ROUTLEDGE HANDBOOK OF BIODIVERSITY AND THE LAW

This volume provides a reference textbook and comprehensive compilation of multifaceted perspectives on the legal issues arising from the conservation and exploitation of non-human biological resources. Contributors include leading academics, policy-makers and practitioners reviewing a range of socio-legal issues concerning the relationships between humankind and the natural world.

The *Routledge Handbook of Biodiversity and the Law* includes chapters on fundamental and cutting-edge issues, including discussion of major legal instruments such as the Convention on Biological Diversity and the Nagoya Protocol.

The book is divided into six distinct parts based around the major objectives which have emerged from legal frameworks concerned with protecting biodiversity. Following introductory chapters, Part II examines issues relating to conservation and sustainable use of biodiversity, with Part III focusing on access and benefit-sharing. Part IV discusses legal issues associated with the protection of traditional knowledge, cultural heritage and indigenous human rights. Parts V and VI focus on a selection of intellectual property issues connected to the commercial exploitation of biological resources, and analyse ethical issues, including viewpoints from economic, ethnobotanical, pharmaceutical and other scientific industry perspectives.

Charles R. McManis is the former Thomas and Karole Green Professor of Law Emeritus and former Director of the Intellectual Property and Technology Law Program at Washington University in St. Louis, Missouri, USA. His book, *Intellectual Property and Unfair Competition in a Nutshell*, is now in its seventh edition. He is also co-author of *Licensing Intellectual Property in the Information Age*, the second edition of which was published in 2005.

Burton Ong is Associate Professor in the Faculty of Law at the National University of Singapore, where he was Deputy Director of the Asia-Pacific Centre for Environmental Law between 2014 and 2017. He teaches and researches in the areas of Competition Law, Intellectual Property and Contract Law. He is the editor of *Intellectual Property and Biological Resources* (2004) and has an interest in the biodiversity and wildlife laws of the ASEAN member countries.

"The *Routledge Handbook of Biodiversity and the Law*, edited by Charles R. McManis and Burton Ong, is required reading for lawyers, scholars and policy-makers for the most recent comprehensive scholarship on a broad spectrum of issues relating to biodiversity. In one single volume world renowned environmental law scholars examine cutting edge issues ranging from genetic resources, biosecurity, access and benefit sharing, synthetic biology, intellectual property, cultural heritage, conservation and sustainable use of marine biodiversity, indigenous peoples' human rights and more. This will clearly become a 'must have' reference book."

– Dr. Nilufer Oral, Law Faculty, Istanbul Bilgi University, Turkey

"This book illuminates the complex set of legal issues surrounding biodiversity by examining them from a wide range of different perspectives. The editors are to be commended for the incredibly rich, varied, and informative scholarship that they have brought together in one volume."

– Professor Graeme B. Dinwoodie, University of Oxford, UK

ROUTLEDGE HANDBOOK OF BIODIVERSITY AND THE LAW

Edited by Charles R. McManis and Burton Ong

Routledge
Taylor & Francis Group
LONDON AND NEW YORK

earthscan
from Routledge

First published 2018 by Routledge

2 Park Square, Milton Park, Abingdon, Oxfordshire OX14 4RN
52 Vanderbilt Avenue, New York, NY 10017

Routledge is an imprint of the Taylor & Francis Group, an informa business

First issued in paperback 2020

British Library Cataloguing-in-Publication Data
A catalogue record for this book is available from the British Library

Library of Congress Cataloging-in-Publication Data
A catalog record for this title has been requested

ISBN: 978-1-138-69330-2 (hbk)
ISBN: 978-0-367-50524-0 (pbk)

Typeset in Bembo
by Out of House Publishing

To Kathleen
– *CM*

To Joyce, Sebastian and Philippa
– *BO*

CONTENTS

CONTRIBUTORS

Regine Andersen holds a doctoral degree in political science and is specialized in the international management of plant genetic resources for food and agriculture and the implementation of relevant international agreements at the national level. She has in particular developed research on the implementation of international provisions regarding farmers' rights related to crop genetic resources and led the "The Farmers' Rights Project" from 2005 until 2013. She has worked with the Fridtjof Nansen Institute as a researcher/senior research fellow since 1995, and on the topic of genetic resources since 2000. From January 2013 until August 2017 she was on leave from the Fridtjof Nansen Institute and working as Executive Director of the organization Oikos – Organic Norway. From August 2017 she is back at the Fridtjof Nansen Institute to continue her research on genetic resources for food and agriculture and farmers' rights, and now also on the politics and management of organic food and farming.

Klaus Angerer is a lecturer at the Institute for the History of Medicine of the Justus-Liebig-University Giessen and a PhD candidate at the Humboldt University Berlin. Based on fieldwork in academic laboratories, botanical gardens and the natural products industry, Angerer is examining the research, transformation and use of collected biological materials in drug discovery. His publications have explored human–animal relations as exemplified with the poison dart frog and the objective of access and benefit-sharing in the Convention on Biological Diversity.

Christoph Antons is Professor of Law in the Newcastle Law School, University of Newcastle, Australia. He is an Affiliated Research Fellow at the Max Planck Institute for Innovation and Competition, Munich; Senior Fellow at the Center for Development Research, University of Bonn; and Project Leader of several Australian Research Council funded projects, including "Intangible Cultural Heritage Across Borders: Laws, Structures and Strategies in China and its ASEAN Neighbours" and "Food Security and the Governance of Local Knowledge in Agriculture in India and Indonesia". His most recent book publication is *The Routledge Handbook of Asian Law* (Routledge, 2017).

Margo A. Bagley is an Asa Griggs Candler Professor of Law at Emory University School of Law. She rejoined the Emory faculty in 2016 after ten years at the University of Virginia

School of Law, where she was most recently the Hardy Cross Dillard Professor of Law and the Joseph C. Carter, Jr. Research Professor of Law. Her scholarship focuses on comparative issues relating to patents and biotechnology, access to medicines, genetic resource appropriation, and technology transfer. Professor Bagley served on the National Academy of Sciences Committee on University Management of Intellectual Property: Lessons from a Generation of Experience, Research, and Dialogue, is a technical expert and advisor to the Government of Mozambique in World Intellectual Property Organization (WIPO) matters and consults with companies, governments and intergovernmental organizations, as well as other entities on a variety of patent-related matters. She currently is Lead Facilitator and Friend of the Chair in the WIPO Intergovernmental Committee on Intellectual Property and Genetic Resources, Traditional Knowledge, and Folklore. Professor Bagley has published numerous articles, book chapters and monographs as well as two books with co-authors: Bagley, Okediji and Erstling, *International Patent Law and Policy* (West Publishing, 2013) and Okediji and Bagley (eds) *Patent Law in Global Perspective* (Oxford University Press, 2014). A chemical engineer by training, Professor Bagley worked in industry for several years before attending law school and is a co-inventor on a patent for reduced fat peanut butter. She is a frequent speaker and writer on patent related topics in the US and abroad and has taught in law schools in China, Cuba, Germany, Israel and Singapore.

Michael J. Balick, an Ethnobotanist, is Vice President for Botanical Science, Director and Philecology Curator, at the New York Botanical Garden Institute of Economic Botany. In his explorations of the diversity and uses of the plant kingdom, and conservation of critical habitats, Dr Balick has led or participated in over 70 scientific field trips to inventory, collect and document plants and how they are traditionally utilized in 24 countries, authored over 150 scientific papers in peer-reviewed journals, and authored or edited 25 scientific and general interest books and monographs. His most recent book is the capstone of a three-decade long collaboration with Dr Rosita Arvigo in Belize, entitled *Messages from the Gods: A Guide to the Useful Plants of Belize* (New York Botanical Garden Press/Oxford University Press, 2015). He received his PhD degree in Biology from Harvard University and was a student of Prof. Richard Evans Schultes, the Amazonian ethnobotanist who is considered the "father" of modern ethnobotany. Balick currently holds adjunct teaching or research appointments at City University of New York, Columbia University, Fordham University, Harvard University and the National Tropical Botanical Garden. Dr Balick has been the recipient of numerous awards and honours in recognition of his accomplishments, including the Distinguished Economic Botanist Award of the Society for Economic Botany; accorded the High Title of *Luhk En Kairoir Dolen Katau Nett*, Kingdom of Nett, Pohnpei, Federated States of Micronesia; the Rachel Carson Award of the Natural Products Association; the John Simon Guggenheim Memorial Foundation Fellowship; the International Scientific Cooperation Award of the American Association for the Advancement of Science (AAAS), of which he is also a Fellow; a Frontiers of Science Awardee, Society of Cosmetics Chemists; and the Janaki Ammal Medal for Ethnobotany by the Society of Ethnobotanists.

Jamie Benidickson teaches Canadian and international environmental law at the University of Ottawa. His publications include *The Culture of Flushing: A Social and Legal History of Sewage* (UBC, 2007). He is the founding Executive Director of the IUCN Academy of Environmental Law.

Michael Blakeney is Winthrop Professor of Law at the University of Western Australia and Visiting Professor in IP at Queen Mary, University of London. He formerly worked in the

Asia Pacific Bureau of the World Intellectual Property Organization. He has directed and been involved in a number of projects concerned with traditional agricultural knowledge with the Asian Development Bank, Consulting Group for International Agricultural Research, European Commission, European Patent Office, Food and Agricultural Organization of the United Nations and in a number of African and Asian developing seeking to become members of the World Trade Organization.

Nuno Pires de Carvalho, Partner, Licks Attorneys, Rio de Janeiro, Brazil. LLM, SJD, Fed. Univ. of Minas Gerais (Brazil); LLM, SJD, Washington University in St. Louis. He served the Secretariats of the World Trade Organization (1996–99) and of the World Intellectual Property Organization (WIPO) (1999–2015). He retired from WIPO in 2015 as Director of the Intellectual Property and Competition Policy Division.

James Ming Chen, an attorney and professor of law with a quarter-century of experience in the law of regulated industries, economics and regulatory policy, holds the Justin Smith Morrill Chair in Law at Michigan State University College of Law. In addition, he is of counsel to the Technology Law Group, of Washington, DC. In November 2012, *National Jurist* magazine named Professor Chen one of the 25 most influential people in legal education. Professor Chen is a highly productive and influential scholar, whose works spans topics such as administrative law, agricultural law, behavioural and mathematical finance, constitutional law, economic regulation, environmental law, industrial policy, legislation and natural resources law. He is the author of *Postmodern Portfolio Theory: Navigating Abnormal Markets and Investor Behavior* (Palgrave Macmillan, 2016) and *Finance and the Behavioral Prospect: Risk, Exuberance, and Abnormal Markets* (Palgrave Macmillan, 2016) and co-author of *Disaster Law and Policy* (3rd edn, Aspen, 2015). Professor Chen is an elected member of the American Law Institute and has served since 2010 as a public member of the Administrative Conference of the United States

Denise Cheong is a senior research fellow and covers developments in two main areas: nuclear law and policy, and ocean law and policy. She is the CIL Project Coordinator for a three-year research project on nuclear safety and security. This project is being undertaken jointly with the Energy Studies Institute of the National University of Singapore and is funded through a National Research Foundation grant as part of the Singapore Government's Nuclear Policy Research Programme. As a member of the Ocean Law and Policy team, she is particularly interested in international law issues that impact on the preservation and protection of the marine environment. Prior to joining CIL, Denise worked with Linklaters (Singapore and Tokyo) for more than ten years, where she was a member of Linklaters' award-winning global capital markets practice. Prior to joining Linklaters, she was a corporate and commercial lawyer with Allens in Sydney, Australia. Denise is an Advocate and Solicitor of the Supreme Court of Singapore as well as a Solicitor of England and Wales. She holds a LLB (Hons) and Master of Science (Environmental Management) (Shell Medal and Prize and Shell Best Dissertation Award winner) from the National University of Singapore.

Carlos M. Correa is Special Advisor on Intellectual Property and Trade of the South Centre and Director of the Center for Interdisciplinary Studies on Industrial Property and Economics at the Law Faculty, University of Buenos Aires. He has been a visiting professor in post-graduate courses of several universities and consultant to various regional and international organizations. He has advised several governments on intellectual property, innovation policy and public health. He was a member of the UK Commission on Intellectual Property, of the Commission on

Intellectual Property, Innovation and Public Health established by the World Health Assembly and of the FAO Panel of Eminent Experts on Ethics in Food and Agriculture. He is the author of several books and numerous articles.

Graham Dutfield is Founding Director of the LLM in Intellectual Property Law, and of the Research Group on Emerging Technologies in Law and Society, both at the University of Leeds, UK. Previously he worked at Queen Mary University of London and at the International Centre for Trade and Sustainable Development where he was Academic Director of a project with UNCTAD on TRIPS and development. He has published numerous articles and several books, including *Global Intellectual Property Law* (Edward Elgar, 2008, with U. Suthersanen), and *Intellectual Property Rights and the Life Science Industries: Past, Present and Future* (World Scientific, 2009).

Rose-Liza Eisma-Osorio is a professor of law at the University of Cebu in the Philippines where she has been teaching for the past 11 years on various subjects, particularly environmental law, legal ethics, public and private international law, legal research and writing, and human rights law. At the same time, she is the Faculty Adviser and Editor-in-Chief of the *Journal of Research in Law and Policy*. She is also one of the faculty members who acts as official representative to the IUCN Academy of Environmental Law (IUCNAEL). She is also a member of IUCN's World Commission on Environmental Law (WCEL). She is also one of the founders and managing trustee of the Philippine Earth Justice Center, Inc. (PEJC), which has filed several environmental cases in the Philippines, including the landmark decision of the Supreme Court in *Resident Marine Mammals and Dolphins v. Reyes*, where she is one of the two lawyers who were recognized as the stewards of the dolphins and whales of Tañon Strait Protected Seascape in Central Philippines against illegal oil exploration in a protected seascape.

Guillaume Futhazar is a PhD student affiliated with the Centre for International and European Research and Study (CERIC) and the "Objectif Terre Méditerranée – Laboratory of Excellence" (OT-Med Labex) based in France. Specialized in international public law and international environmental law, the focus of his current work is on the interactions between law and science, fragmentation and regime complexes in the Mediterranean Region.

Ursula Goodenough is Professor of Biology at Washington University. She received a PhD from Harvard, and her research has focused on the molecular genetics of sexual reproduction in a green alga. Her book, *The Sacred Depths of Nature* (Oxford University Press, 1998), explores the religious potential of our understandings of Nature.

Joseph M. Jez received his BSc in biochemistry from Penn State University (1992), a PhD in biochemistry and molecular biophysics from the University of Pennsylvania (1998), and was an NIH-NRSA postdoctoral fellow at the Salk Institute for Biological Studies (1998–2001). After working as a research scientist at Kosan Biosciences (2001–02), he started his research group at the Donald Danforth Plant Science Center as an assistant member in 2002 and moved to the Department of Biology at Washington University in 2008. He has authored more than 140 papers and received the Presidential Early Career Award for Scientists and Engineers (2005), the Phytochemical Society of North America Arthur Neish Young Investigator Award (2007), and a Fulbright Senior Specialist Award (2012). In 2014, he was named as an HHMI Professor. Research in the Jez lab seeks to understand how environmental changes re-model biochemical pathways in plants at the molecular, cellular and organism levels with the aim of engineering these systems to address agricultural and environmental problems. He has also worked with the

International Life Science Institute (ILSI) to educate regulators and policy-makers in the US and abroad on issues related to food biosafety and genetic-engineering of crops.

Lye Lin-Heng is an Advocate and Solicitor of the Supreme Court, Singapore and Director of the National University of Singapore (NUS) Law Faculty's Asia-Pacific Centre for Environmental Law (APCEL). She teaches and researches in environmental law and property law. She chairs the multi-disciplinary MSc in Environmental Management (MEM) programme, hosted by the School of Design and Environment, which has the collaboration of nine faculties/schools in NUS. She is a member of the Board of Governors, IUCN Academy of Environmental Law and a member of the Board of Governors, WWF Singapore. She is Visiting Associate Professor at Yale University's School of Forestry and Environmental Studies, and a resource person in capacity-building programmes conducted by the IUCN Academy of Environmental Law, the World Bank, the Asian Development Bank, the Ministry of Foreign Affairs, Singapore and the Singapore Environment Institute, Ministry of Environment and Water Resources. She has published and presented papers on environmental law at conferences worldwide. She is a member of the Singapore Land Titles (Strata) Board and a member of the Ministry of Environment and Water Resources' Advisory Committee on Water Quality. She was honorary Legal Advisor to the Nature Society Singapore, for many years.

Porter P. Lowry II, Director of the Africa and Madagascar Program at the Missouri Botanical Garden (MBG) in St. Louis (USA), is a systematist by training, studying the evolution and classification of flowering plants, with a focus on the mostly tropical family *Araliaceae*, and on flora of two biodiversity hotspots, Madagascar (where he began working in 1986) and New Caledonia (where he has worked since 1981). He coordinates the MBG's research and conservation programme in Madagascar, which employs over 100 Malagasy staff members and focuses on the discovery of one of the world's richest and most threatened floras, building local expertise and capacity, and working with local communities to conserve 13 newly established botanical reserves. As a practising plant systematist, Pete conducts research on a wide range of groups in both Madagascar and New Caledonia, making regular visits to both countries (two among the more than 25 countries in which he has conducted field work). Pete also coordinates MBG's consulting programme, which provides expertise on flora and vegetation, primarily to the mining sector. He is permanently based at the *Muséum National d'Histoire Naturelle* in Paris, serving as MBG's liaison to the *Herbier National*. He has authored more than 230 scientific papers and technical reports on a wide diversity of subjects, and serves on numerous committees, councils and boards.

Youna Lyons is a senior research fellow with the Centre for International Law at the National University of Singapore, focusing on marine environmental governance in Southeast Asia. Her primary research areas are the protection of marine sensitive areas in the seas of Southeast Asia and the underlying legal and institutional governance framework, the use of satellite imagery to inform marine environmental management through habitat mapping and maritime boundary disputes on remote features, offshore oil and gas in Southeast Asia including decommissioning and the prospects for rigs to reef, as well as the sustainable use of marine resources in areas beyond national jurisdiction. Prior to this, Youna created and led the private international law practice of the Litigation and Arbitration Group at Clifford Chance (Paris office), one of the world's leading law firms, for more than 12 years. Passionate about the ocean, Youna has been trained in marine ecology, physical oceanography and ocean policy at the College of Ocean and Fisheries of the University of Washington (where she obtained a Masters in Marine Affairs).

Youna's dual expertise in law and marine sciences uniquely position her to integrate these fields for marine policy-making purposes. In addition she focuses on developing industry relevant research and on involving the industry in marine policy proposals.

Sandrine Maljean-Dubois specializes in international environmental law. A researcher at the CNRS, she teaches international environmental law at the Faculty of Law and Political Sciences of Aix-Marseille University. She has edited several books and a large number of articles in this field, focusing in particular on biodiversity and biosecurity, international environmental law and WTO law, climate change negotiations and more generally on the effectiveness of international environmental law.

Charles R. McManis is the former Thomas and Karole Green Professor of Law Emeritus and former Director of the Intellectual Property and Technology Law Program at Washington University in St. Louis, Missouri, USA. He received his BA degree from Birmingham-Southern College in 1964, and both his MA (in Philosophy) and JD degrees from Duke University in 1972. Professor McManis has been a frequent visiting lecturer and paper presenter at universities and academic conferences throughout the United States, Asia, Europe and South America. During 1993 and 1994, Professor McManis was a Fulbright Fellow in Korea, where he lectured and did research at the International Intellectual Property Training Institute in Taejon. He has served as a consultant for the World Intellectual Property Organization, in India, Korea and Oman, and in 2002 he presented a paper at a Joint WIPO/UPOV Symposium, on the Co-existence of Patents and Plant Breeders' Rights, in Geneva, Switzerland. Professor McManis's book, *Intellectual Property and Unfair Competition in a Nutshell*, is now in its seventh edition. He was co-author of *Licensing Intellectual Property in the Information Age*, the second edition of which was published in 2005 by Carolina Academic Press. He is also the editor of a multi-authored volume, entitled *Biodiversity and the Law: Intellectual Property, Biotechnology and Traditional Knowledge* (Earthscan/James & James, 2007).

James S. Miller is the Senior Vice President for Science and Conservation at the Missouri Botanical Garden. He manages a portfolio of international research and conservation projects that all address environmental issues in a world of rapid global change. The Garden's programmes are active throughout the world but focus on tropical parts of the New World, Africa and Madagascar, and Southeast Asia. Through his various research endeavours, Dr Miller has travelled to more than 60 countries, both collecting plants and conducting cooperative research. Dr Miller is an authority on the systematics of the plant group *Boraginales*, a group of ten families and about 2,500 species that are widely scattered in the world. He is also involved with tropical floristics, efforts to explore and catalogue plants of poorly known tropical regions, and his work has focused on Latin America and Tropical Africa and Madagascar. He has worked with numerous plant families in Mexico and Central America, Venezuela, Colombia, Ecuador, Peru, Bolivia, and Madagascar. He is actively involved with using information from these studies to inform conservation initiatives, including which species are most in need of our attention to ensure their future survival and how they can adapt to changing climate. He also has extensive experience in collaborative partnerships to discover new pharmaceutical, agricultural and nutritional products from plants, spanning relationships with government institutions like the National Cancer Institute through corporate and academic partners. Dr Miller earned his PhD though St. Louis University following BS and MS degrees from the University of Maryland. He currently is on the advisory boards for the University of Missouri's Whitney R. Harris World Ecology Center, the University of Mississippi's National Center for Natural Products Research,

Sequoia Sciences (a St. Louis-based pharmaceutical company developing medicines from plants), the University of Missouri's Center for Botanical Interaction Studies and Louisiana State University's NIH Botanicals Center.

John Copeland Nagle is the John N. Mathews Chair at the Notre Dame Law School, where he teaches a variety of environmental law courses, as well as legislation and property law. He is the co-author of three casebooks: *The Practice and Policy of Environmental Law*, *The Law of Biodiversity and Ecosystem Management*, and *Property: Cases and Materials.* His book *Law's Environment: How the Law Affects the Environment*, was published by the Yale University Press in 2010. His other writings address Chinese environmental law, statutory interpretation, the problems associated with lame-duck law-making and how religious teachings influence environmental law. Professor Nagle has lectured on environmental issues at numerous forums in the United States and around the world. He has received two Fulbright awards: first to serve as a Distinguished Lecturer at the Tsinghua University School of Law in Beijing during 2002, where he taught property law and environmental law; and then to serve as a distinguished scholar with the faculty of law of the University of Hong Kong during the spring of 2008. He taught a course on climate change law at Notre Dame's London programme during the fall of 2010. Prior to joining the Notre Dame faculty, Professor Nagle was an associate professor at the Seton Hall University School of Law, from 1994 to 1998. He also worked in the United States Department of Justice, first as an attorney in the Office of Legal Counsel, where he advised other executive branch agencies on a variety of constitutional and statutory issues, and later as a trial attorney conducting environmental litigation. Professor Nagle served as a law clerk to Judge Deanell Reece Tacha of the United States Court of Appeals for the Tenth Circuit, and he was a scientific assistant in the Energy and Environmental Systems Division of Argonne National Laboratory. He is a graduate of Indiana University and the University of Michigan Law School. Professor Nagle has participated in numerous activities outside of the law school. He has served as a member of the executive committee of the section on Legislation of the American Association of Law Schools, and as a vice chair on the Endangered Species Committee of the American Bar Association's environmental section, and he helps organize the annual meeting of the Law Professors' Christian Fellowship.

Omar Oduardo-Sierra is an MA candidate in Linguistics at the University of Puerto Rico-Río Piedras, where he earlier completed an undergraduate degree in political science. Among his half-dozen refereed publications on "access to genetic resources" and "benefit-sharing" is his lead authorship in "Monitoring and Tracking the Economics of Information in the Convention on Biological Diversity: Studied Ignorance (2002–2011)" (*Journal of Politics and Law*, 2012). He is currently exploring how stakeholders and delegates to the Conference of the Parties to the Convention on Biological Diversity can re-frame political metaphors toward "bounded openness".

Burton Ong is an Associate Professor in the Faculty of Law at the National University of Singapore where he teaches and researches in the areas of Competition Law, Contract Law and Intellectual Property Law. Between 2014 and 2017 he was the Deputy Director of the Asia-Pacific Centre for Environmental Law (APCEL), established in 1996 by the World Conservation Union's Commission on Environmental Law (IUCN-CEL), in collaboration with the United Nations Environment Programme (UNEP), to promote environmental legal education within the region. His areas of interest include the intersection between biodiversity and intellectual property law, access and benefit-sharing issues arising from bioprospecting, and the biodiversity and wildlife protection laws of the ASEAN (Association of South East Asian Nations) member states. He is an Advocate and Solicitor of the Supreme Court of Singapore and an Attorney and Counsellor-at-Law of the New York State Bar.

Opi Outhwaite is a Senior Lecturer in Law at the School of Law, University of Greenwich, United Kingdom. Her research focuses on international law and governance, natural resources and the environment. Dr Outhwaite has published widely in her research areas as well as undertaking numerous outreach and dissemination activities. She has a track record of providing research and consultancy for governmental and non-governmental bodies including as a member of the Electronics Watch Advisory Group and for the UK Department for Environment, Food and Rural Affairs (Defra) and equivalent bodies in Tanzania and Belize. Opi is Managing Editor of the IUCN Academy of Environmental Law journal ("e-journal") and participates in several networks for international environmental law and business and human rights. Her current research activities include developing new perspectives on the relationship between animals and the law and ongoing work to address issues of human rights in global supply chains.

Alexander Paterson is a professor of environmental law based in the Institute of Marine and Environmental Law, Faculty of Law, University of Cape Town. He is programme convener of the postgraduate degrees specializing in marine and environmental law on offer through the Institute. His research interests include biodiversity, protected areas and land use planning. He sits on the Governing Board of the IUCN Academy of Environmental Law; is a member of the IUCN World Commission of Environmental Law (CEL), World Commission on Protected Areas (WCPA) and the Association of Environmental Law Lecturers from African Universities (ASSELLAU); and on the Editorial Board of the *South African Journal of Environmental Law and Policy* and the *Chinese Journal of Environmental Law*.

Denis Pesche is a sociologist from the Centre International de Recherche Agronomique pour le Développement (CIRAD), UMR ART-Dev (France) and he specializes in analysing rural interests groups in the policy making, environmental and agricultural policies, at the international level and in developed and developing countries. He coordinates a research programme on ecosystem services from 2009 to 2013. He published a large number of articles in those fields, focusing in particular on biodiversity and ecosystem service approaches.

Colin T. Reid is Professor of Environmental Law at the University of Dundee. He has written widely on environmental issues, including three editions of a book on *Nature Conservation Law*. His recent project examining new approaches to conservation law such as biodiversity offsets and payment for ecosystem services, has led to a book, *The Privatisation of Biodiversity?*, to be published in 2017. He is a founding member of various environmental law bodies in the UK and a member of the IUCN Environmental Law Commission as well as of the editorial boards of several leading academic journals.

Daniel F. Robinson is an Associate Professor and Masters Convenor in Environmental Studies in the Faculty of Arts and Social Sciences, UNSW, Sydney Australia. Daniel is also a Research Fellow with the International Centre for Trade and Sustainable Development (ICTSD) in Geneva. He has worked on Indigenous knowledge, ABS and related issues for more than ten years and for several organizations including the ABS-Initiative, UNDP-GEF, Natural Justice, Pacific Islands Forum Secretariat, Commonwealth Department of Environment, amongst others.

Nicholas A. Robinson is University Professor for the Environment and Kerlin Distinguished Professor Emeritus of Environmental Law at the Elisabeth Haub School of Law at Pace University (New York, USA), and Professor Adjunct at the Yale University School of Forestry and Environmental Studies (New Haven, USA). He chaired the Commission on Environmental Law of the International Union for the Conservation of Nature and Natural Resources (IUCN), and is a Governor of the International Council of Environmental Law. He holds a BA from Brown University (1967) and a JD from Columbia University (1970).

Manuel Ruiz Muller (Peruvian Society for Environmental Law) is an environmental lawyer, with 25 years of experience working in environmental, biodiversity and development law and policy. He has published extensively on issues related to access and benefit-sharing, biotechnology, intellectual property and traditional knowledge. He has worked as a consultant for FAO, Bioversity International, IDB, IUCN, UNEP, UNDP, WIPO, among others, and currently coordinates an International Environmental Law course in the Catholic University of Peru.

Barbara A. Schaal is the Dean of the Faculty of Arts and Sciences, and the Mary Dell Chilton Distinguished Professor, Washington University in St. Louis. Schaal was born in Berlin, Germany and grew up in Chicago, IL, USA. She graduated from the University of Illinois, Chicago with a degree in biology and received a PhD from Yale University. She is a plant evolutionary biologist who uses DNA sequences to understand evolutionary processes such as gene flow, geographical differentiation and the domestication of crop species. Her current research focuses on the evolutionary genomics of rice. She has served as Chair of the Division on Earth and Life Studies at the National Research Council and was a member of President Obama's Council of Advisors for Science and Technology. She has been President of the Botanical Society of America and the Society for the Study of Evolution. She is an elected member of the American Academy of Arts and Sciences and US National Academy of Sciences where she served as Vice President. She was appointed as a US science envoy by former Secretary of State Hillary Clinton. In February 2016 Schaal became the President of the American Association of the Advancement of Science (AAAS).

Brad Sherman is Professor of Law and ARC Australian Laureate Fellow at the TC Beirne School of Law, the University of Queensland, Brisbane, Australia. His research expertise encompasses many aspects of intellectual property law, with particular emphasis on its historical, doctrinal and conceptual development. Brad is currently working on a historical project looking at the materiality of the invention and early historical examples where intellectual property was used to promote food security.

Brendan Tobin, who holds dual Irish and Peruvian citizenship, is an Ashoka fellow and adjunct lecturer with the Irish Centre for Human Rights at the National University of Ireland, Galway. A barrister by training, he has over the past 25 years worked as an activist, consultant and academic, on issues of environmental law, intellectual property and human rights, around the world. In the early 1990s he promoted initial proposals for systems of disclosure of origin and certificates of origin of genetic resources and traditional knowledge and in 2000 he co-chaired final negotiations leading to the adoption of the Bonn Guidelines on Access to Genetic Resources and Benefit-sharing at the Convention on Biological Diversity. He has published widely on issues of access and benefit-sharing, protection of traditional knowledge and customary law, including a recent monograph, *Indigenous Peoples, Customary law and Human Rights – Why Living Law Matters* (Routledge, 2014). He is currently working on issues of natural resource exploitation and cultural genocide.

Joseph Henry Vogel is Professor of Economics at the University of Puerto-Río Piedras. He has served as a technical advisor on the Ecuadorian delegations to the Conference of the Parties to the Convention on Biological Diversity (CBD) and the Framework Convention on Climate Change. Since launching *Genes for Sale* (Oxford, 1994), Vogel has published two dozen refereed texts on the CBD and lectured at 250 venues worldwide. Many of his works have been translated into multiple languages. He is currently preparing an online course, "Bounded Openness as the Global Multilateral Benefit-sharing Mechanism", for stakeholders to the Nagoya Protocol.

Rachel Wynberg holds a Research Chair at the University of Cape Town focused on Social and Environmental Dimensions of the Bio-economy. She has worked on issues related to bioprospecting, biotrade, access and benefit-sharing for more than twenty years, providing policy support and advice on natural resource governance and social justice to a range of governments, civil society organizations, industry and international agencies. Publications include over 140 scientific papers, technical reports and popular articles, and five co-edited books and monographs.

PREFACE

Life on Earth is dependent on the biological bounty of the natural world. The biological diversity of our planet's flora and fauna support our ecosystems, providing shelter, nourishment and medicines to its many inhabitants. Given its vital role in sustaining the existence of mankind, biological diversity deserves a lot more attention from the legal and policy-making communities of this planet – on a myriad of issues encompassing the protection, management and use of this resource. An interdisciplinary approach is needed to fully understand these issues. This book seeks to provide such a perspective by bringing together a diverse group of authors, experts in their own respective fields, to address a broad range of contemporary issues related to biological diversity. Hailing from 12 countries across the world's five main continents, the contributors to this collection of essays include legal scholars, social and physical scientists, environmentalists and an economist.

The purpose of this Handbook is to introduce the various international agreements and implementing domestic legislation concerned with the conservation, sustainable use and the fair and equitable sharing of the benefits deriving from the use of biological diversity. While the heart of what has come to be known as biodiversity law is to be found in the 1992 Convention on Biological Diversity (CBD) and the associated 2014 Nagoya Protocol on Access and Benefit-sharing, the international legal terrain governing biodiversity and the law, as we point out in more detail in Chapter 1, comprises a rich and varied landscape of international agreements and implementing domestic legislation both antedating and postdating the promulgation of the CBD.

Our purpose here is not merely to describe this legal landscape but also (as we will point out in more detail in Chapter 2) to critically evaluate the application and effectiveness of this overlapping skein of laws in promoting the conservation, sustainable use and the fair and equitable sharing of the benefits arising out of the use of biodiversity – the threefold objective of the CBD. Part I of this Handbook maps the legal terrain comprising "biodiversity law" and introduces the various topics discussed in each chapter of this Handbook. Part II of the Handbook focuses on the conservation and sustainable use of biodiversity, while Part III focuses on access and benefit-sharing. Parts IV, V and VI, in turn, discuss the specific topics of traditional knowledge protection, biodiversity and intellectual property protection, and the ethics, economics and the science-policy interface of biodiversity protection.

Preface

This Handbook owes its origin to an earlier book having a similar title and subject matter – namely *Biodiversity and the Law: Intellectual Property, Biotechnology and Traditional Knowledge*, Charles McManis ed. (Earthscan 2007). Although this Handbook is by no means a second edition of the earlier work, the Handbook does contain four chapters that are updated versions of chapters appearing in the earlier volume (namely, Chapters 7, 8, 16 and 24). While the two works have a common objective and can be seen as companion volumes, what distinguishes the current Handbook from the earlier work is the sheer volume of legal developments that have occurred over the past ten years.

This Handbook would never have come into existence without the tireless efforts of Tim Hardwick of Routledge and the dedication of our chapter authors in bringing the Handbook to fruition. We are also grateful to the copyeditor Christopher Long and the project manager Nicola Howcroft. We would also like to acknowledge the support of our families and our home institutions, the Washington University in St. Louis School of Law and the Asia-Pacific Center for Environmental Law at the Faculty of Law, National University of Singapore.

Charles R. McManis and Burton Ong

PART I

Introduction

1

BIODIVERSITY AND THE LAW

Mapping the international legal terrain

Burton Ong

The multi-faceted significance of biological diversity (biodiversity) to the well-being of planet Earth and its inhabitants is reflected in the spectrum of jurisprudence, straddling multiple branches of law, establishing international legal norms relating to the conservation and exploitation of Nature's biological bounty. These legal frameworks are underpinned by a broad range of objectives relating to how biodiversity ought to be protected, managed and utilized by the myriad of stakeholders – including state governments, local authorities, indigenous communities, scientific and commercial researchers, farmers and agricultural practitioners, as well as the biotechnology industry. Biodiversity law is thus embedded within an international legal ecosystem of treaties whose norms are implemented at regional and national levels by countries around the world. This chapter seeks to provide an overview of the principal features of this selection of legal frameworks, explaining their significance in understanding the transnational web of contemporary biodiversity-related issues encountered within the fields of international law, environmental law, trade law and intellectual property law.

The Convention on Biological Diversity: a near-global framework for national biodiversity laws

At the United Nations' Conference on Environment and Development in 1992 (also known as the "Earth Summit") in Rio de Janeiro, the Convention on Biological Diversity (CBD)[1] was launched for members of the global community to sign. The CBD, which embodies the global community's commitment to sustainable development, establishes a broad framework for member states to implement national laws directed at the conservation of biodiversity, the sustainable use of its components, as well as the equitable sharing of the benefits arising from the use of these genetic resources. The CBD came into force on 29 December 1993 with 168 signatory member states. By 2016, 196 parties had ratified the treaty, the United States of America being a notable exception to the near-global reach of the agreement.

The three pillars of the CBD can be found within Articles 8 and 9 (*in situ* and *ex situ* conservation), Article 10 (sustainable use of components of biological diversity) and Article 15 (access to genetic resources). The Conservation, Sustainable Use and Access-and-Benefit-Sharing goals of the CBD require its parties to, "as far as possible and as appropriate", take measures to preserve biodiversity and promote the sustainable use of biological resources within their respective territorial jurisdictions, as well as to "take legislative, administrative or policy measures ... with the aim of sharing in a fair and equitable way the results of research and development and the benefits arising from the commercial and other utilization of genetic resources with the Contracting Party providing such resources ... upon mutually agreed terms" (Article 15(7)).

Administered by a governing body that meets on a biennial basis, the Conference of Parties (COP) comprises government representatives from member states supported by a Subsidiary Body on Scientific, Technical and Technological Advice (SBSTTA) with specialized expertise that provides recommendations to the COP on various technical aspects to the implementation of the CBD. There are also several "ad hoc open-ended Working Groups" established by the COP to address specific issues within the scope of the CBD, including Access and Benefit-Sharing (ABS) and the protection of traditional knowledge under Article 8(j) of the CBD – under which the parties have agreed to "respect, preserve and maintain knowledge, innovations and practices of indigenous and local communities embodying traditional lifestyles relevant for the conservation and sustainable use of biological diversity and promote their wider application with the approval and involvement of the holders of such knowledge, innovations and practices and encourage the equitable sharing of the benefits arising from the utilization of such knowledge, innovations and practices".

In 2010, the COP introduced a Strategic Plan, including 20 headline targets known as the Aichi Biodiversity Targets, to implement the substantive goals of the CBD between 2011 and 2020.[2] Parties are expected to develop and implement their own National Biodiversity Strategies and Action Plans[3] to set their own country-specific targets and translate five goals into reality within their respective jurisdictions: (a) address the underlying causes of biodiversity loss by mainstreaming biodiversity across government and society; (b) reduce the direct pressures on biodiversity and promote sustainable use; (c) improve the status of biodiversity by safeguarding ecosystems, species and genetic diversity; (d) enhance the benefits to all from biodiversity and ecosystem services; (e) enhance implementation through participatory planning, knowledge management and capacity building.

Regulating biosafety concerns arising from transboundary movements of modified living organisms: the Cartagena Protocol

In response to the dangers posed to biodiversity by genetically modified living organisms, the product of modern biotechnology, the Cartagena Protocol on Biosafety to the Convention on Biodiversity was introduced by the COP as a supplementary agreement to regulate transboundary movements of modified genetic resources.[4] The Cartagena Protocol entered into force in 2003, establishing an advance informed agreement procedure that enables countries to make informed decisions before agreeing to import modified living organisms – in accordance with the precautionary principle articulated in Principle 15 of the Rio Declaration on Environment and Development – by establishing a Biosafety Clearing-House to facilitate the exchange of information on such organisms. The Cartagena Protocol operates alongside the World Trade Organization's network of international trade law rules,[5] with explicit recognition in its preamble that trade and environment agreements should be mutually supportive with a view to achieving sustainable development.

Regulating access to, and benefit-sharing from the utilization of, genetic resources: the Bonn Guidelines and the Nagoya Protocol

The scope of the CBD covers a vast range of subject matter, encompassing both *in situ* and *ex situ* biodiversity, as well as their derivatives and by-products. It covers all biological resources containing functional units of heredity, both terrestrial and marine, within the national borders and jurisdiction of all the members of the COP. The subject matter of the CBD also extends beyond biological material – flora and fauna – to encompass "the results of research and development and the benefits"[6] arising from the use of, as well as the "knowledge, innovations and practices of indigenous and local communities embodying traditional lifestyles" (Article 8(j)) associated with these genetic resources. Given the broadly worded nature of the language used in the provisions of the CBD, there is clearly room for divergence between the national laws of individual member states who may, ultimately, interpret their obligations under the treaty in accordance with their respective national policy priorities. Biodiversity-rich parties that are net providers of genetic resources are more likely to implement comprehensive access and benefit-sharing laws that enable them to assert authority over, and to maximize the benefits they can derive from, the exploitation of biological materials within their jurisdictions. Conversely, parties that are net users of genetic resources because of their developed biotechnology industries will be inclined to implement a more facilitative framework that is less onerous to those who seek to benefit from the use of such resources – by developing innovations inspired by or based upon the properties of biological resources, for instance.

The CBD also explicitly recognizes that intellectual property rights may have an impact on the implementation of its objectives – to the extent that patents and other forms of intellectual property involve the assertion of private rights of ownership over biodiversity-generated innovations – and exhorts member states to "cooperate in this regard subject to national legislation and international law in order to ensure that such rights are supportive of and do not run counter to its objectives" (Article 16(5)). This led to calls for the intellectual property treaties administered by the World Trade Organization (the TRIPS Agreement, in particular) and the World Intellectual Property Organization (WIPO) to be more closely aligned to the access and benefit-sharing goals of the CBD. However, despite numerous efforts by developing countries to introduce CBD-inspired reforms to these multilateral intellectual property treaties, little headway appears to have been made on this front over the last two decades.

In 2002, the Sixth meeting of the parties to the CBD adopted the non-binding Bonn Guidelines on Access to Genetic Resources and Fair and Equitable Sharing of Benefits,[7] which explicitly identified various strategies for realizing the benefit-sharing goals of the CBD. These included recognizing IPRs as benefits derived from the utilization of genetic resources which member states could regulate by setting parameters for when such property rights might be granted and exploited. Parties to the CBD were also invited to encourage the disclosure of the country of origin of genetic resources and associated traditional knowledge, utilized to develop patented subject matter, during the patent application process. Such disclosure rules might then be useful in tracking compliance with the CBD's principles of access to genetic resources being granted after prior informed consent has been given on mutually agreed terms. Similarly, the CBD parties were invited to develop *sui generis* regimes for the protection of traditional knowledge,[8] which led to various efforts to develop comprehensive databases of such traditional knowledge that could be utilized by patent authorities as part of the prior art when conducting patent searches and examinations.

These ABS-related facets of the CBD were developed further in the Nagoya Protocol on Access and Benefit-Sharing,[9] which entered into force in 2014. The Protocol aims to facilitate the sharing of benefits derived from the use of genetic resources covered by the CBD in a fair and equitable manner, establishing a transnational Access and Benefit-Sharing Clearing House (ABS Clearing House) mechanism in furtherance of Article 18 of the CBD. This mechanism enhances the CBD's ABS framework by providing the contracting parties with a transparent platform for information exchange, thereby supporting their efforts in enforcing and monitoring compliance with their respective national ABS regulations. Users of the ABS Clearing House can upload national records of information on ABS measures, contact information of national authorities and information on permits and their equivalent constituting internationally recognized certificates of compliance. Other noteworthy features of the Nagoya Protocol include an attempt to elaborate upon ABS-related concepts originating from the CBD, including "prior informed consent" (Articles 5, 6 and 7) and "mutually agreed terms" (Article 18), while contracting parties are exhorted to consider the need for a global multilateral benefit-sharing mechanism (Article 10). However, achieving a unitary system of international governance to give effect to the ABS objectives of the Protocol will be challenging because of the complexity of the relationships between the private sector and public sector stakeholders involved, further complicated by the national sovereignty approach underlying the CBD which gives individual states the final authority to determine conditions for access to, and the value of, genetic resources located within their respective jurisdictions (Oberthür *et al.*, 2014).

In addition, the Nagoya Protocol encourages contracting parties to develop and use model contractual clauses to regulate the use of genetic resources and associated traditional knowledge (Article 19), while explicitly identifying joint ownership of intellectual property rights arising from such use as an example of a non-monetary benefit arising from the utilization of such resources that may be shared with their country of origin and local or indigenous stakeholder communities (Article 5 and Annex). The flexibility of the legal language used in the Nagoya Protocol is both its strength and weakness, giving contracting parties some freedom to tailor their national ABS frameworks to meet their jurisdiction-specific priorities but generating uncertainty over how disputes arising from the asymmetries between users and providers of genetic resources will be resolved by each state (Morgera *et al.*, 2014).

The Ramsar Convention: a pioneer intergovernmental network for promoting conservation and sustainable use of wetlands

Formally titled the Convention on Wetlands of International Importance especially as Waterfowl Habitat, this convention was adopted by a handful of nations in 1971 in the city of Ramsar, Iran, giving rise to the oldest of the world's modern international environmental agreements.[10] Today, 169 states are contracting parties of what is more commonly known as the Convention on Wetlands, or the Ramsar Convention, covering more than 215 million hectares of designated wetlands, and whose purpose is "the conservation and wise use of all wetlands through local, regional and national actions and international cooperation, as a contribution towards achieving sustainable development throughout the world".[11] Wetlands are defined broadly to cover a diverse range of habitats – "areas of marsh, fen, peatland or water, whether natural or artificial, permanent or temporary, with water that is static or flowing, fresh, brackish or salt, including areas of marine water the depth of which at low tide does not exceed six metres" (Article 1.1). Ramsar contracting parties are committed to implementing the "three pillars" of the Convention: (1) designating at least one suitable wetland within their territorial jurisdiction for inclusion into the Convention's List of Wetlands of International Importance (the "Ramsar List", which currently exceeds 2200 sites)[12] and ensuring their effective management; (2) ensuring the "wise use" of wetlands within their territories through national land-use planning laws and policies;[13] and (3) cooperating with other contracting parties on international issues relating to the management of transboundary wetlands, shared water systems and shared species.

The International Treaty on Plant Genetic Resources for Food and Agriculture: a global regime to preserve and protect food crop biodiversity

The Food and Agriculture Organization (FAO) of the United Nations adopted the International Treaty on Plant Genetic Resources for Food and Agriculture (ITPGRFA) in 2001, establishing a global multilateral system to regulate access to a pool of biological samples extracted from 64 of the planet's most important food crops.[14] Since coming into force on 29 June 2004, the ITPGFRA has provided a platform for member states – comprising many of the major crop-producing countries around the world, with 141 parties at the end of 2016 – to collectively manage the exploitation of this *ex situ* repository of genetic resources. Signatory countries which have not ratified the ITPGRFA include the United States of America, Thailand, China, Russia and Mexico.

The ITPGRFA is animated by the spirit of the CBD, applying the CBD's broad objectives within the specific context of plant genetic resources that are valuable to humankind as food sources. As such, the Treaty articulates the same trinity of goals – conservation, sustainable use and fair and equitable benefit-sharing – emphasizing the importance of being "in harmony with the Convention on Biological Diversity, for sustainable agriculture and food security" and how "[t]hese objectives will be attained by closely linking this Treaty to the Food and Agriculture Organization of the United Nations and to the Convention on Biological Diversity" (Article 1).

At its core, the ITPGRFA requires parties to implement measures at the national level that take an "integrated approach to the exploration, conservation and sustainable use of plant genetic resources for food and agriculture", "take steps to minimize or … eliminate threats to plant genetic resources for food and agriculture" and to "develop and maintain appropriate policy and legal measures that promote the sustainable use of plant genetic resources for food and agriculture" (Articles 5 and 6).

In recognition of the contributions of farming communities towards conserving and improving the diversity of plant genetic resources for food crops, Article 9 of the ITPGRFA articulates the concept of Farmers' Rights, requiring parties to take legislative measures that promote the protection of traditional knowledge associated with plant genetic resources used for food and agriculture, as well as the rights of farmers to receive an equitable share of the benefits arising from the use of such resources and to participate in national-level decision-making on matters relating to the conservation and sustainable use of such resources. In addition, Article 9 also makes explicit the absence of any intention in the Treaty "to limit any rights that farmers have to save, use, exchange and sell farm-saved seed/propagating material" – leaving it open to parties to implement appropriate seed-saving systems under their national legal frameworks. Given the central role which farmers play in the conservation and sustainable use of these genetic resources, it has been argued that the international law status of Farmers' Rights ought to be further elevated to give rightholders stronger entitlements to monetary and non-monetary compensation (Girsberger, 1999).

The centrepiece of the ITPGRFA is its Multilateral System of access and benefit-sharing that is applied to plant genetic materials from 64 food crops, listed in Annex I of the Treaty, that are housed in *ex situ* collections managed by the International Agricultural Research Centres (IARC) of the Consultative Group on International Agricultural Research (CGIAR). Article 12 sets out conditions under which access to such genetic resources will be granted ("…solely for the purpose of utilization and conservation for research, breeding and training for food and agriculture, provided that such purpose does not include chemical, pharmaceutical and/or other non-food/feed industrial uses"), including the use of a standard Material Transfer Agreement prescribed by the Treaty's Governing Body. Article 12.3(d) prohibits recipients of such plant genetic resources from claiming "any intellectual property or other rights that limit the facilitated access to the plant genetic resources for food and agriculture, or their genetic parts or components, in the form received from the Multilateral System". Benefit-sharing under Article 13 of the Treaty requires parties to ensure that the benefits derived from the use of resources obtained from the Multilateral System are "shared fairly and equitably" with all contracting parties. Benefit-sharing may take place by way of information exchange, access to and transfer of technology (including technologies protected by intellectual property rights), capacity-building (particularly in developing countries and countries in transition) and the sharing of monetary and other benefits obtained from the commercialization of such resources (Article 13). The standard Material Transfer Agreement prescribed by the Treaty's Governing Body includes "a requirement that a recipient who commercializes a product that is a plant genetic resource for food and agriculture and that incorporates material accessed from the Multilateral System, shall pay … an equitable share of the benefits arising from the commercialization of that product" to a financial mechanism operated by the Treaty's Governing Body.[15]

The Agreement on Trade-Related Aspects of Intellectual Property Rights (TRIPS): biological innovations derived from genetic resources as the subject matter of intellectual property rights

The Agreement on Trade-Related Aspects of Intellectual Property Rights (TRIPS Agreement) is administered by the World Trade Organization, whose members are obliged to implement national intellectual property laws that meet the minimum standards set out in this multilateral treaty.[16] The TRIPS Agreement came into effect on 1 January 1995, establishing a comprehensive global framework of legal standards for all the major branches of intellectual property law (including copyright, trademarks and patents) and general principles governing the enforcement of these intellectual property rights. Every major economy in the world is now party to the TRIPS Agreement, though developing countries have been given a longer transition period to comply with all its provisions. The TRIPS Agreement consolidated and augmented the substantive obligations of the WIPO main conventions, the Paris Convention for the Protection of Industrial Property and the Berne Convention for the Protection of Literary and Artistic Works. The goals of the TRIPS Agreement are directed at the transnational harmonization of intellectual property rights as a means to reduce distortions and impediments to international trade, ensuring the effective and adequate protection of such rights to prevent infringing conduct from becoming a barrier to legitimate trade between member states.

Article 7 ("Objectives") of the TRIPS Agreement declares that the protection and enforcement of intellectual property rights should "contribute to the promotion of technological innovation and to the transfer and dissemination of technology, to the mutual advantage of producers and users of technological knowledge and in a manner conducive to social and economic welfare, and to a balance of rights and obligations". Article 8(1) ("Principles") of the TRIPS Agreement permits member states "in formulating … their laws and regulations, [to] adopt measures necessary to protect public health and nutrition, and to promote the public interest in sectors of vital importance to their socio-economic and technological development, provided that such measures are consistent with the provisions of this Agreement". Taken together, these provisions of the TRIPS Agreement may be read to suggest that there is some latitude for member states to incorporate sustainability and equity policies into their domestic legislative frameworks to regulate if, how and to what extent biological innovations derived from genetic resources should be protected by intellectual property rights.

In the realm of patent law, the TRIPS Agreement provides that "patents shall be available for any inventions, whether products or processes, in all fields of technology" so long as the subject matter of the patent is new, involves an inventive step and is capable of industrial application (Article 27(1)). However, member states are entitled to exclude from patentability patents on grounds of "*ordre public* or morality, including to protect human, animal or plant life or health or to avoid serious prejudice to the environment" (Article 27(2)). More specifically, where biological inventions derived from genetic resources are concerned, Article 27(3)(b) of the TRIPS Agreement permits member states to exclude from patentability:

> plants and animals other than micro-organisms, and essentially biological processes for the production of plants or animals other than non-biological and microbiological processes. However, Members shall provide for the protection of plant varieties either by patents or by an effective sui generis system or by any combination thereof. The provisions of this subparagraph shall be reviewed four years after the date of entry into force of the WTO Agreement.

The mandated review of this provision began in 1999, with numerous discussions raised in the TRIPS Council on issues relating to the patentability of plants and animals, including moral and ethical issues connected to the eligibility of life forms for patent protection and how to deal with the commercial use of traditional knowledge and genetic resources – particularly in patent applications – by parties other than their country (or community) of origin. Paragraph 19 of the 2001 Doha Declaration also contained a statement that the TRIPS Council should look into the relationship between the TRIPS Agreement and the UN Convention on Biological Diversity, the protection of traditional knowledge and folklore.[17] The conflict between the objectives of the CBD and the TRIPS Agreement is most visible from the fact that, as far as the TRIPS Agreement is concerned, patenting the by-products of genetic resources and associated traditional knowledge does not require proof of prior informed consent, neither is it necessary for the patent applicant to demonstrate that benefit-sharing has occurred in accordance with the Nagoya Protocol's ABS framework (Curci, 2009). Patenting biological innovations derived from genetic resources or traditional knowledge does not give rise to any legal obligation on the part of the patent holder to share the revenues generated from the patent with the providers of such resources or knowledge, thereby sowing the seeds of controversy in developing countries, where such conduct is liable to be construed as misappropriative and to be seen as acts of "biopiracy" and "biocolonialism".

The debates in this field have generated divergent views on how the access and benefit-sharing goals of the CBD ought to be integrated into the patent law framework. Some member states proposed that disclosure of the country of origin of genetic resources and traditional knowledge used in inventions, as well as evidence of prior informed consent and benefit-sharing on fair and equitable terms, should be made a requirement under the TRIPS Agreement[18] or within the WIPO Patent Cooperation Treaty. Other member states proposed alternative mechanisms outside the patent law system, including reliance on national legislation and contractual arrangements mandated by legislation that required such disclosure commitments. Despite the numerous attempts at stimulating legislative, regulatory and policy processes on disclosure requirements to support biodiversity conservation and sustainability development objectives (IUCN, 2005), no global consensus has been reached on the most appropriate mechanism for establishing a visible linkage between the CBD and the international intellectual property regime. Similarly, concurrent efforts by the WIPO Intergovernmental Committee on Intellectual Property and Genetic Resources, Traditional Knowledge and Folklore to develop model laws for bioprospecting agreements, *sui generis* protection for traditional knowledge and defensive strategies for preventing the improper acquisition of patent rights over innovations derived from biopiracy, have yet to bear fruit (Ong, 2008).

Other frontiers of biodiversity law: international environmental law, maritime law, human rights law, international law etc.

Apart from the multilateral legal conventions discussed above, the branches of biodiversity law may also be regarded as overlapping other neighbouring fields of environmental and international law. For instance, the regulation of international trade in endangered species under the Convention on International Trade in Endangered Species (CITES)[19] is buttressed by biodiversity conservation policies directed at specific endangered species of fauna and flora (listed in the Appendices of the CITES treaty) that have been commercially over-exploited and threatened with extinction. Biodiversity protection is also closely linked to international environmental laws relating to climate change, where states have adopted "equity"-based approaches in the United Nations' REDD+ (Reducing Emissions from Deforestation and Forest Degradation)[20] framework that are consonant with the ABS principles found in the CBD. Commentators in the international environmental law community have called for a higher degree of "clustering" between the CBD, the United Nations' Climate Change Convention and the United Nations' Convention on the Law of Sea (UNCLOS), to promote closer connectivity between their goals – biodiversity conservation, climate change mitigation and the governance of marine protected areas (Maes *et al.*, 2013). Given their commercial value to the biotechnology industry, maritime genetic resources on the seabed in areas beyond national jurisdiction (ABNJs) have also become a contemporary issue for the international maritime law community whose current agenda includes developing an ABS framework for such resources within the UNCLOS framework.

The nexus between biodiversity and traditional knowledge, and the people with whom they are most closely connected, has also given rise to CBD-related developments in the field of international human rights law, with the United Nations Declaration on the Rights of Indigenous Peoples (UNDRIP) at its core. Leveraging on traditional theories of property rights, law and development and indigenous legal traditions, advocates for giving local and indigenous communities a stronger role to play in the development of an international ABS framework have argued for the emergence of transnational biocultural rights that recognize their contributions as stewards of Nature (Bavikatte, 2014). Similarly, there have been calls to focus on building the socio-economic capacity of indigenous communities to increase their level of engagement with national ABS systems, protecting themselves from being short-changed by regulatory frameworks that favour sophisticated users seeking to extract traditional knowledge from them, while ensuring that the customary cosmologies of these communities are respected as part of the normative basis for any economic engagement with bio-prospectors (Drahos, 2014).

Two prominent features of contemporary biodiversity governance issues have also emerged as focal points for international law and politics theorists. First, while biodiversity is widely regarded as part of the "common heritage of mankind" (*res communis humanis*), the CBD regime is designed around the premise that states have sovereign rights of genetic resources within their boundaries. This has led to the development of an approach that lies between the "free appropriation and use" and "property and market" approaches towards regulating access to genetic resources and traditional knowledge, prompting calls for the entrenchment of a "common pools" approach towards genetic resources and traditional knowledge as a middle-ground for realizing the ABS objectives of the CBD and the Nagoya Protocol (Kamau *et al.*, 2013).

Secondly, given that biodiversity governance issues transcend geographical and spatial boundaries, there is a need for them to be approached from transnational normative perspectives that consider eco-systems holistically, from which a supranational regulatory regime might emerge (Kotzé and Marauhn, 2014).

Interdisciplinary perspectives to biodiversity law have also been injected by philosophers and ethicists who have argued that notions of equity and fairness in ABS systems ought to reflect the value of biodiversity based on Mankind's relationship with Nature. It has been argued, for instance, that the law should reflect human values to ensure human flourishing, in the sense of physical survival and emotional contentment, making it appropriate to approach biodiversity laws as part of an ethically guided framework for achieving meaningful survival (Snape, 1996). Others have advanced the proposition that biodiversity should be defined as an ethical concept, to better understand its value, which differs from disciplinary context to disciplinary context, and that an approach taken towards biodiversity that is embedded in social or natural ethics can be useful in resolving conflicts between scientists, policy-makers and other stakeholders, as well as balancing their conflicting practical interests and goals (Lanzareth *et al.*, 2014). Incorporating ethical foundations into international environmental law might involve embracing multiple considerations such as the ethics of species (maintenance of life and lives), the ethics of animals (protecting animal welfare and sustainable use) and the ethics of areas (preserving aesthetics and cultures) when approaching biodiversity conservation issues (Gillespie, 2011).

Conclusion

Like the subject matter it covers, biodiversity law is an organic and evolving discipline whose reach extends into multiple sub-disciplines and sub-specializations. Unlike many fields of international law, there are no explicit monitoring, compliance or enforcement mechanisms in place to compel states to fulfil their treaty obligations. This is exacerbated further in countries when there is a lack of political will by governments to do what is wise – to conserve and protect the world's natural heritage – when their biodiversity is imperilled by destructive economic policies or otherwise adversely affected by human activity. A multi-factorial perspective is essential to navigating the socio-political landscapes in which these issues are located, in order to fully appreciate the complexities of biodiversity law, which is *sine qua non* to developing effective strategies that might translate the aspirational objectives of these international treaties into reality.

Notes

1 The text of the Convention can be found at www.cbd.int/convention/text/.
2 COP Decision X/2 Strategic Plan for Biodiversity 2011–20.
3 See Art. 6 CBD, which obliges each contracting party to develop national strategies, plans or programmes for the conservation and sustainable use of biological diversity.
4 The text of the Protocol can be found at https://bch.cbd.int/protocol/text/.
5 WTO agreements, such as the agreement on the Application of Sanitary and Phytosanitary Measures (SPS Agreement) and the Agreement on Technical Barriers to Trade (TBT Agreement), and the Agreement on Trade-Related Intellectual Property (TRIPs) contain provisions that are relevant to the modified genetic organisms regulated by the Cartagena Protocol.
6 See Art. 15(7) CBD, read together with Arts 16 and 17, which (respectively) encourage contracting parties to facilitate access to, and transfers of, technologies relevant to the conservation and sustainable use of biological diversity, as well as to exchange the results of technical, scientific and socio-economic research (including indigenous and traditional knowledge) relevant to the conservation and sustainable use of biological diversity.

7 COP Decision VI/24 Part A. The text of the Guidelines can be found at www.cbd.int/abs/bonn/.
8 COP Decision VI/10F para. 34.
9 COP Decision X/1 Access to genetic resources and the fair and equitable sharing of benefits arising from their utilization. The text of the protocol can be found at www.cbd.int/abs/text/.
10 The original and amended text of the convention can be found at http://archive.ramsar.org/cda/en/ramsar-documents-texts/main/ramsar/1-31-38_4000_0.
11 An Introduction to the Ramsar Convention on Wetlands (RAMSAR Handbooks, 5th edn, 2016), 9.
12 Art. 2.2 of the Convention establishes that "wetlands should be selected for the List on account of their international significance in terms of ecology, botany, zoology, limnology or hydrology", criteria which have been subsequently refined by the Conference of Contracting Parties over the years to identify wetlands that are valuable to humanity as a whole – as cradles of biodiversity and providers of "ecosystem services" (including freshwater supply, sources of food and building materials, flood control and climate change mitigation).
13 Art. 3.1 of the Convention requires contracting parties to "formulate and implement their planning so as to promote the conservation of the wetlands included in the List, and as far as possible the wise use of wetlands in their territory". The contracting parties subsequently adopted, in Resolution IX.1 Annex A (2005), the following definition of "wise use", providing the conceptual foundations for a series of guidelines to assist contracting parties in the implementation of this principle: "Wise use of wetlands is the maintenance of their ecological character, achieved through the implementation of ecosystem approaches, within the context of sustainable development".
14 The text of the convention can be found at www.fao.org/3/a-i0510e.pdf.
15 Art. 13.2 (d)(ii) of the ITPGRFA, which directs payments to be made to a "Trust Account" financial mechanism referred to in Art. 19.3f of the Treaty.
16 The text of the latest version of the agreement is available at www.wto.org/english/docs_e/legal_e/31bis_trips_01_e.htm.
17 See "The TRIPS Agreement and the Convention on Biological Diversity" (Paper IP/C/W/368/Rev.1, revised 8 February 2006, WTO, 2006), Review of Art. 27.3(b) (Paper IP/C/369/Rev.1, revised 9 March 2006, WTO, 2006).
18 See Draft Modalities for TRIPS Related Issues (TN/C.W/52, WTO Trade Negotiations Committee, 19 July 2008).
19 The text of the Convention can be found at https://cites.org/eng/disc/text.php.
20 This is a global framework, developed under the auspices of the United Nations' Framework Convention on Climate Change, for reducing carbon emissions from deforestation in developing countries through a system of results-based finance payments funded by developed countries.

References

S.K. Bavikatte, *Stewarding the Earth: Rethinking Property and the Emergence of Biocultural Rights* (New Delhi: Oxford University Press, 2014).

J. Curci, *The Protection of Biodiversity and Traditional Knowledge in International Law of Intellectual Property* (Cambridge: Cambridge University Press, 2009).

P. Drahos, *Intellectual Property, Indigenous People and their Knowledge* (Cambridge: Cambridge University Press, 2014).

A. Gillespie, *Conservation, Biodiversity and International Law* (Cheltenham: Edward Elgar, 2011).

M. Girsberger, *Biodiversity and the Concept of Farmers' Rights in International Law: Factual Background and Legal Analysis* (New York: Lang, 1999).

IUCN (2005), M. Chouchena-Rojas, M. Ruiz Muller, D.Vivas and S.Winkler (eds), *Disclosure Requirements: Ensuring Mutual Supportiveness between the WTO TRIPS Agreement and the CBD* (Gland, Geneva and Cambridge: IUCN and ICTSD, November 2005).

E.C. Kamau and G. Winter (eds), *Common Pools of Genetic Resources: Equity and Innovation in International Biodiversity Law* (Abingdon: Routledge, 2013).

L.J. Kotzé and T. Marauhn (eds), *Transboundary Governance of Biodiversity* (Leiden, Boston: Brill Nijhoff, 2014).

D. Lanzareth and M. Friele (eds), *Concepts and Values in Biodiversity* (New York: Routledge, 2014).

F. Maes, A. Cliquet, W. du Plessis and H. Mcleod-Kilmurray, *Biodiversity and Climate Change: Linkages at International, National and Local Levels* (Cheltenham: Edward Elgar, The IUCN Academy of Environmental Law Series, 2013).

E. Morgera, E. Tsioumani and M. Buck (eds) *Unravelling the Nagoya Protocol: a commentary on the Nagoya Protocol on access and benefit-sharing* (Leiden, Boston: Brill, 2014).

S. Oberthür and G. Kristin Rosendal (eds) *Global Governance of Genetic Resources: Access and Benefit-sharing after the Nagoya Protocol* (Abingdon: Routledge 2014).

B. Ong, 'Biotechnological Innovations, Genetic Resources and Traditional Knowledge: Current Developments at the World Intellectual Property Organisation' in M. Jeffery, J. Firestone, K. Bubna-Litic (eds) *Biodiversity, Conservation, Law and Livelihoods: Bridging the North-South Divide* (Cambridge: Cambridge University Press, IUCN Academy of Environmental Law Research Series, 2008) 553.

W.J. Snape (ed), *Biodiversity and the Law* (Washington D.C.: Island Press, 1996).

2
BIODIVERSITY AND THE LAW IN BRIEF

Charles R. McManis

As Burton Ong points out in Chapter 1, the Convention on Biological Diversity (CBD 1992) states in Article 1 that its threefold objective is (1) the conservation of biological diversity; (2) the sustainable use of its components; and (3) the fair and equitable sharing of the benefits arising out of the utilization of genetic resources. Article 2 defines genetic resources as genetic material of actual or potential value. While Article 3 asserts that states have the sovereign right to exploit their own resources pursuant to their own environmental policies, Article 8(j) imposes an obligation on states to (1) respect, preserve and maintain knowledge, innovations and practices of indigenous and local communities embodying traditional lifestyles relevant for the conservation and sustainable use of biological diversity; (2) promote their wider application with the approval and involvement of the holders of such knowledge, innovations and practices; and (3) encourage the equitable sharing of the benefits arising from the utilization of such knowledge, innovations and practices.

The Nagoya Protocol on Access to Genetic Resources and the Fair and Equitable Sharing of Benefits Arising from their Utilization (ABS) to the Convention on Biological Diversity (Nagoya Protocol 2010), as its name makes clear, is a supplementary agreement to the CBD and is devoted to providing a transparent legal framework for the effective implementation of the third objective of the CBD – namely the grant of access and the fair and equitable sharing of the benefits arising out of the utilization of genetic resources. Article 5 of the Nagoya Protocol states that the benefits arising from the utilization of genetic resources shall be shared in a fair and equitable way with the Party providing such resources—that is, the country of origin of such resources or a Party that has acquired the genetic resources in accordance with the CBD—and specifies that such sharing shall be upon mutually agreed terms. Article 7 of the Protocol specifies that each Party shall take measures, as appropriate, with the aim of ensuring that traditional knowledge associated with genetic resources that is held by indigenous and local communities is accessed with the prior and informed consent or approval and involvement of these indigenous and local communities, and that mutually agreed terms have been established.

Although Articles 5 and 7 of the Nagoya Protocol thus envision essentially bilateral agreements for the fair and equitable sharing of benefits arising from the utilization of genetic resources and associated traditional knowledge, Article 10 specifies that the Parties shall consider the need for and modalities of a global multilateral benefit-sharing mechanism (GMBSM) to address the fair and equitable sharing of benefits derived from the utilization of genetic resources and traditional knowledge associated with genetic resources that occur in transboundary situations or for which it is not possible to grant or obtain prior informed consent. The GMBSM envisioned in Article 10 seems to resemble the multilateral system (MLS) established by the Food and Agriculture Organization's International Treaty on Plant Genetic Resources for Food and Agriculture, which entered into force in 2004 (ITPGRFA 2004), popularly known as the Seed Treaty. The Seed Treaty establishes a multilateral system of access and benefit-sharing to facilitate plant germplasm exchanges of specified plants by means of a Standard Material Transfer Agreement. Among other things, the Seed Treaty specifies that those who access genetic materials through the MLS will either freely share any new developments with others for further research or, if they wish to keep the developments to themselves, pay a percentage of any commercial benefits they derive from their research into a common fund to support conservation and further development of agriculture in the developing world.[1] Among the commercial benefits that might be derived from the utilization of genetic resources are patents, trade secrets, and the licensing agreements and royalties that are based on these two forms of intellectual property. Protection of these and other forms of intellectual property, such as the undisclosed information of indigenous or local communities, is mandated by the 1994 international agreement on Trade-related Aspects of Intellectual Property rights (TRIPS Agreement 1994).

This Handbook is organized around the major objectives and thematic focal points that have emerged from the Convention on Biological Diversity and other legal frameworks concerned with protecting our planet's biodiversity. The chapters contained in Part II address issues pertaining to the conservation and sustainable use of biodiversity, while the chapters contained in Part III address issues relating to access and benefit-sharing. Part IV focuses specifically on traditional knowledge protection as it relates to the preservation and sustainable use of biodiversity and the fair and equitable sharing of benefits arising out of the utilization of genetic resources. Part V discusses specific issues arising out of the intersection between biodiversity and intellectual property protection. Finally, Part VI focuses on the ethics economics and science-policy interface of biodiversity protection.

Part II: Conservation and sustainable use of genetic resources

Opening Part II's discussion of the international and domestic legal frameworks for the conservation and sustainable use of genetic resources, Nicholas Robinson in Chapter 3 examines biodiversity in international environmental law through the lens of the UN Sustainable Development Goals. Robinson provides an overview of the principal substantive and procedural laws by which States seek to conserve and sustain biological diversity. Specifically, Robinson examines (a) the general principles of law applicable to biodiversity; (b) the framework agreements to which most nations have adhered; (c) the sector treaties and generic types of statutes for different aspects of biodiversity; (d) the typical legal stewardship systems of regulatory tools deployed to govern human behavior with respect to biodiversity, including "rational uses of natural resources," and (e) the legal lacunae, where failure to agree on law harms biodiversity and leaves it at risk.

In Chapter 4, Jamie Benedickson and Sandy Paterson survey the role of law in enabling and regulating protected areas to enhance biodiversity. This starting point draws attention to habitat loss as a major driver of biodiversity loss. Following an overview of the international legal framework of relevance to protected areas, including but not limited to the Convention on Biological Diversity, Benedickson and Paterson review key issues which are to be addressed in domestic protected area laws. These include the following: scope and application; objectives and principles; management categories and governance types; systems and management planning; institutional arrangements; regulating potentially harmful activities; compliance and enforcement; and financing. The authors also introduce emerging legal issues relating to protected areas. These include: promoting the principle of non-regression; recognition of diverse protected area governance frameworks; connectivity conservation; scoping and implementing other effective area-based conservation measures; and regulating marine protected areas in areas beyond national jurisdiction.

In Chapter 5, entitled "The international legal framework for the protection and sustainable use of marine biological diversity," Youna Lyons and Denise Cheong discuss biodiversity conservation and sustainable use of genetic resources in habitats particularly in need of international legal protection – namely marine resources found in areas beyond national jurisdiction. They describe the United Nations Convention on the Law of the Sea (UNCLOS) as the cornerstone of the current legal framework concerned with the protection and use of marine biodiversity in areas beyond national jurisdiction. Thus, all other specific conventions that apply to the marine environment, such as those that pertain to specific uses of the sea or to the conservation of specific ecosystems, habitats or species, as well as the Convention on Biological Diversity itself, should be seen as putting into operation the principles and provisions of UNCLOS. Lyons and Cheong first examine the relevant provisions in UNCLOS that seek to protect and preserve marine biodiversity in light of decisions from the International Tribunal for the Law of the Sea. They next examine the criteria set out in the specific conservation conventions to identify ecosystems, habitats or species that should be protected as a matter of priority. They then examine the extent to which conventions that pertain to specific uses of the sea restrict activities in the interest of protecting ecosystems, habitats or species. Finally, they discuss the protection of marine biodiversity in areas beyond national jurisdiction as well as the development of "bioprospecting" of marine genetic resources and the regulatory issues it raises. They also outline the ongoing process of negotiations at the United Nations towards a new internationally binding legal instrument for biodiversity in areas beyond national jurisdiction.

In Chapter 6, entitled "Biosecurity, invasive species and the law," Opi Outhwaite discusses another important driver of biodiversity loss – namely invasive species. Outhwaite points out that as an overarching concept, "biosecurity" can be viewed as incorporating the management of invasive alien species (IAS) – including but not limited to pests, diseases and pathogens – which are now collectively recognized as one of the major drivers of global biodiversity loss. Outhwaite notes that domestic and international legal efforts to address biosecurity have been piecemeal, reactive and sectoral in nature, focusing on agricultural biosecurity, with the United Nations Food and Agriculture Organization (FAO) playing an important role in defining and promoting the concept of biosecurity at the international level. Only recently has the FAO recognized that biosecurity extends beyond risks to human, animal and plant life and health and also includes risks to the environment. After dealing with questions of definition and domestic pre-entry, point-of-entry and post-entry measures, Outhwaite discusses the various international legal and policy measures governing biosecurity and IAS, but noting at the outset that there is no single international instrument for addressing biosecurity. The relevant measures include the sanitary and phytosanitary measures of the World Trade Organization, international standard-setting, the Convention on Biological Diversity, various issue- and pathway-specific instruments, and EU regulation on invasive species.

In Chapter 7, entitled "Biotechnology, biodiversity, and the environment," Barbara Schaal and Joseph Jez address the potential for genetically modified organisms to become a potent form of invasive species. An unfortunate aspect of the highly charged and polarized debate over the impact of the use of genetically modified organisms, both plants and animals, on the environment, say Schaal and Jez, is to view the issue in black or white; biotechnology is either good or bad. However, Schaal and Jez argue that whether or not an application of biotechnology has potential harmful, beneficial or neutral effects on the environment is both species- and context-specific and depends on the genetically modified plant or animal, the geographic region where the organisms are placed, and the local biological environment. Their chapter compares biotechnology and traditional breeding of plants and animals, provides an overview of biotechnology in the tropics, examines the concerns and benefits of biotechnology on biodiversity, and highlights recent advances and new technologies, such as genome editing that will affect future biotechnology applications.

In Chapter 8, James Ming Chen discusses the domestic legal responses in the United States (which is not a party to the Convention on Biological Diversity) to the problems of biodiversity loss and climate change. As Chen explains, the greatest drivers of biodiversity loss today (in order of importance) are climate change, habitat destruction, invasive species, pollution, population, and overkill. The Endangered Species Act of the United States, however, imposes its clearest and harshest sanctions precisely where the drivers of extinction are the weakest – namely when humans consciously capture and kill other living things. Fortunately, says Chen, in recent years the Endangered Species Act has been adapted by the courts to address habitat destruction on private land and to mitigate climate change.

In Chapter 9, entitled "China's biodiversity law," John Nagle discusses the biodiversity law of the single most populous country in the world – namely China. He notes that China is a vast, varied nation that hosts an incredible range of ecosystems and species. However, the rapid economic growth that China has experienced since 1980 strains the nation's ability to preserve ecosystems, species and genetic resources. In his chapter, Nagle discusses the various threats to China's biodiversity and details China's response in the creation of nature reserves and *ex situ* preservation and in educational campaigns, as well as its enactment of environmental and biodiversity laws. He also discusses the difficulties in enforcement of these laws, the role that NGOs have played in efforts to preserve biodiversity, and China's participation in such international efforts embodied in such instruments as the Convention on International Trade in Endangered Species, the Ramsar Convention for the protection of wetlands, and the Convention on Biological Diversity. He concludes that the ultimate success of the measures remains uncertain.

In Chapter 10, entitled "The International Treaty for Plant Genetic Resources for Food and Agriculture: toward the realization of farmers' rights as a means of protecting biodiversity," Regine Andersen concludes Part I by contributing a chapter which spans the topics of biodiversity protection and sustainable use, access and benefit-sharing, traditional knowledge, and the interplay of biodiversity and intellectual property protection with respect to crop genetic resources. To that end, Andersen introduces the International Treaty on Plant Genetic Resources for Food and Agriculture (ITPGRFA) and discusses the extent to which the ITPGRFA furthers the task of realizing farmers' rights as a means of protecting agricultural biodiversity. Andersen first details the historical development of the concept of farmers' rights and offers two conceptual approaches to the development of farmers' rights – the stewardship approach and the ownership approach and explains how each is in evidence in the four elements of farmers' rights: (1) protection of farmers' traditional knowledge; (2) benefit-sharing; (3) participation in decision-making; and (4) the rights to save, use, exchange and sell farm-saved seed. Andersen concludes that, despite its shortcomings, the ITPGRFA is the single most important international instrument that currently exists to ensure the sustainable management of crop genetic resources.

Part III: Access and benefit-sharing

In the first chapter focusing on access and benefit-sharing, Carlos Correa in Chapter 11 expands on a topic earlier introduced by Lyons and Cheong in Chapter 5 – namely the conservation and sustainable use of marine biodiversity – by focusing specifically on access and benefit-sharing in the context of marine biodiversity. In Chapter 11, entitled "Access to and benefit-sharing of marine genetic resources beyond national jurisdiction," Correa examines some of the elements that might be considered in developing a new legally binding instrument on access and benefit-sharing (ABS) derived from the exploitation of marine genetic resources found in areas beyond national jurisdiction (ABNJs). By way of background, Correa first elaborates on some of the problems faced in the implementation of the Convention on Biological Diversity and other relevant international instruments such as the Food and Agriculture Organization's International Treaty on Plant Genetic Resources for Food and Agriculture and the World Health Organization's Influenza Pandemic Influenza Preparedness Framework, which set up specific regimes for ABS. He then details a number of approaches that may be taken to ABS, consistent with the United Nations Convention on the Law of the Sea, and identifies lessons to be learned from current international instruments with regard to establishing definitions, identifying benefits to be shared, and the role, if any, of intellectual property, specifically patents, resulting from access to marine genetic resources in ABNJs.

While the Convention on Biological Diversity established new expectations that direct monetary or research capacity, benefits would accrue to countries from products developed using their biological resources, James Miller and Porter Lowry in Chapter 12, entitled "The impact of natural products discovery programs on our knowledge of the flora of Madagascar," point out another important benefit from natural products discovery programmes that has received less attention; namely that the institutions engaged in "bioprospecting" have been institutions whose mission is to document and understand diversity of some major group of living organisms. Thus, funding of natural products discovery programmes has also resulted in valuable and important natural history collections that have helped advance biological knowledge and in many cases helped promote conservation. Miller and Lowry review two such efforts, both conducted by the Missouri Botanical Garden (MBG) in Madagascar, and outline how these programmes contributed to advancing the biological research and conservation goals of the MBG, its local partners and the Madagascan government.

In Chapter 13, Lye Lin-Heng and Rose-Liza Eisma-Osorio conduct a comparative study of the regulatory measures on access and benefit-sharing in the Philippines and Singapore, as well as recent developments in ASEAN. Both countries are party to the CBD and are members of the Association of Southeast Asian Nations (ASEAN). The Philippines hosts the ASEAN Centre for Biodiversity and is said to have one of the most advanced pieces of legislation regulating access and benefit-sharing (though, discouragingly, despite lowering fees, the Philippines has not issued any permits under this amended legislation). While the Philippines is one of three mega-biodiverse countries (along with Indonesia and Malaysia) in southeast Asia, and explicitly recognizes the rights of indigenous groups, Singapore, by contrast, is a tiny city state having a population of 5.28 million people, with no indigenous population, and with only some 5 per cent of its natural forests remaining. However, Singapore is a repository of genetic materials gathered from the region since colonial times, and seeks to be a leader in greening the urban environment. The Philippines acceded to the Nagoya Protocol in 2015 but Singapore is not yet a member. It is clear that Singapore and the Philippines have very different regimes for access and benefit-sharing, given their differences in socio-economic conditions and geographic landscapes, but both countries have comprehensive laws governing access and benefit-sharing.

Complementing John Nagle's Chapter 8 discussion of the biodiversity law of China with a discussion of the biodiversity law of South Africa, Rachel Wynberg, in Chapter 14, entitled "One step forward, two steps back? Implementing access and benefit-sharing legislation in South Africa," notes that South Africa is one of few African countries to have a comprehensive policy framework in place for access and benefit-sharing (ABS). In her chapter, Wynberg provides a brief history of the development of legal frameworks for access and benefit-sharing in South Africa, discusses the implementation of ABS in practice, and offers suggestions for revitalizing and transforming the regulatory framework for ABS in South Africa. Among the problems she identifies are multiple laws and institutions, impractical permitting, lack of awareness and understanding of science, markets, and industry, and difficulty in identifying traditional knowledge holders and implementing benefit-sharing. Wynberg also offers a number of key lessons that have emerged over the past ten years. Despite these challenges, Wynberg notes that there appears overall to be a positive attitude to ABS in South Africa and a general willingness on the part of companies to comply with ABS laws. She also notes that efforts made by government to establish a Bioprospecting Forum to improve dialogue and trust between different groups have paid off. She concludes by offering a number of key lessons that have emerged over the past ten years, including the need for informed regulators, the need to regulate discovery and commercialization differently, but to recognize that both need regulation to some extent, the need to regulate biotrade in certain circumstances, but not to address these concerns by conflating them with ABS in a single regulatory system, to name only a few of these lessons.

In Chapter 15, Margo Bagley points out that domestic legislation implementing the CBD and Nagoya Protocol, as well as the FAO International Treaty on Plant Genetic Resources for Food and Agriculture, is designed to address the utilization of tangible genetic material and provide for the sharing of benefits to the mutual benefit of users and providers of these materials. However, advances in DNA sequencing and synthetic biology more broadly are resulting in the "de-materialization" of genetic material, with the potential to enable researchers and corporate entities to bypass the obligations to obtain permission to access and to share benefits of use with providers of genetic resources. In her chapter, Bagley considers current challenges associated with this de-materialization, in the context of intellectual property protection for synthetic biology outputs, implementation issues concerning the Nagoya Protocol, and emerging concerns regarding synthetic biology and misappropriation ("biopiracy") of digital genetic information in commercial and non-commercial research projects.

Part IV: Traditional knowledge protection

In the first chapter to discuss the protection of traditional knowledge associated with genetic resources, Michael Balick in Chapter 16, entitled "Traditional knowledge: lessons from the past, lessons for the future," discusses the nature of traditional knowledge and its devolution, emphasizing that, contrary to popular opinion, traditional knowledge is not static but in a constant state of change. Thus, traditional knowledge no longer considered as immediately useful may be lost rather than maintained – a process Balick calls "devolution." Balick points out that the extinction of language is an excellent example of devolution leading to the extinction of cultural knowledge. Concern for the loss of traditional knowledge has led to numerous ethnobotanical and culture-related projects that are essentially "salvage ethnobotany." Balick illustrates his point by describing his ethnobotanical research in Micronesia and his study of traditional healing in Belize. From these examples Balick identifies factors contributing to devolution and strategies for both protecting and letting go of traditional knowledge.

In Chapter 17, Michael Blakeney discusses bioprospecting and traditional knowledge in the "mega-biodiverse" country of Australia. Blakeney identifies Australia's international bioprospecting obligations under the Convention on Biological Diversity, as well as Australian bioprospecting legislation and national and state policy initiatives. He then describes bioprospecting six case studies of bioprospecting projects that have taken place in various Australian states and territories, emphasizing the process of acquiring access to genetic resources and associated traditional knowledge, obtaining prior informed consent from the knowledge holders, and the role played by intellectual property (specifically patents) as a benefit-sharing mechanism.

In Chapter 18, entitled "If we have never been modern, they have never been traditional: 'traditional knowledge', biodiversity, and the flawed ABS paradigm," Graham Dutfield points out that, while traditional knowledge and customary practices relate to all three objectives of the Convention on Biological Diversity (CBD) – conservation, sustainable use and benefit-sharing arising from access (ABS) – most of the attention surrounding traditional knowledge concerns the third objective. Underscoring that concern was the adoption in 2010 of the Nagoya Protocol on Access to Genetic Resources and the Fair and Equitable Sharing of the Benefits Arising from their Utilization. However, according to Dutfield, the conventional wisdom behind the TK aspects of "ABS plus intellectual property" model embodied in the CBD and the Nagoya Protocol is fundamentally flawed. Dutfield identifies four main reasons why this approach generally fails to provide solutions that are fair, equitable and efficient. He names these as (1) the hybrid nature of knowledge systems; (2) the problem of origin and attribution; (3) the over-regulation and corporatization tendency; and (4) the exchange value "distraction." He concludes that international solutions are no substitute for legal and policy initiatives giving more power to local people to make their own decisions.

In Chapter 19, as its title states, Brendan Tobin discusses the fundamental role of customary law in securing protection of traditional knowledge under the Nagoya Protocol. He notes that throughout the negotiation of both the Protocol and the draft texts of the World Intellectual Property Organization (WIPO) Intergovernmental Committee on Intellectual Property, Genetic Resources, Traditional Knowledge and Folklore (IGC), indigenous peoples and local communities have consistently argued that any regime for the protection of their traditional knowledge should be grounded upon their own, primarily customary (i.e. oral, uncodified) legal regimes, in order to ensure that protection accords with their own cultural, social, economic and spiritual priorities. In his chapter, Tobin examines the commitments that states have undertaken with regard to customary law under the Nagoya Protocol, the status of the legal regimes of indigenous peoples and local communities in international law, the significant role customary law already plays in regulation of traditional knowledge in Andean countries, and the significance such recognition has for implementation of the Protocol.

Part V: Biodiversity and intellectual property protection

In Chapter 20, entitled "Biodiversity, intangible cultural heritage and intellectual property," Christoph Antons notes that, while the 1992 Convention on Biological Diversity signified a paradigm shift in thinking about the stewards and custodians of biological conservation, as its Article 8(j) famously included the traditional knowledge and practices of local and indigenous communities in its ambit, Article 3 of the CBD has also reinforced the national sovereignty of governments in controlling their genetic resources. Thus, the customary law of local and indigenous communities remains mostly limited and subordinated to national laws related to environmental protection and resource allocation, and the prior informed consent of and benefit-sharing with local and indigenous communities has brought few tangible results.

Similarly, the attempt to harmonize the principles of the CBD with the strong intellectual property rights granted by the World Trade Organizations Agreement on Trade-related Aspects of Intellectual Property Rights (TRIPS) and the ongoing discussions at the World Intellectual Property Organization concerning international instruments to protect traditional knowledge and traditional cultural expressions have made only modest progress. In this situation of temporary impasse, wider notions of cultural landscapes under the 1972 World Heritage Convention or the safeguarding of the same, as promoted in the UNESCO Convention of 2003, have come to the fore. In this chapter, Antons first describes the paradigm shift in the interpretation of these treaties towards recognition of natural and biocultural heritage that has taken place since the 1970s and then discusses the interplay between national heritage laws and the intellectual property laws in developing countries in Asia, with particular reference to Indonesia.

In Chapter 21, as its title states, Brad Sherman discusses intellectual property, biodiversity and food security. Sherman points out that in developing and developed countries alike, there is an urgent and ongoing need to improve agricultural yields, increase sustainability, enhance and facilitate the breeding and development of new plant varieties, and assist local agriculture in adapting to climate and environmental changes. He notes that over the last decade or so there has been a growing interest in the potential role that intellectual property law might be able to ensure a secure and consistent supply of food. At the same time, there have also been complaints that intellectual property not only undermines attempts to improve agricultural productivity, but that it also threatens local and global food security. The aim of Sherman's chapter is to look at one specific area where, for better or worse, intellectual property law potentially plays a role in shaping food security: namely in relation to the conservation, preservation and use of genetic resources. In his chapter, Sherman first looks at the three international legal regimes that interoperate in the area: namely the CBD, the International Treaty on Plant Genetic Resources for Food and Agriculture, and the Agreement on Trade-related Aspects of Intellectual Property Rights. He then examines how these international regimes operate at the local level and impact the distinctive nature of biological innovation. Reid concludes by discussing the innovative way in which the international research organization, CGIAR, has attempted to resolve the tension between intellectual property law and publicly funded research designed to secure food security.

In Chapter 22, entitled "Sisyphus redivivus? The work of WIPO on genetic resources and traditional knowledge," Nuno Pires de Carvalho discusses the lengthy and thus-far unproductive deliberations of the World Intellectual Property Organization's Intergovernmental Committee on Intellectual Property and Genetic Resources, Traditional Knowledge and Folklore (IGC), the mandate for which was to seek an international consensus on how to create an international mechanism for protecting those intangible outputs of tradition cultures having economic value. After 16-plus years of futile efforts to create this international consensus, the IGC has only obtained the consent of its members to continue to meet and discuss the issue. According to Carvalho, the reason for this lack of progress derives from the divergence between the primary function of intellectual property – which is to differentiate enterprises whenever they exert the rivalry in luring clients from competitors and keeping their clients from being lured by competitors – and the goal of traditional knowledge protection through an intellectual property mechanism whose purpose is not to grant an exclusivity to be enforced against competitors but rather against the traditional knowledge holder's own clients. This fundamental discontinuity, says Carvalho, condemns the IGC to pursuing an endless and futile effort to accomplish an essentially impossible task.

While conceding that the work of the WIPO Intergovernmental Committee has often faltered due to the different aspirations and expectations for the forum, Daniel Robinson in Chapter 23 nevertheless thinks it worthwhile to reflect on the achievements of the IGC and the WIPO Secretariat, which he argues includes the useful materials generated, participation by representatives of indigenous and local communities, and the still evolving draft instruments on intellectual property, genetic resources, traditional knowledge and traditional cultural expressions. In his chapter, Robinson considers the current stalemate in the work of the IGC and possible ways forward, the scale at which traditional knowledge can best be protected, and national legal developments and local projects that are emerging. He also considers the implications and risks for the IGC due to the changing international political environment, characterized by an apparent withdrawal from multilateralism more generally.

Part VI: The ethics, economics and science-policy interface of biodiversity protection

In Chapter 24, entitled "Naturalizing morality," Ursula Goodenough points out that while most arguments for the preservation of biodiversity may be classified as consequentialist or cost/benefit analyses, an alternate ground for preserving biodiversity can be derived from antecedent moral precepts. Moral precepts are commonly assumed to derive from the teachings of traditional religions. But Goodenough argues that the moral precepts for protecting biodiversity can be derived from what she calls "religious naturalism." Her chapter explores the project of articulating an Ecomorality based on a science-based world view.

In Chapter 25, entitled "Making legal use of the valuation of nature," Colin Reid notes that the increasing recognition of the value of ecosystem services creates the potential for new approaches to biodiversity conservation. By recognizing the ways in which human activities are affecting, either positively or negatively, the ecosystem and the services it provides, a basis exists for developing new means of reflecting and influencing our impact on biodiversity, predominantly by applying various economic instruments. Because the effect of our activities can be either positive or negative, both aspects can be incorporated into mechanisms that reward or penalize activities – either a beneficiary pays mechanism, where impacts are positive, or a polluter pays mechanism, where the impacts are negative. Reid discusses various "beneficiary pays" mechanisms for payment of ecosystem services and "polluter pays" mechanisms, such as liability, offsetting schemes and impact fees, and concludes by noting that, notwithstanding significant challenges, such as valuing ecosystem services, there is a growing awareness that simply recognizing that ecosystem services are valuable enables them to become the subject of contractual and other legal arrangements.

In Chapter 26, as its title implies, Joseph Henry Vogel, Klaus Angerer, Manuel Ruiz Muller and Omar Oduardo-Sierra argue that the concept of "bounded openness" should serve as the modality of the Global Multilateral Benefit-Sharing Mechanism specified in Article 10 of the Nagoya Protocol, which envisions such a mechanism for transboundary situations or situations for which it is not possible to grant or obtain prior informed consent to utilize genetic resources. The authors note that the failure of the first 13 Conferences of the Parties (COP) of the CBD to achieve an economically sound policy for access and benefit-sharing can be traced to a mistake in the drafting of Article 2 of the CBD, which other chapter authors in this volume have recognized; namely categorizing "genetic resources" as "material'" rather than as genetic information, the only economics appropriate to which is the economics of information, specifically the concept of "bounded openness." As Vogel *et al.* explain the concept, it is the optimal modality for the transboundary situations referred to in Article 10. As such, natural information would flow unencumbered for RandD, which is the "openness" of bounded openness, while the "boundedness" would be reflected in a significant patent or other royalty (not to exceed 15 per cent) on any subsequent intellectual property acquired, to be shared among the countries of origin, proportional to habitat. After explaining the concept of bounded openness, the authors contrast the economic implications of genetic resources as material with that of genetic resources as natural information, and reproduce passages from the 2013 Online Discussion Groups on Article 10 of the Nagoya Protocol, which display a spirited discussion of bounded openness, but which were ultimately expunged by the Secretariat of the CBD in its official Synthesis of the Online Discussion of genetic resources as natural information.

In Chapter 27, Guillaume Futhazar, Denis Pesche and Sandrine Maljean-Dubois discuss the newly established Intergovernmental Science-Policy Platform on Biodiversity and Ecosystem Services (IPBES). They note that the IPBES was modeled on the Intergovernmental Panel on Climate Change (IPCC) and grew out of a perceived lack of scientific knowledge on how to tackle the loss of biodiversity. The stated objective of the IPBES is to strengthen the science-policy interface for biodiversity and ecosystem services for the conservation and sustainable use of biodiversity, long-term human well-being and sustainable development and thus echoes the objective of the Convention on Biological Diversity. In their chapter the authors provide a general overview of the IPBES in order to make sense of its mandate, structure, rules of procedure and general place in the complex institutional landscape of biodiversity governance.

Note

1 www.fao.org/plant-treaty/areas-of-work/the-multilateral-system/overview/en/.

References

Agreement on Trade-Related Aspects of Intellectual Property Rights, Including Trade in Counterfeit Goods: General Agreement on Tariffs and Trade-Multilateral Trade Negotiations (The Uruguay Round), 33 I.L.M. 81 (December 15, 1993), available at www.wto.int/english/docs_e/legal_e/27-trips.pdf [hereinafter TRIPS Agreement].

Convention on Biological Diversity (CBD), opened for signature June 5, 1992, 31 I.L.M. 818, available at www.cbd.int/convention/text/.

International Treaty on Plant Genetic Resources for Food and Agriculture (ITPGRA) entered into force June 29, 2004, available at www.fao.org/3/a-i0510e.pdf.

Nagoya Protocol on Access to Genetic Resources and the Fair and Equitable Sharing of Benefits Arising from their Utilization to the Convention on Biological Diversity, adopted Oct. 29, 2010, available at www.cbd.int/abs/text/.

PART II

Conservation and sustainable use of genetic resources

3

BIODIVERSITY IN INTERNATIONAL ENVIRONMENTAL LAW THROUGH THE UN SUSTAINABLE DEVELOPMENT GOALS

Nicholas A. Robinson

Biological diversity is an evolving focus of international environmental law. It encompasses three stages of legal development. The first was the diverse nature of conservation laws promulgated since the late 19th century governing fish and game or forestry. Second were the laws which began in the early and mid-20th century that address soil conservation, agricultural activity and fisheries management, and water and drought management. Third came the pollution and environmental quality laws enacted in the mid- and late 20th century, which affect the ways natural systems are impacted by pollution and waste. All have been subsumed within international environmental law, which since 1972 began to elaborate the links between these subjects' association with common threats to biological diversity. As Earth's current "sixth great extinction" has deepened (Kolbert, 2014), the insufficiency of each nation incrementally enacting new statutes or agreeing on new treaties to conserve biodiversity has become evident.

As progenitor of the concept and content of an intergovernmental agreement on biodiversity (Burhenne-Guilmin and Casey-Lefkowitz, 1993),[1] the Environmental Law Programme of the International Union for the Conservation of Nature and Natural Resources (IUCN) had expectations that the United Nations Convention on Biological Diversity (CBD) would provide a unifying framework in which to integrate the various laws on nature (Burhenne-Guilmin and Casey-Lefkowitz 1993). This objective for a strategic plan on global biodiversity was in part attained in a decision of the CBD's 10th Conference of the Parties taken in Japan in 2010,[2] but while these "Aichi Targets" provide invaluable guidance, their goals are far from being embraced by either States or treaty organizations. Especially in their domestic laws, nations have not yet elected to mold the CBD to serve such a function. In 2015, the United Nations General Assembly acknowledged the urgency of Earth's biodiversity crisis in adopting the Sustainable Development Goals, in particular Goals for the terrestrial environment (SDG 15), for the marine environment (SDG 14) and for the Earth's climate (SDG 13).

This chapter provides an overview of the principal substantive and procedural laws by which States seek to conserve and sustain biological diversity. "Biodiversity" generally has come to mean the many different populations and species of organisms, or popularly understood as flora and fauna, together with their evolved reservoirs of generic material, and the ecosystems which sustain these species, popularly seen as landscapes (McNeely 1990; Wilson 1998).[3] Various laws at local, national, regional, or international level focus on different aspects of biodiversity. Just as scientists apply their analysis contextually to the biodiversity within of a rain forest, or on tundra, or within a coral reef or on an alpine glacier, so the drafts of legal texts vary the use of legal terms. The varied legal terminologies used for different biomes or species lead to inconsistencies and sometimes conflicts among laws. Older laws tend to reflect the scientific understanding of their time, and rarely are amended or revised to reflect advances in evolutionary biology or other environmental sciences. Laws for a pragmatic management of natural resources, such as those advanced by the UN Food and Agricultural Organization, will stress goals of reasonable or "wise use," or management for "sustained yield," without providing the procedures by which empirical scientific inquiry is to be applied. Procedures such as "environmental impact assessment," which are required in all national decision-making under Principle 17 of the 1992 UN Declaration on Environment and Development,[4] are frequently perverted into result-oriented studies that do little to assess biological diversity, much less protect it from adverse impacts of a proposed project. In short, there is no single legal standard of performance by which to determine what is the environmental rule of law by which humans govern their relationship with Earth's biodiversity.

As the UN Commission on Environment and Development explained in *Our Common Future* (1987), "The Earth is one, but the world is not." Given this incongruity between the laws of men and the law of nature, this survey of biodiversity in international environmental law examines the following: (a) the general principles of law applicable to biodiversity; (b) the framework agreements to which most nations have adhered; (c) the sector treaties and generic types of statutes for different aspects of biodiversity; (d) the typical legal stewardship systems of regulatory tools deployed to govern human behavior with respect to biodiversity, including "rational uses of natural resources," and (e) the legal lacunae, where failure to agree on law harms biodiversity and leaves it at risk.

As *homo sapiens* procreates, "people" are currently adding two billion more persons to populate the planet. Few places have established sustainable human settlements such as found in Singapore or Costa Rica. At the 1992 Rio Earth Summit, States understood then that "consumption and production" patterns were unsustainable.[5] Trends critically assessed then have spiralled into ever more pronounced patterns of degradation. The Rio Earth Summit produced Agenda 21, which set many nations on the path to designing and implementing reforms to achieve sustainable and resilient socio-economic development (Robinson, 1993).[6] But the slow pace of these reforms prompted nations at the Johannesburg UN World Summit on Sustainable Development (2002) to draft the "Johannesburg Plan of Implementation" for Agenda 21. When this Plan was seen to falter, in 2012 at the "Rio+20" Summit meeting, States reiterated past goals, rather than seeing new paths forward. They adopted "The Future We Want," but soberly also set out to negotiate consensus on new pathways forward. Over three years, they agreed upon the sustainable development goals (SDGs). The pervasiveness of environmental degradation, and its consequent heightened risk of natural disasters, was acknowledged in the summer of 2015 by the Open Working Group of the United Nations General Assembly, which negotiated the text of the UN Sustainable Development Goals.[7] As the Goals were transmitted to the

General Assembly for adoption, the Open Working Group provided the following rationale in support of their rapid adoption and implementation:

> Global health threats, more frequent and intense natural disasters ... and related humanitarian crises and forced displacement of people threaten to reverse much of the development process made in recent decades. Natural resource depletion and adverse impacts of environmental degradation, including desertification, drought, land degradation, freshwater scarcity and loss of biodiversity, add to and exacerbate the list of challenges which humanity faces. Climate change is one of the greatest challenges of our time and its adverse impacts undermine the ability of all counties to achieve sustainable development. Increases in sea level rise, ocean acidification, and other climate change impacts are serious affecting coastal areas and low-lying coastal countries, including many least developed counties and small island developing States. The survival of many societies, and of the biological support systems of the planet, is at risk.[8]

The law of biodiversity ultimately matters because the Earth's natural systems support all human society. Unless human laws safeguard nature, humanity itself reaches the extinction threshold found in other species. Instrumental uses of nature, through ownership of intellectual property in genetic engineering at one end, or of a tract of Amazonian savanna or forest being plowed up for the production of soybeans, tend to ignore the aggregate loss of biodiversity. Similarly, policy debates about the fairness of access to and benefit-sharing of biodiversity will be of little consequence when the species are extinct. This chapter summarizes how laws seek to sustain biological support systems of the planet. The SDGs project that States can bring degradation trends under control by 2030. If these goals are to be realized, all States will need to do much more to strengthen the environmental rule of law, both within their own national legal systems and internationally.

The principles of law for biodiversity

Biological diversity is defined in Article 2 of the CBD as "the variability among living organisms from all sources, including, *inter alia*, terrestrial, marine and other aquatic ecosystems, and the ecological complexes of which they are part; this includes diversity within species, between species, and of ecosystems."[9] Specialists in the IUCN Environmental Law Programme, including Cyrille de Klemm and Françoise Burhenne-Guilmin, among others, prepared a synthesis of many scientific and legal concepts to frame this definition, which the UN's International Negotiating Committee accepted for the CBD (de Klemm and Shine, 1993). When one considers the general principles of law applicable to biodiversity, the principles should be applied with reference to this definition, a holistic concept of life.

Among today's international law system of sovereign states, it is Principle 21 of the UN Stockholm Declaration on the Human Environment that is the foundation for an examination of principles.[10] This Principle was restated in Article 3 of the CBD itself. States shall "ensure that the activities within their jurisdiction and control do not cause damage to the environment of other States, or areas beyond national jurisdiction." Thus, while States have sovereign control over their natural resources, they cannot allow development or alternation of those resources to harm the environmental conditions in other states. Since living resources are globally interconnected, no State can harm the biodiversity of the planet through the extinction of a species or activity risking extinction, or through diminishing populations of species in their habitats or range of migration, or in their relationships within ecosystems.

How can a State know if it is observing Principle 21 (Article 3 of CBD)? A State is in compliance with Principle 21 when it conscientiously seeks to observer the environmental rule of law. Responding to this basic question requires a State to observe a set of well-defined principles guiding all state conduct. These have been restated many times (Kurukulasuriya and Robinson, 2006),[11] most recently by the IUCN World Congress on Environmental law in Rio de Janeiro in 2016.[12]

Duty to cooperate – Under International law, and the UN Charter, all States shall cooperate. This is true for neighboring states, as seen in *droit de voisinage*, or law of good neighborly relations. It is also true for acts affecting the high seas or atmosphere or the natural systems working across the planet (Sucharitkul, 1996).

Duty for sustainable development – States have a duty to observe the norms of sustainable development, and constrain their use of resources by respecting the expected need of future generations. In short, States have a fiduciary obligation as trustees today for the needs of tomorrow.[13] They shall place limits on the use and exploitation of nature sufficient to sustain biodiversity.

Duty to protect nature – States have a duty to prevent harm to, and to care for and sustain, biodiversity, regardless of its worth to humans,[14] as a common concern of humankind.

Right to the environment – States have an obligation to respect and advance the right of humans to a safe, clean, healthy, and sustainable environment.[15] This right necessarily entails the ecological function of property, by which those who control biodiversity or land, water and other natural systems, have the duty to refrain from actions that harm basic ecological functions.[16]

Duty of precaution – Where there are threats of serious or irreversible damage to biodiversity or the environment, States shall apply a precautionary approach, whereby lack of full scientific certainty shall not be used as a reason for postponing effective measures to prevent degradation.[17]

Duty of environmental impact assessment – States shall exercise precaution and prevent harm to biodiversity by undertaking a prior assessment of potentially harmful impacts, and enable public participation and access to environmental information, and access to justice.[18]

Duty to indigenous people – States shall recognize, respect, preserve, and maintain the knowledge, innovations, and practices of indigenous and local communities embodying knowledge of biodiversity, and encourage equitable sharing of the benefits of biological diversity.[19]

Polluter pays principle – States shall endeavor to internalize the environmental costs of sustaining biodiversity, taking into account that the party harming biodiversity should bear the cost of the harm.[20]

Common but differentiated responsibilities principle – States shall cooperate in a spirit of global partnership to conserve, protect and restore the health and integrity of biodiversity. In view of the different contributions to global environmental degradation, States have common but differentiated responsibilities.[21]

These general principles of international law are to be applied whenever State conduct affects biodiversity. From a scientific perspective, in the case of biodiversity duties, all actions taken by states affecting biodiversity within their territory impact, at the same time, the Earth's universal biological diversity. There are no purely domestic actions when it comes to biodiversity.

Application of these principles is evident in the many treaties that govern aspects of biodiversity. National legislation also illustrates adherence to these principles, as for example in statutes for establishing protected areas, protecting species in danger of extinction, regulating fish and game, managing silvaculture and timber forestry, or the mitigation of greenhouse gas emissions. When specific legal issues are addressed, the over-arching duties to ensure sustainability should be applied. This is well illustrated by a court decision interpreting the scientific whaling rules of the International Whaling Commission (IWC).[22] The 2014 International Court of Justice judgment on *Whaling in the Antarctic* interpreted the IWC's decisions regarding Japanese whaling. The Court addressed biodiversity principles as follows:[23]

> Taking into account the Preamble and other relevant provisions of the Convention … the Court notes that programmes for purposes of scientific research should foster scientific knowledge; they may pursue an aim other than either conservation or sustainable exploitation of whale stocks. This is also reflected in the Guidelines issued by the IWC for the review of scientific permit proposals by the Scientific Committee. In particular, the Guidelines … referred not only to programmes that "contribute information essential for rational management of the stock" or those that are relevant for "conduct[ing] the comprehensive assessment" of the moratorium on commercial whaling, but also those responding to "other critically important research needs". The current Guidelines, Annex P, list three broad categories of objectives. Besides programmes aimed at "improv[ing] the conservation and management of whale stocks", they envisage programmes which have as an objective to "improve the conservation and management of other living marine resources or the ecosystem of which the whale stocks are an integral part" and those directed at "test[ing] hypotheses not directly related to the management of living marine resources".

The general principles of law mandate decision-makers to examine how biodiversity may be affected, and apply applicable laws to sustain biodiversity.

UN agencies and multilateral environmental agreements on biodiversity

Several framework agreements function globally and can be deployed to sustain biodiversity. Virtually all nations have adhered to these multilateral environmental agreements (MEAs). The Conferences of the Parties (COP) to each agreement meet in both plenary gatherings, and through working groups or subsidiary bodies, which each COP establishes. For example, most MEAs have scientific advisory subsidiary bodies, with States nominating biodiversity experts to serve on these bodies. Each MEA and its COP have their own distinct and independent rules of procedure and work plans. The older agreements follow the practice of the United Nations General Assembly's Rules of Procedure, and provide for decisions by majority vote. The more recently established COPs generally prefer rules of procedure that require a consensus of all (or virtually all) parties to make a decision. When the State Parties to an MEA also agree to a Protocol, or an agreement related to the obligations under the MEA, the States that adhere to the Protocol deliberate through a Meeting of the Parties (MOP). These MOPs often are scheduled to take place in the same timeframe as the COP.[24]

It is beyond the scope of this chapter to examine each of the MEAs. Typically, the secretariat for each MEA will maintain a web page which includes the treaty or other international agreements establishing the MEA, a record of each COP meeting and its decisions, and the ongoing work plans for the MEA, and documentation about the issues addressed by the MEA, much of this submitted by the State Members of the MEA, or produced at their request by the Secretariat. These websites are invaluable references for assessing the work of each MEA. Each State Member designates a national agency or office to be the official link to the MEA or UN body. These national focal points are often not connected to the foreign ministries, and often are poorly funded and staffed, and unable to coordinate with other national offices of MEAs.

These inevitably lead to a fragmentation in both national and international decision-making about biodiversity, and all other common environmental concerns.

Since each MEA is autonomous in its decision-making, it is important for States to adhere to the general principles of law governing biodiversity as they deliberate in any COP. Likewise, the Secretariats are bound by the same general principles.

United Nations agencies

United Nations – General Assembly and Economic and Social Council – based in New York (USA), with offices in Geneva (Switzerland), the Second Committee of the UN General Assembly, for each annual UN Session, has an agenda of usually some 20 to 40 resolutions that touch on biodiversity. The UN Economic and Social Council (ECOSOC) oversees the work of the UN Environmental Programme (UNEP, as of 2017 branded "UN Environment") and the UN Development Programme (UNDP), with its extensive work on the Sustainable Development Goals. In Nairobi, the UN Environment Assembly, composed of all UN Member States, meets to advance an agenda of environmental protection activities. UN Environment published periodic reports on the state of the Earth's environment, which includes an assessment of deteriorating biodiversity trends.[25] The UN General Assembly (UNGA), with a Secretariat Office, is responsible for the UN Convention on the Law of the Sea, including its protection of marine biodiversity. In 2016–18, the UN General Assembly is conducting a negotiation to prepare a new international agreement on biodiversity in areas beyond national jurisdiction (BBNJ); this proposed new treaty will address biodiversity in the high seas, areas beyond 200 miles from each coast. The UN Human Rights Committee, based in Geneva, also has programs on the human right to the environment. The UN Permanent Forum in Indigenous Issues meets in the UNGA every spring. The UN General Assembly conducts an annual examination in the spring, by Ministers of the Members States, about how the SDGs are being implemented, and every fifth year it schedules a summit meeting of Heads of Member States to assess the state of implementation of the 17 SDGs, overall and especially in the prior four-year period.

UN Educational, Scientific and Cultural Organization (UNESCO) – based in Paris (France), UNESCO's "Man and the Biosphere" program and its duties under the World Heritage Convention for conserving cultural and natural monuments are important for building knowledge about and action to conserve biodiversity. UNESCO is established under an international agreement separate from that of the UN Charter.[26]

UN Food and Agricultural Organization (FAO) – based in Rome (Italy), FAO has extensive programs on conservation of fisheries, forests, soils, wetlands, and the biodiversity related to these topics. For example, the FAO has a non-binding Soils Charter, a Code of Conduct for Responsible Fisheries, and a global systems for conservation and sustainable use of plant genetic resources. FAO is established under an agreement separate from the UN Charter.[27]

World Health Organization – the WHO secretariat is based in Geneva (Switzerland). Many WHO activities on public health come into contact with biodiversity issues, such as zoonotic diseases.[28]

"World Bank" and regional development banks – the Asian Development Bank (ADB) and the European bank for Reconstruction and Development (EBRD), and other regional banks, plus the complex of institutions constituting the World Bank, have a wide range of environmental protection programs, which can and do include conservation of biological diversity. All have provisions that require environmental impact assessments.

Multilateral environmental agreements

1948 Statutes of the International Union for the Conservation of Nature and Natural Resources – IUCN is a hybrid intergovernmental organization established by Statutes adopted at a diplomatic conference in Fontainebleau, convened by France, in 1948. Its headquarters are in Gland (Switzerland) and the UN General Assembly has invited IUCN to maintain a Permanent Observer Mission at the UN. IUCN is governed by a bicameral World Conservation Congress, in which its State Members vote in one chamber and its non-governmental members vote in a second chamber. It has five commissions, and the World Commission on Protected Areas is an international forum for all national park agencies. IUCN has a Species Survival Commission and an Ecosystem Management Commission. The IUCN World Commission on Environmental Law developed the agreements for CITES and the CBD, among others. IUCN is focused primarily on biodiversity issues.[29]

1971 Ramsar Convention on Wetlands of International Importance, especially as Waterfowl Habitat – the Ramsar Convention has its Secretariat in Gland (Switzerland), together with the IUCN. IUCN had facilitated the establishment of the Ramsar Convention. States nominate large wetlands areas, situated in their nation, to be listed as Ramsar sites. It has vastly expanded the scope of wetlands protection beyond the original focus on water birds, to encompass hydrology and biodiversity broadly.[30]

1972 Convention Concerning the Protection of the World Cultural and Natural Heritage – the "UNESCO World Heritage Convention" is headquartered with UNESCO in Paris (France). States nominate sites of cultural or natural importance to be listed as world heritage sites. IUCN provides scientific and technical advice to the Convention in relation to nominated sites of natural importance.

Convention on the International Trade in Endangered Species – CITES was established by a diplomatic conference held in Washington DC (USA) in 1973, and its Secretariat is based in Geneva (Switzerland). It maintains the rosters of species that are threatened with extinction, and oversees an international ban on trade in any of the listed species or their parts. CITES collaborates with Interpol and TRAFFIC and other governmental and non-governmental bodies to combat illegal trade in endangered species.[31]

1979 Bonn Convention on the Conservation of Migratory Species of Wild Animals – the Bonn CMS Convention Secretariat is located in Bonn (Germany). It is a framework agreement to facilitate the negotiation and adoption of agreements among the states within the range of migration of any species. It has concluded a number of species-specific agreements. No State in the Western Hemisphere has, as yet, adhered to this Convention.

Vienna Convention on the Protection of the Stratospheric Ozone Layer, and the Montreal Protocol – with Secretariat units in New York and Nairobi, this MEA is the most successful in terms of establishing a global network of national and international specialists that reclaim chlofluorocarbons and phase out their use, in order to protect and restore the Earth's stratospheric ozone layer. All life on Earth is affected by solar ultraviolet rays that reach the surface of the planet because of the loss of the ozone layer. The COP for the Vienna Convention and MOP for the Montreal Protocol, and other sister protocols, are independent of the other UN MEAs.[32]

UN Convention on Biodiversity – signed at the 1992 UN Rio Earth Summit, the CBD took several years to become operational. The CBD Secretariat is based in Montreal (Canada), and holds its COPs in different cities around the world each year. Under the CBD, the States Parties have agreed to the 2000 Cartagena Protocol on Biosafety, which addresses genetically modified organisms, or "Living Modified Organisms" (LMOs) under the Protocol; advanced

Informed Agreement procedures exist for introducing LMOs into use. The CBD's Strategic Plan for biodiversity provides for national biodiversity plans, which include development of national legislation to implement the CBD (Belbase, 1999).

UN Framework Convention on Climate Change – signed at the 1992 Rio Earth Summit, the UNFCCC Secretariat is based in Bonn (Germany) and its COP meets annually, now also with the MOPs for the 2015 Paris Agreement and the Kyoto Protocol. The UNFCCC deliberations touch upon biodiversity in several aspects, for instance: ways in which the marine environment absorbs carbon dioxide from the atmosphere; the value of wetlands in continuing to sequester the carbon dioxide in wet peat deposits; or the role of "Reducing Emissions from Deforestation and Degradation, with conservation" (REDD+) and promoting photosynthesis.

UN Convention to Combat Desertification in Countries experiencing Serious Drought and/or Desertification – the CCDD aims to advance international cooperation to assist countries to restore degraded lands. National Action Programs (NAPs) are the means by which States implement the CCDD.[33]

UN Stockholm Convention on Persistent Organic Pollutants – the POPs Convention seeks to prevent the use or release of toxic chemicals that bioaccumulate in species, including in humans. It bans the use of twelve chemicals and provides for its Member States adding other dangerous substances also. Although this agreement is critical for the integrity of biodiversity, States have not as yet expanded its list of controlled substances.[34]

There are many other international organizations whose activities affect biodiversity. The international environmental agreements and organizations outlined above contain the broad legal pathways through which States and intergovernmental organizations cooperate, or fail to cooperate, to govern human impacts on biodiversity. These activities constitute a global matrix, with large gaps or sectors that do not relate well to one another.[35] At regional or national levels, some of these gaps are filled, however, since biodiversity is a universal concern, global stewardship cannot be effective until a comprehensive system is established.

Regional and sectorial treaties

In each region, governments have established ways to coordinate and cooperate to sustain shared natural resources and biodiversity assets. In many cases, the agreements are defined in terms of geography, such as the many "regional seas" agreements encouraged by UNEP, which aim to monitor and sustain selected biological assets. Others are specific to conserving an animal or plant found locally. There are between 3,000 and 10,000 such agreements. Many have been compiled in a database created by the IUCN, FAO, and UNEP, known as "Ecolex."[36] The CBD has encouraged use of ECOLEX for accessing biodiversity law.

Not all regions have the same types of intergovernmental cooperation systems for biodiversity, and most are very different in their effectiveness. Nonetheless, when working to assess biodiversity law in any region, it is necessary to identify the salient treaties and agencies which are part of this dimension of international environmental law. Here follows a representative set of such laws and agencies:

African Union – since 2002 the AU,[37] which was built on the foundation of the 1963 Organization of African Unity, has been the regional economic and political integration organization, with a secretariat in Addis Ababa (Ethiopia). Within the African Union, there are regional courts and other inter-governmental organizations. There are many regional wildlife treaties in Africa, including the pan-African and IUCN-facilitated Convention on the Conservation of Nature and Natural Resources.

Alpine Convention – the European Alps are governed by an international agreement on protection of the Alps. It has many components and is well elaborated. A similar agreement is in place for the Carpathian mountains in Central Europe.

Arab League – founded in 1945, the Arab League includes 22 States across North African and West Asia. It has a Council of Environment Ministers, but pan-Arab programs on biodiversity have yet to be developed. Arab States participate in the Barcelona Convention, and other sub-regional programs such as the Red Sea or Kuwait Regional Seas agreements.

Arctic Council – the Arctic States and Permanent Representatives of the Inuit and other indigenous peoples of the Arctic region, in 1996 formed an Arctic Council, with a Secretariat in Tromsø (Norway). There are also Polar Bear treaties, dating from 1973 but updated more recently, among Arctic States.

Antarctica and Antarctic Ocean – the 1959 Antarctic Treaty reserves the Antarctic for scientific study as "the common heritage of mankind." States' duties are further set out in the 1991 Madrid Protocol on Environmental Protection in the Antarctic Treaty, which incorporates the Precautionary Principle. The 1980 Convention on the Conservation of Antarctic Marine Living Resources (CCAMLR) governs the biodiversity of the marine areas adjacent to the continent.

Association of South East Asian Nations – ASEAN is a regional economic and political integration organization for 10 states, established in 1967, with a secretariat in Jakarta (Indonesia). It has an extensive system of ASEAN Heritage Parks and a regional biodiversity institute in LOS BAÑOS, Philippines. ASEAN has a rolling five-year plan to enhance biodiversity, but does not yet have a regional seas program or marine protected areas (Koh, Robinson and Lye, 2016).[38]

Barcelona Convention for the Protection of the Marine Environment and for Coastal Region of the Mediterranean – in 1976, all States around the Mediterranean Sea, agreed to cooperate to abate pollution and protect the environment through the Barcelona Convention. In 1982 a Protocol concerning Mediterranean Specially Protected Areas was adopted, and in 1999 renamed to include Biodiversity.

The Caribbean Community Market – CARICOM is composed of 15 Caribbean nations, with a secretariat in Guyana. Established in 1973, CARICOM's objectives are to promote economic integration and cooperation among its members and share the benefits of integration equitably.[39] CARICOM has launched environmental cooperation activities, including cooperation among terrestrial and marine protected areas, which conserve biodiversity. In addition, with UNEP, the Caribbean States with Meso America have a Convention on the Protection and Development of the Marine Environment in the Wider Caribbean Region.

European Union – the EU is a regional economic cooperation and integration organization of States in Europe.[40] It has a secretariat in Brussels, with agencies in different EU member States (the Environment Agency is in Demark), and a well-elaborated set of EU-wide regulations and directives that harmonize the environmental protection across all the Member States. The Wild Bird and Habitat Directives, for example, have been extremely important in protecting biodiversity (Jans and Vedder 2012). The EU is a member of many of the multilateral environmental agreements described above, which also harmonizes biodiversity law across the Member States.

South Asian Cooperative Environmental Programme – SACEP was established to encourage environmental cooperation across the Indian subcontinent. It has a secretariat in Colombo (Sri Lanka). Because there are profoundly different political perspectives among the South Asian

States, they do not find consensus on many issues and SACEP has been constrained in facilitating even basic aspects of environmental cooperation.

South Pacific Regional Environmental Programme – SPREP was established to promote environmental cooperation across the States of the Pacific Ocean, including biodiversity. It has a secretariat in Vanuatu.

There are many other regional or bilateral agreements, such as the Great Lakes Compact between the Provinces of Canada and the States of the United States for the protection of the watersheds and ecology of the Great Lakes in North America, or agreements in international rivers basins such as the Zambezi River in Southern Africa, or the Amazon Basin agreement among all States of the Amazon in South America. These legal agreements have great potential for biodiversity stewardship, but much will need to be done to align them behind that objective. They are mostly all at early stages of being elaborated, funded, and implemented.

Characteristic national biodiversity laws

Even when a State has enacted a new law to implement the Convention on Biological Diversity, virtually all nations have kept in place their 19th century customs and laws on wildlife (hunting, fishing) and water allocation practices, and for agriculture and silvaculture. Few national environmental laws have been restated or codified, as has the New Zealand Natural Resources Management Act (1991). Thus, virtually all national laws governing biodiversity laws are parts of an amalgam of older statutes, which by default must be construed and applied together regardless of inconsistencies. Even the holistic application of environmental impact assessment laws has more often than not left biodiversity assessment out of decisions on the scoping of environmental studies. Biodiversity takes time to study and protect, and economic development schedules seek rapid and often expedient decision-making. Economic growth in most regions still favors exploitation of nature without regard to biodiversity conservation. Since many vested interests rely on resource exploitation rights, they tend to oppose law reform.

Earth's environment is pervasive, and almost all legal settings affect environmental interests. Biodiversity is universal. States have not had, or made,[41] the time to establish consistent approaches to biodiversity. In 1982, the UN General Assembly adopted the World Charter For Nature, which proclaimed fundamental principles and recommended that States take specific actions to safeguard nature.[42] For example, paragraph 19 of the World Charter for Nature states: "The status of natural processes, ecosystems and species shall be closely monitored to enable early detection of degradation or threat, ensure timely intervention and facilitate the evaluation of conservation policies and methods."[43] During the decades since adopting this Charter, few states have succeeded in establishing monitoring systems needed to track degradation of biodiversity.

The CBD provides guidelines for the design of national biodiversity laws, but each nation is unique in how its legal systems already governs ecological and other natural systems. In the World Charter For Nature, the UN General Assembly recommend specific implementation measures for national laws. Since then, UNEP's "Montevideo Programme" also has made recommendations about appropriate environmental laws for all nations. UNEP and the World Commission on Environmental Law have advanced norms for the "environmental rule of law," to further compliance and enforcement. In 2016, the First World Congress on Environmental Law took into account the 2015 UN Sustainable Development Goals, and sought to further SDG 16[44] when it adopted recommendations for national law-makers in the World Declaration on the Environmental Rule of Law. The Declaration posited several measures that States should address to establish the environmental rule of law:[45]

(a) Monitoring and reporting systems that enable accurate assessments of the state of the environment and the pressures on it.
(b) Anti-corruption measures, including those that address ethical conduct and oversight.
(c) Legally supported environmental management systems that take due consideration of environmental risk and the vulnerability of social and economic systems in the face of ecological deterioration.
(d) Environmental assessment, incorporating multidimensional, polycentric perspectives and the complexity of social-ecological relationships.
(e) Quantitative and qualitative modeling and visioning tools that enable planning based on science, environmental ethics, strategies, and options that remain robust under multiple plausible futures.
(f) Collaborative and adaptive management and governance which involves stakeholders from a range of socio-economic and cultural backgrounds including historically marginalized groups, including local and indigenous peoples, women, and the poor.
(g) Coordination mechanisms such as regional enforcement networks, intelligence-sharing, and judicial cooperation.
(h) Environmental legal education and capacity-building for all people, and especially for women, girls, and traditional leaders and governance bodies of indigenous people, focusing on exchange of knowledge on best practices, taking into account the relevant legal, political, socio-economic, cultural, and religious aspects, as well as recognizing common features founded on international norms and standards.
(i) Harnessing new technologies and media for promoting environmental law education and access to information, as well as complementary tools which draw on and respect customary laws and practice.
(j) Communication systems enabling the production and dissemination of guidelines, tool-kits, checklists, and associated technical and legal implementation assistance.
(k) Strengthening civil society, environmental law associations, and other non-state actors that fill gaps in state-based environmental governance systems.
(l) Linking environmental crimes to other types of crime such as money laundering.
(m) Enabling public interest dispute resolution concerning environmental conservation and protection, and upholding the rights of future generations.
(n) Strengthening the capacity and independence of the courts in the effective application and interpretation of environmental law and in acting as guarantors of the environmental rule of law.

Whether national legislatures and environmental decision-making systems have the discipline to adhere to environmental law remains in doubt. Little heed was paid to the recommendations of the World Charter for Nature. The SDGs themselves illustrate the difficulties that States have in making consistent decisions to safeguard biodiversity.

SDG 15 reflects an understanding that terrestrial biodiversity is in grave danger. In 2018, the UN General Assembly is to review the progress that States have made with respect to this SDG.[46]

> Protect, restore and promote sustainable use of terrestrial ecosystems, sustainably manage forests, combat desertification, and halt and reverse land degradation and halt biodiversity loss.

It is not possible to fulfil this SDG unless States act to halt biodiversity loss, and this entails a reordering of all national development economics. Courts in South America have begun to invoke the principle "*In dubio pro natura*" to oblige parties to err on the side of protecting the environment.[47] The Philippines has gone further, and adopted court rules that shift the burden of proof onto the party, which is harming nature through its economic activity, to demonstrate that it complies with all applicable environmental law.[48] These legal measure hint at the transformations needed to attain SDG 15.

SDG 14 deals with the marine environment.[49] It provides that States are to "Conserve and sustainably use the oceans, seas and marine resources for sustainable development."

SGD 14 regrettably ignores the command of the UN Convention on the Law of the Sea, to protect and preserve the marine environment. Unless one reads into sustainable use and development the sort of strict obligation of SGD 15, "to halt biodiversity loss" in the seas as States must on land, this SDG will be regressive in allowing current marine degradation to continue. The UN negotiations to provide a new agreement on biodiversity beyond areas of national jurisdiction (BBNJ) may help produce terms consistent with Article 192 of UNCLOS.[50] In any event, States must take legal measures in their coastal waters and exclusive economic zones "to protect and preserve" the marine environment. The designation of Marine Protected Areas (MPAs) is a positive development in this regard.

Finally, the SDG 13 focuses on how States should address the impacts and causes of climate change in activities other than those dealt with through the UN Framework Convention on Climate Change.[51] SDG 13 provides that States are to "Take urgent action to combat climate change and its impacts."

Unless climate change is slowed, arrested, and reversed by sequestering greenhouse gases and allowing atmospheric temperature increases to be contained and reversed over the next two centuries, or less, the prospects for halting biodiversity loss are dim. One "urgent" action needed is to enact reforms to establish the environmental rule of law, as described above. It remains to be seen how the SDGs will stimulate better legal protection of Earth's biodiversity.

Legal lacunae and biodiversity

International environmental law has demonstrated its success in many instances. It has the tools to protect Earth's biodiversity, but they are not deployed in all nations or at a global scale. Can they be?

Edward O. Wilson has proposed that nations set aside half the Earth primarily for nature. (Wilson, 2016). Doing so would allow human civilization to continue to advance in reliance on Earth's ongoing natural systems. He argues that "only by setting aside half of the planet in reserve, or more, can we save the living part of the environment and achieve the stabilization required for our own survival."[52] By scaling up the terrestrial systems of protected areas, and adding the marine protected areas, States can attain this objective. "Reasonable" use of all natural areas will need to yield to minimal use and enhancing stewardship or trusteeship for natural areas. The legal basis exists for requiring such actions needs to be better articulated.

There are many gaps in environmental legislation. Beyond enacting laws to implement each of the treaties described above – something most states have not done – new laws are needed. For example there is virtually no protection for peat, which is found on 3 per cent of the planet's surface but sequesters 30 per cent or more of the Earth's carbon dioxide. Peat is being destroyed rapidly in Kalimantan and Sumatra (Indonesia), releasing huge greenhouse gas emissions. The need for peat legislation is acute.[53] Most nations still have antiquated laws requiring generation of electricity by burning fossil fuels (some burn peat), thereby exacerbating climate change. Laws are needed to require electricity to be conserved and used efficiently, and laws to encourage solar and wind energy use need to be enacted. The monitoring of biodiversity needs to be as routine as the monitoring of public health conditions, or compiling statistics about employment and other economic indicators. What a society does not measure, it ignores. Laws are needed to establish national inventories of biodiversity and mandate the study and surveillance of trends. Biodiversity exists everywhere. Concern by conservation biologists for biodiversity in exotic locations may have caused neglect of biodiversity where most humans are found. For instance, making preservation of urban biodiversity a priority could sensitize national leaders to the need to save biodiversity everywhere (Alfsen-Norodom, *et al.*, 2004).

Biodiversity law – both internationally and nationally – is manifestly inadequate to conserve the Earth's legacy of biodiversity. Biodiversity losses are likely to continue. The SDGs represent an opportunity to educate decision-makers globally about how to shift from unsustainable legal practices to those that are characterized by the environmental rule of law.

Biodiversity law is still evolving as this book goes to print, and as climate change induces species and ecosystems to adapt to survive. Humans will adapt as well. The challenge for biodiversity law is to synchronize these life systems. To sustain civilization, human legal constructs will need to accommodate the laws of nature. Whether or how this happens will be determined by how biodiversity law will be shaped in the coming decades.

Notes

1 See the Resolutions of the IUCN General Assemblies, e.g. the 17th GA held in Costa Rica in 1988, at https://portals.iucn.org/library/node/6083. See Burhenne Guilmin and Casey-Lefkowitz (1993).

2 In decision X/2, the tenth meeting of the Conference of the Parties, held from 18 to 29 October 2010, in Nagoya, Aichi Prefecture, Japan, adopted a revised and updated Strategic Plan for Biodiversity, including the Aichi Biodiversity Targets, for the 2011–20 period. See the "CBD Strategic Plan for Biodiversity 2011–2020," at www.sbd.int/sp.

3 All species of animals, plants, insect, microorganisms, and the ecosystems and ecological processes of which they are a part, are subsumed within the concept of biodiversity. Scientific discourse, policy and law tend to focus on biodiversity in terms of genetic diversity, species diversity, and ecosystem diversity. See McNeely (1990) at 17. See generally Wilson (1988).

4 UN Declaration on Environment and Development, Rio de Janeiro, 14 June 1992, at www.unep.org/documents.multilingual/default.asp?documentid=78andarticleid.

5 UN Conference on Environment and Development, Agenda 21, Chapters 3 and 4 (1992).

6 Report of the 1992 UN Conference on Environment and Development (UNCED), available at https://sustainabledevelopment.un.org/content/documents/Agenda21.pdf; for an annotated edition of Agenda 21, with the full *traveaux préparatoires* for the negotiation of Agenda 21, see Robinson (1993).

7 For background on the United Nations Sustainable Development Goals, see the UN Department of Social and Economic Affairs website: https://sustainabledevelopment.un.org/.

8 Paragraph 14, "Transforming Our World: The 2030 Agenda for Sustainable Development." Contained in A 68/970, "Report of the Open Working Group of the General Assembly on Sustainable Development Goals" (see also A. 68/970 Add. 1), adopted in UNGA Res. A/Res/70/1 (25 September 2015), distributed 21 October 2015, for Agenda Items 15 and 116, 70th Session, UN GA.

9 UN Convention on Biological Diversity, at www.cbd.int.

10 Declaration of the UN Conference on the Human Environment, Stockholm, 1972, at www.unep.org/documents.multilingual/default.asp?documentid=97&articleid.

11 See the restatement in Kurukulasuriya and Robinson (2006), in Chapter 3 "Principles and Concepts of Environmental Law" (UN Environment Programme, 2006), online at http://digitalcommons.pace.edu/lawfaculty/791/.

12 IUCN World Declaration on the Environmental Rule of Law, adopted at the World Congress on Environmental Law (Rio de Janeiro, 26–29 April 2016), at www.iucn.org.

13 This is akin to the concept of a trust in common law, or the *waqf* in Islamic law. The preamble to the CBD concludes by observing that inter-generational rights are recognized through observing CBD duties.

14 Sub-section 7 of Chapter 3, in Kurukulasuriya and Robinson (2006).
15 The right to the environment is now recognized in the national constitutions of 174 States, see Boyd (2012), as well as in the 2016 Rio World Congress Declaration on the Environmental Rule of Law.
16 Principle 6 of the Rio Declaration on the Environmental Rule of Law provides as follows: "Any natural or legal person or group of people, in possession or control of land, water or other resources, has the duty to maintain the essential ecological functions associated with those resources and to refrain from activities that would impair such functions. Legal obligations to restore ecological conditions of land, water or other resources are binding on all owners, occupiers, and uses of a site, and liability is not terminated by the transfer of title to others."
17 See Principle 15 of the 1992 Rio Declaration on Environment and Development, found also in the preamble to the CBD, and in Art. 3(3) of the UN Framework Convention on Climate Change.
18 This duty of "EIA" is Principle 17 of the 1992 Rio Declaration on Environment and Development, and appears in the 1991 Espoo Agreement on EIA in a Tranboundary Context, and other agreements, and the duty to provide public participation is Principle 10 of the 1992 Rio Declaration, and appears in the Aarhus Agreement.
19 This duty appears in Principle 22 of the 1992 Rio Declaration on Environment and development, and in Art. 8(j) of the 1992 CBD, and elsewhere, such as in Principle 7 of the 1993 Nuuk Declaration on Environment and Development in the Arctic States. See subsection 9 of chapter 3 in Kurukulasuriya and Robinson (2006).
20 Principle 16 of the 1992 Rio Declaration on Environment and Development, widely accepted based on studies in the 1970s by the Organization for Economic Cooperation and development (OECD). The science of ecology has documented how pollution can and does harm ecosystems, and biodiversity, and thus the polluter pays principle is now applied to more than a narrow concept of pollution. For example, ecosystem restoration and payments and assessing natural resources damages are part of hazardous waste remediation. See sub-section 7 in chapter 3 of Kurukulasuriya and Robinson (2006).
21 Principle 7 of the 1992 Rio Declaration on Environmental and Development provides in this regard that "The developed countries acknowledge the responsibility that they bear in the international pursuit of sustainable development in view of the pressures their societies place on the global environment and of the technologies and financial resources that they command." This Principle appears in Art. 20(4) of the CBD and in Art. 5 of the Convention to Combat Desertification, and in other international environmental law instruments.
22 For the International Whaling Convention and its work, see https://iwc.int/home.
23 Paragraph 58 of Judgment, *Australia v. Japan: New Zealand intervening* (2014), at www.icj-cij.org/docket/files/148/18136.pdf.
24 On the administration of biodiversity agreements generally see chapter 4, "Compliance and enforcement of multilateral environmental agreements," in Kurukulasuriya and Robinson (2006).
25 See, e.g. the UNEP Fifth *Global Environmental Outlook Assessment Report*, GEO-5 Report, at www.google.com/webhp?sourceid=chrome-instant&ion=1&espv=2&ie=UTF-8#q=unep+geo+5+pdf&*.
26 For UNESCO, see http://en.unesco.org/.
27 For FAO, see www.fao.org/home/en/.
28 For WHO, see www.who.int/en/.
29 See www.iucn.org.
30 For the Ramsar Convention see www.ramsar.org/.
31 For CITES, see www.cites.org/.
32 For the Vienna Convention and Montreal Protocol see http://ozone.unep.org/.
33 See www.unccd.int/main.php.
34 For the Stockholm Convention on POPs, see www.pops.int, and http://chm.pops.int/default.aspx.
35 The gaps and short-comings of biodiversity law and governance have been ably described in the 2005 Millennium Ecosystem Assessment, see www.millenniumassessment.org/.
36 See www.ecolex.org/.
37 See www.au.int/.
38 See Koh, Robinson and Lye (2016).
39 See www.caricom.org/.
40 See the EU website: https://europa.eu/european-union/index_en.
41 For implementation of this Act, see www.mfe.govt.nz/rma.
42 UNGA Res. 37/7, at www.un.org/documents/ga/res/37/a37r007.htm.

43 *Ibid.*, Para 19.
44 SDG 16 encompasses promoting the rule of law. See https://sustainabledevelopment.un.org/sdg16.
45 These Recommendations, "a) to n)" are set forth in Part III, "Means of Implementation of the Environmental Rule of Law," in the IUCN World Declaration on the Environmental Rule of Law, available at www.iucn.org.
46 SDG 15 is discussed at https://sustainabledevelopment.un.org/sdg15.
47 In cases of any doubts, courts and administrative agencies are mandated to make their decisions in ways that favor the protection and conservation of the environment. See the Brazilian High Court decisions in Bryner (2012).
48 See the Philippine Supreme Court Rules for the Writ of Kalikasan (Nature), A.M. No. 09-6-8 (13 April 2010), at www.lawphil.net/courts/supreme/am/am_09-6-8-sc_2010.html.13.
49 www.un.org/sustainabledevelopment/oceans/.
50 UNCLOS is at www.un.org/depts/los/convention_agreements/texts/unclos/unclos_e.pdf. See Art. 192, "General obligation," which provides that "States have the obligation to protect and preserve the marine environment."
51 See https://sustainabledevelopment.un.org/sdg16.
52 Wilson (2016) at p. 3.
53 See IUCN World Conservation Congress Resolution on Peat, WCC-216-Res-043,EN, adopted in at Hawaii (September 2016), available at https://portals.iucn.org/congress/assembly/motions?field_motion_id_value=&title=peat.

References

Alfsen-Norodom, C.D., Lane, B.D., and Corry, M. (Eds), "Urban Biosphere and Society – Partnership of Cities", *Annals of the NYU Academy of Sciences* 1023 (2004).

Belbase, N., *National Implementation of the Convention on Biological Diversity – Policy and Legislative Requirements* (IUCN, 1999).

Boyd, D., *The Environmental Rights Revolution* (UBC Press, 2012).

Bryner, N., "Brazil's Green Court: Environmental Law in the *Superior Tribunal de Justiça*," 29 *Pace Environmental Law Review* 470 (2012).

Burhenne-Guilman, F., and Casey-Lefkowitz, S., "The Convention on Biological Diversity: A Hard Won Global Achievement," 3 *Yearbook of International Environmental Law* 42 (1993).

de Klemm, C., and Shine, C., *Biological Diversity Conservation and the Law: Legal Mechanisms for Conserving Species and Ecosystems* (IUCN EPL Paper 29, 1993).

Jans, J.H., and Vedder, H.H.B., *European Environmental Law* (Europa, 2012).

Koh K.L, Robinson, N.A. and Lye, L.H., *ASEAN Environmental Legal Integration – Sustainable Goals?* (Cambridge University Press, 2016).

Kolbert, E. *The Sixth Great Extinction – An Unnatural History* (Henry Holt, 2014).

Kurukulasuyriya, L., and Robinson, N.A., *UNEP Manual on International Environmental Law* (UNEP, 2006), on line at http://digitalcommons.pace.edu/lawfaculty/791/.

McNealy, J.A., *Conserving the World Biological Diversity* (IUCN, 1990).

Robinson, N.A., *Agenda 21 and the UNCED Proceedings* (Oceana Publications, 1993), 6 Vols.

Sucharitkul, S., *The Principles of Good-Neighborliness in International Law* (Golden Gate University School of Law Digital Commons (1996), Paper 559), at http://digitalcommons.law.ggu.edu/cgi/viewcontent.cgi?article=1559andcontext=pubs.

UN World Commission on Environment and Development ("Brundtlund Commission"), *Our Common Future* (Oxford University Press, 1987).

Wilson, E.O., *Biodiversity* (National Academy Press, 1988).

Wilson, E.O., *Half Earth* (Liveright, 2016).

4

BIODIVERSITY, PROTECTED AREAS AND THE LAW

Jamie Benidickson and Alexander Paterson

Introduction

> Protected areas are critical for maintaining a healthy environment for people and nature. They are essential for biodiversity conservation and vital to the cultures and livelihoods of indigenous peoples and local communities. They also deliver clean air and water, bring benefits to millions of people through tourism, and provide protection from climate change and natural disasters.
>
> (UNEP-WCMC and IUCN, 2016, p. 1)

As the above opening words from the *Protected Planet Report 2016* suggest, protected areas fulfill essential ecological, climatic, cultural and socio-economic functions (Dudley *et al.*, 2015; Dudley *et al.*, 2010; Stolton and Dudley, 2010). Still more broadly, protected areas contribute to goals contained in the Strategic Plan on Biodiversity (2011–20) and the associated Aichi Biodiversity Targets (CBD COP 10, 2010) agreed to by parties to the Convention on Biological Diversity (CBD) in 2010, and the Sustainable Development Goals (UNGA, 2015) adopted by members of the United Nations in 2015. This potential is comprehensively canvassed in the *Protected Planet Report 2016* and other recent texts (Joppa *et al.*, 2016), and requires no repetition here.

At the last available count, there were 202,467 recorded protected areas, covering just under 15% of the Earth's terrestrial environment and inland waters (UNEP-WCMC and IUCN, 2016). This is complemented in the coastal and marine environment by a further 14,688 marine protected areas covering approximately 4% and 10% of these environments under national jurisdiction respectively (UNEP-WCMC and IUCN, 2016). Significant further expansion is nevertheless needed to attain the numerical targets prescribed in Aichi Target 11, specifically:

> By 2020, at least 17% of terrestrial and inland water areas and 10% of coastal and marine areas, especially areas of particular importance for biodiversity and ecosystem services, are conserved through effectively and equitably managed, ecologically representative and well-connected systems of protected areas and other effective area-based conservation measures, and integrated into the wider landscape and seascape.

Recent assessments also indicate that there is a significant way to go in ensuring that protected areas are 'effectively and equitably managed', 'ecologically representative', 'well-connected', complemented by 'other effective area-based conservation measures' and 'integrated into the wider landscape and seascape' (UNEP-WCMC and IUCN, 2016).

So what are these protected areas that serve essential functions for all species, that are so central to global biodiversity and sustainable development goals, and which demand additional global attention? In terms of the CBD, a protected area is defined as 'a geographically defined area which is designated or regulated and managed to achieve specific conservation objectives' (CBD, Art. 2). Alternatively, in a widely accepted formulation developed by the IUCN World Commission on Protected Areas (WCPA), a protected area is 'a clearly defined geographical space recognized, dedicated and managed, through legal and other effective means, to achieve the long-term conservation of nature with associated ecosystem services and cultural values'.

These definitions recognize – explicitly and implicitly – the role of law in enabling and regulating protected areas. References to 'legal' mechanisms or to the characteristic of being 'regulated' constitute explicit recognition. Implicit recognition of the law can be garnered through references to: a 'defined' space or area (which implies a legal demarcation of borders); 'recognized', 'dedicated' and 'long-term' (which suggest elements of consent, agreement and ongoing, indeed potentially perpetual, commitment provided for through the law for different types and categories of protected areas); and management (which implies, amongst other things, the establishment of management authorities, the prescription of management objectives, the preparation of management plans, adherence to reporting obligations, the regulation of activities and enforcement of the relevant scheme through law). The role of law in enabling and regulating protected areas has been comprehensively canvassed in the *IUCN Guidelines for Protected Areas Legislation* (Lausche, 2011), which outlines legal issues respecting protected areas, provides guidance on underlying principles and describes approaches to statutory drafting.

Within the above context, this chapter has two broad purposes. First, the chapter provides a succinct survey of the current role of law in enabling and regulating protected areas, with an overview of the international legal framework of relevance to protected areas which has informed the development of domestic protected area laws. The chapter then briefly reviews key issues which should be canvassed in domestic protected area laws, such as: scope and application; objectives and principles; management categories and governance types; systems planning and management planning; institutional arrangements; regulating activities; compliance and enforcement; and financing. The chapter's second purpose is to introduce some key emerging legal issues relating to protected areas. These include: promoting the principle of non-regression; recognizing the diversity of protected areas governance; fostering connectivity; scoping and implementing other effective area-based conservation measures; and regulating marine protected areas in areas beyond national jurisdiction.

International legal framework of relevance to protected areas

This section of the chapter surveys global international legal initiatives respecting protected areas with reference to three eras – early developments; the Convention on Biological Diversity and its immediate aftermath to 2004; the Strategic Plan and Aichi Targets – before briefly noting other more specialized measures.

Although relevant initiatives may be traced to the early twentieth century, arrangements of current significance date back half a century to the creation of biosphere reserves under UNESCO's Man and the Biosphere Programme (UNESCO; Gillespie, 2007). Shortly thereafter, the World Heritage Convention (WHC), a further UNESCO initiative, constituted an enduring effort to protect both cultural and natural heritage underpinned by the standard of 'outstanding universal value'. In the case of natural sites, outstanding universal value may be demonstrated with reference to 'science, conservation or natural beauty' (WHC, Art. 2). The standard of outstanding universal value has been expressed as being 'so exceptional as to transcend national boundaries and to be of common importance for present and future generations of all humanity' (WHC, 2015, para. 49). Although the convention itself contains no precise reference to biodiversity or ecology, subsequent elaboration in the form of operational guidelines (WHC, 2015, para. 77) makes reference to two specific criteria for outstanding universal value that incorporate these considerations.

As of 2016, about 200 natural sites along with three dozen sites containing mixed cultural and natural heritage were designated as World Heritage sites under the convention. It is noteworthy that arrangements for monitoring and surveillance exist under the convention, including provision for a List of World Heritage in Danger (WHC, 2015, Art. 11(4)).

Without explicit reference to protected areas, the 1972 Stockholm Declaration on the Human Environment asserted in Principle 2 that 'the natural resources of the earth, including the air, water, land, flora and fauna and especially representative samples of natural ecosystems, must be safeguarded for the benefit of present and future generations' (Stockholm Declaration, 1972). Recommendation 38, in particular, called on governments to 'take steps to set aside areas representing ecosystems of international significance for protection under international agreement' (Stockholm Report, 1972).

The Convention on International Trade in Endangered Species (CITES), though oriented around species protection through trade regulation, represents a further level of potential safeguards for species that may often rely heavily on protected areas for habitat. The World Charter for Nature, a non-binding instrument taking the form of a UN General Assembly resolution (UNGA 1982), sets out general principles which are broadly supportive of protected area development. Fundamentally, nature is to be respected and its essential processes are not to be impaired (UNGA, 1982, para. I.1). Earth's genetic viability must not be compromised, and the populations of both wild and domesticated life forms need to be preserved at levels sufficient for their survival, through mechanisms to safeguard 'necessary habitats' (UNGA, 1982, para. I.2). In applying these principles to terrestrial and marine environments, 'special protection shall be given to unique areas, to representative samples of all the different types of ecosystems and to the habitats of rare or endangered species' (UNGA, 1982, para. I.3).

While by no means a central consideration in the work of the WCED on sustainable development, the Brundtland Report, *Our Common Future*, considered protected areas in the context of its concerns for species and ecosystems. Noting that 'even the most effectively managed parks and protected areas are unlikely to provide a sufficient answer' to the preservation of species (WCED, 1987, p. 152), the report nevertheless echoed a call to expand existing networks of protected areas by some 300% in order to encompass representative ecosystems.

The Earth Charter is intended as an ethical framework for global sustainability. With reference to protected areas, it calls for actions to '[e]stablish and safeguard viable nature and biosphere reserves, including wild lands and marine areas, to protect Earth's life support systems, maintain biodiversity, and preserve our natural heritage' (Earth Charter, s. 5(b)).

The CBD built upon these legal and intellectual foundations, with protected areas figuring prominently amongst measures that were contemplated. The CBD sets out specific obligations respecting *in-situ* conservation of biological diversity, calling on parties 'as far as possible and as appropriate' to, among other measures, establish a protected area system requiring special measures to conserve biological diversity (CBD, Art.8). Guidelines for selecting, establishing and managing protected areas are also called for along with measures to regulate or manage biological resources important for the conservation of biological diversity whether within or outside protected areas, with a view to ensuring their conservation and sustainable use. CBD parties must also protect ecosystems and natural habitats, as well as maintain viable populations of species in their natural surroundings. Moreover, parties are expected to enhance safeguards for protected areas by pursuing environmentally sound and sustainable development in adjacent areas. In conjunction with the promotion of sustainable development, including the CBD at the time of its initial formulation, Agenda 21 provided further direction respecting the conservation of biological diversity, often with specific reference to protected areas (UNCED, 1992, ch.15).

The CBD provisions were subsequently elaborated through the work of the Conference of the Parties (COP). A Programme of Work on Protected Areas (PoWPA) was adopted at the 7th COP in 2004 and is considered to represent 'a defining framework' or 'blueprint' for protected areas. The PoWPA offers guidance on cooperation between governments, donors, NGOs and local communities in order to promote programmes that are sustainable in the long-term (CBD Overview).

Significantly, the CBD's PoWPA endorsed 10% of each country's territory as a national protected area target for 2010 and contains four inter-linked elements: (1) direct actions for planning, selecting, establishing, strengthening and managing protected area systems and sites; (2) governance, participation, equity and benefit-sharing; (3) enabling measures such as legislative and policy initiatives and capacity-building; and (4) the development of standards, assessment procedures and monitoring arrangements (CBD COP 7, 2004).

The Strategic Plan on Biodiversity (2011–2020) and the associated Aichi Biodiversity Targets later emerged from CBD deliberations at COP 10, held in Nagoya, Japan, in 2010 (CBD COP 10, 2010). Amongst the range of initiatives required in pursuit of its stated mission – 'to halt the loss of biodiversity in order to ensure that by 2020 ecosystems are resilient and continue to provide essential services' (CBD COP 10, 2010, para. 12) – the Strategic Plan specifically noted the necessity for direct interventions – including protected areas – to safeguard or restore biodiversity and ecosystem services (CBD COP 10, 2010, para. 10(c)).

As one means to promote and encourage the establishment of protected areas for biodiversity protection, a target-based approach was endorsed and the PoWPA ambition was extended as reflected in the Aichi Target 11 quoted above. While formulated in global terms rather than as explicit national obligations, Aichi's numeric targets have become widely accepted as benchmarks for assessing performance with CBD COP 13 recently acknowledging progress (CBD COP 13, 2016).

In keeping with the CBD's preferred approach to national implementation (CBD, Art. 6) and as explicitly reinforced in Aichi Target 17 (CBD COP 10, 2010), parties have formulated or revised National Biodiversity Strategy and Action Plans (NBSAPs), frequently incorporating protected areas initiatives (CBD National Biodiversity Strategy).

Apart from global frameworks of the kind illustrated above, international legal initiatives to safeguard biodiversity by means of protected areas have also emerged at the regional level, or with reference to the needs of particular species or in relation to distinctive landscapes (Head, 2012). In the case of wetlands, the Ramsar Convention dating from 1971 encourages protection of 'the fundamental functions of wetlands as regulators of water regimes and as habitats supporting a characteristic flora and fauna, especially waterfowl'. To date, more than 2,200 sites covering over 215 million ha have been designated. Although the Ramsar Convention does not formally call for protected areas to be established, it endorses 'the conservation of wetlands and waterfowl by establishing nature reserves on wetlands whether or not they are specifically listed' (Ramsar, Art. 4(1)).

Through its Regional Seas Programme, UNEP has comprehensively promoted coastal and marine protected areas across 18 identified regional seas. Protocols respecting specially protected marine areas have been successfully concluded in several regions, notably the Mediterranean and the Caribbean. These, in general, advance the CBD objective adopted in the Jakarta Mandate on Marine and Coastal Biological Diversity (1995) and elaborated in the Aichi Targets. Pursuant to the Barcelona Protocol on Specially Protected Areas, for example, parties commit to take measures to 'protect, preserve and manage in a sustainable and environmentally sound way areas of particular natural or cultural value, notably by the establishment of specially protected areas' (Barcelona Convention, 1995a). In its objectives, the protocol calls upon signatories to safeguard: 'representative types of coastal and marine ecosystems of adequate size to ensure their long-term viability and to maintain their biological diversity' along with critical and endangered habitats and scientifically or culturally important sites (Barcelona Convention, 1995b).

Also on a broad regional level, protected areas were recognized in 1964 pursuant to an agreement on measures for conservation of Antarctic fauna and flora within the context of the 1959 Antarctic Treaty. More detailed and comprehensive arrangements were subsequently set out in Annex V to the 1991 Environment Protocol. Antarctic Specially Protected Areas may now be designated in order to protect outstanding values, including environmental values such as major terrestrial and marine ecosystems, and major colonies of breeding native birds, mammals or habitats (Antarctic Treaty, 1991).

Key issues regulated in domestic protected areas laws

Guided by the myriad of international commitments highlighted above, domestic lawmakers pursue the task of forging effective and equitable protected area regimes. This is no easy feat given the diversity issues at play and the need to ensure that protected area laws are tailored to the ecological, climatic, social and economic context of a particular jurisdiction. Protected area laws are often also influenced by the legal traditions, processes and hierarchy, and the institutional framework of each country. However, commentators have identified certain generic elements which are inherent in most protected area law frameworks. These generic elements have most recently been canvassed in the IUCN Guidelines on Protected Areas Legislation (Lausche, 2011) and the associated Building Capacity on Protected Areas Law and Governance – Training Modules (Paterson *et al.*, 2015). These elements include: scope and application; objectives and principles; institutional arrangements; systems planning and management planning; management categories and governance types; establishment, disestablishment and amending boundaries; regulating activities; compliance and enforcement; and financing. We highlight below some of the main considerations underpinning the formulation of each with a view to clearly illustrating the key role of law in both enabling and regulating protected areas.

Every protected area law needs to clearly define its geographical scope of application in order to ensure certainty. Certain laws delimit their geographical scope to the entire country or to a particular region, state or province. Other laws delimit their geographical scope to either the terrestrial environment or the marine environment, drawing a distinction between laws regulating terrestrial protected areas and marine protected areas. Others, recognizing the need for a seamless regulatory and institutional framework to span the complex marine/terrestrial interface, introduce a single protected areas law covering both environments.

Protected area laws also need to clearly outline their substantive purposes, ordinarily through provisions expressly setting out the principles and/or objectives underpinning the law. Principle provisions, which reflect the fundamental principles governing the law, can reflect both generally accepted and emerging international norms (such as the principle of non-regression, precautionary principle, social equity and justice, public participation, prior-informed consent and science-based decision-making) and norms relevant to specific domestic contexts. Objective provisions, which set out the purpose or intent of the law and often link to the management objectives underlying the protected area management categories referred to below, can span both conservation and sustainable use agendas. In respect of both principle and objective provisions, protected areas legislation should prescribe their substantive content, and the manner in which they should/may inform decision-making under the protected area law and the implementation of other laws which may impact on protected areas.

The task of implementing protected area law may be allocated among various institutions and authorities including government ministries, departments, statutory authorities, non-governmental organizations, local communities and indigenous peoples, private citizens and corporate entities. These institutions and authorities fulfil an array of functions relating to policy formulation, advice provision, general administration and management. In the case of government authorities, they can span different spheres, levels and sectors of government. Protected area law needs to identify relevant institutions, set out their respective powers and functions, provide mechanisms for promoting cooperation and coordination between them, and outline reporting and accountability procedures. Where it is anticipated that more than one authority will be involved in managing a protected area, provision should be made for co-management, feasibly through the conclusion of co-management agreements, with the form, nature, content and process preceding the conclusion of such agreements being clearly prescribed in the law.

Planning is an essential component of all protected area laws. There are generally two forms of planning relevant to protected areas: system planning and management planning. System planning recognizes that individual protected areas are part of an overall protected areas system and need to be understood in this context. In its simplest sense, a system plan represents a strategy aimed at informing decisions about where to establish protected areas, where to undertake complementary biodiversity conservation initiatives and how to coordinate the different areas and initiatives. Where provision for a system plan is incorporated within protected area law, as opposed to in broader overarching resource conservation legislation, provision should be made for the following general aspects: the form, nature and content to be included in the plan; which authority/ies is/are tasked with preparing it and within what timeframe; the process to be followed in doing so; implementation mechanisms; and provision for renewal and coordination with other relevant planning instruments. Management planning on the other hand, relates to a particular protected area. A management plan is characteristically developed by the appointed management authority, and its purpose is to guide the operations and management of a particular protected area. The protected area law should regulate the same issues in the context of management plans as are highlighted above in the context of system plans.

Efforts to develop a global system to categorize, plan for, accurately record, consistently report and compare protected areas across the world are longstanding, with the most recent iteration appearing in the IUCN Guidelines for Applying Protected Area Management Categories (Dudley, 2008; Stolton *et al.*, 2013). These Guidelines combine a set of management categories with a range of governance types, both of which are important in the context of national and international planning and reporting. Management categories are primarily concerned with the management objectives underpinning different protected areas, notably in relation to biodiversity. These management categories include: strict nature reserve (Ia); wilderness area (Ib); national park (II); natural monument (III); habitat/species management (IV); protected landscape/seascape (V); and managed resource protected area (VI). The different management

Table 4.1 IUCN Protected Area Management Categories

Category	*Definition of Management Objective*
Category Ia Strict Nature Reserve	Strictly protected areas set aside to protect biodiversity and also possibly geological or landform features, where human visitation, use and impacts are strictly controlled and limited to ensure protection of the conservation values. Such protected areas can serve as indispensable reference areas for scientific research and monitoring.
Category Ib Wilderness Area	Protected areas are usually large unmodified or slightly modified areas, retaining their natural character and influence, without permanent or significant human habitation, which are protected and managed so as to preserve their natural condition.
Category II National Park	Protected areas are large natural or near natural areas set aside to protect large-scale ecological processes, along with the complement of species and ecosystems characteristic of the area, which also provide a foundation for environmentally and culturally compatible spiritual, scientific, educational, recreational and visitor opportunities.
Category III Natural Monument/ Feature	Protected areas are set aside to protect a specific natural monument, which can be a landform, sea mount, submarine cavern, geological feature such as a cave or even a living feature such as an ancient grove. They are generally quite small protected areas and often have high visitor value.
Category IV Habitat/Species Management Area	Protected areas aim to protect particular species or habitats and management reflects this priority. Many category IV protected areas will need regular, active interventions to address the requirements of particular species or to maintain habitats, but this is not a requirement of the category,
Category V Protected Landscape/ Seascape	A protected area where the interaction of people and nature over time has produced an area of distinct character with significant ecological, biological, cultural and scenic value: and where safeguarding the integrity of this interaction is vital to protecting and sustaining the area and its associated nature conservation/other values.
Category VI Protected Area with Sustainable Use of Nature Resources	Protected areas conserve ecosystems and habitats, together with associated cultural values and traditional natural resource management systems. They are generally large, with most of the area in a natural condition, where a proportion is under sustainable natural resource management and where low-level non-industrial use of natural resources compatible with nature conservation is seen as one of the main aims of the area.

Source: Lausche (2011). Copyright IUCN.

objectives related to each of these management categories are reflected in Table 4.1, drawn from the IUCN Guidelines for Applying Protected Area Management Categories.

Governance types, which are primarily concerned with governance arrangements applicable to the protected area (namely who exercises authority and how), include: governance by government (A); shared governance (B); private governance (C); and governance by indigenous peoples and local communities (D). The different governance arrangements related to each of these main governance types are reflected in Table 4.2, which is similarly drawn from the IUCN Guidelines for Applying Protected Area Management Categories.

Many domestic protected area laws directly reflect management categories within their domestic categorization, while others do not. Many protected areas laws similarly provide

Table 4.2 IUCN Protected Area Governance Types

Governance Type	*Sub-types*
Type A. Governance by government	Federal or national ministry or agency in charge Sub-national ministry or agency in charge (e.g., at regional, provincial, municipal level) Government-delegated management (e.g., to an NGO)
Type B. Shared governance	Transboundary governance (formal arrangements between one or more sovereign States or Territories) Collaborative governance (through various ways in which diverse actors and institutions work together) Joint governance (pluralist board or other multi-party governing body)
Type C. Private governance	Conserved areas established and run by: individual landowners non-profit organizations (e.g., NGOs, universities) for-profit organizations (e.g., corporate owners, cooperatives)
Type D. Governance by indigenous peoples and local communities	Indigenous Peoples' conserved territories and areas – established and run by indigenous peoples Community conserved areas and territories – established and run by local communities

Source: Borrini-Feyerabend *et al.* (2013). Copyright IUCN.

tangible legal tools for promoting all governance types. In seeking to give effect and recognition to both management categories and governance types, countries should not be precluded or discouraged from tailoring the categories and types reflected in their protected area laws to suit their domestic context. Using different legal terminology and categories does not necessarily preclude the promotion of the IUCN management categories and governance types, but may make it more complicated when seeking to report on their performance globally.

Protected area laws need to prescribe clear criteria and procedures for establishing, disestablishing and/or altering the boundaries of protected areas. The following issues should be incorporated in any such scheme: who can initiate the process and how; criteria informing any decision; the process to be followed in making the decision (including provision for relevant assessments, public participation, cross-consultation with relevant authorities and interim protection pending declaration); land acquisition procedures; demarcation of boundaries and zones; and the formal declaration and registration of the protected area. The relevant legal framework should also prescribe a detailed management regime, which would include: the appointment of a management authority; the development of a management plan (including the prescription of mandatory/discretionary content, the process to be followed in developing it, and who should prepare and approve it); the demarcation of internal zones and external buffers if applicable; provision for co-management arrangements where relevant; and monitoring, reporting and reviewing obligations and timeframes. With growing global emphasis placed on assessing and reporting on protected area management effectiveness (PAME) (Stolton *et al.*, 2016; Hockings *et al.*, 2016), the framing of clear and rigorous management provisions in protected area laws becomes even more imperative (CBD COP 7, 2004; CBD COP 10, 2010).

Legal frameworks play an essential role in promoting the different management objectives underpinning protected areas by giving management authorities clear and adequate powers to regulate activities inside and adjacent to (particularly in identified buffer zones) the protected area. Protected area laws should set out frameworks for regulating access to and use of amenities and resources and for the prevention and control of potentially harmful activities that may threaten conservation objectives. A common legal approach is to identify or list different types of activities in the protected area law and enable the relevant authority (generally the appropriate ministry or other high-level body) to make additional regulations as need be. There are three main legal techniques used for regulating activities inside a protected area, namely: prohibiting certain activities under most or all situations (prohibited activities); requiring written permission to undertake certain activities, which are otherwise prohibited (permitted activities); and allowing certain activities without written permission, as long as certain general rules/conditions are followed (generally authorized activities). In addition to identifying the array of prohibited, permitted and generally authorized activities, the protected area law needs to prescribe: general principles and/or decision-making criteria governing the regulation of activities; procedures which must be followed in securing any relevant permission (including relevant impact assessments where necessary); procedures to deal with emergency incidents threatening the protected area; and offences and sanctions in the case of non-compliance (which would include the identification of authorities responsible for compliance and enforcement, and the designation of their powers and functions).

Finally, current spending on protected areas is grossly inadequate to meet the needs of existing protected areas and to expand coverage to achieve international goals for biodiversity conservation. Two financial mechanisms have traditionally been used to support protected areas: financing allocated from the central/regional/local government budget; and revenues generated by protected areas, predominantly through tourism and other user fees. While these two financial mechanisms remain important, other innovative and complementary financial mechanisms for protected areas are increasingly being explored, such as: external sources (funds generated from sources external to government protected area budgets, such as from private voluntary donations, foreign assistance and debt-for-nature swaps); environmental funds (funds set up to attract and administer financing secured from these external sources); incentives for conservation (such as income tax deductions for expenses incurred in undertaking management activities in a protected area or property tax deductions for land contracted into a protected area); market-based charges for goods and services (such as entrance fees, development charges and resource extraction fees); and payment for ecosystem service schemes (in terms of which the financial value of certain ecosystems services provided by the protected area – such as carbon sequestration, watershed protection and/or wetland conservation – are quantified and allocated in the form of funding to the protected area). Protected area legislation should recognize the full range of financial options that may be feasible in the particular country. Several of the innovative financial mechanisms referred to above normally fall under the jurisdiction of other legal frameworks, particularly revenue, income and property tax legislation. Protected area law needs to be aligned with this legislation – so as to ensure that the protected area authorities have the power to pursue their use.

Key emerging legal issues

As highlighted by the 2016 winner of the Nobel Prize for Literature, Bob Dylan, the 'times they are a changin'', and the law is frequently compelled to respond and adapt to new issues and challenges. These new issues and challenges are perhaps as prolific as the number of songs produced by Bob Dylan, and the prescribed scope of this chapter unfortunately confines us to only deal with a handful of those most relevant to the intended readership.

Promoting the principle of non-regression

Emerging from the international human rights discourse, the principle of non-regression proposes that existing environmental laws should not be modified in a manner which reduces, erodes or compromises the level of protection they accord to the environment (Prieur, 2011, 2012). Aligned with other pre-existing principles of environmental law (such as the duty to prevent harm, public participation, intergenerational equity and the precautionary principle), the principle of non-regression effectively seeks to characterize the protective norms contained in environmental laws as non-revocable on the basis that they protect the common interest of humanity. Originating in California in 2010, the principle has increasingly become the focus of international and domestic academics and policymakers and is partially reflected in the UN Rio+20 outcome document, *The Future We Want*, which states that 'it is critical that we do not backtrack from our commitment to the outcome of the United Nations Conference on Environment and Development' (UNGA, 2012, para. 20). Most recently, the principle of non-regression was expressly recognized in the recommendations emerging from the IUCN World Conservation Congress held in 2016, which amongst other things urge 'all governments to take steps to implement effectively and to reinforce the principle of non-regression in the areas of environmental policy and law…' (IUCN WCC, 2016).

The principle of non-regression may soon influence the form and nature of protected area laws. As highlighted by proponents of the principle, it does not preclude amending existing legislation, but urges lawmakers to proceed with care and subject any proposed amendment to proper informed scrutiny prior to adoption (Prieur, 2012). Some examples of how the principle of non-regression may impact protected areas laws are notable. Firstly, the principle of non-regression might be embedded into the preamble, objective and/or principle provisions of the law thereby ensuring that it informs implementation. Secondly, legislative provisions may be inserted to ensure that any process regulating the de-gazetting or alteration of the boundaries of a protected area is clear, rigorous, open, transparent and subject to a high level of authority. Thirdly, non-regression suggests the importance of introducing rigorous mechanisms in the law providing for environmentally deleterious activities undertaken within or adjacent to protected areas to be phased out. Fourthly, prohibitions against similar harmful activities (notably mining operations) from taking place within existing protected areas may be strengthened by the non-regression principle. Fifthly, subjecting all permitted activities to rigorous assessment and associated public participatory processes prior to their commencement would be consistent with the procedural caution aspects of the principle.

Recognizing the diversity of protected areas governance types

Governance is a rich, complex and multifaceted concept (Mitchell, 2007). In the context of protected areas, it has been defined as 'the interactions among structures, processes and traditions that determine how power and responsibilities are exercised, how decisions are taken and how citizens or other stakeholders have their say' (Graham, 2003, pp. 2–3). Protected areas governance is accordingly about who makes decisions, how such decisions are made, how authority is shared by relevant actors, and who is held accountable for decisions and how? The concept has been comprehensively canvassed in the IUCN Guidelines on Governance of Protected Areas and in the *Primer on Governance for Protected and Conserved Areas*, and is generally acknowledged as comprising of two broad dimensions, namely types of governance and quality of governance. Types of governance, referring to the main actors that hold authority for making key decisions relating to the protected area, include: governance by government; shared governance; governance by private individuals and organizations; and governance by indigenous peoples and/or local communities. The four broad types, which effectively constitute a continuum from state-centered governance to entirely decentralized governance by indigenous peoples and local communities, are further broken down into 11 sub-types. Quality of governance, while packaged in a slightly different form, relates generally to the effective and equitable exercise of authority and endorses several principles associated with the concept of good governance, namely: public participation; transparency; predictability; equity; and justice (Borrini-Feyerabend *et al.*, 2013, 2014).

In the context of protected areas, authority can be held and exercised in two different ways: *de iure* (where the authority is prescribed or recognized by the law); or *de facto* (where the authority is not prescribed or recognized by law but nonetheless exists in practice). Our focus here is naturally on the former, the role of law in promoting recognition and implementation of a diverse array of protected areas types; and in providing for the effective and equitable exercise of authority relating to protected areas. Both dimensions require attention by international and domestic law and policymakers given the relative novelty of the protected areas governance discourse.

The following key issues inform protected areas governance in both dimensions and are often regulated by the law. Firstly, the types of decisions that can or should be undertaken respecting protected areas are often prescribed in the law. Secondly, the key actors (including both government and non-government actors) authorized to exercise authority within a protected area are often constituted and/or empowered by the law. Thirdly, the instruments that impact upon, provide or influence the exercise of authority in the protected area are often created by the law. Fourthly, the nature of authority accorded to the actors through these instruments, together with the procedures they are required to adhere to when exercising such authority, are often prescribed in the law. Fifthly, the different levels or spheres within which this decision-making takes place, and the coordination across these different levels or spheres, are often identified and regulated by the law.

The interaction between the law and these different governance issues is too complex to deal with here. Fortunately, the issues have been comprehensively canvassed in the Guidelines for Protected Areas Legislation (Lausche, 2011) and more recently in the *Building Capacity on Protected Areas Law and Governance – Training Modules* (Paterson *et al.*, 2015). However, it is worth distilling here certain general lessons relating to the role of the law in promoting the IUCN protected areas governance types and the quality of governance generally.

Protected area laws play an essential role in giving domestic effect to the IUCN governance types. This is essential in two main respects. Firstly, it ensures that the necessary legal framework is in place to promote a diversity of protected area governance types, which in turn ensures that authorities can feasibly promote greater participation from a more diverse set of actors in respect of a greater proportion of land/natural resources within protected areas. Secondly, it ensures that the authority operating within the protected areas is not only feasibly more effective, but also more open, transparent and accountable.

The law should however be seen and used as an enabler and not a straightjacket. Protected area laws should recognize and promote all forms of IUCN governance types, especially given their current distribution as reflected in recent assessments. According to the *Protected Planet Report* (2016), the different governance types constitute the following percentage of protected areas reflected in the World Database of Protected Areas: governance by government (84%); private governance (4.5%); shared governance (1.8%); and governance by indigenous peoples and local communities (0.6%) (UNEP-WCMC and IUCN, 2016). For their potential to safeguard biodiversity, particularly the three latter forms of governance types could be promoted through the law by, for example: providing for the incorporation of private and communal land within the formal protected area system; incorporating mechanisms to facilitate voluntary conservation initiatives (such as agreements with government, conservation easements/covenants/servitudes and land trusts); appointing non-government actors as management authorities; and making provision for co-management arrangements. The governance arrangements within a protected area may shift along the 'governance continuum' (in either direction) over time as circumstances change. Therefore, the relevant legal framework regulating the governance arrangement needs to be sufficiently flexible to allow for it to adapt to changing circumstances. Finally, protected area laws could make periodic reporting against the IUCN governance types a mandatory requirement, given the international trend towards reporting on both IUCN management category and governance type.

Fostering connectivity

One call that has consistently echoed across the conservation discourse over the past few decades is that 'protected areas, as part of increasingly fragmented and degraded natural ecosystems, will become isolated "ecological islands", less able to stem the accelerating loss of terrestrial and marine biodiversity and less able to maintain ecosystem functions, such as species migration and hydrological flows, that operate at the landscape/seascape level' (Lausche *et al.*, 2013, p. xi). Accordingly, much attention has been placed on improving linkages within and between protected areas, with a view to promoting and/or maintaining connectivity between landscapes, seascapes, habitats, ecological and evolutionary processes (Aune *et al.*, 2011; Worboys *et al.*, 2010). The scale and form of mechanisms seeking to promote such connectivity vary, and the associated terminology used by landscape ecologists is diverse including patches, matrices, corridors, linkages, buffers and mosaics. Similarly the many benefits associated with improved connectivity span not only biodiversity conservation and climate change adaptation, but also climate change mitigation (Lausche *et al.*, 2013).

Although the concept of connectivity, in an ecological sense, arose in the 1960s and has been the focus of increasing scientific attention ever since (see most recently Gross *et al.*, 2016; Vasilijević *et al.*, 2015; Worboys *et al.*, 2015), policymakers have only recently sought to incorporate and give recognition to connectivity in the international, regional and domestic legal instruments and policy. Perhaps the clearest global reflection in an international legal instrument is *Aichi Target 11* that identifies the need for 'well-connected systems' of protected areas complemented by 'other effective area-based conservation measures' that are 'integrated' into 'wider landscapes and seascapes'. Connectivity now permeates the recent decisions emerging from the *CBD* COPs (CBD COP 10, 2010), other relevant international conventions' COPs (Ramsar Convention, 2008; CMS Convention, 2011) and the IUCN World Conservation Congresses (IUCN WCC, 2008; IUCN WCC, 2012).

A concept paper on *The Legal Aspects of Connectivity Conservation: Concept Paper* comprises the first effort to advance conceptual thinking and legal understanding about important law and policy tools and options for supporting the connectivity of protected area systems (Lausche *et al.*, 2013). The concept paper outlines a framework allowing domestic law and policymakers to analyse the feasibility of promoting connectivity through their existing legal frameworks, and where deficient, possible options for reform. Certain generally applicable lessons are noted here.

Firstly, many domestic protected area laws contain (or should contain) legal mechanisms for promoting connectivity within and between protected areas. These include: reference to connectivity in principles and objectives informing the implementation of the law; provision for comprehensive protected area system planning which promotes linkages, connectivity and the notion of ecological networks; recognizing a diversity of management categories and governance types as contributors to connectivity; providing for voluntary conservation initiatives; formally recognizing buffer zones and landscape initiatives; embedding connectivity issues within site-specific management planning; providing for environmental impact assessment relating to activities within or adjacent to the protected area which may impact on connectivity; blending connectivity into applicable decision-making criteria, monitoring and reporting obligations; and prescribing economic and other incentives in respect of activities which promote connectivity.

Connectivity is not however only the purview of protected areas legislation, and many natural resource laws, land use planning laws, development control laws and fiscal laws contain mechanisms which need to be developed and applied in a complementary manner. Given the prevalence of a range of laws, administered by a variety of authorities containing diverse legal mechanisms to promote connectivity within and between protected areas, cross-sectoral integration and cooperation is often key. Furthermore, the creative, coordinated and phased implementation of existing legal mechanisms for promoting legal reform may be more expedient than wholesale legal reform aimed at creating a 'one-stop legal connectivity shop'. Examples of the former type of approach could include: embedding connectivity considerations within spatial/land use planning initiatives; prescribing connectivity issues as mandatory considerations in environmental impact assessment processes informing development proposals within and adjacent to protected areas; imposing buffers around protected areas; or enabling private landowners to contract land of high conservation value between protected areas under a range of stewardship options. Legal approaches will need to be tailored to match the nuanced characteristics of a particular jurisdiction and context, specifically taking into account conservation priorities, land tenure arrangements, institutional arrangements, development priorities and patterns, and resource issues.

Scoping and implementing other effective area-based conservation measures

As highlighted above in the context of connectivity, the concept of 'other effective area-based conservation measures' (OECMs) is embedded within Aichi Target 11. OECMs constitute an essential complementary mechanism for achieving the prescribed numerical targets. Accordingly, legal clarity is required regarding their definition, scope, nature, purpose and objective. However, notwithstanding six years having passed since the approval of the target and calls by the international community for clarity on the above aspects, such guidance is still absent (Jonas *et al.*, 2014). Thus several important matters remain unresolved: what constitute these measures; how these measures relate to protected areas; what areas/regions should be the subject of/qualify for recognition as OECMs; how such recognition should be accorded to them; what criteria should inform such recognition; who should be accorded authority to provide such recognition; what procedures should be followed in according such recognition; what is the temporal scale of these OECMs; and who should be tasked with managing these areas and reporting on management effectiveness. Underlying each of these is the question of the role of law.

Moves are fortunately afoot to provide some clarity. At the IUCN World Conservation Congress held in 2012, parties recognized the need to 'interpret the implications of Target 11 at national and regional scales to enhance action and accountability and to clarify terms and measures, including the use of the term "other effective area-based conservation measures"' (IUCN WCC, 2012). In 2015, the IUCN WCPA established a task force to specifically provide guidance on the nature, form and scope of OECM for both the IUCN and parties to the CBD. In the context of the CBD, parties attending the 20th Subsidiary Body on Scientific, Technical and Technological Advice in 2016 called on the Executive Secretary of the CBD to provide similar guidance and support the efforts of the task force. The task force has already produced a range of guidance documentation, such as *Framing the Issues: Discussion Paper 1* (Jonas and McKinnon, 2015; 2016a; 2016b). Most recently the task force published draft *Guidance for Recognition and Reporting of OECM under Aichi Target 11* (IUCN WCPA, 2016). The last of these documents provides the first tentative definition of OECMs as a 'clearly defined geographical space, beyond the protected areas network, governed and managed in ways that deliver the long-term and effective *in-situ* conservation of biodiversity and associated ecosystem services and cultural values, regardless of its management objectives'. The concept of OECM would therefore appear to encapsulate area-based conservation initiatives which do not meet the IUCN and CBD's formal requirements to be regarded as a protected area, but which significantly contribute nonetheless to biodiversity conservation. These conservation initiatives could feasibly include conservancies, stewardship programmes and area-based conservation schemes undertaken by indigenous peoples and local communities. Once finalized, the guidance currently being developed under the auspices of the IUCN may well inform domestic lawmakers in their efforts to frame or amend legal frameworks seeking to give tangible legal recognition and effect to OECMs.

Regulating marine protected areas beyond national jurisdiction

While approximately 4.12% of the global ocean and 10.2% of coastal and marine areas under national jurisdiction are currently incorporated within marine protected areas, only 0.25% of the marine environment situated in areas beyond national jurisdiction enjoy a similar status (UNEP-WCMC and IUCN, 2016). Given the ecological importance of these areas and their low percentage representation in the global protected areas estate, the United National General Assembly adopted a resolution in June 2015 calling for the development of 'an international legally binding instrument under the United Nations Convention on the Law of the Sea on the conservation and sustainable use of marine biological diversity in areas beyond national jurisdiction' (UNGA, 2015). The proposed instrument will address high seas marine protected areas.

The challenges inherent in formulating a legally binding instrument of this nature have been canvassed by several commentators (Wright *et al.*, 2015; Druel *et al.*, 2012; Nordtvedt, 2012), and include: the fragmented state of the current international regime governing marine areas beyond national jurisdiction (MABNJs); the dearth of a comprehensive set of overarching governance principles; a disjointed institutional framework; the absence of a global framework to establish marine protected areas in MABNJs; legal uncertainty surrounding the status of marine genetic resources in MABNJs; a lack of global rules for EIAs and SEAs in MABNJs; limited capacity-building and technology transfer; uneven and ineffective governance of high seas fisheries; and compliance and enforcement issues relating to flag state responsibility and the use of flags of convenience. At the September 2016 discussions of the Preparatory Committee on Marine Biodiversity beyond Areas of National Jurisdiction, which is responsible for making recommendations to the UN General Assembly on the form and content of the legally binding instrument, the following aspects were noted as essential elements for the regime: definitions; management objectives; strategic planning; stakeholder consultation; open and transparent decision-making; delineation of spatial boundaries of the MPAs; management planning, monitoring and review; institutional arrangements; regulation of activities in MPAs; and protection measures (Earth Negotiations Bulletin, 2016). Encouragingly, these match elements frequently found in legislation regulating marine protected areas under national jurisdiction (Lausche, 2011, pp. 209–64; Day *et al.*, 2012), and it will be interesting to see what adjustments are made to grapple with an entirely different jurisdictional setting.

Conclusion

With specific reference to biodiversity, the 2016 *Protected Planet Report* documents that protected areas have contributed to biodiversity conservation in several ways, notably by reducing habitat loss, by benefitting a broad set of species and, for those species whose most critical and important sites were protected, by lowering the risk of extinction (UNEP-WCMC and IUCN, 2016). Indeed, effectively managed protected areas have the potential to contribute to the recovery of threatened populations. As expressed in the TEEB Report where the global significance of biodiversity is comprehensively addressed, protected areas serve as the 'cornerstone' of conservation (TEEB, 2009, p. 20).

In this chapter we have surveyed legal dimensions of the contribution of protected areas to biodiversity. The international legal foundations of biodiversity conservation, including the elaboration of protected areas networks, now serve more broadly to guide national and international action. The *Strategic Plan for Biodiversity*, for example, as elaborated within the CBD context, was subsequently adopted as a framework for sustainability and biodiversity action across the UN system and has been endorsed by the governing bodies of five other biodiversity-related conventions. Moreover, the 2015 *Sustainable Development Goals* once again demonstrate the significance of biodiversity as a global policy objective and the critical contribution of protected areas within the overall agenda comprised of seventeen SDGs incorporating 169 specific targets.

At the same time, however, it is acknowledged that no formula dictates or prescribes the 'correct' allocation of lands and waters to the conservation of biodiversity. As recently expressed, 'there is no scientific answer to how much space we should leave for other forms of life. This is a moral and ethical discussion' (Baillie *et al.*, 2016, p. 7). These ethical dimensions of global biodiversity challenges alongside practical complexities and the dynamic context in which they must be addressed underscore the ongoing contribution of the legal framework (TEEB, 2009, p. 21). While long-established considerations associated with the rule of law principle help to secure the legal foundations of permanent and accountable protected area frameworks, emerging legal approaches in the form of non-regression, precautionary decision-making, participatory governance and mechanisms to address distinctive threats to biodiversity beyond national marine jurisdiction offer new opportunities for protected areas law to advance effective and equitable biodiversity conservation initiatives.

References

Antarctic Treaty (1991), 'Area Protection and Management' Annex V to the Protocol on Environmental Protection (Madrid), www.ats.aq/documents/recatt/Att004_e.pdf.

Aune, K., Beier, P., Hilty, J. and Shilling, F. (2001) *Assessment and Planning for Ecological Connectivity: A Practical Guide* (Wildlife Conservation Society, Bozeman).

Baillie, J., Joppa L. and Robinson J., 'Introduction: Do Protected Areas Safeguard Biodiversity?' in Joppa, L., Baillie, J. and Robinson, J. (eds) (2016) *Protected Areas: Are They Safeguarding Biodiversity?* (Wiley Blackwell, Chichester).

Barcelona Convention (1995a) Protocol Concerning Specially Protected Areas and Biological Diversity in the Mediterranean, www.rac-spa.org/sites/default/files/protocole_aspdb/protocol_eng.pdf.

Barcelona Convention (1995b) 'Action Plan for the Protection of the Marine Environment and the Sustainable Development of the Coastal Areas of the Mediterranean (MAP II)', http://ec.europa.eu/environment/marine/international-cooperation/regional-sea-conventions/barcelona-convention/index_en.htm.

Borrini-Feyerabend, G., Bueno, P., Hay-Edie, T., Lang, B., Rastogi, A. and Sandwith, T. (2014) *A Primer on Governance for Protected and Conserved Areas* (IUCN, Gland).

Borrini-Feyerabend, G., Dudley, N., Jaeger, T., Lassen, B., Pathak Broome, N., Phillips, A. and Sandwith, T. (2013) *Governance of Protected Areas: From Understanding to Action* (IUCN Best Practice Protected Area Guidelines Series No. 20, Gland).

Chacon, M. (ed) (2015) *El Principio de No Regresión Ambiental en Iberoamérica* (IUCN, Gland).

Convention on Biological Diversity (1992) (CBD) 31 *ILM* (1992) 818.

CBD (CBD National Biodiversity Strategy) 'National Biodiversity Strategy and Action Plans', www.cbd.int/protected/implementation/actionplans/.

CBD (CBD Overview) 'Protected Areas – An Overview', www.cbd.int/protected/overview/default.shtml.

CBD COP 7 (2004) Decision VII/28 (2004) Protected Areas (Articles 8(a) to (e), 'Annex'.

CBD COP 10 (2010) Decision X/2 (2010) Strategic Plan on Biodiversity (2011–20).

CBD COP 13 (2016) Decision XIII/2 Progress towards the achievement of Aichi Biodiversity Targets 11 and 12.

Convention on International Trade in Endangered Species (1973) (CITES) 12 *ILM* 1085 (1973).

Convention on Migratory Species of Wild Animals (1980) (CMS Convention) 19 *ILM* 11 (2011).

Convention on Wetlands of International Importance Especially as Waterfowl Habitat (Ramsar Convention 1971) (1971) 11 *ILM* 963 (1972).

Convention on Wetlands of International Importance Especially as Waterfowl Habitat (Ramsar Convention 2008) (2008) Resolution X.24/2008.

Day, J., Dudley, N., Hockings, M., Holmes, G., Laffoley, D., Stolton, S. and Wells, S. (2012) *Guidelines for Applying the IUCN Protected Area Management Categories to Marine Protected Areas* (IUCN, Gland).

Druel, E., Rochette, J., Bille, R. and Chiarolla C. (2012) *A Long and Winding Road: International Discussions on the Governance of Marine Biodiversity in Areas beyond National Jurisdiction*, Study No. 07/13 (IDDRI, Paris).

Dudley, N. (ed) (2008) *Guidelines for Applying Protected Area Management Categories* (IUCN, Gland).

Dudley, N., Allen, D. and Campbell, K. (2015) *Natural Solutions: Protected Areas are Vital for Human Health and Well-being* (IUCN, Gland).

Dudley, N., Stolton, S., Belokurov, A., Krueger, L., Lopoukhine, N., MacKinnon, K., Sandwith, T. and Sekhran, N. (eds) (2010) *Natural Solutions: Protected Areas Helping People Cope with Climate Change* (IUCN-WCPA, TNC, UNDP, WCS, World Bank and WWF).

Earth Charter, http://earthcharter.org/invent/images/uploads/echarter_english.pdf.

Earth Negotiations Bulletin (2016) 'Summary of the Second Session of the Preparatory Committee on Marine Biodiversity beyond Areas of National Jurisdiction' Vol. 25 no. 118.

Gillespie, A. (2007) *Protected Areas and International Environmental Law* (Martinus Nijhoff).

Graham J., Amos, B. and Plumptre, T. (2003) 'Governance Principles for Protected Areas in the 21st Century A Discussion Paper' (Institute on Governance in collaboration with Parks Canada and the Canadian International Development Agency, Ottawa).

Gross, J., Woodley, S., Welling, L. and Watson, J. (eds) (2016) *Adapting to Climate Change: Guidance for Protected Area Managers and Planners* (Best Practice Protected Area Guidelines Series No. 24, IUCN, Gland).

Head, J. (2012) *Global Legal Regimes to Protect the World's Grasslands* (Carolina Academic Press, Durham, NC).

Hockings, M., Stolton, S., Leverington, F., Dudley, N. and Courrau, J. (2016) *Evaluating Effectiveness: A Framework for Assessing Management Effectiveness of Protected Areas*, 2nd edn (IUCN, Gland and Cambridge, UK).

IUCN WCC (2008) Resolution 4.062 (2008).

IUCN WCC (2012) Resolution 5.056 (2012).

IUCN WCC (2012) Resolution 035.

IUCN WCC (2016) Motion 082 – Reinforcing the principle of non-regression in environmental law and policy (adopted).

IUCN WCPA (2016) *Guidance for Recognition and Reporting of OECM under Aichi Target 11*, www.iucn.org/protected-areas/world-commission-protected-areas/wcpa/what-we-do/other-effective-area-based.

Jonas, H. and Mckinnon, K. (2015) *Framing the Issues: Discussion Paper 1 of the Task Force on Other Effective Area-Based Conservation Measures* (IUCN, Gland).
Jonas, H. and Mckinnon, K. (2016a) *Co-Chairs' Report of the First Meeting of International Experts: Task Force on Other Effective Area-Based Conservation Measures* (IUCN/WCPA, Gland).
Jonas, H. and Mckinnon, K. (eds) (2016b) *Advancing Guidance on Other Effective Area-Based Conservation Measures: Report of the 2nd Meeting of the IUCN-WCPA Task Force on OECM* (Bundesamt fur Naturshutz, Bonn).
Jonas, H., Barbuto, V., Jonas, H., Kothari, A. and Nelson, F. (2014) 'New Steps of Change: Looking Beyond Protected Areas to Consider Other Effective Area-Based Conservation Measures' *Parks* Vol. 20 no. 2, pp. 111–28.
Joppa, L., Baillie, J. and Robinson, J. (eds) (2016) *Protected Areas: Are They Safeguarding Biodiversity?* (Wiley Blackwell, Chichester).
Krueger, L. (2016) 'Government commitments for protected areas, status of implementation and sources of leverage to enhance ambition' in Joppa, L., Baillie, J. and Robinson, J. (eds) (2016) *Protected Areas: Are they Safeguarding Biodiversity?* (Wiley Blackwell, Chichester).
Lausche, B. (2011) *IUCN Guidelines for Protected Areas Legislation* (IUCN, Gland).
Lausche, B., Farrier, D., Verschuuren, J., La Viña, A., Trouwborst, A., Born, C. and Aug L. (2013) *The Legal Aspects of Connectivity Conservation. A Concept Paper* (IUCN, Gland).
MacArthur, R. and Wilson, E. (1967) *The Theory of Island Biogeography* (Princeton University Press, Princeton, NJ).
Mitchell, B. (2007) 'Who's Doing the Protecting in Protected Areas? A Global Perspective on Protected Area Governance' *The George Wright Forum*, Vol. 24 no. 3, pp. 81–91.
Nordtvedt Reeve, L., Rulska-Domino, A. and Gjerde, K. (2012) 'The Future of High Seas Marine Protected Areas' *Ocean Yearbook*, Vol. 26, pp. 265–89.
Paterson, A., Lausche, B., Benidickson J. and Moore P. (2015) *Building Capacity on Protected Areas Law and Governance – Training Modules* (IUCN, Gland), www.protectedareaslaw.org.
Prieur, M. (2011) 'De L'urgente Nécessité De Reconnaître Le Principe De "Non Régression" En Droit De L'Environnement' *IUCN Academy of Environmental Law e-Journal*, Vol. 1, pp. 26–40.
Prieur, M. (2012) 'Non-Regression in Environmental Law' *Surveys and Perspectives Integrating Environment and Society*, Vol. 5, no. 2, pp. 1–7.
Stolton, S. and Dudley, N. (eds) (2010) *Arguments for Protected Areas: Multiple Benefits for Conservation and Use* (Earthscan, London).
Stolton, S. and Dudley, N. (2016) *METT Handbook: A Guide to Using the Management Effectiveness Tracking Tool (METT)*, WWF-UK, Woking.
Stolton, S., Shadie P. and Dudley, N. (2013) *IUCN WCPA Best Practice Guidance on Recognising Protected Areas and Assigning Management Categories and Governance Types* Best Practice Protected Area Guidelines Series No. 21 (IUCN, Gland).
TEEB (2009) *The Economics of Ecosystems and Biodiversity for National and International Policy-Makers: Responding to the Value of Nature*, http://img.teebweb.org/wp-content/uploads/Study%20and%20Reports/Reports/National%20and%20International%20Policy%20Making/TEEB%20for%20National%20Policy%20Makers%20report/TEEB%20for%20National.pdf.
United Nations Conference on Environment and Development (UNCED) (1992) Agenda 21, https://sustainabledevelopment.un.org/content/documents/Agenda21.pdf.
United Nations Conference on the Human Environment (1972) (Stockholm Declaration) *Declaration of the United Nations Conference on the Human Environment (Stockholm)* (1972) UN Doc.A/CONF/48/14 Rev.1.
United Nations Conference on the Human Environment (1972) (Stockholm Report) *Report*, Stockholm 5–16 June 1972, A/Conf.48/14/Rev.1, www.un-documents.net/aconf48-14r1.pdf.
UNEP-WCMC and IUCN (2016) *Protected Planet Report 2016* (UNEP WCMC and IUCN, Gland).
UNGA (1982) *World Charter for Nature* A/RES/37/7 (28 October 1982).
UNGA (2012) Resolution 66/288 (2012) *The Future We Want*.
UNGA (2015) Resolution 69/292 (19 June 2015).
UNGA (2015) Resolution 70/1 (2015).
Vasilijević, M., Zunckel, K., McKinney, M., Erg, B., Schoon, M. and Rosen Michel, T. (2015) *Transboundary Conservation: A Systematic and Integrated Approach*, Best Practice Protected Area Guidelines Series No. 23 (IUCN, Gland).

Worboys, L., Francis, W. and Lockwood, M. (2010) *Connectivity Conservation Management: A Global Guide* (Earthscan, London).

Worboys, G., Lockwood, M., Kothari, A., Feary, S. and Pulsford I. (eds) (2015) *Protected Area Governance and Management* (ANU Press, Canberra).

World Commission on Environment and Development (WCED) (1987) *Our Common Future* (Oxford University Press).

WHC (2015) *Operational Guidelines for the Implementation of the World Heritage Convention*, WHC 15/01 8 July 2015.

WHC *World Heritage List,* http://whc.unesco.org/en/list/.

Wright, G., Rochette, J., Druel, E. and Gjerde, K. (2015) *The Long and Winding Road Continues: Towards a New Agreement on High Seas Governance*, Study No. 01/16, IDDRI, Paris.

5

THE INTERNATIONAL LEGAL FRAMEWORK FOR THE PROTECTION AND SUSTAINABLE USE OF MARINE BIOLOGICAL DIVERSITY

Youna Lyons and Denise Cheong

Introduction

The last two decades have seen an intensification of projects and programmes designed to map, investigate and better understand marine biological diversity (biodiversity, for short) at local and global scales.[1] Scientific publications also report tirelessly on threatened, near-extinct and extinct species as well as newly discovered species: a common feature of under-studied marine ecosystems such as the deep seabed or other systems found in areas which are difficult to access. However, efforts to protect marine biodiversity are slowed by the tension that exists between conservation efforts, which aim to set aside 10% of all wildernesses as 'natural reserves',[2] and development efforts led by industries and many governments, which favour economic growth and the use of oceans and their resources. The existing legal framework for the protection and sustainable use of marine biodiversity reflects this tension as it includes rules from conservation treaties as well as from treaties relating to uses of the sea. The perspective presented in this chapter is that the protection of marine biodiversity should not be framed solely as a conservation issue but that it should also be considered in the context of pollution of the sea and generally all regulations designed to protect the marine environment through the limitation of environmental impacts from activities at sea. This is because pollution thresholds and other restrictions in the use of marine ecosystems are necessary to ensure the protection of biodiversity in areas that have been identified as warranting protection under the conservation treaties.

This chapter, which presents an overview of the current legal framework applicable to the protection and use of marine biodiversity, discusses, in its first section, how the United Nations Convention on the Law of the Sea (UNCLOS)[3] is the cornerstone of this legal framework and the Convention on Biological Diversity (CBD),[4] which can be deemed to be incorporated into UNCLOS. The criteria adopted by the Conference of the Parties (COP) to the CBD to identify areas of ecological and biological significance are then discussed. The second section investigates the extent to which these criteria include those found in more specific conservation treaties such as the 1971 Ramsar Convention on the Protection of Wetlands of International Importance (Ramsar Convention),[5] the 1972 UNESCO Convention concerning the Protection of the World Natural and Cultural Heritage (the UNESCO World Heritage Convention),[6] the 1979 Bonn Convention on the Conservation of Migratory Species of Wild Animals (Convention on Migratory Species)[7] and in other more recent soft law instruments. The third section examines the way in which conventions and international regulations and guidelines that pertain to specific uses of the sea restrict activities in the interest of protecting ecosystems, habitats or species and therefore biodiversity. The discussion involves an examination of international regulations relating to the exploitation of marine living resources (1995 Agreement for the Implementation of the Provisions of the UNCLOS relating to the Conservation and Management of Straddling Fish Stocks and Highly Migratory Fish Stocks, the FAO Code of Conduct for Responsible Fisheries[8] and Whaling Sanctuaries established by the International Whaling Commission[9]), the trade of wildlife (the Convention in International Trade in Endangered Species of Wild Fauna and Flora – CITES[10]), shipping ('Special Areas' under the 1973/1978 MARPOL Convention,[11] the Ballast Water and Sediment Management Convention,[12] as well as IMO Guidelines on Particularly Sensitive Sea Areas (PSSA) and Routeing Measures) and deep seabed mining (Mining Code adopted by the International Seabed Authority[13]). The fourth section of this chapter discusses the protection of marine biodiversity in Areas Beyond National Jurisdiction (ABNJ) and overall gaps in the protection of marine biodiversity. First, it discusses the development of bioprospecting of marine genetic resources and the regulatory issues it raises; second, it analyses the deficiencies of the overall existing legal regime with respect to the protection of marine biodiversity in ABNJ as well as within national jurisdiction; and third, it outlines the ongoing process of negotiation at the United Nations towards a new Internationally Binding Legal Instrument (IBLI) for biodiversity in ABNJ.

UNCLOS, the cornerstone of the protection of marine biodiversity

The marine environment in UNCLOS encompasses marine biodiversity

UNCLOS, which includes an entire section of 45 articles on the protection and preservation of the marine environment (Part XII), does not define the term 'marine environment', nor does it refer literally to biological diversity. However, the definition of 'pollution of the marine environment' is very wide, as it includes 'living resources and marine life' as well as quality of seawater and all properties that are necessary for human activities. Furthermore, Article 194(5) specifically refers to the protection and preservation of the marine environment as including the protection and preservation of rare or fragile ecosystems as well as the habitat of depleted, threatened or endangered species and other forms of marine life. Finally, the provisions on monitoring and environmental assessment direct states to 'keep under surveillance the effects of any activities … in order to determine whether these activities are likely to pollute the marine environment'. In circumstances where such activities 'may cause substantial pollution of or significant and harmful changes to the marine environment, [States] shall, as far as practical, assess the potential effects of such activities on the marine environment and shall communicate reports of the results'.[14] The term 'marine environment' in UNCLOS appears to include all naturally occurring living and non-living parts of the oceans.[15] This has also been the approach of the United Nations General Assembly, which, in its resolution on Oceans and Law of the Sea, has been including marine biodiversity within its section on 'Marine Environment, Marine Resources and the Protection of Vulnerable Marine Ecosystems' since 2006.[16]

The most comprehensive definition of biological diversity contained in a binding treaty is that of the Article 2 of the CBD: 'the variability among living organisms … and the ecological complexes of which they are part; this includes diversity within species, between species and of ecosystems'. The main difference between protecting and preserving the marine environment and protecting marine biodiversity is therefore that the latter refers to a particular aspect of the marine environment, which is the 'variability within species, between species and of ecosystems'. The emphasis is on the number of different species and ecosystems, rather than the number of individuals of the same specie (although this may also be commercially valuable and therefore desirable). However the implementation guidelines developed by the COP to the CBD (to identify Ecologically and Biologically Significant Areas (EBSAs)) do not focus solely on species richness and distribution. They take a more functional approach to biodiversity, where biodiversity serves as a basis for building environmental resilience, and as such have developed criteria that include (in addition to criteria of biological diversity and threatened, endangered or declining species and/or habitats): (i) uniqueness and rarity; (ii) special importance for life history of species; (iii) vulnerability, fragility, sensitivity and slow recovery; (iv) biological productivity; and (v) naturalness.[17]

Given that the living ocean is composed of interacting ecosystems made up of numerous species from different taxa, there is no doubt that the protection of the marine environment envisaged in UNCLOS includes the protection of marine biodiversity and that the CBD should therefore be read as further articulating aspects of the UNCLOS obligation to protect the marine environment.

The pre-eminence of UNCLOS and the obligation to protect and preserve marine biodiversity in UNCLOS and the CBD

As suggested by its best descriptor of 'constitution for the oceans',[18] and in its preamble where it is described as the needed 'legal order for the seas and oceans', the pre-eminence of UNCLOS for all issues relating to the oceans has been highlighted by many (see for instance, Boyle, 2006 and Redgwell, 2006). This view is primarily based on the fact that UNCLOS is comprehensive (in that it seeks to regulate all the uses of the sea[19]), is approaching universality (with 168 parties as at 23 September 2016[20]) and was designed based on an intertemporal approach.[21] Furthermore many of its provisions are commonly referred to as customary international law.[22]

UNCLOS's pre-eminence vis-à-vis other treaties is further established in Article 237(2) of UNCLOS which provides that 'Specific obligations assumed by States under special conventions, with respect to the marine environment, should be carried out in a manner consistent with the general principles set forth in this Convention.'[23] Article 22(2) of the CBD also reinforces this pre-eminence as it provides that 'Contracting Parties shall implement this Convention with respect to the marine environment consistently with the rights and obligations of States under the Law of the Sea'. A key design feature of the ocean governance regime established by UNCLOS which ensures its pre-eminence is its in-built mechanisms that allow sub-regimes to be developed and integrated into UNCLOS. For example UNCLOS places the responsibility to develop and adopt specific measures (including for the protection of the marine environment) on Member States acting through 'competent organizations' (such as the International Maritime Organization (IMO) for shipping matters[24] and the Food and Agricultural Organization (FAO) as well as Regional Fisheries Management Organizations for fisheries.[25] Another example of such an in-built mechanism is the incorporation by reference into UNCLOS of generally accepted rules, standards and recommended practices developed and adopted by competent organizations for various sub-regimes.[26] In the recent arbitration case between the Philippines and China, the Tribunal considered the relationship between UNCLOS and the CBD. It found that the general obligation to protect and preserve the marine environment under Article 192 of UNCLOS is broad enough to include the protection of biodiversity and that Article 194 also includes the protection and preservation of the biological diversity, in the form of coral reefs, giant clams and sea turtles. It further decided that the content of the obligation provided in Article 192 of UNCLOS is informed by other applicable rules of international law, including the CBD.[27]

Based on this approach, the following paragraphs discuss the content of the obligation to protect and preserve the marine environment provided for in UNCLOS and the way in which the CBD informs it. UNCLOS places the responsibility to manage the marine environment on all States who have jurisdiction or control over activities that may have an adverse environmental impact. However, like in the CBD, the obligations of States to protect the marine environment in UNCLOS are subject to their rights to use the ocean and exploit living and non-living resources. Article 193 of UNCLOS provides that States have sovereign rights to exploit their natural resources pursuant to their environmental policies and in accordance with their duty to protect and preserve the marine environment. Similarly, the CBD's main aim is the conservation *and* sustainable use of biological diversity.[28]

As such, while the obligation to protect and preserve the environment is worded in general terms, there are further provisions in UNCLOS that elaborate on this general obligation by seeking to prevent environmental harms from specific ocean uses such as shipping, seabed mining, dumping etc., for which states are encouraged to adopt further international and/or regional agreements to detail their specific obligations and expected to adopt domestic laws, regulations and measures to implement these international regulations (de la Fayette, 2004). For pollution from seabed mining and dumping at sea, these regulations and measures must be at least as effective as international rules and standards. Under UNCLOS, States also have a direct obligation to 'conduct an environmental impact assessment' for activities 'they engage in or permit' in order to determine whether they are likely to pollute the marine environment, including marine biodiversity. They must then publish their report.[29]

Furthermore, international courts and tribunals have had the opportunity to clarify States' obligations to ensure that activities within their jurisdiction or control do not harm the marine environment in the specific circumstances brought to them.[30] They have found that the obligation to 'ensure' imposes a duty to act with 'due diligence',[31] in the sense that it is an obligation 'to deploy adequate means, to exercise best possible efforts, to do the utmost, to obtain this result.'[32] They have further indicated that:

> it is an obligation which entails not only the adoption of appropriate rules and measures but also a certain level of vigilance in their enforcement and the exercise of administrative control … such as the monitoring of activities undertaken.[33]

In another case involving a question of responsibility of flag States with respect to vessels flying its flag that engage in illegal fisheries, the tribunal determined that States must not only adopt appropriate rules and measures to prohibit such activities but are also 'obliged to investigate and, if appropriate, take any action necessary to remedy the situation' if violations nevertheless occur and are reported by other States; that flag States are under an obligation of due diligence, which includes an 'obligation to exercise effectively its jurisdiction and control in administrative matters over fishing vessels flying its flag'. This further requires that domestic laws, regulations and measures include 'enforcement mechanisms to monitor and secure compliance … Sanctions … must be sufficient to deter violations and to deprive offenders of the benefits accruing from their [illegal] activities'.[34] In the context of the harvesting of giant clams by means that destroy their coral habitats and associated species, the arbitral tribunal decided that the due diligence obligation under UNCLOS Article 192, read in the context of Article 194(5), is an obligation to take those measures 'necessary to protect and preserve rare or fragile ecosystems as well as the habitat of depleted, threatened or endangered species and other forms of marine life'. The obligation includes first a duty to prevent harvesting of species that are recognized internationally as being at risk of extinction and requiring international protection and, second, a duty to prevent harm that would affect depleted, threatened or endangered species indirectly through the destruction of their habitat.[35] Also noteworthy is the fact that to determine that the species and ecosystems concerned were at risk, the tribunal relied on the listing in CITES Appendix II (species that may become threatened by extinction if subject to uncontrolled trade)[36] and scientific expert evidence which also refers to the IUCN Red List.[37]

In the same way, it is proposed that the obligation of identification and monitoring of components of biodiversity (important for their conservation and sustainable use) under the CBD should be read as forming part of the obligation of due diligence under UNCLOS, the CBD forming part of the corpus of international law that informs the content of UNCLOS provisions on the protection of the marine environment, including marine biodiversity (along with CITES and other environmental conventions).[38]

Obligation of identification and monitoring under the CBD and the identification of ecologically and biologically significant areas

While it encompasses both terrestrial and marine biodiversity, the CBD does not distinguish between them in terms of the measures it provides. That said, it seems to have been drafted with terrestrial biodiversity in mind. For instance, the convention focuses on species diversity and areas of high endemism to determine areas where protection would be desirable (Freestone, 1995). This makes less sense in the context of marine biodiversity than it does for terrestrial biodiversity, because only 20% of known species are marine and endemism is generally low[39] due to the ability of marine organisms to disperse over large geographic scales (unlike terrestrial animals) (Palumbi, 1992; Angel, 1993). Other important differences which result in different geographic representations of biodiversity (and a different approach to geographic scale) include the presence of higher orders of the taxonomic classification being more represented in the ocean: 28 phyla of animals, compared to 11 on land. The need to develop implementation measures for the CBD that are specific to marine and coastal biodiversity has led to the creation of a Programme of Work on Marine and Coastal Biodiversity adopted at the COP 5 (held in 1998).[40]

In 2004, the UNGA's call for area-based approaches and tools for the conservation and sustainable use of marine biodiversity in ABNJ further stimulated the development of the work undertaken in this programme, triggering expert meetings on the topic and eventually the adoption (Bax *et al.*, 2016; Dunn *et al.*, 2014)[41] at the COP 9 (held in 2008) of the following seven criteria for the identification of EBSAs: (1) uniqueness or rarity, (2) special importance for life-history stages of species, (3) importance for threatened, endangered or declining species and/or habitats, (4) vulnerability, fragility, sensitivity or slow recovery, (5) biological productivity, (6) biological diversity and (7) naturalness (Dunn, 2014).[42] COP 9 also adopted five separate criteria for the identification of suitable areas with a view to establishing networks of Marine Protected Areas, EBSAs being one such.[43] This decision was driven by a focus 'on provision of scientific and, as appropriate, technical information and advice relating to marine biological diversity, the application of the ecosystem approach and the precautionary approach'.[44]

Application of these criteria relies primarily on biological and ecological information but also on oceanographic information as an indicator of missing biological and ecological information. They demonstrate a holistic and partly qualitative approach to marine biodiversity that does not focus solely on species abundance and distribution. However, they do not take into account the extent of potential or actual vulnerability to ongoing human activities, natural events or their impact. Criterion (4) on 'vulnerability, fragility, sensitivity or slow recovery' focuses on functionally or inherently sensitive or vulnerable habitats, but does not include the extent of exposure to potential or actual disruption, degradation or depletion by human activities or natural events.

Nevertheless, COP 10 which adopted the Aichi Biodiversity targets called for '10% of coastal and marine areas, especially areas of particular importance for biodiversity and ecosystem services [to be] conserved through *effectively and equitably managed*, ecologically representative and well-connected systems of protected areas and other effective area-based conservation measures'. As such, this call incorporates the element of sustainable use (also included in the objectives of the CBD) which is not reflected in the EBSAs criteria which focus solely on scientific parameters. The decision was also made to hold a series of regional workshops in order to describe potential areas for protection in ABNJ that meet the EBSAs criteria; this led to the inclusion of areas of Exclusive Economic Zones as proposed by the relevant States. A repository for storing and sharing scientific and technical information was further established which may include data from areas within national jurisdiction, in addition to data from ABNJ.[45] The intention of the COP to the CBD is that areas that are identified as EBSAs should also be considered as 'sensitive areas' under other instruments and that maritime and marine uses that are taking place within these areas, as well as pollution from these activities, should be regulated within the regime applicable to each activity (e.g. shipping through the International Maritime Organization (IMO), fisheries through the Food and Agricultural Organization (FAO) and relevant Regional Fisheries Management Organizations (RFMOs), deep seabed mining through the International Seabed Authority (ISA), etc.) (Dunn *et al.*, 2014).[46] The identification of an area as an EBSA is not a first step towards declaring it a Marine Protected Area; it is solely a scientific exercise to shortlist areas that present particular characteristics and for which management solutions should be considered. Such management solutions could include (but are not limited to) the establishment of Marine Protected Areas associated with management measures for marine and maritime activities occurring within those areas.

Conservation regimes developed under the auspices of conventions pre-dating UNCLOS and subsequent soft law instruments

Wetlands of international significance

The objective of the 1971 Ramsar Convention is the wise use of wetlands with particular conservation and management of wetlands of international significance in terms of ecology, botany, zoology, limnology or hydrology;[47] the term 'wetland' includes areas of marine waters that are up to six metres deep at low tide; deeper sea areas may also be included in the regime if they are riparian and coastal zones adjacent to wetlands and islands or bodies of marine water, especially if they support important waterfowl habitat.[48]

In order to fulfil the objective of designating and listing wetlands of international significance, the COP to the Ramsar Convention has adopted sets of criteria to identify such wetlands. The last set of alternative criteria was adopted at COP 9 (held in 2005) and is made up of nine criteria, five of which are clearly included in the EBSAs criteria. The other four criteria are wetland-specific and could possibly, in some circumstances, not qualify under EBSAs. They relate to significant numbers of individuals or populations of water birds and non-avian species from wetlands as well as wetland fish that would be representative of wetland benefits. Examples of marine areas that have been listed as wetlands of international significance under the Ramsar Convention include the Tubbataha Reefs Natural Park in the Philippines, and part of the Great Barrier of Reef of Australia.

Importantly, neither the convention nor subsequent guidelines and resolutions of the COP provide guidance on activities that may qualify as 'wise use' of the wetlands. Instead, the emphasis is on the identification of wetlands which present certain characteristics, including the provision of ecosystem services, and 'wise use' is defined as 'the maintenance of [the] ecological character [of the wetlands], achieved through the implementation of ecosystem approaches'.[49] The guidelines for reviewing laws and institutions to promote the wise use of wetlands focus on the development of national and subnational policies, regulations and institutional responsibilities to ensure the wise use of wetlands, starting with a review of the existing framework. They do not prescribe thresholds of acceptability nor activities that should be avoided.[50]

Protection of world cultural and natural heritage

The 1972 UNESCO World Heritage Convention aims to protect, conserve and preserve world cultural and natural heritage, which includes

> natural features consisting of physical and biological formations or groups of such formations which are of Outstanding Universal Values [OUV] from the aesthetic or scientific point of view; geological and physiographical formations and precisely delineated areas which constitute the habitat of threatened species of animals and plants of [OUV] from the point of view of science or conservation; natural sites or precisely delineated natural areas of [OUV] from the point of view of science, conservation or natural beauty.[51]

A particularity of this convention is that it created an intergovernmental committee for the protection of the world cultural and natural heritage which is the main body in charge of the implementation of the Convention. In this capacity, it developed criteria, for the inscription of sites of OUV on the World Heritage List, which include 10 criteria, two of which are concerned with marine biodiversity. Criteria 9 and 10 aim to protect sites with significant ongoing ecological and biological processes in the evolution and development of … coastal and marine ecosystems and communities of plants and animals; or contain the most important and significant natural habitats for *in-situ* conservation of biological diversity, including those containing threatened species of [OUV] from the point of view of science or conservation.[52]

While neither the Convention itself, nor subsequent guidelines, contain specific indications of activities that may or may not take place within a World Heritage Site, the Convention requires a monitoring of the characteristics, which earned the site its status (and listing) as a world heritage of OUV so that should these key characteristics be lost, the eligibility of the site (as being of 'OUV') will be reconsidered, and if necessary, deleted from the list.[53] The most recent guidelines set out, in detail, the methodology by which to justify that the relevant criteria are met. Nevertheless, it still leaves a margin of interpretation as demonstrated by the controversy that surrounded the degradation of the environmental condition of the Great Barrier of Reef and the determination as to whether it should be listed as 'being in danger'.[54]

Convention on the Conservation of Migratory Species of Wild Animals

The 1970 Convention on Migratory Species (CMS) focuses first on the protection of endangered migratory species (listed in Appendix 1) for which Range States (those within the jurisdiction of which endangered migratory species may cross or stay in temporarily) have an obligation to conserve and restore the habitat of these species and to limit and mitigate potential impacts to migratory corridors.[55] It also lists (in Appendix II) migratory species which have an unfavourable conservation status and require or would benefit from international cooperation through international agreements for their conservation and management.[56] Such agreements may for instance include provisions for the maintenance of a network of suitable migratory routes and sound management of the taking of these species.

To the extent that this convention concerns the identification and protection of certain categories (migratory species) of threatened and endangered migratory species, it contributes to the protection of marine biodiversity and overlaps with the objectives of the CBD and some of the EBSA criteria, while being more specialized. It seeks to identify threatened and endangered migratory species (EBSA criterion 3) as well as the nature of the threats they face, their migration routes, and ecological characteristics, in order to set a framework for protection.

Species-specific memoranda of understanding and agreements aim to set out the framework and invite Range States to adopt measures that will control activities that may have adverse impacts on the species concerned while they are within their jurisdiction (e.g. limit fishing or shipping within migration corridors). Should activities be occurring beyond national jurisdiction, States and the COP of this convention would seek cooperation with relevant international authorities (e.g. the relevant RFMO in the context of large migratory species that may be caught as by-catch, such as protected sharks under the 2012 MoU on the Conservation of Migratory Sharks).[57]

Global soft law instruments that aim to protect marine biodiversity

In addition to resolutions of the United Nations General Assembly,[58] the declarations of the United Nations Environmental Assembly of the United Nations Environment Programme from 23–27 May 2016 (UNEA-2)[59] also contain expressions of concern and the need for action, without specifying the actions to be taken nor identifying ecosystems or species in need of immediate protection. However, other soft law instruments such as the IUCN Red List[60] are an important part of the current international governance regime of marine biodiversity. The IUCN, originally the International Union for Conservation of Nature, which is a membership union composed of governments and civil society organizations, has been assessing the conservation status of species, sub-species varieties and other subpopulations on a global scale for the past 50 years in order to identify critically endangered, endangered, vulnerable and near threatened species according to a detailed methodology[61] and including marine species. As such, in practice, the IUCN is seeking to identify species and their habitats that fall under EBSA criterion 3 on the importance for threatened, endangered or declining species and/or habitats. The global recognition of the IUCN as an expert observer in most international fora, and its constituency, provide legitimacy to the data it publishes which are used by COPs and States to implement treaties that seek to protect marine biodiversity.

The conservation treaties discussed in this section overlap in scope with the CBD and its EBSAs criteria in that their ultimate purpose is to identify and protect components of marine biodiversity. Some areas such as Tubbataha reef system in the Sulu Sea even benefit from protection from all three pre-UNCLOS conservation instruments, illustrating their complementarity. In so far as threatened and endangered species are concerned, the conservation treaties and the IUCN Red List overlap, for those that are listed or qualify for protection under the relevant treaty. However, when considered together, these bodies of hard and soft laws have the same overall purpose of protecting and preserving the marine environment and favouring sustainable use of resources; they are generally consistent, complementary and can be implemented together. The difficulty resides in a multiplicity of fora, a coordination challenge. Recent COPs reports from these treaties acknowledge the issue and report on coordination efforts.[62]

International treaties, regulations and guidelines that regulate ocean activities

Operational mechanisms to preserve and protect marine biodiversity can be divided into two broad categories: (1) operational management of the use of marine living and non-living resources in order to limit the impact on biodiversity, and (2) permanent and temporary (e.g. seasonal) closure of specific areas to some or all uses of the sea, depending on the resources to be protected and the nature of the activity (e.g. shipping may be acceptable with specific discharge requirement, whereas extractive activities may be banned). These mechanisms are mostly provided in sectoral regulations developed to meet the needs and impact of different activities (exploitation of living resources, trade of wildlife, shipping, seabed activities, etc.). However, in reality, the measures adopted in each sector often do not fall neatly into either category but instead serve dual purposes.

The main international instruments which seek to regulate activities at sea in order to protect marine biodiversity are presented below. However, not all maritime activities, whether within or beyond national jurisdiction, are regulated internationally. Within national jurisdiction, offshore oil and gas activities and other types of mineral extractions are well-known examples of activities for which States have failed to fulfil their obligation to agree on global rules and standards.[63] As such, there are no global rules on the protection of the marine biodiversity in this context although some regional seas agreements have filled this gap.[64] The gap in the international normative framework as regards bioprospecting in ABNJ is another example which will be discussed further below in this chapter. It is nevertheless the responsibility of each State to take adequate domestic measures to ensure the protection of marine biodiversity against all activities.

Protection of marine biodiversity threatened by the exploitation of living resources (fisheries)

UNCLOS's provisions on fisheries concern both the conservation as well as the exploitation of living resources. The overriding objective for exploitation is that of 'maximum sustainable yield' through the determination of allowable catch based on best scientific evidence, in order to avoid over-exploitation. Commercial species are the primary focus of these provisions; effects on species associated with or dependent upon harvested species must also be taken into account to the extent that their depletion could impact on commercial species. These provisions therefore protect marine biodiversity only insofar as they impact on commercial species.

However, the 1995 United Nations Fish Stock Agreement (UNFSA)[65] further implements the obligation in UNCLOS to protect and preserve the marine environment, which as mentioned earlier, encompasses the protection of marine biodiversity. Specifically mentioned in the preamble to UNFSA is 'the need to avoid adverse impacts on the marine environment, preserve biodiversity [and] maintain the integrity of marine ecosystems'. With these objectives in mind, the UNFSA also explicitly advocates the adoption of the precautionary approach in the implementation of its provisions.[66] The 1995 Code of Conduct for Responsible Fisheries adopted by the FAO Conference (composed of representatives of Member States) further emphasizes the need to maintain biodiversity and conserve the population structure and aquatic ecosystems[67] and to conserve genetic biodiversity of endangered species,[68] setting as an objective that fisheries policies (be they at national, sub-national, regional or other level) should aim to conserve biodiversity of aquatic habitats and ecosystems and protect endangered species.[69] It recommends that States should (i) protect and rehabilitate 'all critical fisheries habitats in marine … ecosystems'; and (ii) use selective and environmentally safe fishing gear and practices (including non-destructive fishing gear and closed seasons and areas) to minimize waste, discards, bycatch and negative impacts on associated or dependent species and habitats, in particular endangered species.[70]

In the context of the depletion of deep-sea fish stocks in the high seas, the FAO also developed the concept of the Vulnerable Marine Environment (VME) and the International Guidelines for the Management of Deep-Sea Fisheries in the High Seas to facilitate and encourage the efforts of States and RFMOs towards sustainable use of marine living resources exploited by deep-sea fisheries; the prevention of significant adverse impacts on deep-sea VMEs; and the protection of marine biodiversity that these ecosystems contain.[71] VMEs constitute areas that may be vulnerable to impacts from fishing activities and present one of five criteria: (i) uniqueness or rarity; (ii) functional significance of habitat; (iii) fragility; (iv) life history traits of component species that make recovery difficult; (v) structural complexity. The overlap with the EBSAs criteria, which were developed in parallel, is clear. The FAO has since also developed a VME database in collaboration with regional bodies with mandates to manage deep-sea fisheries in ABNJ. It maps VMEs occurring in ABNJ and includes information on management measures taken for these areas.[72]

More recently, in 2011, the FAO adopted New Technical Guidelines for Responsible Fisheries No. 4: Marine Protected Areas (MPAs) and Fisheries, to provide information and guidance on the use of marine protected areas in the context of fisheries. These guidelines are focused on fisheries management but also include the protection of biodiversity. This focus on fisheries management is not contemplated within the MPA design advocated by the CBD, which is driven solely by the protection of marine biodiversity.[73]

In the context of the exploitation of marine living resources, another instrument which has established protected areas to conserve endangered species (and therefore biodiversity) is the International Whaling Convention, whose mission is the conservation and management of whale stocks.[74] Sanctuaries where commercial whaling is prohibited were designated by the Commission in 1979 in the Indian Ocean and in 1994 in the Southern Ocean around Antarctica.[75] However, there are no clear criteria such as those developed by the COPs to the CBD, the Ramsar Convention, the Convention on Migratory Species or the FAO that appear to have been developed by the Whaling Commission for the designation of whale sanctuaries, despite discussions on this topic.[76]

Protection of biodiversity from wildlife trade

Another challenge in the protection and sustainable use of marine biological diversity is the threat posed by illegal trade in wildlife. Many marine species are in decline due to the illegal trade in wildlife, which has an estimated value of US$19 billion each year.[77] The traded wildlife is put to a range of uses including as food, traditional medicines, accessories, pets, curios and trophies. The Convention in International Trade in Endangered Species (CITES),[78] seeks to protect both terrestrial and marine biodiversity by regulating international trade in specimens of CITES-listed species with certain controls. All import, export, re-export and introduction from the sea (meaning the transportation into a State of any CITES-listed specimens taken from ABNJ) of species covered by the Convention have to be authorized through a national permit and institutional system contemplated by CITES. The level of protection given to a species depends on which of the three appendices it is listed within CITES. Appendix I lists species threatened with extinction that are or may be affected by international trade and such trade is forbidden except with some exceptions. Appendix II lists species that although not necessarily threatened with extinction at present, may become so unless trade in such species is strictly controlled and trade in such species is permitted subject to control. Appendix III contains species that Parties have identified as being subject to regulation in their jurisdiction for purposes of preventing or restricting exploitation and needing the cooperation of other Parties in the control of trade.[79] CITES requires that each Party has to establish at least one Management Authority to administer the permit system and one Scientific Authority to advise on the effects of trade on the status of the species.[80] Collectively, these national institutions have the responsibility for verifying that the required conditions for the issue of permits for species listed in Appendices I, II and III have been fulfilled and for granting the applicable permit(s) only when such conditions have been complied with.

CITES lists many marine species and these include numerous species of baleen whales (e.g. bowhead and right whales), toothed whales (e.g. beaked whales, species of dolphins and porpoises), species of marine turtles and dugong in Appendix I and giant clams, certain species of sharks (e.g. hammerhead, basking and whale sharks), manta rays, species of soft and hard coral and seahorses in Appendix II. Certain CITES-listed marine species are also protected by the CMS. Marine species that are protected by both CITES and CMS include marine turtles and sharks and rays. It is worth highlighting that where a CITES-listed organism (e.g. a marine turtle) is taken from a geographic area that benefits from the protection of one or several conservation treaties (for example, Tubbataha reef ecosystem, a RAMSAR and UNESCO site in the Philippines) and subsequently traded, it may not necessarily mean that an offence has been committed. On the assumption that the State where the site is located is a party to CITES, whether an offence has been committed would depend on national legislation (which may or may not have prohibited the initial taking of the specimen regardless of the site's status as a RAMSAR site) as well as the quality of the CITES-implementing legislation of the State in question (which, if implemented properly, would make any illegal international trade in that specimen an offence).

Protection of marine biodiversity from the impact of shipping

The IMO is the United Nations specialized agency responsible for the promotion of safe, secure, environmentally sound, efficient and sustainable shipping through cooperation and the adoption of guidelines and standards. In this context, several shipping instruments negotiated under the auspices of the IMO include special protection provisions for areas that are considered to be particularly sensitive. The first such instrument was the 1973/78 International Convention for the Prevention of Pollution from Ships (MARPOL). This defines certain sea areas as 'special areas' where for technical reasons relating to their oceanographic and ecological conditions, and sea traffic in those areas, the adoption of a higher level of protection than in other areas of the sea is necessary. These 'Special Areas under MARPOL' concern increased tolerance thresholds for the discharge of oil, noxious liquid substances, sewage and garbage. 'Emission Control Areas' have also been designated, with more stringent restrictions on specific emissions from ships (sulphur emissions and nitrogen oxides) in the context of the prevention of air pollution from ships.[81] The conditions to be met for an area to qualify under this protection scheme may be grouped into three categories of oceanographic conditions, ecological criteria and vessel traffic characteristics. In particular, the scientific ecological criteria that need to be satisfied are set out in the IMO Guidelines for the Designation of Special Areas under MARPOL and are as follows: (i) depleted, threatened and endangered marine species; (ii) high productivity; (iii) spawning, breeding, nursery areas and migratory routes; (iv) rare or fragile ecosystems; and (v) critical habitats for marine resources.[82]

In order to protect areas that need special protection against other forms of ship-source pollution and environmental hazards associated with shipping such as physical damage to marine habitats or organisms, the IMO also adopted Guidelines for the Designation of Particularly Sensitive Sea Areas (PSSA Guidelines). These guidelines aim to grant a greater degree of protection to such areas based on their significance for recognized ecological, socio-economic or scientific attributes. First adopted in 1991, they were subsequently amended in 1999, 2001, 2005 and 2015.[83] While the protection of marine biodiversity is not the only ground on which an area may qualify as a PSSA, the 11 scientific criteria set out in the PSSA Guidelines to demonstrate the ecological significance of the area concerned include most of the criteria to identify EBSAs and designate MPAs under the CBD, and seem to share the objective of protecting marine biodiversity, despite differences in the way the criteria are defined.[84] To be successful in an application to designate an area as a PSSA, the applicant States must also demonstrate that this area is vulnerable to impacts from international shipping, describe the measure proposed (called an Associated Protective Measure or APM) and describe how this APM protects the area from the identified vulnerability. Possible APMs include ship routeing measures, reporting systems, and areas to be avoided. The 2004 International Convention for the Control and Management of Ships' Ballast Water and Sediments, when it enters into force on 8 September 2017, could arguably create a new type of APM; this would limit the risk of transfer of harmful aquatic organisms and pathogens through ships' ballast water and sediment in particularly vulnerable areas, through compliance with additional standards and requirements.[85]

States have also successfully proposed voluntary or mandatory routeing measures designed to protect the marine environment (including marine biodiversity), such as areas to be avoided, non-anchoring areas and ship reporting systems. Examples include areas to be avoided to protect migrating whales without going through an application for a PSSA. These applications are made on the basis of the General Provisions on Ships' Routeing adopted by the IMO on 20 November 1985 (as amended) (Roberts, 2005).[86]

Several other IMO instruments, such as the 1990 International Convention on Oil Pollution Preparedness, Response and Cooperation (OPRC), have developed measures designed to protect the marine environment For example, through the Guidance on sensitivity mapping for oil spill response adopted by the IMO for the implementation of OPRC, the IMO has provided a methodology to map sensitive marine habitats including marine biodiversity, in order to mitigate the impact of oil pollution incidents.[87] The 1972 London Convention on the Prevention of Marine Pollution by Dumping of Wastes and Other Matter and its 1996 Protocol have also factored the protection of marine biodiversity into the criteria for dump-site selection which include the avoidance of areas of important biodiversity, especially benthic species and habitats, vulnerable ecosystems, seasonal and critical habitats and sensitive areas, protected areas, EBSAs, spawning nursery and recruitment areas, as well as migration routes.[88] However, it is noted that the site-selection criteria do not seek to protect solely biodiversity; they focus on all valuable marine resources. The site-selection criteria also focus on not interfering with, or impacting on, other lawful uses of the sea such as commercial fisheries or areas of beauty or significant cultural or historical importance.

Protection of biodiversity from the impact of deep seabed mining activities in the Area (seabed beyond national jurisdiction)[89]

Rights to the Area and its resources are 'vested in mankind as a whole, on whose behalf the Authority shall act'. The International Seabed Authority (ISA) is a body created for this purpose by the UNCLOS and is charged with the responsibility of organizing and controlling exploration and exploitation of seabed minerals in the Area.[90] Its mandate includes the adoption of necessary measures to ensure effective protection of the marine environment from harmful effects which may arise from deep seabed mining.[91] In this capacity, the ISA has been developing a mining code which includes provisions for the protection of the marine environment, one of the components of which is marine biodiversity. As such, applicants for deep seabed exploration must submit a programme for oceanographic and environmental baseline studies and the assessment of impact on biodiversity would be based on these studies.[92] The ISA must also ensure that the proposed plan of work for exploration provides for effective protection and preservation of the marine environment including, but not restricted to, the impact on biodiversity and that adequate monitoring is set in place (in Impact and Preservation Reference Zones) as a means of controlling and managing the outcome.[93]

In 2012, the ISA also adopted the first regional-scale environmental management plan for the deep seabed of the Clarion-Clipperton Zone, which includes the designation of Areas of Particular Environmental Interest (APEIs) (Lodge, 2014).[94] One of the six guiding principles on which the plan is based is the conservation and sustainable use of biodiversity. The scientific indicators used to guide the design of the network of nine APEIs in the CCZ are: (i) spatial variations to ensure the protection of areas thought to be representative of the full range of habitats; (ii) of a size to cover large areas of self-sustaining populations and a broad range of habitat variability, biodiversity and ecosystem structure and function within the management area (200km by 200km, surrounded by a buffer of 100km); (iii) scientific design that relies on generally accepted and widely applied principles for the design of marine protected area networks and includes the protection of 30% to 50% of the total management area.[95] Although no scientific criteria appear to have yet been formalized for the designation of APEIs in other geographic parts of the Area and for other minerals, the management plan also mentions that it is mostly consistent with the criteria for VMEs (FAO), EBSAs, and MPA networks (CBD). However, areas of special significance valued for their uniqueness, biological diversity or productivity, as well as areas of special importance to the life histories of non-fish species referred to in the criteria of EBSAs have not been incorporated in the scientific design. However, it adds that they may be included in the future, as more information becomes available. The decision which adopted these APEIs also indicates that no application for exploration work should be granted in APEIs for a period of five years, or until further review by the Legal and Technical Commission.[96]

Marine biodiversity in areas beyond national jurisdiction and gaps in the overall management of marine biodiversity

The realization by developing States of the perceived value of marine genetic resources (MGRs) found in ABNJ, and the exploitation by developed States of these resources, led to the debate on whether bioprospecting is an activity regulated by UNCLOS. The development of this discussion within the UNGA over the years brought into focus another concern to some States, namely, the weakness of the protection granted to the marine environment in ABNJ, whether due to the fragmentation of the international governance regime in place or gaps within it. Ongoing negotiations within the UN in this area reflect these concerns and centre around the four following elements: (1) MGRs, including questions on the access and sharing of benefits); (2) capacity-building and transfer of marine technology; (3) sustainable use of marine biological diversity of ABNJ and the role of Environmental Impact Assessments (EIAs); and (4) area-based management tools including MPAs.[97] Steps are currently being taken by the PrepCom to help identify elements of the draft text of an IBLI. Parts of these discussions that focus on attempting to remedy the adverse effects of the fragmented ocean governance regime in ABNJ have the potential to also improve (at least in part) the governance and management of the marine environment and marine biodiversity within national jurisdiction. This section also discusses different types of existing deficiencies in the global regime for the governance of oceans and the extent to which they may be addressed in the ongoing negotiations.

Bio-prospecting: the main unregulated activity in ABNJ that was not contemplated during the negotiations of UNCLOS

As there is currently no international regime setting out clear rules on access and benefit-sharing for bioprospecting of MGRs in ABNJ, developing states are concerned not to be disadvantaged, given that the wealthier countries tend to be those with the resources to engage in such activity. The main commercial interest in deep seabed biodiversity is the exploitation of commercially useful MGRs through biotechnology. For example, MGRs found in organisms that are able to survive in extreme environments (e.g. in high pressure at great depth, or in extreme temperatures) are regarded as sources of great potential for scientific advancement and commercial application, and enzymes derived from extremophiles have been used in a variety of industrial applications. When negotiations for the regime to regulate deep seabed mining in ABNJ at the Third Law of the Sea Conference started, it was generally considered that the deep seabed was rich only in mineral resources, and bioprospecting was not an activity that was contemplated. This may possibly explain why the definition of 'resources of the Area' is limited to 'mineral resources' (Heidar, in press).

Discussions early on in the Working Group revealed differing views on the legal regime applicable to MGRs. Some states consider that the principle of the Common Heritage of Mankind (CHM) is the basis for a future specific legal regime, similar to the one governing the Area (i.e., controlled access where any financial and non-financial gain associated with the appropriation of resources are subject to a benefit-sharing mechanism, based on the view that the resources belong to, and are for, the benefit of all mankind) (Millicay, 2007; T. Heidar, in press). Yet others believe that MGRs fall under the legal regime governing the freedom of the high seas (i.e., freedom of access to all to engage in lawful activities permissible under UNCLOS and subject to the respect for the interests of other States and their activities on the high seas and certain conditions and other detailed regulations including the duty to protect the marine environment) (de la Fayette, 2009), and are opposed to negotiating a new benefit-sharing regime for the use of such MGRs. Still others were focused on the need to address all issues relating to marine biodiversity, especially conservation, in a comprehensive implementing agreement to UNCLOS.

Ongoing discussions have questioned the distinction between bioprospecting and Marine Scientific Research (MSR); in dispute is the determination of the regime applicable to bioprospecting in ABNJ and, more specifically, whether it should be treated as a freedom of the high seas. Although the concept of MSR is not defined in UNCLOS (despite an entire chapter setting out its regime, see Birnie, 1995), it seems that the main difference between bioprospecting and typical MSR in the high seas, being that which is driven by 'pure research' objectives, is one of intent. Some research cruises might begin with the intention to develop commercial applications, but most will start as 'pure research' MSR and may later develop into bioprospecting for MGR, although the exact point when this occurs may not be clear. All MSR (whether pure research or bioprospecting) is treated as a freedom of the high seas and there is currently no mechanism to ensure that the data gathered through MSR are made publicly available[98] (except where gathered within the context of deep seabed mining and under the purview of the ISA).

For States that lack the financial resources to conduct MSR, this raises first and foremost the issue of accessibility; second, the question arises as to whether appropriation or the right to commercially exploit the MGR based on this data is permissible under the existing framework. Be that as it may, underlying the call of developing States for MGRs to be the CHM is the expectation that not only the financial benefits arising from the exploitation of the MGRs, but the non-financial benefits also, should be subject to any benefit-sharing mechanism.[99] A key area for concern is the intellectual property rights accruing to commercial and/or industrial applications of MGRs, especially the extent to which naturally occurring components of MGRs can be appropriated through patenting. How this issue is to be addressed is currently also being discussed within the World Trade Organization (WTO), the World Intellectual Property Organization (WIPO) and the Food and Agricultural Organization (FAO).[100]

Systemic deficiencies in the fragmented ocean governance regime

Deficiencies in the existing ocean governance regime applicable to the management of the marine environment, especially marine biodiversity, can be divided into three categories: (i) deficiencies that concern impacts from activities that were not envisaged during the negotiations of UNCLOS and are therefore linked to a clear gap in the regime; (ii) impacts from activities that were envisaged but required implementation agreements between States or other measures to be taken by member States who have failed to do so; these are linked to clear gaps in implementation; and (iii) impacts from activities that fall in between two or more sectoral regimes such as combined impacts from several activities that cannot be addressed by the sectoral regimes despite the goodwill of regional bodies and competent international organizations due to limitations in their original mandate; these are often linked to combined gaps in coordination and implementation.

Deficiencies from the first category include bioprospecting for MGRs (discussed in the previous section), geo-engineering activities (Scott, 2013; Verlaan, 2011) and the commercial harvesting of sedentary species (such as bivalves and sponges) in and on the seabed beyond national jurisdiction.[101] Other new activities in ABNJ that can be contemplated are the development of aquaculture or other commercial activities in ABNJ for which no strict environmental measures have yet been devised. Solutions to such issues require amendments to existing international instruments or negotiation of new ones.

Examples of clear gaps in implementation that would characterize the second category of deficiencies include the management of fisheries in general, the low ratification status of the 1995 Fish Stock Agreement, and the lack of management of impacts of destructive fisheries practices on biodiversity in sea areas with weak RFMOs; with RFMOs that only deal with a small selection of commercially exploited fish stocks; or with no RFMOs. Furthermore, RFMOs are typically focused on optimizing fisheries yield, rather than directly concerned with impacts on biodiversity. Some even lack the authority to prevent fishing in areas within their geographic scope that may be recognized as being areas with particular characteristics or vulnerability that justify protection. Another example is the lack of implementation of UNCLOS Article 208, which calls on States to establish global and regional rules, standards and recommended practices and procedures to prevent, reduce and control pollution of the marine environment arising from, or in connection with, seabed activities within national jurisdiction. No such global instrument has been successfully negotiated, and there is to date no international organization clearly in charge of seabed mining within national jurisdiction (including offshore oil and gas activities).

The third category concerns deficiencies that result from inconsistencies, gaps and lack of coordination between regimes. A critical example is the lack of recognized international thresholds and standards for Environmental Impact Assessments (EIAs), the absence of a methodology for cumulative EIAs that would be globally accepted as well as for Strategic Environmental Assessments that aim to identify and evaluate environmental consequences of proposed policies, plans or programmes and take into consideration these consequences in the decision-making process. Furthermore, the EIA processes developed for different activities that are regulated or managed by international or regional organizations (as the case may be) are specific to those activities undertaken within a particular geographic or environmental context. For example, EIAs or equivalent mechanisms developed for fishing activities by the FAO and NEAFC are different from those developed by the IMO regulations and treaties for shipping and the ISA regulations for deep seabed mining. Provisions for the establishment of reference environmental baselines and to assess and monitor the potential and actual impacts of such activities also vary (to the extent that they are provided for). The sectoral approach also results in different criteria having been developed for the designation of areas that may require protective measures (including restricting marine/maritime activities) as discussed earlier in this chapter. Examples include EBSAs developed under the CBD; PSSAs, Special Areas under MARPOL and routeing measures developed by the IMO; VMEs developed by the FAO; and APEIs developed by the ISA. The case of organisms that move between the water column and the seabed (as part of their ecological or biological cycles), thereby triggering the application of two distinct regimes, provides another example of the need to ensure consistency and coordination between regimes.

These issues are being discussed in the ongoing UN discussions on biodiversity beyond national jurisdiction, especially in the context of discussions on MPAs and EIAs. Some of the proposals to resolve these issues involve the idea of having an overall body charged with coordinating the actions and processes of the various international, regional and sectoral bodies and governance frameworks in order to overcome these deficiencies and promote more effective management of ABNJ. This does not necessarily require the establishment of a new body; one alternative could be to officially extend the mandate of existing organizations such as the ISA. Such a body could also be responsible for coordinating MPA proposals under different instruments and international organizations and proposing areas for protection as well as ensuring that EIAs standards are implemented consistently for all activities taking place in ABNJ and include an assessment of their cumulative impact.

It must be pointed out that these deficiencies are observed in ABNJ as well as within national jurisdiction. As such, similarly to the way EBSAs criteria are now being used within national jurisdiction, mechanisms developed to improve coordination between ocean governance regimes and protection of the marine environment in ABNJ (including marine biodiversity), especially with respect to MPAs and EIAs, may also benefit the protection of the marine environment within national jurisdiction.

UNGA and ongoing negotiation process

For around a decade now, discussions on marine biodiversity of areas beyond national jurisdiction, frequently referred to as 'BBNJ', have been taking place at the United Nations (UN). The UN General Assembly, pursuant to resolution 59/24,[102] established the Ad Hoc Open-ended Informal Working Group to study issues relating to the conservation and sustainable use of BBNJ (Working Group) and called upon states and international organizations to take action urgently to address, in accordance with international law, destructive practices that have adverse impacts on marine biodiversity and ecosystems. The Working Group held nine meetings between 2006 to 2015, culminating in a consensus during its ninth meeting on recommendations for a decision to be taken at the 69th session of the UN General Assembly to develop a legally binding instrument (ILBI) on BBNJ under UNCLOS, and to start a negotiating process to that end.

Pursuant to resolution 69/292,[103] the General Assembly decided to develop such an ILBI and to that end, the Assembly established a Preparatory Committee (PrepCom) whose mandate is (i) to make substantive recommendations to the General Assembly on the elements of a draft text of an ILBI under UNCLOS, taking into account the various reports of the Co-Chairs on the Working Group's work; and (ii) for the Assembly to decide at its 72nd session whether to convene an intergovernmental conference to elaborate the text of the agreement. The resolution made clear that negotiations are to address topics identified in the 2011 'package'; exhorted the PrepCom to 'exhaust every effort' to reach agreement on substantive matters by consensus,[104] and made clear that the negotiating process must not undermine existing legal frameworks and institutional arrangements.

The PrepCom met twice in 2016[105] under the chairmanship of Eden Charles (Trinidad and Tobago) and, in accordance with resolution 69/292, was to meet for two sessions in 2017 and by the end of 2017, report to the General Assembly with substantive recommendations on the 'elements of a draft text' of an ILBI.[106]

Conclusion

This chapter demonstrates the complexity of the international legal framework in place for the protection of marine biodiversity and shows that it can only be effectively protected (i) as a component of the protection of the marine environment of which it is an integral part; and (ii) through the regulation of activities that may adversely impact it. It is clear that UNCLOS supplemented by the CBD is the legal cornerstone for its protection. Other treaties that seek to promote marine biodiversity conservation and protect the marine environment from ocean activities also work together to implement UNCLOS. As such, while appearing complex and fragmented, the legal framework can be seen as a coherent framework consisting of two complementary parts, namely (i) international rules that relate to the identification of areas that would meet scientific criteria and therefore are eligible for enhanced protection and management measures; and (ii) international rules that regulate activities that adversely impact the marine environment, including marine biodiversity.

Deficiencies in the overall ocean governance framework have been the subject of discussions for decades and highlighted in the UNGA Resolutions on Oceans and Law of Sea without significant progress. However, the ongoing negotiations on BBNJ, while centred on MGRs, have reopened the discussions relating to coordination and implementation of the existing governance framework (in particular with respect to EIAs and MPAs). This has presented a new opportunity for the international community to work towards strengthening the overall ocean governance framework to ultimately benefit the protection of the marine environment and its biodiversity, not just in ABNJ but also in parts of the sea within national jurisdiction. The fact remains that it is incumbent on each and every state to adopt the necessary implementing laws and to take all measures that are necessary to ensure that activities under their jurisdiction or control comply with these laws.

Notes

1 See for instance the Ocean Biogeographic Information System platform (OBIS), www.iobis.org/, or the UNEP-WCMC Ocean Data Viewer, http://data.unep-wcmc.org/.
2 Conservation goal of 10% of coastal and marine areas, especially areas of particular importance for biodiversity and ecosystem services through effective area-based conservation measures: see 'Decisions Adopted by the Conference of the Parties to the Convention on Biological Diversity at its Eighth Meeting' (20–31 March 2016), www.cbd.int/doc/decisions/cop-08/full/cop-08-dec-en.pdf, Annex IV of Decision VIII/15, and further reiterated in the revised Aichi Biodiversity Target for 2011–20 in 'Decision Adopted by the Conference of the Parties to the Convention on Biological Diversity at its Tenth Meeting: Decision X/2 – The Strategic Plan for Biodiversity 2011–2020 and the Aichi Biodiversity Targets' (29 October 2010), www.cbd.int/doc/decisions/cop-10/cop-10-dec-02-en.pdf, Strategic Goal C, Target 11. See also International Union for Conservation of Nature (IUCN) and UNEP World Conservation Monitoring Centre (UNEP WCMC) (October 2013) 'The Official MPA Map', www.protectplanetocean.org/official_mpa_map, or MPAtlas.org (28 September 2016) 'Map Gallery', www.mpatlas.org/data/. Note the IUCN World Park Congress recommendation of a target of 30% no-take MPA since the 2003 Durban Congress in Laffoley, D. *et al.* (2008) 'Progress with Marine Protected Areas since Durban, and future directions', www.protectplanetocean.org/resources/docs/Progress_with_MPAs_paper_Parks_17_1.pdf.
3 United Nations Convention on the Law of the Sea, 10 December 1982, 1833 U.N.T.S. 3, 21 I.L.M. 1261 (entered into force 16 November 1994) [UNCLOS].
4 Convention on Biological Diversity, 5 June 1992, 1760 U.N.T.S. 79, 31 I.L.M. 818 (entered into force 29 December 1993) [CBD].
5 Convention on Wetlands of International Importance especially as Waterfowl Habitat, 2 February 1971, 996 U.N.T.S. 245, 11 I.L.M. 963 (entered into force 21 December 1975) [Ramsar Convention], also available at Ramsar.org, www.ramsar.org/sites/default/files/documents/library/scan_certified_e.pdf.
6 Convention Concerning the Protection of the World Cultural and Natural Heritage, 23 November 1972, 1037 U.N.T.S. 151, 11 I.L.M. 1358 (entered into force 15 December 1975) [UNESCO World Heritage Convention].
7 Convention on the Conservation of Migratory Species of Wild Animals, 23 June 1979, 1651 U.N.T.S. 333, 19 I.L.M. 15 (entered into force 1 November 1983) [Convention on Migratory Species], also available at CMS.org, www.cms.int/en/convention-text.
8 Code of Conduct for Responsible Fisheries, 31 October 1995, FAO Doc. 95/20/Rev/1, 1995 W.T.S. 3 [FAO Code of Conduct for Responsible Fisheries], also available at FAO.org, www.fao.org/3/a-v9878e.pdf.
9 International Convention for the Regulation of Whaling, 2 December 1946, 161 U.N.T.S. 72, 62 Stat. 1716 (entered into force 10 November 1948) [1946 Whaling Convention].
10 3 March 1973, 993 UNTS 243Convention on International Trade in Endangered Species of Wild Fauna and Flora, 3 March 1973, 993 U.N.T.S. 243, 27 U.S.T. 1087 (entered into force 1 July 1975) [CITES].
11 International Convention for the Prevention of Pollution from Ships, 2 November 1973, 1340 U.N.T.S. 184, 12 I.L.M. 1319 (entered into force 2 October 1983) [1973/1978 MARPOL Convention].
12 International Convention for the Control and Management of Ships' Ballast Water and Sediments, 13 February 2004, IMO Doc. BWM/CONF/36 [Ballast Water and Sediment Management Convention].
13 The mining code for activities in the Area is available at ISA.org, www.isa.org.jm/mining-code.
14 UNCLOS, above note 3, Arts 204 and 206.
15 See definition of 'marine environment' in the Regulations on Prospecting and Exploration for Polymetallic Nodules in the Area (adopted 13 July 2000) which was later updated and adopted 25 July 2013; the Regulations on Prospecting and Exploration for Polymetallic Sulphides in the Area (adopted

7 May 2010); and the Regulations on Prospecting and Exploration for Cobalt-Rich Crusts (adopted 27 July 2012). These regulations are available online at www.isa.org.jm/mining-code.

16 See 'A/RES/58/240 – Resolution adopted by the General Assembly on 23 December 2003 – Oceans and the Law of the Sea' (5 March 2004), https://documents-dds-ny.un.org/doc/UNDOC/GEN/N03/508/92/PDF/N0350892.pdf?OpenElement, and note in 'A/RES/59/24 – Resolution adopted by the General Assembly on 17 November 2004 – Oceans and the Law of the Sea' (4 February 2005), https://documents-dds-ny.un.org/doc/UNDOC/GEN/N04/477/64/PDF/N0447764.pdf?OpenElement, the inclusion of 'marine biodiversity' in the title of Chapter X 'Marine Environment, Marine Resources, Marine Biodiversity and the Protection of Marine Ecosystems'.

17 See 'Decision Adopted by the Conference of the Parties to the Convention on Biological Diversity at its Ninth Meeting' (9 October 2008), www.cbd.int/doc/decisions/cop-09/cop-09-dec-20-en.pdf, para. 14.

18 Remarks by T. Koh reproduced in UN, *The Law of the Sea: Official text of the UNCLOS* (United Nations, New York 1983), p. xxxiii.

19 The fourth paragraph of UNCLOS' preamble provides: 'Recognizing the desirability of establishing through this Convention, with due regard for the sovereignty of all States, a legal order for the seas and oceans which will facilitate international communication, and will promote the peaceful uses of the seas and oceans, the equitable and efficient utilization of their resources, the conservation of their living resources, and the study, protection and preservation of the marine environment'.

20 See www.un.org/depts/los/reference_files/chronological_lists_of_ratifications.htm.

21 Manifestations of this intertemporal approach include the use of terms and concepts that allow for evolution through their dynamic interpretation to ensure that they keep reflecting current international law (Boyle, 2006).

22 This is the position frequently taken by the USA to justify its compliance with UNCLOS (see for instance www.unclosdebate.org/argument/855/us-already-abides-unclos-matter-customary-international-law-and-domestic-policy). See also Boyle (2006) and Roach (2014).

23 PCA Case No. 2013–19 – South China Sea Arbitration (*Philippines v. China*), Arbitral Award of 12 July 2016 (SCS Arbitration Award), www.pcacases.com/pcadocs/PH-CN%20-%2020160712%20-%20Award.pdf, at para.942.

24 E.g. UNCLOS, above note 3, Arts 22, 41, 53, 211, 217 and 220.

25 E.g. UNCLOS, above note 3, Arts 61, 64, 118, 119.

26 E.g.UNCLOS, above note 3, Arts 60(3), 197, 208 and 210.

27 SCS Arbitration Award, above note 25 at 941 and 945.

28 CBD, above note 4, Arts 1 and 6.

29 UNCLOS, above note 3, Arts 204–206.

30 This duty to 'ensure' is for instance described in the context of transboundary pollution (UNCLOS Art. 194(2)). Nevertheless, it is noted that these cases do not necessarily rely on an interpretation of Art. 192 of the UNCLOS as they each concern different activities that adversely impact the aquatic (marine or riverine) environment and therefore involve the application of different legal provisions. However, it appears that the concept of due diligence relied on in these different cases is being consciously developed as a common principle that seeks to operationalize the meaning of the obligation to protect and preserve the aquatic environment by elaborating on the elements of this obligation.

31 ICJ Reports 2010 – Case Concerning Pulp Mills on the River Uruguay (*Argentina v. Uruguay*), Judgment of 20 April 2010, para. 197, www.icj-cij.org/docket/files/135/15877.pdf.

32 ITLOS Case No. 17 – Responsibilities and Obligations of States Sponsoring Persons and Entities with Respect to Activities in the Area, paras 110–16, www.itlos.org/fileadmin/itlos/documents/cases/case_no_17/17_adv_op_010211_en.pdf.

33 Pulp Mills Case, above note 31, para. 197.

34 ITLOS Case No. 21 – Request for an Advisory Opinion Submitted by the Sub-Regional Fisheries Commission (SFRC) (Request for Advisory Opinion Submitted to the Tribunal), Advisory Opinion of 2 April 2015, paras 119, 125 and 138, www.itlos.org/fileadmin/itlos/documents/cases/case_no.21/advisory_opinion/C21_AdvOp_02.04.pdf.

35 PCA Case No. 2013–19 – South China Sea Arbitration (*Philippines v. China*), Arbitral Award of 12 July 2016, paras 955–60, www.pcacases.com/pcadocs/PH-CN%20-%2020160712%20-%20Award.pdf.

36 CITES, above note 10, Appendix II.

37 South China Arbitration, above note 35, paras 823, 945 and 956.
38 South China Arbitration, above note 35, para. 956.
39 Except for benthic communities.
40 See 'Decisions Adopted by the Conference of the Parties to the Convention on Biological Diversity at its Fourth Meeting' (4–15 May 1998), www.cbd.int/doc/decisions/cop-04/full/cop-04-dec-en.pdf, Decision IV/5. This was building on the draft programme attached to Annex II of Decision II/10 the second COP in 1995, www.cbd.int/decision/cop/default.shtml?id=7083.
41 See Resolution of UNGA 59/24, above note 18, paras 70–73 and 'A/RES/59/25 – Resolution adopted by the General Assembly on 17 November 2004 – Sustainable fisheries, including through the 1995 Agreement for the Implementation of the Provisions of the United Nations Convention on the Law of the Sea of 10 December 1982 relating to the Conservation and Management of Straddling Fish Stocks and Highly Migratory Fish Stocks, and related instruments' (14 January 2005), https://documents-dds-ny.un.org/doc/UNDOC/GEN/N04/477/70/PDF/N0447770.pdf?OpenElement, para. 66. These approaches and tools were developed primarily for the management of the threats to biodiversity beyond national jurisdiction on the basis that the CBD has a key role to play in supporting the work of the GA with regard to marine protected areas beyond national jurisdiction by focusing on provision of scientific and, as appropriate, technical information and advice relating to marine biodiversity (CBD COP 8, above note 2, para. 42 of Decision VIII/24).
42 CBD COP 9, above note 16, section 18 and Annex 1 of Decision IX/20. See also on this topic Dunn *et al.*, *ibid.*
43 The other four are (1) representativity (2) connectivity (3) replicated ecological features and (4) adequate viable site (CBD COP 9, above note 16, Annex II of Decision IX/20).
44 CBD COP 8, above note 2, para. 42 of Decision VIII/24.
45 See 'Decision Adopted by the Conference of the Parties to the Convention on Biological Diversity at its Tenth Meeting' (29 October 2010), www.cbd.int/doc/decisions/cop-10/cop-10-dec-29-en.pdf, para. 39.
46 CBD COP 9, above note 16, especially paras 11, 17 and 18 of Decision IX/20, which adopts the scientific criteria to identify EBSAs emphasizing the importance of coordination with all the relevant international organizations such as the FAO and IMO and invite them to apply the EBSA criteria and provide their feedback to assist in the implementation of conservation and management measures.
47 Ramsar Convention, above note 5, Arts 2(2), 2(6), 3(1) and 4(1).
48 Ramsar Convention, above note 5, Arts 1(1) and 2(1).
49 Ramsar Convention Secretariat (2010) *Handbook 1 – Wise use of wetlands: Concepts and approaches for the wise use of wetlands*, 4th edn, vol. 1, www.ramsar.org/sites/default/files/documents/library/hbk4-01.pdf, p. 16.
50 Ramsar Convention Secretariat (2010) *Handbook 3 – Laws and institutions: Reviewing laws and institutions to promote the conservation and wise use of wetlands*, 4th edn, vol. 3, www.ramsar.org/sites/default/files/documents/pdf/lib/hbk4-03.pdf.
51 UNESCO World Heritage Convention, above note 6, Art. 2.
52 See UNESCO, 'The Operational Guidelines for the Implementation of the World Heritage Convention' available in their successive versions from 1977 to 2015, http://whc.unesco.org/en/guidelines/, especially the last version (8 July 2015).
53 UNESCO (30 June 1977) 'World Heritage Committee – First Session', http://whc.unesco.org/archive/opguide77a.pdf, section I(A)(5)(iv).
54 See for instance O. Milman (2015) 'Great Barrier Reef: Australia says Unesco decision shows it is a "world leader"', www.theguardian.com/environment/2015/jul/02/great-barrier-reef-australia-says-unesco-decision-shows-it-is-a-world-leader.
55 Convention on Migratory Species, above note 7, Art. 3(4).
56 Convention on Migratory Species, above note 7, Art. 4(1). Appendices I and II (effective 8 February 2015) are available online at CMS.int, 'Appendices I and II of the Convention on the Conservation of Migratory Species of Wild Animals (CMS)', www.cms.int/sites/default/files/document/Appendices_COP11_E_version5June2015.pdf.
57 See CMS.int (4–9 November 2014) 'Proceedings of the 11th Meeting of the Conference of the Parties', www.cms.int/sites/default/files/publication/cms_cop11_proceedings_e.pdf, and CMS.int, 'Memorandum of Understanding (MOU) on the Conservation of Migratory Sharks', www.cms.int/en/legalinstrument/sharks-mou.

58 See for instance UNGA Resolutions on Oceans and the Law of the Sea (see above, notes 18 and 34) or 'A/RES/66/288 – Resolution adopted by the General Assembly on 27 July 2012 – The future we want' (11 September 2012), www.un.org/en/ga/search/view_doc.asp?symbol= A/RES/66/288.

59 See UNEA, 'List of Resolutions adopted at UNEA-2', http://web.unep.org/unea/list-resolutions-adopted-unea-2, No. 2/10 – Oceans and sea.

60 Available online at The IUCN Red List of Threatened Species, www.iucnredlist.org.

61 IUCN Species Survival Commission (9 February 2009) 'IUCN Red List Categories and Criteria – Version 3.1, Second edition', http://s3.amazonaws.com/iucnredlist-newcms/staging/public/attachments/3192/redlist_cats_crit_en.pdf, and IUCN Species Survival Commission (January 2010) 'Guidelines for Application of IUCN Red List Criteria at Regional and National Levels', http://s3.amazonaws.com/iucnredlist-newcms/staging/public/attachments/3101/reg_guidelines_en.pdf.

62 Above, notes 47 and 58.

63 UNCLOS, above note 3, Art. 208.

64 See UNEP Regional Seas Marine Biodiversity Assessment and Outlook Series, http://icriforum.org/sites/default/files/UNEP%20Global%20Synthesis%20Marine%20Biodiversity%20Series.pdf, and Regional Seas Programmes and other UNEP Activities Relevant to Marine Biodiversity in Areas beyond National Jurisdiction, submitted to the UN in the context of the Development of an international legally binding instrument on the conservation and sustainable use of marine biological diversity of areas beyond national jurisdiction under the United Nations Convention on the Law of the Sea, www.un.org/depts/los/biodiversity/prepcom_files/UNEP_and_BBNJ_PrepCom2.pdf.

65 United Nations Agreement for the Implementation of the Provisions of the United Nations Convention on the Law of the Sea of 10 December 1982 relating to the Conservation and Management of Straddling Fish Stocks and Highly Migratory Fish Stocks, 4 August 1995, 2167 U.N.T.S. 88, 34 I.L.M. 1542 (entered into force 11 December 2001) [UN Fish Stock Agreement].

66 UN Fish Stock Agreement, *ibid.*, Art. 5(c) and (g).

67 FAO Code of Conduct for Responsible Fisheries, above note 8, Art. 6(6).

68 FAO Code of Conduct for Responsible Fisheries, above note 8, Art. 9(3)(5).

69 FAO Code of Conduct for Responsible Fisheries, above note 8, Art. 7(2)(2)(d).

70 FAO Code of Conduct for Responsible Fisheries, above note 8, Arts 6(8), 8(4)(8) and 12(10).

71 FAO.org, 'International Guidelines for the Management of Deep-Sea Fisheries in the High Seas', www.fao.org/docrep/011/i0816t/i0816t00.HTM, para. 6.

72 FAO.org, 'More about the VME Database', www.fao.org/in-action/vulnerable-marine-ecosystems/about-vme-database/en/.

73 FAO.org (2011) 'Fisheries Management – 4. Marine protected areas and fisheries', www.fao.org/docrep/015/i2090e/i2090e.pdf, Abstract and text of the New Technical Guidelines for Responsible Fisheries. See also note EBSAs and MPA Design CBD COP9, above section 1.3 and note 16.

74 1946 Whaling Convention, above note 9, Art. 5.

75 See IWC/66/08, 'The South Atlantic: A Sanctuary for Whales', https://iwc.int/sanctuaries.

76 See the joint proposal for a Whale Sanctuary in the South Atlantic by Argentina, Brazil, South Africa and Uruguay to the 65th Annual Meeting of the International Whaling Commission (September 2014). Available online.

77 R. Morelle (11 February 2014) 'Animal trafficking: Wildlife trade crisis talks held', www.bbc.com/news/science-environment-26122438.

78 Above note 10.

79 CITES, above note 10, Art. II.

80 CITES, above note 10, Art. X.

81 1973/1978 MARPOL Convention, above note 11, Annex VI. Existing special areas and Emission Control Areas can be consulted online at IMO.org, 'Special Areas under MARPOL', www.imo.org/en/OurWork/Environment/SpecialAreasUnderMARPOL/Pages/Default.aspx.

82 The last guidelines have been adopted in 2013 and supersede the previous version. See IMO Assembly Resolution 1087(28).

83 See IMO (6 February 2006) 'Resolution A.982(24) – Adopted on 1 December 2005 (Agenda item 11) – Revised Guidelines for the Identification and Designation of Particularly Sensitive Sea Areas', www.imo.org/blast/blastDataHelper.asp?data_id=25322&filename=A982(24).pdf, and MEPC 267(68).

84 PSSA criteria to demonstrate ecological significance: (i) uniqueness or rarity, (ii) critical habitat (incl. those of endangered species), (iii) dependency, (iv) representativity, (v) diversity, (vi) productivity,

(vii) spawning or breeding ground, (viii) naturalness, (ix) integrity, (x) fragility, (xi) bio-geographic importance. To compare with EBSAs criteria, above section 1.3 and note 16.

85 Ballast Water and Sediment Management Convention, above note 12, Regulation C-1.

86 IMO.org (6 January 2003) 'MSC/Circ.1060 – Guidance Note on the Preparation of Proposals on Ships' Routeing Systems and Ship Reporting Systems for Submission to the Sub-Committee on Safety of Navigation', www.imo.org/en/OurWork/Safety/Navigation/Documents/1060.pdf.

87 Adopted at MEPC 63 held in March 2012, 2011 IMO/IPIECA Guidance on Sensitivity Mapping for Oil Spill Response (Doc MEPC 62/8).

88 Convention on the Prevention of Marine Pollution by Dumping Wastes and Other Matter, 29 December 1972, 1046 U.N.T.S. 120, 11 I.L.M 1294 (entered into force 30 August 1975) [1972 London Convention], Art. 4 and Annex III (Art. 4(2)); the 1996 Protocol to the London Convention, 7 November 1996, 36 I.L.M. 1 (entered into force 24 March 2006) [1996 London Protocol], Annex II; and IMO.org (revised 9 December 2005) '1997 Guidelines for the Assessment of Wastes or Other Matter', www.imo.org/blast/mainframemenu.asp?topic_id=1503&doc_id=7550, Arts 6(4) and 7(4).

89 The 'Area' is a maritime zone defined in UNCLOS Art. 1(1) as 'the seabed and ocean floor and subsoil thereof, beyond national jurisdiction'. Access to resources is governed by Part XI of UNCLOS titled 'the Area'.

90 UNCLOS, above note 3, Arts 136, 137 and 153, as amended by the 1994 Agreement Relating to the Implementation of Part XI of the United Nations Convention on the Law of the Sea of 10 December 1982, 28 July 1994, 1836 U.N.T.S. 3.

91 UNCLOS, above note 3, Arts 145 and 162, as amended by the 1994 Agreement Relating to the Implementation of Part XI of the United Nations Convention on the Law of the Sea of 10 December 1982, 28 July 1994, 1836 U.N.T.S. 3.

92 International Seabed Authority (22 July 2013) 'ISBA/19/C/17 – Decision of the Council of the International Seabed Authority relating to amendments to the Regulations on Prospecting and Exploration for Polymetallic Nodules in the Area and related matters', www.isa.org.jm/sites/default/files/files/documents/isba-19c-17_0.pdf, reg. 18(b).

93 ISBA/19/C/17, *ibid.*, regs 21(4)(d), 31 and 32.

94 International Seabed Authority (26 July 2012) 'ISBA/18/C/22 – Decision of the Council relating to an environmental management plan for the Clarion-Clipperton Zone', www.isa.org.jm/files/documents/EN/18Sess/Council/ISBA-18C-22.pdf, and International Seabed Authority, 'ISBA/17/LTC/7 – Environmental Management Plan for the Clarion-Clippertone Zone', www.isa.org.jm/sites/default/files/files/documents/isba-17ltc-7_0.pdf.

95 ISBA/17/LTC/7, above note 82.

96 The criteria for EBSAs under the CBD and VME under the FAO had not been fully developed when work on this Environmental Management Plan started in 2007.

97 For further background information on this area see United Nations University (2005) 'UNU-IAS Report – Bioprospecting of Genetic Resources in the Deep Seabed: Scientific, Legal and Policy Aspects', http://collections.unu.edu/eserv/UNU:3101/DeepSeabed1.pdf.

98 Another deficiency in the regime of MSR in ABNJ is the lack of a framework to ensure that they are so conducted as to preserve and protect the marine environment (Glowka, 2003).

99 *Ibid.*, p. 273.

100 For WIPO, see work of the Intergovernmental Committee on Intellectual Property and Genetic Resources, Traditional Knowledge and Folklore (www.wipo.int/meetings/en/topic.jsp?group_id=110) and summary by the IUCN, www.iucn.org/news/traditional-knowledge-genetic-resources-and-wipo. See also Kirchner-Freis and Kirchner (2014).

101 UNCLOS Part XI on the Area (the seabed beyond national jurisdiction) only applies to mineral resources. There are no provisions in UNCLOS with respect to the harvesting of living resources in the Area although the ISA is the Competent Organization with respect to the protection of the marine environment including marine living resources (UNCLOS Art. 145, above note 3).

102 A/RES/59/24, above note 18.

103 See 'A/RES/69/292 – Resolution adopted by the General Assembly on 19 June 2015 – Development of an international legally binding instrument under the United Nations Convention on the Law of the Sea on the conservation and sustainable use of marine biological diversity of areas beyond national jurisdiction' (6 July 2015), www.un.org/en/ga/search/view_doc.asp?symbol=A/RES/69/292.

104 Where there is a lack of consensus in the development of certain elements of the draft text of the ILBI, such elements may also be noted in the recommendations of the PrepCom to the General Assembly.

105 The two sessions of the PrepCom in 2016 were held from 28 March to 8 April and from 29 August to 12 September.

106 See A/RES/69/292, above note 95, para. 1(a).

References

M.V. Angel (1993) 'Biodiversity of the Pelagic Ocean', *Conservation Biology*, Vol. 7, no. 4, pp.760–72.

N.J. Bax *et al.* (2016) 'Results and Implications of the First Global Effort to Identify Ecologically or Biologically Significant Marine Areas', *Conservation Biology*, Vol. 30, no. 3, pp. 571–81.

P. Birnie (1995) 'Law of the Sea and Ocean Resources: Implications for Marine Scientific Research', *International Journal of Marine and Coastal Law*, Vol. 10, no. 2.

A. Boyle (2006) Further Developments of the 1982 Convention on the Law of the Sea, in D. Freestone *et al.* (eds), *The Law of the Sea: Progress and Prospects* (Oxford University Press), pp. 40–62.

D. Dunn *et al.* (2014) 'The Convention on Biological Diversity's Ecologically or Biologically Significant Areas: Origins, Developments and Current Status', *Marine Policy*, Vol. 49, no. 1, pp. 137–45.

L.A. de La Fayette (2006) 'The Role of the United Nations in International Oceans Governance' in D. Freestone *et al.* (eds), *The Law of the Sea: Progress and Prospects* (Oxford University Press), pp. 63–75.

L.A. de La Fayette (2009) 'A New Regime for the Conservation and Sustainable Use of Marine Biodiversity and Genetic Resources Beyond Limits of National Jurisdiction', *International Journal of Marine and Coastal Law*, Vol. 24, no. 2, p. 236.

D. Freestone (1995) 'The Conservation of Marine Ecosystems under International Law', in M. Bowman and C. Redgwell (eds) *The Convention on Biological Diversity* (Kluwer, The Netherlands), pp. 91–107.

L. Glowka (2003) 'Putting marine scientific research on a sustainable footing at hydrothermal vents', *Marine Policy*, Vol. 27, pp. 303–12.

T. Heidar (In press), 'Conservation and Sustainable Use of Marine Biological Diversity Beyond Areas of National Jurisdiction: A Third Implementing Agreement under the Law of the Sea Convention', in *Liber Amicorum for Gudmundur Eiriksson* [Copy from the author].

I. Kirchner-Fries and A. Kirchner (2014) *The IMLI Manual on International Maritime Law*: Volume I: The Law of the Sea (Oxford University Press).

M. Lodge *et al.* (2014) 'Seabed mining: International Seabed Authority Environmental Management plan for the Clarion-Clipperton Zone – A partnership approach', *Marine Policy*, Vol. 49, no. 1, pp. 66–72.

F. Millicay (2007) 'A Legal Regime for the Biodiversity of the Area', in T. Heidar *et al.* (eds) *Law, Science and Ocean Management* (Brill, The Netherlands), p. 795.

S.R. Palumbi (1992) 'Marine Speciation on a Small Planet', *Trends in Ecology and Evolution*, Vol. 7, no. 4, pp.114–18.

K. Scott (2013) 'International Law in the Anthropocene: Responding to the Geoengineering Challenge', *Michigan Journal of International Law*, Vol. 34, no. 2, http://repository.law.umich.edu/cgi/viewcontent.cgi?article=1004&context=mjil.

A. Roach (2014) 'Today's Customary International Law of the Sea', *Ocean Development and International Law*, Vol. 45, pp. 239–59.

J. Roberts (2005) 'Protecting Sensitive Marine Environments: The Role and Application of Ships' Routeing Measures', *The International Journal of Marine and Coastal Law*, Vol. 20, no. 1, pp. 135–59.

C. Redgwell (2006) 'From Permission to Prohibitions: The 1982 Convention on the Law of the Sea and Protection of the Marine Environment', in D. Freestone *et al.* (eds), *The Law of the Sea: Progress and Prospects* (Oxford University Press).

P. Verlaan (2009) 'Geo-engineering, the Law of the Sea and Climate Change', *Carbon and Climate Law Review*, www.jura.uni-freiburg.de/institute/ioeffr2/downloads/verlaan.

P. Verlaan (2011) 'Current Legal Developments London Convention and London Protocol', *International Journal of Coastal and Marine Law*, Vol. 26, no. 1, pp. 185–94.

6

BIOSECURITY, INVASIVE SPECIES AND THE LAW

Opi Outhwaite

Introduction: biosecurity and invasive species as an environmental problem

The impact of pests, diseases and pathogens on agriculture and human health has long been recognized. In an increasingly globalized world the nature of the threats that these pose has changed – international travel and trade have provided novel pathways for their movement, allowing diseases and pests to spread to parts of the world and populations that they might not have previously reached. The natural boundaries presented by mountain ranges and oceans, for instance, no longer restrict the range of a disease outbreak or the movement of a mosquito carrying a particular disease (Waugh, 2009). Consequently, the need arose to develop systems to mitigate and manage the risks posed, and the concept of *biosecurity* became increasingly important in domestic and international frameworks. The negative impacts of pests and diseases and of biosecurity failings on biodiversity and the environment more broadly are now also recognized. Animal and plant diseases can affect wild animal or native plant populations for instance, with devastating consequences (Meyerson and Reaser, 2002; Leibler *et al.*, 2009; Lowe *et al.*, 2004). It is also estimated that the process of climate change will lead to an increase in unwanted introductions including in the range of pathogens that may be introduced and the pathways by which those introductions could occur (Masters and Norgrove, 2010). Negative impacts on biodiversity render other species and ecosystems more susceptible to pest and disease threats (Outhwaite, 2013(a)).

The scope of biosecurity is discussed in greater detail below, but as an overarching concept it can also be viewed as incorporating the management of invasive alien species (IAS). IAS have in fact tended to receive greater explicit attention than biosecurity has in the field of environmental law, with IAS conceived of as an environmental problem, compared with the agricultural emphasis and more encompassing scope of biosecurity. The legal frameworks to manage IAS have in many cases been developed more recently and are less well established than the historical frameworks to manage plant and animal diseases from an agricultural or health perspective but legal and political attention has nevertheless been given to the problem of IAS. The presence of such species can present problems where they act as a direct pest to naturalized or indigenous plants or animals and local ecosystems (Waugh, 2009) and through the consequences of their presence in the form of predation, herbivory, competition for resources, disease transmission, hybridization and habitat or ecosystem change. The impact of of IAS is such that they are now recognized as one of the major drivers of global biodiversity loss (Jay *et al.*, 2003; Meyerson and Reaser, 2002).

Questions of definition

Before proceeding to examine the nature of the legal regime for biosecurity and IAS, some further examination of the scope and meaning of these concepts is warranted. Although modern biosecurity concerns require the adoption of extensive legal measures, attempts to address aspects of biosecurity through legislative means are not new. Such measures have been used to control plant pests and animal diseases for over a century and aimed to prevent their entry or to control the spread of pests or diseases even where scientific understanding of the basis for this was lacking (see further Ebbels, 2003; Waage and Mumford, 2008). Early efforts in domestic law were, however, piecemeal, reactive and sectoral in nature, with a focus on 'agricultural biosecurity' and in particular on responses to economic threats associated with the agriculture sector as well as on major public health issues (especially food safety) (Outhwaite, 2013(b)). This emphasis on 'agricultural biosecurity' was largely reflected in developments in biosecurity at the international level (see below). Although its scope is not always precisely defined, however, a strategic and integrated approach to the management of pest and disease risks is crucial to contemporary conceptions of biosecurity.

The UN Food and Agriculture Organization (FAO) has played an important role in promoting the concept of biosecurity at the international level, emphasizing both the limitations that can arise, particularly for developing countries, in continuing to regulate on a sectoral basis and the benefits to be derived from a more integrated and strategic 'biosecurity' approach. The FAO's earlier work on biosecurity had however a relatively narrow focus, framing biosecurity principally in terms of the historically delineated agricultural sectors – food safety, plant health, and animal health and life, with limited reference to the environmental significance of pests and diseases. In 2001 it provided the following definition:

> Biosecurity is composed of three sectors, namely food safety, plant health and life, and animal life and health. These sectors include food production in relation to food safety, the introduction of plant pests, animal pests and diseases, and zoonoses, the introduction and release of Genetically Modified Organisms (GMOs) and their products, and the introduction and safe management of invasive alien species and genotypes.
>
> (UN FAO, 2001)

This focus was subsequently expanded. Maintaining the defining characteristics of a strategic and integrated approach the FAO's 2007 'Biosecurity Toolkit' extended their earlier definition, describing biosecurity as:

> a strategic and integrated approach to analysing and managing relevant risks to human, animal and plant life and health and associated risks to the environment ... Thus biosecurity is a holistic concept of direct relevance to the sustainability of agriculture, and wide-ranging aspects of public health and protection of the environment, including biological diversity.

Jay *et al.* (2003) similarly recognize that 'from an earlier biosecurity focus on economically significant pests, weeds and diseases, there has developed a wider concern with threats from bioinvasives for native plants, animals and ecosystems'.[1] Despite movement towards a more comprehensive definition, inconsistency in the use of this and other key terms has presented challenges to the consistent conceptualization of biosecurity and IAS and their incorporation into legal frameworks.

Few international legal instruments include the term 'biosecurity', it having developed after most of these were drafted. The Terrestrial Animal Health Code (TAHC) of the Office International des Epizooties (OIE, also known as the World Animal Health Organization) introduced a definition in its 2016 edition, providing that biosecurity

> means a set of management and physical measures designed to reduce the *risk* of introduction, establishment and spread of animal *diseases*, *infections* or *infestations* to, from and within an animal population [original emphasis].
>
> (OIE, 2016)

This definition attempts to overcome some of the variation found at the level of its member countries, where biosecurity is sometimes referred to as an overarching management or regulatory approach and sometimes as a set of practical measures, of varying scope. In the UK for instance biosecurity is often used to describe 'on-farm' measures, such as disinfecting vehicles and regulating local movement.[2] In the American context it has more often been linked with bioterrorism, including deliberate introductions of pests and diseases to attack food security, the agricultural economy or public health. In Australia and New Zealand, a more comprehensive approach, in line with the modern definitions described above, has been pursued (Donaldson, 2008; Hinchcliffe and Bingham, 2008; Jay *et al.*, 2003, Fletcher and Stack, 2008). These approaches reflect different ideas about the processes, aims and outcomes of biosecurity, and in the first two cases do not necessarily reflect the integrated nature of biosecurity which is key to its relevance. While the OIE has incorporated the term biosecurity directly into the TAHC this is not the case for the equivalent documents in international plant health.[3]

Concerning IAS specifically, the use of numerous terms, sometimes as synonyms, can similarly give rise to a lack of clarity. In this case, the terms 'invasive species', 'alien invasive species', 'non-native species', 'invasive alien species' and other associated terms appear variously in applicable instruments (CBD, 2016; OIE, 2011).[4] Problems with competing or inadequate definitions have presented challenges to the development and coordination of legal instruments in this area (see also Shine, 2010, p. 71). Key to this term however is the understanding that the species is both *alien*, meaning that it is occurring outside of its 'natural' distribution, and that it is *invasive*, meaning that the species' establishment or spread threatens biological diversity and the disruption of local ecosystems and ecosystem functions (CBD, 2002; Riley, 2009).

Other important terms are also sometimes used inconsistently. For instance, terms such as 'introduction', 'establishment', 'precautionary approach' or 'risk analysis' are not used consistently by the UN-FAO International Plant Protection Convention (IPPC) and the Convention on Biological Diversity (CBD), despite the clear overlap and potential for coordination in this area. Efforts to improve international coordination are being undertaken but a lack of consistency remains (see STDF, 2013). Attempts by the IPPC to harmonize terminology related to IAS with the CBD had limited success, with the IPPC reporting that differing terms were based on different concepts and therefore had different meanings and could not easily be harmonized (IPPC, 2010). Again, this means that the basis for action at the domestic level can be unclear or contradictory.

Managing biosecurity risks and IAS

Before examining the nature and extent of obligations and opportunities presented within the international legal regime it is also necessary to understand the types of measure that a country may wish to adopt, since the form and content of these will to an extent be dictated by the international framework. There is a range of actions which a government may wish to take to prevent the entry into the country of pests, diseases, invasive species or pathogens, or to limit their spread or attempt to eradicate them. Such measures may apply to species themselves, to derivatives of those species (for instance, animal and food products) or to the vectors and pathways which facilitate their movement (such as vehicles, packaging materials or farm equipment). In many instances these measures will apply to unintentional introductions or movement, but legislation may also establish a legal basis for intentional introductions (for instance, the introduction of exotic pests as part of a strategy for integrated pest management). From a regulatory perspective, the classic approach to managing these risks revolves around prevention, eradication and control, with each stage tending to present progressively greater costs and risks (because of reduced likelihood of success and increased negative impacts). At the domestic level a variety of responses may be adopted in this context, and these can be usefully categorized as pre-entry, point-of-entry and post-entry controls.

For imported produce – a common source of introductions – measures might include some of the following.[5]

Pre-entry

- Prohibiting the entry of a commodity (for example a type of fruit or vegetable) which presents a high risk for the introduction of a particular pest.
- Requiring that other imported commodities have been produced in an area that is free from a particular pest, or have been subject to appropriate treatment to prevent the presence of the pest. This might in turn lead to measures for verifying compliance with these requirements, for instance through in-country (pre-export) inspections by authorities of the exporting country and/or the importing country.

Point-of-entry

- Documentary and/or physical inspections of the commodities (or a sample thereof) at the point-of-entry, usually a border point.
- Requiring treatment, return to origin or destruction of the commodity in the event of non-compliance with requirements or detection of the pest.

Post-entry

- Quarantine.
- Surveillance and inspection to detect outbreaks of the relevant pest by a designated regulatory authority.
- Surveillance and/or reporting obligations for relevant stakeholders (for example, those most likely to come into contact with the pest).
- Treatment or destruction in the event that the pest is detected.
- The imposition of sanctions and penalties in the event of non-compliance with obligations or restrictions.

The extent to which controls of this type are on the one hand permitted or on the other hand required will of course have implications for the effective management of biosecurity risks and the control of IAS and consequently for their impacts on biodiversity and the environment.

International law and governance for biosecurity and invasive alien species

To begin with it should be noted that there is no single instrument for addressing biosecurity and no comprehensive approach has been adopted at the international level. Instead a number of instruments and agreements are applicable. While broader commitments and obligations are found in some hard law instruments, more specific guidance and rules are often to be found in the patchwork of relevant soft law instruments.

Sanitary and phytosanitary measures within the World Trade Organization

The regulation of international trade is a key issue for biosecurity since international trade is one of the principal drivers for the movement of species, diseases and pathogens. Measures which might be used to protect against the arrival or spread of such things (as described above) might, in practice, have the result of restricting trade (see also Jack, 2009).

The general principles of the WTO (such as non-discrimination and 'national treatment') apply to national biosecurity measures just as they do to other domestic measures which affect trade (WTO GATT, 1994; WTO, 2015, pp. 10–11). More specifically the Agreement on Technical Barriers to Trade and, more directly, the Agreement on the Application of Sanitary and Phytosanitary Measures (SPS Agreement) have direct implications for the adoption of domestic legal measures for biosecurity since they govern the adoption by members of rules relating to plant and animal health and food safety. As already noted, some countries have been addressing particular biosecurity concerns for decades. Consequently, in some of these countries legislation was established long before the development of the modern multilateral trading system. Legislation in developing countries often dates to colonial times or was established upon independence (see also Ikin, 2002). One effect of the WTO framework is that member countries have needed to (or still need to) revise relevant domestic legislation. Whereas a country previously legislated according to its own needs and priorities, that legislation should now reflect WTO obligations. As well as being justified on the basis of risk, being transparent, non-discriminatory and least-trade-restrictive, technical terms and requirements should also be incorporated into legal measures. This may mean changing the content of the legislation – revising definitions, for instance, and introducing institutional changes (such as the designation of official contact points) – as well as changing the subject and nature of the restrictions that are permissible.

One of many examples of such revisions can be seen in the plant health law of Mauritius. Until relatively recently, the key statute was the Plants Act 1976. This Act did not reflect modern plant health requirements (in line with international standards and principles) in a number of ways. Key terms (as established by the IPPC) such as 'quarantine pest', 'phytosanitary certificate' and 'pest risks analysis' were not defined in the Act. 'Officers' were empowered to detain, examine, remove, treat, destroy (etc.) any article which they had 'reasonable grounds' to suspect may be 'infected'. Such actions do not reflect principles of transparency and risk analysis. There were no provisions for the use or issuing of phytosanitary certificates and no schedules of pests subject to control, as established in international standards. While this legislation may or may not have served Mauritius well with respect to that country's own plant health concerns, it did not reflect international requirements in the light of the WTO (1994), though Mauritius became a member of the WTO as of 1 January 1995. The Mauritius Plant Protection Act 2006 set out a revised framework including new definitions which reflect IPPC standards, such as a revised definition of 'pest' and new definitions for 'phytosanitary certificate', 'quarantine' and other terms which had been missing. A National Plant Protection Office was established and other measures enacted, such as powers to designate 'pest free areas'. A list of quarantine pests and powers to revise the list was also adopted. Although there is no obligation to revise 'pre-1994' legislation *per se*, a failure to do so may mean that it does not give effect to technical requirements set out in international standards or to broader WTO principles. In turn this may damage trade opportunities or even lead to a WTO dispute.

With respect to the nature and extent of measures that may be imposed, disputes have arisen between member countries regarding the 'appropriate level of protection' and the application of key concepts including scientific justification and precaution, in the context of the requirement for risk analysis. These disputes demonstrate that measures such as import bans and detailed sanitary or phytosanitary requirements can and will be challenged at the WTO. One such high-profile case was the 'Australia – Apples' dispute. Australia had banned the importation of New Zealand apples in 1921, following the entry and establishment of fire blight in New Zealand in 1919. New Zealand had sought to gain access to the Australian apple market from 1986, eventually initiating a dispute through the WTO dispute settlement process. New Zealand claimed that the Australian phytosanitary measures were not justified, challenging the import risk management measures that were then being imposed by Australia and the methodology on which the risk assessment was based. One of the arguments raised by Australia was that the measures were justified on the basis of the *available* evidence. The Dispute Settlement Panel (August 2010) found that the import risk analysis was not supported by scientific evidence, and that less trade-restrictive measures were available. On appeal, the Appellate Body (November 2010) also found that the measures were not justified. Australia subsequently replaced the existing measures with detailed controls and quarantine requirements.

Against this background the frameworks of the international standard-setting bodies for plant health, animal health and food safety (see below) have assumed even greater significance because they are formally recognized in Annex A(3) of the SPS Agreement and the standards and guidelines issued by them are assumed to be compatible with it and to reflect the general principles of the WTO. Although countries are free to adopt their own standards they must, in that case, be able to justify them based on risk analysis (Articles 2 and 5, SPS). Consequently, the standards and guidance issued by these bodies have significant regulatory power.

Intergovernmental standard-setting

International governance of biosecurity reflects in most cases the traditional sectoral divisions seen historically and at the domestic level. Separate international standard-setting bodies are accordingly found for animal health, plant health and food safety. These bodies are responsible for the international standards which form the basis of national measures for member countries and are therefore a crucial influence on the form and content of biosecurity measures. Although the standards and codes published by these intergovernmental bodies are not legally binding, their status within the framework of the SPS Agreement gives them considerable regulatory power. These standards form the basis for domestic legislative and regulatory measures for their members and member states of the WTO.

The establishment of international bodies in these areas principally took place in the early twentieth century. The Office International des Epizooties (OIE, which, as we have seen, is sometimes referred to as the World Animal Health Organization) was established in 1924 with the aim of introducing international measures in order to protect animal health. The OIE produces the Terrestrial Animal Health Code (and an equivalent for aquaculture), which sets out international standards for animal health (OIE, 2016). Although the OIE has a general mandate to address animal health, it has tended to focus on diseases of livestock and those which would have direct economic implications, and not on wild animals or the relationship of animal diseases to biological diversity. This is evident from the categorization of most notifiable diseases (as diseases of cattle, sheep, horses, etc.). This approach of listing diseases also highlights the historical development of the framework, in contrast with modern perspectives which may recognize more explicitly that pathogens can be relevant to a range of species (including those of both 'kept' and 'wild' animals) and that they may cause a number of additional and indirect effects (Perrings *et al.*, 2010a). More recently, the OIE has extended its efforts to address diseases affecting wildlife and to consider animal diseases in a more holistic sense including the impact of pathogens on human health and ecosystems health. Specifically, the OIE has promoted the 'One Health' approach, along with the FAO and the World Health Organization (UN FAO-OIE-WHO). The One Health partnership envisages the possible use of tripartite protocols and increased coordination in standard-setting (Mackenzie *et al.*, 2013). Concerning IAS, the OIE has produced separate Guidelines for Assessing the Risk of Non-native Animals Becoming Invasive (2011).[6] These are to act separately from the OIE standard for risk assessment for the introduction of pathogens. Despite these developments in policy and technical work the TAHC maintains its focus on animals of agricultural importance.

For plant health, a group of European countries took action in 1881 to control the spread of grape phylloxera, following the extensive damage caused to European vineyards after its accidental introduction from North America (Ebbels, 2003). The International Plant Protection Convention (IPPC) eventually followed in 1951. The current international plant health framework is based on the IPPC 1997 and on the numerous specific standards developed through the IPPC governing body, the Commission on Phytosanitary Measures. These standards (International Standards for Phytosanitary Measures, or ISPMs) address numerous specific and technical matters associated with the international movement of plants, plant material and plant-based packaging.[7] The framework of the IPPC and ISPMS aims to prevent the movement, establishment and negative impacts of plant pests through the adoption of harmonized rules and procedures at the national level. The IPPC framework does not explicitly address biosecurity or IAS and the use of differing terms and concepts (as above) has sometimes been problematic. Nevertheless since some IAS will qualify as plant pests (Shine, 2007) the framework of the IPPC is an important aspect of the international framework both for biosecurity and for IAS specifically. Although the specific remit of the IPPC is not environmental protection *per se*, it is also noteworthy that supplementary guidance has sought to address this aspect of plant protection: ISPM 11 for instance was revised to include direct and indirect environmental effects within the Pest Risk Analysis process. ISPM 15, which addresses wood packaging material, is also especially important in the control of IAS since it deals with a significant pathway for the movement of plants and plant pests and diseases.

The Codex Alimentarius Commission (Codex) was established in 1963 by the UN Food and Agriculture Organization (FAO) and the World Health Organization (WHO). This body seeks to establish food safety standards in order to protect human health and was developed out of concern to harmonize food laws and standards, especially in the light of international trade. It also sought to address consumer food safety concerns, which were of an increasingly technical nature (WHO–FAO, 2006). Numerous Codex standards have been developed and may deal with all characteristics of a particular commodity or with a specific characteristic, such as Maximum Residue Limits for pesticides or veterinary drugs. Although of less obvious relevance, food safety issues form an aspect of biosecurity particularly where they involve matters of infestation or biological contamination and because food products and food production can provide pathways for the introduction of IAS.

The Convention on Biological Diversity

Discussion of the environmental aspects of biosecurity, including matters related to biodiversity and ecosystems, and particularly to the management of invasive alien species (IAS), is limited in the SPS Agreement and in the standards set by the applicable sectoral bodies reflecting the greater emphasis that had been afforded to 'agricultural biosecurity' and the scope of the international trade regime. When considering environmental agreements it is IAS, rather than biosecurity, that has received greatest attention.

The Convention on Biological Diversity (CBD) is the most broadly applicable of the relevant instruments of international environmental law, providing an important basis in binding treaty obligations. As noted by Shine (2000), the CBD 'is the only globally applicable, legally binding instrument to address generally alien species introduction, control and eradication across all biological taxa and ecosystems'. Most pertinently, Article 8 sets out obligations for *in situ* conservation, including:

- the establishment or maintenance of means to regulate and manage the risks associated with LMOs (living modified organisms) and biotechnology 'which are likely to have adverse environmental impacts that could affect the conservation and sustainable use of biological diversity, taking also into account the risks to human health' (paragraph g);
- preventing the introduction of, controlling or eradicating, alien species 'which threaten ecosystems, habitats or species' (paragraph h); and
- developing and/or maintaining legislation and regulatory provisions for the protection of threatened species and populations (paragraph k).

Concerning LMOs, the Cartagena Protocol to the CBD sets out further measures relating to their regulation. The protocol requires 'advance informed agreement' between exporting and importing states for the movement of LMOs intended for release into the environment and the preamble and several of the substantive provisions make it clear that the precautionary approach[8] should be applied.

Recognizing their significance to biological diversity, the CBD has developed further guidance on tackling IAS, with the Conference of the Parties (COP) having consistently addressed the question of IAS. COP 6 (2002) adopted the (non-binding) Guiding Principles for the Prevention, Introduction and Mitigation of Impacts of Alien Species that Threaten Ecosystems, Habitats or Species. These principles call, among other things, for the adoption of a precautionary approach, the implementation of appropriate State border controls and quarantine measures, adoption of appropriate regulatory measures for both intentional and unintentional introductions in domestic frameworks and increased coordination at the international level. The principles adopt the prevention – eradication – control approach, described above (Guiding Principle 2).

In 2011 the CBD published its draft 'Considerations for implementing international standards and codes of conduct in national invasive alien species strategies and plans' (Reaser, 2011). Importantly, the document recognizes the role that is, or could be, played by other key instruments and organizations such as the OIE, IPPC, Codex and the Convention on International Trade in Endangered Species (CITES). It recognizes also that the international legal framework is simultaneously overly complex, with several overlapping agreements, and inadequate in its scope, with numerous pathways for introduction not addressed by any of the applicable instruments (see also CBD COP Decision VIII/27). CBD COP XII (2014) adopted voluntary Guidance on devising and implementing measures to address the risks associated with the introduction of alien species as pets, aquarium and terrarium species, and as live bait and live food, addressing a specific gap in the existing framework.

Significantly, as part of its strategic plan for Biodiversity 2011–2020 the CBD adopted the Aichi Biodiversity Targets. Target 9 provides:

> By 2020, invasive alien species and pathways are identified and prioritized, priority species are controlled or eradicated, and measures are in place to manage pathways to prevent their introduction and establishment.
>
> (CBD Decision X/2)

Collectively these decisions and guidelines provide significant impetus and direction for state parties to act to prevent and manage the threat that IAS pose to biological diversity. All however fall short of forming binding obligations (beyond those set out in the Convention), but contribute to the expanding body of soft law in which the significance of IAS to biodiversity is recognised.

Issue and pathway-specific instruments

While the CBD provides the most comprehensive international instrument relevant to the environmental aspects of biosecurity and IAS, numerous other instruments in environmental law are relevant. Based on the definitions described above, FAO-Norway (UN FAO, undated) cites the following non-exhaustive list of 'sectoral instruments' related to biosecurity: the Rotterdam Convention on the Prior Informed Consent Procedure for Certain Hazardous Chemicals and Pesticides in International Trade; the Convention on Persistent Organic Pollutants; the FAO International Code of Conduct on the Use and Distribution of Pesticides; the Biological and Toxin Weapons Convention; the FAO International Code of Conduct on Responsible Fisheries; the Ramsar Convention on Wetlands; the Protocol to the Antarctic Treaty on Environmental Protection; the Convention on the Conservation of Migratory Species of Wild Animals; the Global Programme of Action for the Protection of the Marine Environment from Land-Based Activities. These agreements address specific environmental issues and in doing so impose obligations or guidance on implementing countries which may affect their biosecurity strategies. For instance, controls related to the presence of IAS may be required for the conservation of specific habitats such as wetlands.

In some instances, the measures are more explicitly relevant. The Convention on International Trade in Endangered Species (CITES) is important to biosecurity because it concerns international trade which, as we have seen, is a key pathway for the global spread of species, pests and pathogens. Its significance is limited insofar as it aims to protect *endangered* species by preventing their illegal trade and not to limit trade for the purpose of protecting species more generally. Nevertheless, CITES articles place further obligations on contracting states by introducing an import and export permit system for the movement and trade in specimens listed in the appendices. For example, Article III (2) requires that any species listed in Appendix I (species that are the most endangered) shall only be exported if accompanied by an export permit. Consequently, states that have ratified CITES must build these obligations and requirements into domestic measures, which place controls on the trade of plants and animals. Further, Resolution 13.10 addressed IAS and recommended that they be considered by Parties when developing national legislation concerning trade in endangered species and that they consult importing country to determine whether appropriate import controls are in place but specific guidelines have not been developed.

The Ramsar Convention adopted Resolution III.18 (2002) on invasive species and wetlands encouraging Parties to take action to address IAS in a holistic manner. This includes undertaking risk assessments of species that may pose a threat to the ecological character of wetlands and taking action to identify and eradicate invasive species including through cooperation where relevant to transboundary wetlands. Identification and management of IAS forms part of the Ramsar Strategic Plan 2016–24. The Strategic Plan identifies as a baseline that 36% of Parties have established national policies or guidelines on IAS and 20% have a national inventory of invasive species relevant to wetlands (4th Strategic Plan, 2015, p. 25), indicating that there is a considerable way to go before comprehensive management of IAS is achieved at the State level.

Recognizing that activities associated with shipping can be a major problem in transferring aquatic IAS, the International Maritime Organization (IMO) and the UN have developed measures to address this particular pathway. The IMO adopted the International Convention for the Control and Management of Ships' Ballast Water and Sediments (2004, due to enter into force in 2017). Under the Convention, all ships will be required to implement a Ballast Water and Sediments Management Plan, as well as complying with further record-keeping and management requirements including specified practices (see also Young, 2006, pp. 14–16). The United Nations Convention on the Law of the Sea provides a more general obligation, requiring that States 'take all measures necessary to prevent, reduce and control pollution of the marine environment resulting from the use of technologies under their jurisdiction or control, or the intentional or accidental introduction of species, alien or new, to a particular part of the marine environment, which may cause significant and harmful changes thereto' (Art. 196).

EU Regulation on Invasive Species

As a regional instrument, the new EU IAS Regulation is noteworthy both because of the impact that it will have on the way that all EU States regulate IAS and because of its broad and preventative approach. Regulation 1143/2014 entered into force on 1 January 2015 having been promised as part of the EU 2020 Biodiversity Strategy (European Commission, 2011, p. 20). The adoption of a Regulation is obviously significant in that, in contrast with a Directive, it has direct effect on the EU Member States and is also expected to address some of the gaps left by the limited ECJ case law in this area (Shine *et al.*, 2010). The Regulation applies to listed species and emphasizes that prevention is preferable, both in terms of cost and effectiveness, to control and eradication.[9]

As noted earlier, a key characteristic of IAS is their presence outside of their natural range. The EU Regulation takes an unusual step in distinguishing the extension of a 'natural range' attributable to environmental change: the preamble notes that 'Some species migrate naturally in response to environmental changes. They should not be considered as alien species in their new environment and should be excluded from the scope of this Regulation. This Regulation should focus only on species introduced into the Union as a consequence of human intervention.' This is to allow for changes including those resulting from adaptation to climate change (Art. 2). Article 2 also makes a clear distinction between IAS and pathogens causing animal diseases and plant pests as regulated under separate EU instruments. Reflecting the language of the CBD, invasive species are conceived of as 'species, subspecies or lower taxon of animals, plants, fungi or micro-organisms' and as such could include pathogens in other capacities but these measures ensure a strict separation from a more encompassing biosecurity approach (Art. 3(1) and (2)). Incorporating a relatively strong preventative element, the Regulation requires *inter alia* that Member States carry out an analysis of pathways of unintentional introductions and that they adopt a corresponding action plan (Art. 13).[10] The Regulation potentially goes further than the specific guidance of the CBD in requiring the restoration of ecosystems that have been damaged by IAS, though this is based on a cost-benefit assessment (Art. 20). Crucial to the impact of this regulation will be the thresholds established for key requirements for listing IAS (Art. 4) including the interpretation of 'available scientific evidence' (see above concerning discussions in the context of the WTO) and 'significant adverse impact on biodiversity'.[11]

Conclusion

International law related to biosecurity and IAS has developed steadily, particularly over the past 20 years, with more detailed guidelines emerging as the impact and threat of pests, diseases and invasive species becomes increasingly visible. This is not to say that there is a full and coherent applicable body of international law. It is clear from the preceding exploration that this is not the case. In most cases international instruments have not managed to move away from the sectoral divisions which mirror the early development of aspects of biosecurity. Consequently, there exists an often fragmented patchwork of guidance and commitments, and it has been recognized often that this leads to gaps, overlaps and inconsistencies (Shine *et al.*, 2010; Outhwaite, 2010; Riley, 2005). The sometimes conflicting concepts and priorities found in international agreements have been discussed elsewhere but issues such as the lack of clarity over application of the precautionary principle are yet to be settled despite being on the international agenda for a number of years. The Standards and Trade Development Facility (STDF, 2013) – a partnership between the WTO, WHO, FAO and World Bank which aims to develop cooperation with respect to SPS matters – highlighted the ongoing inconsistencies between the CBD and the SPS Agreement including measures related to precaution and risk assessment as they relate to IAS and the use of differing terms. Its recommendations included the development of guidance by the SPS Committee on the relationship between IAS and the SPS Agreement, but this has not yet been produced.

Although the conceptualization of biosecurity has evolved, in theory, to include the clear environmental implications of this umbrella concept, IAS are usually addressed as a separate issue, and in the context of biodiversity specifically. Genovasi (2015) notes for instance that the new EU Regulation on invasive species falls short of the more encompassing biosecurity approach found in New Zealand and Australia. In the EU, other aspects of biosecurity including plant and animal health continue to be regulated under separate legislative frameworks. Hence, the calls for greater international collaboration between key institutions, including the WTO, CBD, International Union for the Conservation of Nature, OIE, IPPC, IMO, WHO and others, have maintained their relevance. Of some encouragement perhaps is the inclusion of IAS in the newly adopted Sustainable Development Goals (SDGs), providing broad global political support for development of this issue (General Assembly of the United Nations, 2015). SDG 15 (Life on Land) includes the target of introducing, by 2020 'measures to prevent the introduction and significantly reduce the impact of invasive alien species on land and water ecosystems and control or eradicate priority species'.

Beyond this, there are further challenges including the extent to which customary international law does or might provide assistance to states seeking to reduce the harm caused by pathogens and IAS or establish liability where it has occurred. As discussed by Riley (2009), IAS could be considered to fall within the definition of transboundary harm. However, although customary international law does impose obligations on States as regards environmental harm and transboundary environmental risks it is by no means certain that the scope of customary law extends to harm arising from IAS and, further, the possibility of requiring transboundary environmental impact assessment to prevent harm caused by IAS is also based on soft law commitments rather than binding obligations. Perrings *et al.* (2010) note also the absence of mechanisms to internalize the costs of (in that case) IAS within the international institutional framework, undermining the extent to which the polluter can be made to pay.

At the domestic level, as regimes for biosecurity and the control and management of IAS continue to be developed, further issues of cost and liability will arise with responsibilities and liabilities to be balanced between those directly responsible for, for example, the international and local movement of animals, and the general public, given that protecting health and biodiversity in this regard can be considered a 'public good' (see Outhwaite *et al.*, 2008 and Jenkins, 2013). Such matters may develop more quickly for Member States subject to the new EU Regulation. Implementing States also face significant challenges in terms of resources and institutional frameworks even where appropriate legislation has been adopted (Outhwaite, 2010).

Notes

1 See also Kezia Barker (2008) and Glynn Maynard and David Nowell (2009).

2 Or in the context of plants it is described as a set of 'precautions', similarly aimed to 'prevent the introduction and spread of harmful organisms'. See Defra (2014).

3 See IPPC ISPM 05: *Glossary of Phytosanitary Terms.*

4 Further, the CBD Guiding Principles on IAS note that '[i]t must be reiterated that the terminology surrounding the issue of impact arising from alien species is interpreted differently by different Parties and other Governments, and that additional terminology problems arise in the translation'.

5 This example emphasizes measures related to plants and plant products but equivalent measures for animals and animal products are equally relevant.

6 Welcomed by the CBD: Decision/XI/28, 5 December 2012.

7 See list of ISPMs at www.ippc.int/en/publications/626/.

8 The precautionary approach to risk management states that if an action or policy has a suspected risk of causing harm to the public or the environment, in the absence of scientific consensus that the action or policy is not harmful, the burden of proof that it is *not* harmful falls on those taking an action that may or may not be a risk. See https://en.wikipedia.org/wiki/Precautionary_principle.

9 On the relationship of the Regulation with other key frameworks and instruments within the EU and on some of the legislative gaps existing within Member States see Shine *et al.* (2010). A blacklisting approach is adopted; on listing options (such as grey and white listing) see also Young (2006).

10 Further, Art. 14 makes provision for systems of surveillance.

11 See further Genovesi *et al.* (2015).

References

Agreement on the Application of Sanitary and Phytosanitary Measures (SPS Agreement) 1867 U.N.T.S. 493.

Australia – Measures Affecting the Importation of Apples from New Zealand, Report of the Appellate Body, WT/DS367/AB/R, 29 November 2010.

Australia – Measures Affecting the Importation of Apples from New Zealand, Report of the Panel, WT/DS367/R, 9 August 2010.

Barker, K. (2008) 'Flexible boundaries in biosecurity: accommodating gorse in Aotearoa New Zealand', *Environment and Planning*, Vol. 40, p. 1598.

Biological and Toxin Weapons Convention (London, Moscow and Washington, 10 April 1972).

Cartagena Protocol on Biosafety (Montreal, 29 January 2000).

Convention on Biological Diversity (CBD) (Rio de Janeiro, 5 June 1992).

Convention on Biological Diversity, COP 8 Decision VIII/27, Alien species that threaten ecosystems, habitats or species (Article 8 (h)): further consideration of gaps and inconsistencies in the international regulatory framework, UNEP/CBD/COP/DEC/VIII/27, 15 June 2006.

Convention on Biological Diversity (2002) Guiding Principles for the Prevention, Introduction and Mitigation of Impacts of Alien Species that Threaten Ecosystems, Habitats or Species (UNEP/CBD/COP/6/23), Secretariat of the Convention on Biological Diversity, United Nations Environment Programme, Montreal.

Convention on Biological Diversity (2014) COP 12 Decision XII/16, Invasive alien species: management of risks associated with introduction of alien species as pets, aquarium and terrarium species, and as live bait and live food, and related issues, UNEP/CBD/COP/DEC/XII/16, Pyeongchang, Republic of Korea, 6–17 October 2014, Agenda item 22.

Convention on Biological Diversity, Decision X/2, The Strategic Plan for Biodiversity 2011–2020 and the Aichi Biodiversity Targets, UNEP/CBD/COP/DEC/X/2 29 October 2010.

Convention on Biological Diversity, Glossary of Terms at www.cbd.int/invasive/terms.shtml.

Convention on International Trade in Endangered Species of Wild Fauna and Flora (CITES) (Washington DC, 3 March 1973).

Convention on Persistent Organic Pollutants (Stockholm, 22 May 2001).

Convention on Wetlands of International Importance especially as Waterfowl Habitat – the 'Ramsar Convention', The 4th Strategic Plan 2016–2024, Adopted by the 12th Meeting of the Conference of the Parties at Punta del Este, Uruguay, 1–9 June 2015, through Resolution XII.2.

Convention on Wetlands (Ramsar, Iran, 1971) 'Wetlands: water, life, and culture', 8th Meeting of the Conference of the Contracting Parties, Valencia, Spain, 18–26 November 2002, Resolution VIII.6, A Ramsar Framework for Wetland Inventory.

Defra (2014) *Protecting Plant Health: A Plant Biosecurity Strategy for Great Britain*, www.gov.uk/government/uploads/system/uploads/attachment_data/file/307355/pb14168-plant-health-strategy.pdf.

Donaldson, A. (2008) 'Biosecurity After the Event: Risk Politics and Animal Disease', *Environment and Planning*, Vol. 40, pp. 1552–67.

Ebbels, D.L. (2003) *Principles of Plant Health and Quarantine*, CABI Publishing, Wallingford, UK.

European Commission (2011) *The EU Biodiversity Strategy to 2020*, European Union: Belgium.

FAO – United Nations Food and Agriculture Organization – *see* UN FAO.

Fletcher, J. and Stack, J.P. (2008) 'Crop Biosecurity: Definitions and Role in Food Safety and Food Security', in M.L. Gullino, J. Fletcher, A. Gamliel and J.P. Stack (eds), *Crop Biosecurity: Assuring Our Global Food Supply*, NATO–Springer, Dordrecht.

General Assembly of the United Nations (21 October 2015), Resolution 70/1 Transforming our world: the 2030 Agenda for Sustainable Development, adopted at the seventieth session, 25 September 2015.

Genovasi, P., Carboneras, C., Vila, M., and Walton, P. (2015) 'EU adopts innovative legislation on invasive species: a step towards a global response to biological invasions?' *Biol Invasions*, Vol. 17, pp.1307–11.

Global Programme of Action for the Protection of the Marine Environment from Land-Based Activities (Washington, 3 November 1995). 47 United Nations Framework Convention on Climate Change (New York, 9 May 1992).

Hinchcliffe, S. and Bingham, N. (2008) 'Securing Life: The Emerging Practices of Biosecurity', *Environment and Planning*, Vol. 40, pp. 1534–51.

Ikin, R. (2002) 'International Conventions, National Policy and Legislative Responsibility for Alien Invasive Species in the Pacific Islands', *Micronesia Supp.*, Vol. 6, pp. 123–28.

International Maritime Organization (IMO), International Convention for the Control and Management of Ships' Ballast Water and Sediments, Adoption: 13 February 2004.

International Plant Protection Convention, Rome, 6 December 1951.

International Plant Protection Convention (Revised) Rome, 17 November 1997.

IPPC, ISPM 05: Glossary of Phytosanitary Terms (2010).

IPPC, ISPM 11: Pest Risk Analysis for Quarantine Pests Including Analysis of Environment Risks and Living Modified Organisms (2004).

IPPC, ISPM 15: Regulation of Wood Packaging Material in International Trade (2009).

ISPM – International Standards for Phytosanitary Measures. See note 7 above.

Jack, B. (2009) *Agriculture and EU Environmental Law*, Ashgate, Farnham, UK.

Jay, M., Morad, M. and Bell, A. (2003) 'Biosecurity, A Policy Dilemma for New Zealand', *Land Use Policy*, Vol. 20, pp. 121–29.

Jenkins, P.T. (2013) 'Invasive animals and wildlife pathogens in the United States: the economic case for more risk assessments and regulation', *Biol. Invasions*, Vol. 15, pp.243–48.

Leibler, J.H., Otte, J., Roland-Holst, D., Pfeiffer, D.U., Soares Magalhaes, R., Rushton, J., Graham, J.P., and Silbergeld, E.K. (2009) 'Industrial Food Animal Production and Global Health Risks: Exploring the Ecosystems and Economics of Avian Influenza', *EcoHealth*, Vol. 6, p. 58.

Lowe, S., Browne M., Boudjelas S., and De Poorter M. (2004) *100 of the World's Worst Invasive Alien Species A selection from the Global Invasive Species Database*, The Invasive Species Specialist Group (ISSG) a specialist group of the Species Survival Commission (SSC) of the World Conservation Union (IUCN).

Mackenzie, J., Jeggo, M., Daszak, P., Richt, J.A. (eds) (2013) *One Health: The Human-Animal-Environment Interfaces in Emerging Infectious Diseases: The Concept and Examples of a One Health Approach*, Springer.

Masters, G. and Norgrove, L. (2010) *Climate Change and Invasive Alien Species*, CABI Working Paper 1, p. 30.

Mauritius Plant Protection Act (Act No. 10 of 2006) (Mauritius).

Maynard, G. and Nowell, D. (2009), 'Biosecurity and Quarantine for Preventing Invasive Species', in Clout, M.N. and Williams, P.A. (eds), *Invasive Species Management: A Handbook of Techniques*, OUP.

Meyerson, L.A. and Reaser, J.K. (2002) 'Biosecurity: Moving Toward a Comprehensive Approach', *BioScience*, Vol. 52, no. 7, pp. 593–600.

OIE (2011) Guidelines for Assessing the Risk of Non-Native Animals Becoming Invasive, OIE: Paris.

OIE (2016) Terrestrial Animal Health Code 2016, www.oie.int/international-standard-setting/terrestrial-code/access-online/.

Outhwaite, O. (2013a) 'Preventing and mitigating the impacts of climate change and biodiversity loss through biosecurity', in F. Maes, A. Cliquet, W. du Plessis, H. McLeod-Kilmurray (eds), *Biodiversity and Climate Change*, Edward Elgar/IUCN.

Outhwaite, O. (2013b) 'Legal Frameworks for Biosecurity', in A. Dobson, K. Barker and S. Taylor (eds), *Biosecurity: Theory and Politics in Practice*, Routledge/Earthscan.

Outhwaite, O. (2010) The International Legal Framework for Biosecurity and the Challenges Ahead, *Review of EC and International Environmental Law (RECIEL)*, Vol. 19, no. 2, pp. 207–26.

Outhwaite, O., Black, R., and Laycock, A. (2008) 'The Significance of Cost Recovery for the Regulation of Agricultural Health', *Journal of Law and Society*, Vol. 35 Issue 1(s).

Perrings, C., Burgiel, S., and Lonsdale, M. (2010a) 'Globalisations and Bioinvasions: The International Policy Problem', in C. Perrings, H. Mooney and M. Williamson (eds), *Bioinvasions and Globalisation*, OUP, Oxford.

Perrings, C., Burgiel, S., and Lonsdale, M. (2010b) 'International cooperation in the solution to trade-related invasive species risks', *Ann. N.Y. Acad. Sci.*, Vol. 1195, pp. 198–212.

Plants Act 1976 (Act No. 12, 1976) (Mauritius).

Protocol to the Antarctic Treaty on Environmental Protection (Madrid, 4 October 1991).

Ramsar Convention on Wetlands of International Importance especially as Waterfowl Habitat (Iran, 2 February 1971) – see also Convention on Wetlands.

Reaser, K. and the Secretariat of the Convention on Biological Diversity, *Considerations for Implementing International Standards and Codes of Conduct in National Invasive Alien Species Strategies and Plans* (Draft) (29 November 2011), UNEP: Montreal.

Regulation (EU) No. 1143/2014 of the European Parliament and of the Council of 22 October 2014 on the prevention and management of the introduction and spread of invasive alien species.

Riley, S. (2005) 'Invasive Alien Species and the Protection of Biodiversity: The Role of Quarantine Laws in Resolving Inadequacies in the International Legal Regime', *Journal of Environmental Law*, Vol. 17 issue 3: 323.

Riley, S. (2009) 'Preventing Transboundary Harm From Invasive Alien Species', *RECIEL*, Vol. 18 issue 2.

Rotterdam Convention on the Prior Informed Consent Procedure for Certain Hazardous Chemicals and Pesticides in International Trade (Rotterdam, 10 September 1998).

Shine, C. (2007) 'Invasive species in an international context: IPPC, CBD, European Strategy on Invasive Alien Species and other legal instruments', *Bulletin EPPO Bulletin*, Vol. 37, pp.103–13.

Shine, C., Williams, N., and Gündling, L. (2000) *A Guide to Designing Legal and Institutional Frameworks on Alien Invasive Species*, Environmental Policy and Law Paper No. 40, IUCN, Gland.

Shine, C., Kettunen, M., Genovesi, P., Essl, F., Gollasch, S., Rabitsch, W., Scalera, R., Starfinger, U., and ten Brink, P. (2010) *Assessment to Support Continued Development of the EU Strategy to Combat Invasive Alien Species*, Final Report for the European Commission. Institute for European Environmental Policy (IEEP), Brussels, Belgium.

SPS Agreement – see Agreement on the Application of Sanitary and Phytosanitary Measures.

Standards and Trade Development Facility (STDF) (2013) *International Trade and Invasive Alien Species*, WTO: Geneva.
UN FAO, Committee on Agriculture, *Biosecurity in Food and Agriculture, Item 8 of the Provisional Agenda*, Sixteenth Session (Rome, 26–30 March 2001 COAG/01/8).
UN FAO (2007) *FAO Biosecurity Toolkit*, Rome: FAO.
UN FAO *International Code of Conduct on Responsible Fisheries* (Rome, 31 October 1995).
UN FAO *International Code of Conduct on the Use and Distribution of Pesticides* (Rome, November 2002), as revised.
UN FAO International Plant Protection Convention (IPPC) (Rome, 6 December 1951), as revised.
UN FAO-Norway PCA, *Legal Frameworks for Biosecurity* (FAO, undated), http://km.fao.org/biosecwiki/index.php/Biosecurity_Resources_from_Norway_PCA#Legal_frameworks_for_Biosecurity.
UN FAO-OIE-WHO Collaboration *Sharing Responsibilities and Coordinating Global Activities to Address Health Risks at the Animal-Human-Ecosystems Interfaces, A Tripartite Concept Note*, April 2010, www.who.int/influenza/resources/documents/tripartite_concept_note_hanoi_042011_en.pdf.
United Nations Convention on the Law of the Sea (Montego Bay, 10 December 1982).
United Nations, Sustainable Development Goals [summary], www.un.org/sustainabledevelopment/sustainable-development-goals/.
Waage, J.K. and Mumford, J.D. (2008) 'Agricultural Biosecurity', *Philosophical Transactions of the Royal Society*, Vol. 363, pp. 863–76.
Waugh, J.D. (2009) *Neighborhood Watch: Early Detection and Rapid Response to Biological Invasion Along US Trade Pathways* IUCN, Gland, Switzerland.
WHO–FAO (2006) *Understanding the Codex Alimentarius*, www.who.int/ foodsafety/publications/codex/understanding_codex/en/index.html.
World Trade Organization Agreement on the Application of Sanitary and Phytosanitary Measures (Marrakech, 15 April 1994).
WTO (1998) *Understanding the WTO Agreement on Sanitary and Phytosanitary Measures*, available at www.wto.org/english/tratop_e/sps_e/spsund_e.htm.
WTO, General Agreement on Tariffs and Trade 1994.
WTO (2015) Understanding the WTO, 5th edn, Geneva.
Young, T.R. (2006) *National and Regional Legislation for Promotion and Support to the Prevention, Control, and Eradication of Invasive Species*, World Bank Environment Department, Biodiversity Series, Paper No. 108, IBRD: Washington DC.
Zavaleta, E.S., Hobbs, R.J., and Mooney, H.A. (2001) 'Viewing invasive species removal in a whole-ecosystem context', *TRENDS in Ecology and Evolution*, Vol. 16 issue 8, p. 454.

7

BIOTECHNOLOGY, BIODIVERSITY, AND THE ENVIRONMENT

Barbara A. Schaal and Joseph M. Jez

The topic of biotechnology and genetically engineered (GE) organisms, both plants and animals, generates highly polarized debate around such issues as health, the environment, economics, international relations, the business practices of large corporations, and ethics. One of the most active areas of this debate is the potential effect of agricultural biotechnology on the environment (NRC Board on Agriculture Report, 2002). At one extreme is the claim that GE organisms will greatly harm global agriculture, human health, and the environment. Whereas, advocates maintain that there are few, if any, new risks and that biotechnology benefits both global agriculture and the environment. However, there is a vast middle ground that acknowledges the great potential of agricultural biotechnology but also raises science-based concerns. An unfortunate aspect of the controversy is the tendency to see biotechnology as either good or bad. Biotechnology involves many species, both plants and animals, with a wide range of modifications that are placed in diverse agricultural and natural systems worldwide. Whether or not an application of biotechnology has potential harmful, beneficial, or neutral effects on the environment is both species- and context-specific and depends on the GE plant (or animal), the geographical region where the organisms are placed, and the local biological environment.

Breeding versus biotechnology

Before we go on to examine the effects of biotechnology on biodiversity, our topic here, we need to define what is a 'genetically engineered' (GE) organism. And, we need to determine how GE varieties differ from conventionally bred plants and animals. First, the term 'genetically modified organism' or GMO, often used to describe a variant produced by biotechnology, is misleading and inaccurate. All of our domesticated plants and animals are the result of significant and deliberate genetic modification from their original wild ancestors, first during the course of domestication by either early farmers or herders and then by modern breeding (Shi and Lai, 2015; Wang *et al.*, 2014). In contrast to traditional breeding, biotechnology provides a way of genetically modifying organisms using molecular biology methods to manipulate DNA with the goal of inserting a gene (or genes) from one species into the genome of another. To distinguish the modifications generated from the application of biotechnology versus those from breeding, we use the term 'genetically engineered' (GE) in this chapter. For example, crops with specific genes introduced into them by modern methods of biotechnology (i.e., the organisms are genetically transformed) are GE crops.

How does traditional plant and animal breeding compare to the production of new varieties by biotechnology? Modification of wild species to make them more useful or compatible to humans is an ancient process. Humans from the earliest times have interacted with native biodiversity and have used this biodiversity for their own benefit. Early farmers in the Middle East, Asia, South America and Africa began to grow plants near their villages that they had first collected for food or fiber, in the wild. They chose plants with traits that were most useful—the individual with bigger seeds or longer and tougher fibers, and they used the seeds of these plants to begin the next generation of plants. Over many generations of selection, genetic differences accumulated between the domesticated crop and its wild relative. These genetic differences coded for the morphological, physiological and biochemical traits that distinguish the crop from its wild ancestor. In some species, such as corn, the process so changed the crop that the wild parent species is no longer obvious by morphology alone. In the development of other crops, such as wheat or kales, different species have been crossed, to incorporate genes from one species into the genome of another (e.g. Simpson and Ogorzaly, 2003). The concept of using genes from different species as a basis for improvement is a well-established principle of plant and animal breeding. Early farmers developed plant varieties for their local region, and when the new varieties were useful, they traded seeds and animals over vast geographical scales. Often these new, introduced varieties crossed on their own with local landraces and native species. The introduction of either a new species or new variety into different geographical regions in many cases had a profound effect on biodiversity, by altering agricultural practices, by introducing species which displaced native species, or by altering community dynamics. Agriculture has a long history of impacting both native biodiversity and the environment.

What are some of the characteristics of traditional crop breeding today? The first requirement is a source of new genes or traits. This source is from either other varieties of the same crop, wild relatives, or closely related species. Having identified a new source of desired traits, crosses between the crop and source of new genes allows for recombination of genetic information. Because of recombination, traditional crop breeding is inexact and the progeny contain many genes beyond those for the selected trait. This process can result in the transfer of whole sections of chromosomes, which may introduce genes that produce an undesirable trait, such as early dropping of seeds or reduced crop yield (Simmonds and Smartt, 1999). After the initial cross, repeated crossing of the progeny and their progeny over several generations eliminates these undesirable genes and concentrates the desirable traits. This is a time-consuming process, particularly in the case of perennial crops, such as bananas or cassava, where the generation length (i.e., the time to first flowering) may be several years. Even with annual crops, the process is slow. For example, rice breeding can take 10–15 years before commercial release of a new variety (Anthony and Ferroni, 2012).

Recent advances in DNA sequencing and genome analysis tools are accelerating the pace of traditional breeding. Genomes for multiple crops (alfalfa, apple, banana, barley, cacao, canola, cassava, castor bean, chickpea, chocolate, common bean, cotton, cucumber, date palm, grape, maize, millet, mung bean, orange, papaya, peach, pea, potato, rice, sorghum, soybean, strawberry, sugar beet, sugarcane, tomato, watermelon, wheat) are foundational for marker-assisted breeding of traits, for guiding backcrossing, and recurrent selection (Anthony and Ferroni, 2012; Morrell *et al.*, 2012; Dhanapal and Govindaraj, 2015). Using such information, private sector seed companies, national plant breeding programs, and agribusiness can increase selection efficiency to reduce the number of breeding cycles, which shaves years from the breeding process, and to provide consistent evaluation during the breeding of new crop varieties.

The success of traditional breeding cannot be understated. Of the ~150 domesticated food plants, 30 species produce 95% of human food needs, with four crops (rice, wheat, maize, and potato) accounting for 60% of global food production (www.fao.org/nr/cgrfa/cthemes/plants/en). Traditional breeding generated all of these crops, including those that led to the Green Revolution, which increased the yield of wheat and rice. Regardless of technological advances, plant breeding remains an important source of new crop varieties. Moreover, traditional breeding provides the background stock for new crops produced by genetic engineering, as the choice of which variety to use for genetic transformation is critical. Some of the earliest efforts at producing GE crops were unsuccessful because a relatively poor variety was chosen as the stock for transformation; this happened in tomatoes, making the GE lineage commercially unviable.

Biotechnology presents a complementary approach to traditional plant breeding. Although the overall concept of genetic engineering is actually straightforward and parallels traditional crop breeding with its rounds of selection, the molecular techniques are technologically complex (see Chrispeels and Sadava, 2003). Using molecular biology, a single gene that codes for a desired trait, such as insect resistance, increased protein content, or tolerance to drought, is isolated and combined with a promoter sequence that allows for expression of the gene. This combination of DNA is then introduced directly into the plant genome either by physical means, such as particle bombardment, or biologically by the Ti plasmid of the bacterium *Agrobacterium tumefaciens*, which causes crown gall disease in plants. Following introduction of the DNA encoding the desired trait into the genome, these genetically transformed cells are grown by tissue culture into adult plants that now contain the foreign gene. These plants produce seeds by standard cross-pollination of one plant by another. The resulting seeds produce the next generation of plants that contain the newly inserted gene. Current molecular biology approaches offer defined DNA for insertion, but lack control of where that DNA inserts into a plant genome. Thus, GE crops (just like traditionally bred crops) undergo multiple rounds of selection and evaluation to find useful genetic transformation events. This eliminates unwanted traits and results in a GE plant that shares greater than 99% genome identity with the traditionally bred plant originally used for transformation. In contrast to traditional breeding and its reliance on recombination, which can lead to insertion of large regions of foreign DNA, genetic engineering introduces only a targeted DNA sequence into the plant genome.

As with traditional breeding, new discoveries and technologies offer greater precision for genetic engineering. Recent advances in genome editing have the potential for developing an entirely new generation of GE crops. Genome editing relies on a set of proteins that can be guided to specific DNA sequences to make targeted changes. In particular, the development of programmable nucleases radically changes how scientists can approach genetic modifications in plants and animals (Kim and Kim, 2014). Programmable nucleases, such as zinc-finger nucleases (ZFNs), transcription activator-like effector nucleases (TALENs), and RNA-guided nucleases based on the bacterial clustered regularly interspaced short palindromic repeat (CRISPR)–Cas (CRISPR-associated) system, can produce genetic modifications in cultured cells, as well as plants and animals. Importantly, ZFNs, TALENs, and CRISPR/Cas enzymes are site-specific and can be guided to particular target DNA sequences in a genome. These nucleases share a common mode of action. Each nuclease cleaves DNA, which triggers DNA repair mechanisms that result in genome modifications. The resulting modifications depend on the nuclease system and targeting approach used, but include single base-pair changes (i.e., point mutations), insertion of new genes, small and large deletions of sequence that silence endogenous genes, or even translocations of regions between different chromosomes. Combined with genome data, these methods offer precision undreamt of using either traditional breeding or first generation transgenic technologies.

The potential of genome editing using programmable nucleases for engineering biochemical pathways, changing expression of endogenous genes for the development of a particular trait, for removal of allergens, for the introduction of new traits in a targeted (instead of random) manner, and for a variety of biomedical, agricultural, and biotechnology applications, is unprecedented. Although still at an early stage, experiments in a variety of organisms, including viruses, bacteria, yeast, nematodes, frogs, plants, insects, fish, mice, rats, rabbits, pigs, cows, primates, and cultured mammalian and avian cells, demonstrate the feasibility of genome editing for various purposes. Each nuclease system has pros and cons for its use, including various off-target effects that result in unwanted changes and the inability to control the DNA repair systems that actually introduce the genome modifications. At this stage, extensive screening is required to obtain specific outcomes.

How do plants produced by genetic engineering differ from those produced by traditional breeding? First, as noted above, the generation of GE crops only introduces specific genes to the target species, in contrast to the many genes introduced by traditional breeding. Second, genes used for genetic engineering can come from a wide variety of organisms. Genetic compatibility limits traditional breeding to using only closely related species, usually within the same plant genus. Genetic engineering can use genes from across kingdoms. GE plants can contain genes from bacteria, fungi, and animals, which in turn increases the range of traits available to improve crop performance. For example, expression of a bacterial protein (an RNA-chaperone) in maize and other plants maintains crop productivity under drought conditions (Castiglioni *et al.*, 2008). Likewise, domestic animals can be genetically transformed. Pet stores sell glow-in-the-dark GE fish that express green fluorescence protein from jellyfish (www.glofish.com). Salmon engineered to produce higher levels of growth hormone grow twice as fast and to double the size of normal salmon (Du *et al.*, 1992). Although approved by the US Food and Drug Administration for human consumption in 2015, GE salmon are particularly controversial because they are highly mobile and thus have a possibility of escaping into native marine environments. Plants are also currently being engineered to serve as factories to produce useful compounds that are not found in plants in nature, such as pharmaceuticals (De Luca *et al.*, 2012), plastics (Snell *et al.*, 2015), and human vaccines, including the Ebola virus vaccine (Qiu *et al.*, 2014). A final difference between traditional breeding and genetic transformation to produce new varieties is the time scale. Breeding projects take many years, whereas GE transformation is relatively quick. Genetic transformation is also more efficient. Breeding of a perennial crop, such as cassava or bananas, also requires vast amounts of space and labor to grow the large numbers of individuals to screen for selected traits. Genetic transformation occurs in the laboratory and after it is successful plants are transferred to the greenhouse and ultimately field grown.

Genetically engineered plants and animals

Since the introduction of GE crops in the early 1990s, the use of agricultural biotechnology expanded worldwide, from 1.7 million hectares in six countries to 185.1 million hectares in 26 countries, between 1996 and 2016 (James, 2016). In 2016, developing countries planted more GE crops than industrial countries, with farmers in Latin America, Asia, and Africa planting 54% of global GE crops. Moreover, 90% of the 15 million farmers planting GE crops are small, resource-poor farmers in developing countries.

Currently, the most widely used varieties of GE crops carry introduced genes either for insect resistance or for herbicide resistance. Insect resistance comes from a natural insecticide gene found in the soil bacterium *Bacillus thuringiensis* (Ananda Kumar *et al.*, 1996). *B. thuringiensis* produces a family of crystalline proteins (i.e., cry proteins) that inhibit insect growth. Cry proteins are environmentally friendly insecticides; organic farmers routinely use the bacterium as a natural insecticide. Instead of external application of dried bacteria, GE crops such as soybean, corn, and cotton produce one of these cry proteins and are resistant to several major insect crop pests. Herbicide resistance is the other major GE crop trait (Dekker and Duke, 1995). Competition with weeds reduces crop yields. Killing of weeds with herbicides such as glyphosate (Roundup) that leave the GE plant unaffected greatly enhances productivity and yield. In addition to insect and herbicide resistance, GE crops with enhanced drought tolerance (maize, soybean, sugarcane), faster growth and yield (eucalyptus, maize, soybean), disease resistance (bean, papaya, plum, potato, squash, sweet pepper, tomato), modified product quality, and pollination control traits are available (www.isaaa.org/gmapprovaldatabase).

The development of the next generation of GE crops is actively proceeding and we can expect a diversity of new approved crop varieties (Jez *et al.*, 2016). These crops will expand the range of agricultural biotechnology for different species, for the geographical regions where GE crops are grown, and for the type of trait engineered into the crop. Such traits focus disease and pathogen resistance, increased protein content, production of health-promoting lipids, and engineered production pharmaceutical compounds, among others. Moreover, new traits achieved through combinations of breeding and GE are also aiming to reduce inputs required for crop growth and promise to lead to plants that reduce environmental impact and contribute to sustainability (Jez *et al.*, 2016). Development of GE varieties is not limited to row crops such corn, soy, and cotton. Work is being conducted on producing new varieties of trees for wood and pulp, ornamental plants for gardening and landscaping, and new forage grasses. Major efforts are underway to engineer new crops for the developing world, to provide food security and alleviate nutritional inadequacies that are found so often in the developing world. At the same time, animal biotechnology is rapidly advancing. For example, many Asian countries have large aquaculture industries and efforts are underway to produce genetically transformed fish and crustaceans that are resistant to disease, grow rapidly, and are adapted to aquaculture conditions. These applications of biotechnology present particular challenges since these animals are highly mobile. While it is outside of our discussion here, there are also well-established efforts to genetically transform insects such as mosquitoes, to eliminate them as vectors of disease (Waltz, 2016).

Agricultural biotechnology in the tropics: issues

The use of GE plant and animal varieties for the developing world presents challenges for assessing environmental impact. Why do we need to specifically assess the environmental impact of GE agriculture in tropical regions? Why are the lessons already learned from GE agriculture in the developed world inadequate? Both the type of agriculture and the environmental context of agriculture in tropical developing countries differ from those in the temperate developed world. In developed countries, fields of a crop grown in monoculture with large inputs of fossil fuel in the form of agrochemicals, fertilizers, pesticides, and herbicides characterize modern agriculture. In contrast, farmers in developing and tropical countries use a greater variety of agricultural practices. Indigenous people can use traditional intercropping and/or swidden agriculture (i.e., crop rotation) that employ many plant species and varieties with little to no agrochemical use. Many crops are grown in small gardens, orchards, or fields and come in close contact with local native biodiversity. Yet, increasingly, modern agricultural methods are employed for major crops, like maize and soybean.

For our discussion of biodiversity and the environment, the most significant difference between agricultural systems of the developed and developing world is the ambient levels of biodiversity, both in natural habitats and agricultural ecosystems. The tropics contain the greatest natural biodiversity on Earth. Moreover, the biological relationships among species are complex. Species often have specialized ecological niches and are closely tied to other species in the community by feeding relationships, by competition, parasitism, or mutualisms. These intricate connections between species potentially make tropical species and communities vulnerable when changes occur. The major concern is that tropical communities may be highly sensitive to perturbations and ripple effects because of these elaborate interrelationships. However, the relative resistance of tropical versus temperate communities is a long-standing debate in ecology (e.g., Tillman and Downing, 1994). The combination of high species diversity and potential sensitivity to disturbance requires careful evaluation of the potential environmental effects of GE agriculture in tropical regions.

Another important aspect of biodiversity in tropical regions needs to be considered. In the US, most major crops were imported from other regions of the globe and are not grown in contact with their wild ancestors. Thus, maize, wheat, rice and soybean are all crops of either the Old World (wheat, rice, soy) or Mesoamerica (maize). In many cases, there are no close relatives to the imported crop, and the crop is grown in genetic isolation from the native biodiversity. Thus, the environmental concerns regarding gene flow between crop and wild relative and its effect on biodiversity are not major ones.

As GE plants and animals are developed for tropical species and their use incorporated into the agriculture of developing nations, the effects of gene migration between GE species and wild relatives will have increasing importance. We might expect that for many species the contact between crop or GE animal and wild ancestor will be more frequent in regions of high biodiversity. Close contact, which raises the possibility of gene flow, is more likely in some tropical regions for several reasons. First, many genera are species-rich in the tropics and offer many more native candidates for gene flow (i.e., cross-pollination) between wild and domesticated species. Second, many tropical crops are not as highly domesticated as the major crops of the industrial world. These local varieties may be genetically more similar to their wild ancestors or relatives that live nearby and hence more likely to produce fertile offspring when crossed. Finally, many regions in the developing world still use locally adapted landraces of a crop; these landraces are of great importance because they contain valuable agricultural biodiversity, and are a genetic resource for future crop improvement. It is important to consider the effects of GE crops on this aspect of agricultural biodiversity, as well as the potential effects on native biodiversity.

Up to now, we distinguished agriculture and biodiversity between developed and less developed countries. This distinction is far from complete. In the US, several crops are grown in close association with either their wild ancestors (e.g., sunflowers) or weedy relatives have been introduced (rice, sorghum, pannicum). In addition, large monoculture fields of GE crops are increasingly common in developing countries. While the environmental issues that center on biotechnology are the same globally, their relative importance varies with crop, geographical region, and community context.

Before considering the specific effects of biotechnology on biodiversity, we need to address two points. First, many concerns with agricultural biotechnology, such as harm to non-target organisms from pesticides and herbicides, gene flow, and the production of weeds, also plague traditional agriculture. Thus, neither should these issues be ignored for GE crops (because they are the same ones for non-GE crops) nor GE crops avoided because they affect the environment (*all* agriculture affects the environment). Second, the debate regarding biotechnology often focuses on whether there is harm from GE crops. All agriculture is two-sided, with potential liability and benefits for biodiversity. Any assessment of the effects of biotechnology must consider the context of use and be compared to current agricultural practices. Given that we are not going to stop the practice of agriculture, we need to determine the relative risk of GE plants and animals compared to the risk associated with current varieties.

Effects of biotechnology on biodiversity: potential concerns

What are the concerns about the effect of GE crops on biodiversity and the environment? First, we consider the effect that biotechnology-derived species might have on non-target organisms. A study of Monarch butterflies and maize containing a gene encoding a *B. thuringiensis* cry protein (i.e., Bt maize) brought this issue to light in 1999 (Losey *et al.*, 1999). Feeding of Bt maize pollen to Monarch caterpillars in a laboratory experiment led to larvae that displayed stunted growth or were killed (Bt is particularly effective against lepidopterons). However, the study did not ask if this mortality actually occurred in the field. Subsequent scientific risk assessment showed a negligible effect of Bt pollen in the field due to a number of factors (Sears *et al.*, 2001). In the lab experiment, the variety of Bt maize used expressed high levels of the toxin in the pollen, whereas new varieties of Bt maize contain little cry protein in pollen. Other studies showed that the timing of pollen release, the dispersal curve of pollen over distance, and the proximity of milkweed (the larval food source) to farm fields all were such that Bt maize would have a minimal effect on the mortality of Monarch larva. Such extensive risk assessment studies have been done for only a few organisms.

Another issue is the cross-pollination between crop and closely related species (Ellstrand *et al.*, 1999). Gene flow is the migration of genes from one population or taxon to another. Gene flow has a homogenizing effect, making populations that exchange genetically similar genes. Why is such gene flow or cross-pollination a concern? First, it can alter the gene pool of native species. When the native species are either wild relatives or ancestors of domesticated species, homogenization of populations can result in the loss of critical genetic biodiversity. One of the hallmarks of domestication is a genetic bottleneck that results in a decline in genetic variability within the domesticated plant or animal species. In some cases, up to 80% of the genetic variation originally in the wild species was lost during domestication (Olsen and Schaal, 2001). Thus, populations of wild ancestors are extremely important for future crop improvement because they can contain potentially useful genes. As an example, the green revolution in Asia was fostered by new high-yield varieties of rice. Genes were incorporated from rice's wild ancestor, *Oryza rufipogon*, including such traits as disease resistance, small stature, and response to fertilizers. Another concern with gene flow from GE crops into the wild ancestor is that GE traits may cause selective changes that sweep through wild populations and result in a decline in variation. Any loss of variation would include some useful traits. Such loss of variation could also compromise the ability of wild populations to adapt to environmental change, either biological or physical.

Numerous studies of rice in Asia indicate significant gene migration between crop and wild ancestor. The gene flow curve for rice is leptokurtic – most genes migrate at small or moderate distances, but there is a long tail of low levels of gene dispersal across large distances. In the case of rice, identifying plants that are morphologically intermediate between cultivated and wild species detected hybridization between the crop and wild ancestor. This work also illustrates another concern – the production of weedy hybrids. The worry is that when a GE crop hybridizes with a wild ancestor, the hybrid offspring will be a vigorous weed (called super weeds by some). This is again a situation found in conventional agriculture, where there are many crop-weed systems. Such hybridization is of particular concern in Asia, where the wild ancestor of rice grows in close contact with cultivated rice. In Thailand, gene flow results from changing agricultural practices and results in plant hybrids that are very aggressive in growth, interfere with rice cultivation, and cause a decline in yield. The concern for biodiversity is that these weeds will then spread outside of the fields and negatively affect native species. Work by Allison Snow and colleagues on hybrids of Bt sunflowers and native sunflowers has indicated that hybrids may have an enhanced fitness relative to the wild sunflowers (Snow *et al.*, 2003). The hybrid sunflowers incorporated the Bt gene from transgenic sunflowers and are resistant to attack by some lepidopterons. Bt hybrids display greater seed production than the wild sunflowers, thus raising the specter of gene flow altering both the gene pool of the native sunflowers and producing a new, weedy taxon. But whether or not these negative affects actually occur remains to be assessed.

In global regions with high biodiversity, we expect that many related species will grow in close proximity to crops. The likelihood of gene migration between closely related taxa is an issue that needs to be carefully evaluated. We expect that the results of such evaluations will vary, depending on the crop. In some cases, where the crop grows adjacent to the wild ancestor, where the crop has not accumulated major genetic differences that isolate it from the wild ancestor, and where there is no reproductive isolation or lack of pollinators, gene flow is likely. However, for some species there will be no gene migration between crop and wild relatives due to lack of compatibility, variation in flowering time, or spatial isolation of the crop from wild relatives. This conclusion is both encouraging and discouraging, since either the detection of risks or the absence of risks in one species does not bear on risk assessment of gene flow in other agricultural species. Each species needs to be carefully assessed separately, and any generalizations need to be drawn with great care.

Effects of biotechnology on biodiversity: potential benefits

Up to this point, this chapter has addressed potential negative consequences of GE agriculture on biodiversity; however, there are also potential positive aspects. These benefits frequently stem from a mitigation of current agricultural practices such as pesticide and herbicide application. Different regions of the globe use different kinds of chemicals and in vastly different amounts, with tropical agriculture of developing countries often having very high rates of pesticide application. Some rice fields in southeast Asia are sprayed with pesticide several times a week, jeopardizing farmers, their families, and the entire ecosystem with pesticides (Phipps and Park, 2002). Bt crops such as maize and cotton produce their own pesticides by genetic modification and require less insecticide spray. Data from cotton fields show a clear reduction in pesticide use over conventional agriculture, but possible reductions for some other crops are not always well documented. For example, use of crops engineered with cry proteins (i.e., Bt) reduces pesticide applications by 500 million kg, and as a result of less machinery use, lowered CO_2 emissions by 28 million kg (the equivalent of 12.4 million cars being removed from the roads). Reductions in insecticide use would also benefit human health and mitigate non-target organism exposure to maintain diversity. Reductions in agrochemical use exposes species to less pesticide, either by direct contact or through sequestration in the food chain.

Another major concern is that herbicides commonly used in Western agriculture are increasingly used in developing countries. Some herbicides can be toxic, degrade slowly, or are difficult to assay. Glyphosate (RoundUp) is environmentally benign with little if any toxicity and rapidly degrades. RoundUp-ready crops use applications of glyphosate as an alternative to more toxic herbicides. Thus, the switch to glyphosate-resistant GE crops mitigates or reduces potential effects of more toxic herbicides. This simple change in agrochemical use can enhance biodiversity. Moreover, herbicides reduce plant biodiversity, and thus indirectly affect other species in a food chain. Less diverse plant communities may lead to less diverse arthropod, mammal, bacterial, etc. populations. Such changes can then have a ripple effect through the food chain.

New varieties of GE crops currently in development are engineered to respond more readily to fertilizers or for drought resistance. Such crops afford the possibility of reducing fertilizer application and irrigation, both processes that significantly modify native habitats and lessen biodiversity. Other potential benefits include providing alternative, cash-generating crops for local farmers in the developing world. In many regions of the developing world with low agricultural production, local farmers subsidize their diets by hunting animals. Such 'bush meat' may often be species that are rare or even endangered. Economic stability from new cash crops can lessen the harvest pressure on native biodiversity. One of most intriguing aspects of biotechnology for environmental benefit is the use of GE plants modified to take up and sequester toxic substances, such as heavy metals or chemicals (Bizily *et al.*, 2000). These specialized plants, developed for bioremediation, are sown as a lawn on a toxic spill site, grown, and the resulting plants then harvested and disposed of as toxic waste. Several years of treatment can effectively remove contaminants and dramatically reduce the levels of toxins in the soil.

GE crops offer the hope of reducing agrochemical use by developing plants that produce their own insecticides to reduce pesticide application, by developing plants that are resistant to herbicides to modify application schedules, and by developing plants that require less fertilizer. Such potential benefits are particularly important in tropical regions where pest pressure on crops and the corresponding use of pesticides is exceedingly high. In a study of potential use of GE crops in developing countries, Qaim and Zilberman (2003) illustrated that the demand for GE crops could be high in developing countries due to their expected enhancement of yield. At the same time, data from India on cotton indicates that Bt cotton greatly reduces the use of pesticides to reach the same yield. To prevent a loss of 20% yield, Bt cotton requires pesticide application of 0.8 kg/ha, but non-Bt cotton requires an application of 4.8 kg/ha (Qaim and Zilberman, 2003). Not only are such reductions in pesticide good for biodiversity, they are critical for the health of local farmers, who often suffer from the effects of frequent applications of toxic pesticides, pesticides whose use is often banned in the US.

While many studies speculate that any reduction of agrochemical use would enhance biodiversity, few supporting data are available. One study examined the effect of the timing and use of herbicides on arthropod community diversity in forage beet populations in Denmark (Standberg and Pedersen, 2002). The study compared arthropod biodiversity in fields treated with conventional herbicide application (non-GE crop) to a RoundUp-ready (GE) crop with applications of herbicide according to label recommendations and with a late application of Roundup. There was no significant difference between the arthropod communities for the conventional crop and the RoundUp-ready beets treated according to label directions. Interestingly, late application of the herbicide led to nearly a doubling in the number of arthropod species. The authors speculated that letting weeds remain longer in the field enhanced arthropod species diversity. Such research demonstrates not only that GE agriculture can enhance species diversity relative to conventional agricultural practices, but also shows the necessity of fine tuning agricultural practices for specific crops and location.

Such studies will be criticized, with the observation that if no herbicides were used at all, then there would be an even greater biodiversity. This is of course correct, but the assessment of agricultural practices needs to be made realistically and in comparison with current practices. Biodiversity would be greatest if we had no agriculture at all; agriculture since the time of the earliest plant domestication has reduced native biodiversity. Such arguments ignore the global requirements of human populations. We need agriculture to feed populations in cities and the expanding populations of the developing world. The best way to minimize the negative effects of agriculture, both GE and non-GE, is to carefully apply the learned scientific principals from ecology, genetics, molecular biology, and agronomy to each agricultural situation.

Conclusion

It is clear that many of the issues that relate to the potential environmental effects and biodiversity of GE agriculture are location- and crop-specific. For example, there is no risk of gene flow between GE maize and its wild ancestors in the US. But, in central Mexico such gene flow may be a threat to the few remaining populations of teosinte, maize's wild ancestor. The wealth of biodiversity in tropical regions is a particular challenge for agricultural biotechnology. In the tropics, many species are cultivated in contact with their wild ancestors, and some tropical crops may have little genetic differentiation from their wild ancestor, thereby increasing the chances of gene flow. Moreover, environmental interactions in the tropics are complex, with food chains and connections between species often intricate. Thus, one might expect perturbations of local species to pass through other components of the ecosystem. At the same time, pesticide use is high in the tropics, with a cost to humans, the environment and biodiversity.

The only way to determine the effect of biotechnology on the environment and on biodiversity is to conduct appropriate scientific studies that assess relative risk, measure gene flow, determine the fitness of hybrids, examine the effects on non-target species, and monitor ecosystems for things gone wrong (Kjellsson and Strandberg, 2001). This is not a well-received answer to the general question: is biotechnology harmful, neutral, or beneficial to the environment?

This question can only be answered for a specific case and depends on the GE plant or animal, the geographical region where the organisms are placed, and the local biological environment. Moreover, the effect of a GE organism needs to place in the context of current local agricultural practices on the environment and biodiversity. While such work is complex and often tedious, careful scientific assessment of the environmental risks of biotechnology will assure that biotechnology develops in concert with local biodiversity and will ultimately help in gaining the public's confidence in and acceptance of these technologies.

References

Anananda Kumar, P., Sharma, R., and Malik, V. (1996) 'The insecticidal proteins of *Bacillus thuringiensis*', *Advances in Applied Microbiology* 42: 1–43.

Anthony, V.M. and Ferroni, M. (2012) 'Agricultural biotechnology and smallholder farmers in developing countries', *Current Opinion in Biotechnology* 23: 278–85.

Bizily, S., Rugh, C., and Meagher, R. (2000) 'Efficient phytodetoxification of the environmental pollutant methylmercury by engineered plants', *Nature Biotechnology* 18: 213–14.

Castiglioni, P., Warner, D., Bensen, R.J., Anstrom, D.C., Harrison, J., Stoecker, M., Abad, M., Kumar, G., Salvador, S., D'Ordine, R., Navarro, S., Back, S., Fernandes, M., Targolli, J., Dasgupta, S., Bonin, C., Luethy, M.H., and Heard, J.E. (2008) 'Bacterial RNA chaperones confer abiotic stress tolerance in plants and improved grain yield in maize under water-limited conditions', *Plant Physiology* 147: 446–55.

Chrispeels, M. and Sadava, D. (2003) *Plants, Genes, and Crop Biotechnology*, Jones and Bartlett, Sudbury, Massachusetts.

De Luca, V., Salim, V., Atsumi, S.M., and Yu, F. (2012) 'Mining the biodiversity of plants: a revolution in the making', *Science* 336: 1658–61.

Dekker, J. and Duke, S. (1995) 'Herbicide resistant field crops', *Advances in Agronomy* 54: 69–116.

Dhanapal, A.P. and Govindaraj, M. (2015) 'Unlimited thirst for genome sequencing, data interpretation, and database usage in genomic era: the road towards fast-track crop plant improvement', *Genetics Research International* 2015: 684321.

Du, S.J., Gong, Z.Y., Fletcher, G.L., Shears, M.A., King, M.J., Idler, D.R., and Hew, C.L. (1992) 'Growth enhancement in transgenic Atlantic salmon by the use of an "all fish" chimeric growth hormone gene construct', *Biotechnology* 10: 176–81.

Ellstrand, N., Hand, S., and Handcock, J. (1999) 'Gene flow and introgression from domesticated plants into their wild relatives', *Annual Review of Ecology and Systematics* 30: 539–63.

James, C. (2014) 'Global Status of Commercialized Biotech/GM Crops', *ISAAA Brief No. 51*, International Service for the Acquisition of Agri-Biotech Applications, Ithaca, New York.

James, C. (2016) 'Global Status of Commercialized Biotech/GM Crops: 2016', *ISAAA Brief No. 52*, International Service for the Acquisition of Agri-Biotech Applications, Ithaca, New York.

Jez, J.M., Lee, S.G., and Sherp, A.M. (2016) 'The next green movement: plant biology for the environment and sustainability', *Science* 353: 1241–44.

Kim, H. and Kim, J.S. (2014) 'A guide to genome engineering with programmable nucleases', *Nature Reviews Genetics* 15: 321334.

Kjellsson, G. and Strandberg, M. (2001) *Monitoring and Surveillance of Genetically Modified Higher Plants,* BirkhauserVerlag, Basel.

Losey, J., Rayor, L., and Carter, M. (1999) 'Transgenic pollen harms Monarch larvae', *Nature* 399: 214–16.

Morrell, P.L., Buckler, E.S., and Ross-Ibarra, J. (2012) 'Crop genomics: advances and applications', *Nature Reviews Genetics* 13: 85–96.

NRC Board on Agriculture, Committee on Environmental Impacts associated with Commercialization of Transgenic Plants (2002) *Environmental Effects of Transgenic Plants*, National Academy Press, Washington DC.

Olsen, K. and Schaal, B. (2001) 'Microsatellite variation in cassava (*Manihot esculenta, Euphorbiaceae*) and its wild relatives: further evidence for a southern Amazonia origin of domestication', *American Journal of Botany* 88: 131–42.

Phipps, R. and Park, J. (2002) 'Environmental benefits of genetically modified crops: global and European perspectives on their ability to reduce pesticide use', *Journal of Animal and Feed Sciences* 11: 1–18.

Qaim, M. and Zilberman, D. (2003) 'Yield effects of genetically modified crops in developing countries', *Science* 299: 900–02.

Qiu, X., Wong, G., Audet, J., Bello, A., Fernando, L., Alimonti, J.B., Fausther-Bovendo, H., Wei, H., Aviles, J., Hiatt, E., Johnson, A., Morton, J., Swope, K., Bohorov, O., Bohorova, N., Goodman, C., Kim, D., Pauly, M.H., Velasco, J., Pettitt, J., Olinger, G.G., Whaley, K., Xu, B., Strong, J.E., Zeitlin, L., and Kobinger, G.P. (2014) 'Reversion of advanced Ebola virus disease in nonhuman primates with ZMapp', *Nature* 514: 47–53.

Sears, M.K., Hellmich, R.L., Stanley-Horn, D.E., Oberhauser, K.S., Pleasants, J.M., Mattila, H.R., Siegfried, B.D., and Dively, G.P. (2001) 'Impact of Bt corn pollen on monarch butterfly populations: a risk assessment', *Proceedings of the National Academy of Sciences USA* 98: 11937–42.

Shi, J. and Lai, J. (2015) 'Patterns of genomic changes with crop domestication and breeding', *Current Opinion in Plant Biology* 24: 47–53.

Simmonds, N. and Smartt, J. (1999) *Principles of Crop Improvement*, Blackwell Science, Oxford, UK.

Simpson, B. and Ogorzaly, M. (2003) *Economic Botany: Plants in Our World*, McGraw Hill, New York.

Snell, K.D., Singh, V., and Brumbley, S.M. (2015) 'Production of novel biopolymers in plants: recent technological advances and future prospects', *Current Opinion in Biotechnology* 32: 68–75.

Snow, A.A., Pilson, D., Rieseberg, L.H., Paulsen, M.J., Pleskac, N., Reagon, M.R., Wolf, D.E., and Selco, S.M. (2003) 'A Bt transgene reduces herbivory and enhances fecundity in wild sunflowers', *Ecological Applications* 13: 279–86.

Strandberg, B. and Pedersen, M. (2002) 'Biodiversity in glyphosate tolerant fodder beet fields', *NERI Technical Report* No. 410. Ministry of the Environment, Denmark.

Tillman, D. and Downing, J. (1994) 'Biodiversity and stability in grasslands', *Nature* 367: 363–65.

Waltz, E. (2016) 'GM mosquitoes fire first salvo against Zika virus', *Nature Biotechnology* 34: 221–22.

Wang, G.D., Xie, H.B., Peng, M.S., Irwin, D., and Zhang, Y.P. (2014) 'Domestication genomics: evidence from animals', *Annual Review of Animal Bioscience* 2: 65–84.

8

LEGAL RESPONSES IN THE UNITED STATES TO BIODIVERSITY LOSS AND CLIMATE CHANGE

James Ming Chen[1]

The greatest causes of biodiversity loss today are climate change, habitat destruction, invasive species, pollution, population and overkill. Climate change, habitat destruction and alien invasive species should figure more prominently than overkill and the marketing of products derived from endangered species. The law of the United States, however, imposes its clearest, harshest sanctions precisely where the drivers of extinction are weakest: when humans consciously capture or kill other living things. More helpfully, the Endangered Species Act has been adapted to address habitat destruction on private land and to mitigate climate change.

Across the Apocalypse on horseback

Humanity is living through the sixth mass extinction of the Phanerozoic Eon, a span of 542 million years from the initial emergence of hard-shelled animals to the present (Kolbert, 2014). Because full restoration of biodiversity after a mass extinction event requires 10 million to 100 million years (Wilson, 1992, p. 330), "the loss of genetic and species diversity" is probably the contemporary crisis "our descendants [will] most regret" and "are least likely to forgive" (ESA Oversight Hearings, 1981, p. 366).

To this concern, we must add anthropogenic climate change (Intergovernmental Panel on Climate Change, 2014). Biodiversity loss and climate change are closely related biological catastrophes. Indeed, within the framework of biodiversity loss, climate change represents an extreme variation on the theme of habitat destruction. Large-scale habitat destruction through climate change puts many plant and animal species at the risk of extinction (Parmesan, 2006).

If indeed biodiversity loss and climate change have reached apocalyptic proportions, it is perhaps fitting to describe the drivers of extinction in equine terms. To describe these drivers, Edward Wilson uses an acronym derived from the Greek word for horse. HIPPO represents **H**abitat destruction, **I**nvasive species, **P**ollution, **P**opulation and **O**verharvesting (Wilson, 2002, pp. 50–51). Metaphorically, this chapter's review of legal responses to biodiversity loss and climate change rides across the apocalypse on horseback.

Before the rise of Neolithic agriculture and the spread of sedentary human settlements across much of the globe's surface, Wilson's deadly HIPPO took the reverse sequence: OPPIH. In Paleolithic times, the overharvesting of large mammals and flightless birds had a greater ecological impact than what was then "a still proportionately small amount of habitat destruction" (Wilson, 2002, p. 50). In North America, the sudden disappearance of large mammals such as mammoths and ground sloths 11,000 to 12,000 years ago, after the continent's megafauna had survived 22 glacial cycles, strongly suggests that this mass extinction was attributable to "blitzkrieg" (Diamond, 1989). Today, "the principal cause of biodiversity loss is the fragmentation, degradation, and destruction of ecosystems and habitats through conversion of land to economically productive uses, especially agriculture, forestry, mineral and fossil fuel extraction, and urban development" (Karkkainen, 1997, p. 7). Global climate change represents an even more potent driver of ecological ruin and evolutionary change (Parmesan and Yohe, 2003).

Legal responses to biodiversity loss and climate change typically fail to calibrate remedies according to the severity of the biological threat. Perversely enough, the legal understanding of extinction mechanisms remains frozen in time, like a cave dweller in ice. The legal enterprise of preventing extinctions would seem likelier to succeed if it addressed the most powerful causes of biodiversity loss today. Climate change, habitat destruction and alien invasive species should figure more prominently than overkill in the law of biodiversity protection.

The following section of this chapter describes how the law of the United States seeks to preserve biodiversity by deterring overkill, habitat destruction and the introduction of alien invasive species. The law imposes its clearest and harshest sanctions precisely where the drivers of extinction are weakest: when humans take conscious steps to capture or kill other living things for human gain. The chapter then proceeds to examine more closely the use of the Endangered Species Act to address habitat destruction on private land and to mitigate climate change.

Legal responses to biodiversity loss

Overkill

The second decade of the new millennium marks the centenary of the Great War – and two of the most spectacular extinctions in cultural memory. Humanity has never emotionally recovered from witnessing the extermination of the Carolina parakeet and the passenger pigeon. These birds, respectively the only parrot native to North America and what is thought to have been not only the most abundant bird but also the most abundant terrestrial vertebrate, became extinct at the Cincinnati Zoo four years apart. Martha, the last passenger pigeon, died on September 1, 1914; Incas, a male Carolina parakeet and the last of his kind, died on February 21, 1918 (Cokinos, 2000; Schorger, 1955). The 1916 treaty at issue in *Missouri v. Holland*, perhaps one of the first legal enactments in the United States (or anywhere else in the world) to treat biodiversity conservation as "a national interest of very nearly the first magnitude," focused exclusively on "the killing, capturing or selling … of … migratory birds."[2]

The paradigmatic act of converting wildlife to personal property through capture or slaughter remains the central focus of laws designed to protect endangered species.[3] In the United States, section 9 of the Endangered Species Act of 1973[4] (ESA) flatly prohibits the "tak[ing]" of any protected species.[5] "The term 'take,'" in turn, "means to harass, harm, pursue, hunt, shoot, wound, kill, trap, capture, or collect, or to attempt to engage in any such conduct."[6] Section 9 so unequivocally condemns the harvesting of protected organisms that few litigated ESA cases discuss this aspect of the statute.[7]

The Endangered Species Act reveals a rather overt bias in favor of preventing direct takings of large, charismatic fauna over all other threats to biodiversity. The Act excludes certain insects from its protective aegis,[8] even though insects are so essential to human welfare that if they "and other land-dwelling arthropods ... were to disappear, humanity probably could not last more than a few months" (Wilson, 1992, p. 133). Moreover, even though "[t]he biological differences between animals and plants ... offer no scientific reason for lesser protection of plants," the Act significantly undervalues plants (National Research Council, 1995, p. 90). Threatened and endangered plants are protected only insofar as they appear on federal land or are destroyed in knowing violation of state law.[9] Plants receive far fewer critical habitat designations than do threatened and endangered animals.[10] The ESA perpetuates rather than corrects the common law's baneful practice of treating plants as private property merely by virtue of dwelling on private land (Rolston, 1990, p. 293).

Traffic in goods derived from endangered species remains the single act of biodiversity destruction on which international law has reached a punitive consensus. The Convention on International Trade in Endangered Species (CITES)[11] would represent a major step toward conserving biodiversity, as long as one is willing to overlook the fact that it does not work. The extension of CITES during the 1980s to "all aspects of trade and research" in orchids "immediately increased the desire for the plants, raised their market value dramatically, and led to even more collecting of rare orchid species from the wild" (Hanson, 2000, p. 67). Nothing in CITES stops developers and farmers who would "flood [critical] habitat with a hydroelectric dam, log it, level the hillsides of a road, build a golf course on the site, or burn the jungle to the ground for agricultural purposes" (Hansen, 2000, p. 17). Not surprisingly, "no reliable data [show] that CITES and similar efforts ha[ve] reduced smuggling, saved any orchid species from extinction, helped protect orchid habitats, or even salvaged orchid plants facing ... certain destruction" (Hansen, 2000, pp. 262–63). The focus on politically visible but environmentally secondary acts of overkill and commercial exploitation has rendered CITES tragically impotent.

Alien invasive species

Human ecological mismanagement often takes the form of introducing an invasive species (Cox, 1999; Elton, 1958, Williamson, 1996). "[M]ost invasions have a weak impact," but on occasion "an invasive species [is] capable of precipitating monumental changes to an ecosystem" (McCann, 2000, p. 232). For example, introducing the Nile perch into Lake Victoria devastated endemic cichlids (Goldschmidt, 1996). Starlings, a scourge to many native birds, entered North America by virtue of a single man's perverse obsession to import all birds mentioned by Shakespeare (Dillard, 1999, p. 37).

As overall biological diversity decreases, the environmental impact of invasive species will probably increase. If "simplified communities are more vulnerable to invasion," then "we should also expect an increase in frequency of successful invaders as well as an increase in their impact" (McCann, 2000, p. 233). Repeated cycles of extirpation and invasion, whether intentional or inadvertent, "can, and eventually will, invoke major shifts in community structure and dynamics" (McCann, 2000, p. 233). In this game of ecological roulette, the disturbances with the "greatest ecological impact frequently incur high societal costs" (Chapin *et al.*, 2000, p. 239).

Existing law offers few, if any, ways to address the problem of invasive species. Laws targeting the animal and plant pests do enable the Department of Agriculture to constrict the movement of organisms known or suspected to have an adverse effect on agriculture.[12] Such laws, however, serve more to regulate the proposed releases of genetically modified crops than to provide broad-based authority to restrain the diffusion of invasive species.[13]

The National Environmental Policy Act of 1970 (NEPA)[14] – a statute whose "procedural requirements ... are analogous" to those of the ESA[15] – provides a somewhat broader platform for legal intervention. One federal court of appeals has used NEPA to require a federal agency to address how dam construction could introduce zebra mussels into previously uninfested waters.[16]

More typically, however, NEPA proves impotent to curb invasions. Rejecting arguments that airport expansion could dramatically increase the rate at which commercial flights (especially from Asia) would introduce alien species into Maui, the Ninth Circuit declined to find a NEPA violation. That court took refuge in the vagaries of airport demand projections, the multiplicity of invasion vectors and the impossibility of determining *ex ante* which species would become established and, among those, which would become "economic pests."[17]

No single country can contain the menace posed by alien invasive species. Within the inherently global project of biodiversity conservation, any hope of addressing the scourge of alien invasive species demands especially vigorous dedication to international cooperation (Glowka, 2000, pp. 333–49). The Convention on Biological Diversity[18] exhorts its contracting parties, "as far as possible and as appropriate," to "[p]revent the introduction of, control or eradicate those alien species which threaten ecosystems, habitats or species." The United States' persistent refusal to sign the Convention, however, effectively short-circuits international law's potential to spur domestic legal change (Blomquist, 2000).

Habitat destruction and public land management

Habitat destruction threatens even deeper biodiversity loss (Ehrlich, 1988; Matson *et al.*, 1997). Contracting the physical range of endangered species spurs their extinction (Channell and Lomolino, 2000). Island biogeography posits that a 90 per cent reduction in the area of a biological island – which may consist of an island in the geographic sense, or merely an isolated patch of wildlife habitat – dictates a 50 per cent reduction in biological carrying capacity as measured by the number of distinct species that can be sustained (MacArthur and Wilson, 1967; Simberloff, 1976; Whitehead and Jones, 1969).

Biologically catastrophic episodes of habitat destruction recur with alarming frequency. An area as large and diverse as Centinela, a diverse forest ridge in Ecuador, can fall to victim to cacao cultivation (Dodson and Gentry, 1991). Large-scale damming, as typified by California's Hetch Hetchy Reservoir (White, 1991, pp. 412–15), Egypt's Aswan High Dam (White, 1988) and China's Three Gorges Dam (Wu *et al.*, 2003), can likewise erase multiple ecological niches in a single blow. Destroying large chunks of the earth's physical infrastructure within a temporal frame that by geological standards is effectively instantaneous significantly accelerates the rate of evolutionary change attributable to human activity.

Traditionally, much of the legal apparatus of the United States for habitat conservation has focused on public lands. Although "[t]he Endangered Species Act of 1973 was motivated in part by the need to [regulate] beyond the limited confines of federal land,"[19] a significant degree of habitat conservation takes place under the aegis of public land management. The law of public lands rests on the primary premise of "multiple use,"[20] defined as a range of uses "including, but not limited to, recreation, range, timber, minerals, watershed, wildlife and fish, and natural scenic, scientific and historical values."[21] Because "[m]ultiple use posits that all uses from commodity extraction and production to biodiversity are equal," this principle "both supports and hinders biodiversity conservation" (Tarlock, 2002, pp. 10, 540–41).

When it first appeared, the concept of "multiple use" represented a substantial improvement in federal land management policy. "[I]ncreased competition for forage" among cattle and sheep ranchers during the nineteenth and early twentieth centuries "led … to overgrazing, diminished profits, and open hostility among forage competitors."[22] The Federal Land Policy and Management Act of 1976[23] (FLPMA) explicitly adopted two statutory principles: "multiple use" for recreation, range, timber, mineral extraction, wildlife and fish habitat, and natural, scenic, scientific and historical uses;[24] and "sustained yield" of renewable resources.[25] At the same time, FLPMA retained "first priority" for existing grazing permitholders as long as federal land-use planning continued to leave land "available for domestic livestock grazing."[26]

Although a statutory commitment to multiple use may theoretically "provide … the legal foundation for a management decision to preserve biodiversity" (Tarlock, 2002, p. 10,541), disputes over federal land management expose a bias favoring commercialization over conservation. When the Interior Department tried in 1995 to "accelerate restoration" of rangelands by making its managerial approach "more compatible with ecosystem management,"[27] incumbent ranchers argued in response that the Interior Department was legally obliged to "safeguard" livestock interests' reliance on the perpetuation of grazing privileges.[28] This argument ran squarely against an explicit statutory proviso that neither "the creation of a grazing district [n]or the issuance of a permit … shall … create any right, title, interest, or estate in or to the lands."[29]

Other decisions have demonstrated the willingness of federal land management agencies to favor grazing and other historically privileged land uses. A federal district court was forced to remind federal land managers in 1985 that grazing "[p]ermittees must be kept under a sufficiently real threat of cancellation or modification in order to adequately protect the public lands from overgrazing or other forms of mismanagement."[30] In spite of its statutory mandate to maintain "final control and decisionmaking authority over livestock grazing practices on the public lands," the federal government had all but ceded jurisdiction over grazing permits.[31]

On the whole, federal land management policy concentrates its habitat preservation efforts on tracts designated as "wilderness." "A wilderness, in contrast with those areas where man and his own works dominate the landscape, is … an area where the earth and its community of life are untrammeled by man, where man himself is a visitor who does not remain."[32] Unlike other public lands, wilderness areas fulfill their function solely by virtue of remaining "in their natural condition."[33] Wilderness preservation helps ensure "that an increasing population, accompanied by expanding settlement and growing mechanization, does not occupy and modify" the entire physical surface of the earth.[34]

Cold and high-elevation wilderness areas, however, cannot anchor a comprehensive and effective biodiversity program (Tarlock, 2002, p. 10,542). Biodiverse "hot spots," rich in species, typically live up to their name: most such locales lie in the tropics (Kunich, 2001, pp. 1157–58). The National Park Service – which is directed to "conserve the scenery and the natural and historic objects and the wild life" in the most spectacular federal lands and "to leave them unimpaired for the enjoyment of future generations"[35] – was designed to preserve geological wonders, not to serve broader ecological purposes (Sellars, 1997, pp. 2–3). Wilderness policy, in microcosm, reveals the overall weakness of laws responding to biodiversity loss.

The Endangered Species Act

Having examined legal tools of varying effectiveness in addressing overkill, alien invasive species and habitat destruction on public lands, I will now focus on the Endangered Species Act.[36] Whatever its shortcomings, the Act deserves credit for "preventing the ultimate extinction of the vast majority of protected species" (Ruhl, 2008, p. 5). After outlining the Act's mechanics, I will address the application of this statute to prevent habitat destruction on private lands and to protect biodiversity from the effects of climate change.

Endangered Species Act mechanics

Listing endangered and threatened species

The Fish and Wildlife Service (FWS) and the National Marine Fisheries Service (NMFS) (known together as the "Services") collectively enforce the Endangered Species Act. The FWS administers the Act for terrestrial and freshwater species, while the NMFS administers the act for most marine species.[37] A species is defined as *endangered* if it "is in danger of extinction throughout all or a significant portion of its range."[38] A *threatened species* is one "which is likely to become an endangered species within the foreseeable future."[39]

Of particular interest in the context of climate change is the time frame deemed *foreseeable*. Because neither the Act nor its implementing regulations define the term *foreseeable future*, the Services determine foreseeability on a case-by-case basis.[40] Definitions of foreseeability have varied considerably. One federal district court has declined to decide whether a risk that the coho salmon might become endangered within "30 or 100 years" satisfied the statutory definition of *foreseeable future*, because an administrative determination that this species "would not become endangered within the next two years" would "[fall] far short of any reasonable definition of the 'foreseeable future.'"[41] Another court has noted (but neither endorsed nor rejected) the assumption that 24 years constitutes the "foreseeable future" for purposes of predicting the likelihood of endangerment.[42] In listing decisions involving salamanders, foreseeability has been set at 40 years.[43] By contrast, one court has held that the same 40-year time horizon, from 2010 to 2050, as identified in projections of deleterious effects from climate change, was not sufficiently foreseeable to warrant the listing of the ribbon seal as a threatened species.[44] The FWS, of its own accord, has declined to list the American pika as threatened or endangered on the basis of climate change risks beyond 2050.[45]

Listing as an endangered or threatened species is a prerequisite to protection under the Act. The Services must base their listing decisions on five factors:[46]

(1) The present or threatened destruction, modification or curtailment of a species' habitat or range.
(2) The overutilization of a species for commercial, recreational, scientific or educational purposes.
(3) Disease or predation.
(4) The inadequacy of existing regulatory mechanisms.
(5) Other natural or manmade factors affecting a species' continued existence.

The decision to list rests solely on biological grounds and must be made "without reference to possible economic or other impacts of [that] determination."[47] Moreover, listing decisions must be made "solely on the basis of the best scientific and commercial data available."[48] The requirement to use the best *available* data is not tantamount to a command to seek and apply "the best … possible" data.[49] Rather, this requirement prevents the Services from disregarding evidence that is better than the scientific basis on which the Services do base their listing decisions.[50]

Critical habitat

After listing a species as endangered or threatened, the Services must also designate critical habitat "to the maximum extent prudent and determinable."[51] Critical habitat includes areas containing "physical and biological features" that are "essential to the conservation of the species and which may require special management considerations or protection."[52] Critical habitat may also include areas outside a species' current range if such habitat is essential to the conservation of that species.[53] Although the designation of critical habitat must "tak[e] into consideration the economic impact" of designating any particular area, the Services may not deny the critical habitat designation to any area where the "best scientific and commercial data available" indicate that "the failure to designate such area as critical habitat will result in the extinction of the species."[54]

Because the Act aspires not merely to "halt" but also to "reverse the trend towards" biodiversity loss,[55] the Act directs the Services to develop a recovery plan aimed at improving the status of each listed species so that listing is no longer necessary.[56] A recovery plan must identify "management actions necessary ... for the conservation and survival of the species," to the point of either "recommend[ing] corrective action" or explaining why such action "is impracticable or unnecessary."[57] Although a recovery plan need not specify a precise timetable, it must include estimates for the time needed to perform recovery measures.[58] The ultimate factors for delisting a species are the same as those that inform the decision to list a species as endangered or threatened.[59]

Interagency consultation

Section 7 of the Act requires each federal agency to ensure that its actions are "not likely to jeopardize the continued existence of any endangered or threatened species or result in the destruction or adverse modification of [critical] habitat."[60] The jeopardy prong of section 7 addresses the impact of agency action on the survival and recovery of a listed species.[61] By contrast, the adverse modification prong concerns critical habitat. A determination that proposed agency action "may affect" a listed species or its critical habitat triggers the obligation to formally consult the FWS or NMFS, as appropriate.[62]

Formal consultation under section 7 typically results in the issuance of a biological opinion evaluating jeopardy to a listed species' continued existence and adverse modification of its habitat.[63] At the very least, where a biological opinion has found that proposed federal action will directly affect a listed species for reasons independent of climate change, that biological opinion must also address the cumulative effects of climate change on that species.[64]

It is not statutory language, but administrative practice and judicial review that have infused the Endangered Species Act with the power to address climate change. As a textual matter, the Act does not obligate the Services, in their discharge of their obligations under sections 4 and 7, to consider the impact of climate change.[65] The Act contains "no statutory requirement" compelling the Services "to consider climate change in [their] listing decisions."[66] Nor does the Act require that the Services account for climate change in their critical habitat designation decisions.[67]

More recent judicial decisions have breathed new power into the Endangered Species Act as a legal tool for addressing the effects of climate change. The Ninth Circuit decided in 2011 to invalidate the Fish and Wildlife Service's attempt to delist Yellowstone grizzly bears as a threatened species, on the grounds that the Service had failed to properly account for the impact of climate change on the whitebark pine, a primary source of food for grizzlies. The climate-driven loss of whitebark pine trees could foreseeably increase conflicts between bears and humans and thereby harm the bears' prospects for reproductive success and overall survival.[68]

Thanks to its breadth, section 7's requirement that federal agencies must consult the FWS or NMFS if proposed action "may affect" a listed species or its critical habitat has the potential to cover "any action that results in non-trivial net increases" in greenhouse gases.[69] As between administrative discretion and judicial review, more aggressive enforcement of the Endangered Species Act by the Services will have greater impact on efforts to mitigate climate change. Because reviewing courts are admonished "not to substitute [their] judgment for that of [an] agency," especially where disputed matters involve "a high level of technical expertise,"[70] courts will hesitate to reverse agency action on the basis of challenges "amount[ing] to nothing more than competing views about policy and science."[71]

Habitat conservation on private lands

"[I]t is unlawful for any person subject to the jurisdiction of the United States to … take any [endangered] species within the United States or the territorial sea of the United States [or] take any such species upon the high seas."[72] This prohibition against the "tak[ing]" of endangered species, contained in section 9 of the Endangered Species Act, dramatically expands the scope of the statute from agencies of the federal government to all actors, including the entire private sector. Although the Act does not directly prohibit the taking of *threatened* species, section 9 punishes the "violat[ion of] any regulation pertaining … to any threatened species of fish or wildlife listed pursuant to" section 4 of the Act.[73] "The Secretary may by regulation prohibit with respect to any threatened species any act prohibited under section 1538(a)(1) of this title, in the case of fish or wildlife, or section 1538(a)(2) of this title, in the case of plants, with respect to endangered species."[74] By regulation, the Services have defined the taking of a threatened species as a violation of section 9.[75]

The Act's definition of *take* and its administrative interpretation are the true source of legal power in section 9's prohibition against the taking of endangered species. The Act defines *take* to mean "harass, harm, pursue, hunt, shoot, wound, kill, trap, capture, or collect" a member of an endangered species.[76] In turn, regulations issued by the Services have defined the term *harm* as including "significant habitat modification or degradation" adversely affecting an endangered or threatened species, with no regard to intent to injure any individual specimen. The celebrated 1995 Supreme Court case of *Babbitt v. Sweet Home Chapter of Communities for a Great Oregon*[77] upheld the application of these expanded definitions of *take* and *harm* to the destruction or significant modification of critical habitat.[78]

The use of section 9 against habitat destruction triggers other provisions of the Endangered Species Act. Section 10 authorizes incidental take permits upon submission and approval of a habitat conservation plan (HCP).[79] In turn, approval of an HCP triggers the federal government's obligation under section 7 to "insure that any action" it undertakes "is not likely to jeopardize the continued existence of any endangered species or threatened species or result in the destruction or adverse modification" of critical habitat.[80] Section 4(d) of the Act may also be used to establish the functional equivalent of HCPs for threatened species (16 U.S.C. § 1533(d)). These provisions have been interpreted as imposing an affirmative obligation to pursue an active species conservation policy.[81]

Before HCPs became a familiar fixture of ESA enforcement, developers and farmers facing potential section 9 liability often resorted to "the 'scorched earth' technique" of pre-emptively clearing wildlife habitat (Bean, 2002, p. 415). ESA enforcement from the 1990s onward transformed "the previously obscure and rarely used permit provision" of section 10 into "the centerpiece of endangered species and ecosystem conservation policy" (Karkkainen, 2003, p. 970). Threatened section 9 liability became merely "the opening gambit in a prolonged bargaining process" (Farber, 1997, pp. 316–17). HCPs today represent "perhaps the most visible example of a consensus-based, multi-stakeholder approach to resource management" (Freeman, 2000, p. 194).

The strategy has its limits. Like the ESA as a whole, HCPs proceed species by species, and only after an individual species has begun to decline. Despite well-founded doubts about the territorial and institutional suitability of states as participants in ecosystem management (Karkkainen, 2002, p. 216), state-law restrictions on land use can enhance the effectiveness of federal HCPs.[82] California law, for instance, facilitates natural community conservation plans that provide "large-scale, multispecies equivalents of HCPs."[83] Ultimately, however, the ESA only indirectly addresses habitat loss and altogether ignores "other causes" of biodiversity loss "such as the invasion of exotic species and air and water pollution" (Tarlock, 2002, p. 10,537). The Act as a whole falls far short of "promot[ing] the conservation of ecosystems on the geographic scale necessary to promote biodiversity generally" (Tarlock, 2002, p. 10,540).

Αρκτούρος: climate change in the "last great wilderness"

Though the polar regions may be poor in biodiversity, they exhibit some of the most dramatic effects of global warming. Climatic impacts on Arctic Ocean sea ice are among the most alarming harbingers of rising temperatures worldwide (Alexander *et al.*, 2014; Cassano *et al.*, 2014). The application of the Endangered Species Act to species most immediately menaced by climate change offers a promising set of remedies for the seemingly relentless emission of greenhouse gases and the anthropogenic contribution to global climate change. With a reach that exceeds that of sections 4 and 7, section 9 of the Act may yet be construed to treat greenhouse gas emissions as a legally critical link in a causal chain leading to the unlawful "taking" of an endangered species (Cumming and Siegel, 2007, pp. 4, 7).

The application of section 9 to climate change would represent a significant step beyond Justice O'Connor's *Sweet Home* concurrence, which emphasized limitations imposed "by ordinary principles of proximate causation," including embedded "notions of foreseeability,"[84] in order to curb the perceived excesses of the Ninth Circuit's 1988 *Palila* decision.[85] In 1995 Justice O'Connor questioned whether section 9 could be lawfully construed to reach destruction of the palila bird's habitat in Hawaii through sheep-grazing. The question today is whether section 9 may be applied to significant modification or degradation of habitat traceable to anthropogenic climate change.

Climate change has figured prominently in both listing and critical habitat designation decisions for species ranging from subtropical elkhorn and staghorn coral,[86] to sage grouse and wolverine on the North American mainland,[87] and bearded and ringed seals in northern seas.[88] Nevertheless, Endangered Species Act cases addressing climate change have extended the longstanding legal preference for large, charismatic fauna over all other forms of biodiversity (Verissimo *et al.*, 2011).

The signature battle over the application of the Act to climate change has involved, quite unsurprisingly, the polar bear (Chen, 2015). The English word *Arctic*, after all, stems from the Greek word for *bear*, in honor of the constellation that other ancient people called *Ursa Major* (Harper, n.d.). *Arcturus* (*Αρκτούρος*), the celebrated northern star, means the "guardian of the bear" (Harper, n.d.). Other ancient sources have drawn inspiration from these northern asterisms: "He is wise in heart and mighty in strength … Which maketh Arcturus, Orion, and Pleiades, and the chambers of the south." Job 9: 4, 9 (King James Version).

Litigation has swamped all aspects of the FWS's efforts to protect the polar bear, from its listing as a threatened species,[89] to the designation of large portions of the Arctic as critical habitat,[90] and the application of section 9's prohibition against takings of polar bears.[91]

The FWS elected in 2008 to list the polar bear as threatened by the effects of climate change on the bear's Arctic habitat.[92] Although the FWS initially declined to designate critical habitat for the polar bear, it dramatically reversed course in 2010 by designating 187,157 square miles in Alaska and adjacent waters of the United States and its territories (see Figure 8.1).[93]

Of central importance to the listing of the polar bear and to the designation of its habitat as critical is the existential threat that climate change poses to Arctic sea ice. A sympathetic D.C.

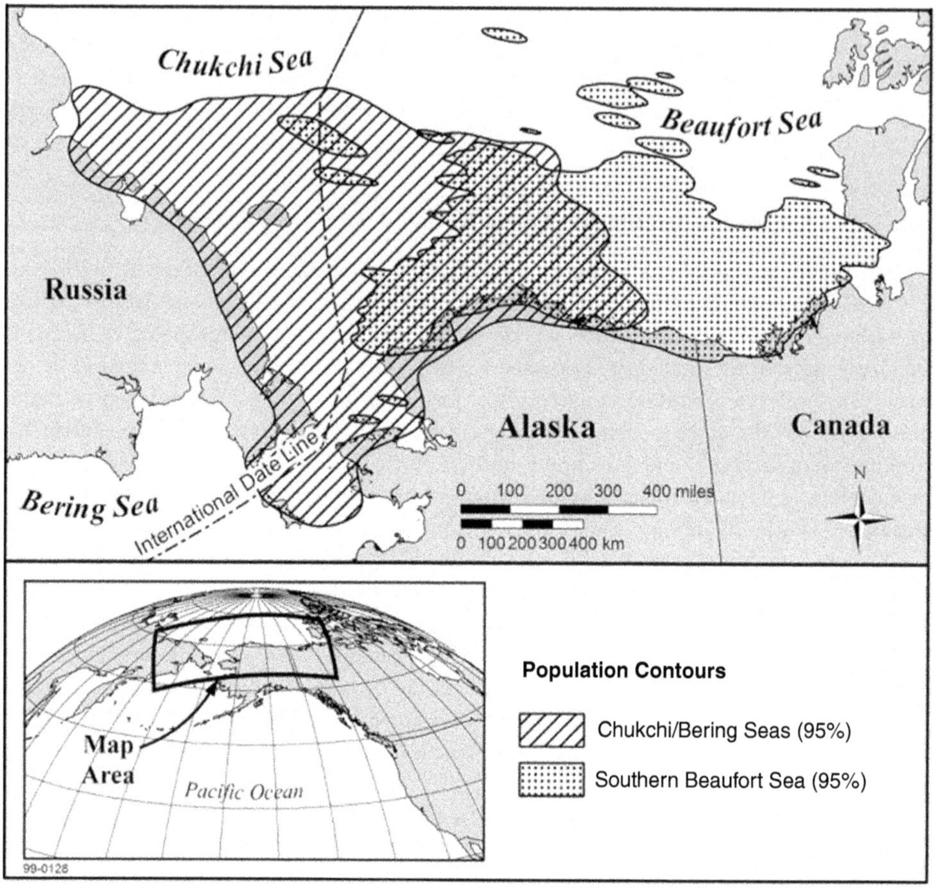

Figure 8.1 Ranges of the Chukchi/Bering Seas and Southern Beaufort Sea Polar Bear (*Ursus maritimus*) Populations
Source: 75 Fed. Reg. (Dec. 7, 2010)

Circuit recognized that irreversible "changes to the polar bear's habitat will soon pose an existential threat to the species":

> Productivity, abundance, and availability of ice seals, the polar bear's primary prey base, would be diminished by the projected loss of sea ice, and energetic requirements of polar bears for movement and obtaining food would increase. Access to traditional denning areas would be affected. In turn, these factors would cause declines in the condition of polar bears from nutritional stress and reduced productivity. … [W]ithin the foreseeable future, all populations would be affected, and the species is likely to become in danger of extinction throughout all of its range due to declining sea ice habitat.[94]

Federal courts have upheld most aspects of the FWS's polar bear decisions. The United States District Court for the District of Alaska did invalidate the FWS's designation of Unit 2, a stretch of northern Alaska spanning the Canadian border and the town of Barrow, because the

FWS used its finding of a need to isolate polar bear dens from humans and human activities, an "essential feature" of Unit 2 that constituted only "approximately one per cent of the entire area" as an improper basis for "designat[ing] a large swath of land … as 'critical habitat.'"[95]

For its part, the United States District Court for the District of Columbia has rejected a challenge to the FWS's decision to confine the protection of polar bears under section 9 of the Endangered Species Act according to exemptions granted by the Marine Mammal Protection Act[96] and the Convention on International Trade in Endangered Species,[97] and to refrain from enforcing section 9 to activities outside the polar bears' range, notwithstanding those activities' incidental impact on polar bears.[98] Using its authority under the Marine Mammal Protection Act,[99] the FWS has routinely authorized nonlethal, incidental takings of polar bears and Pacific walruses.[100] The authority to permit incidental takings requires the Service to determine that such takings will have no more than a "negligible impact" on the affected population.[101]

The brief legal record of applying the Endangered Species Act to climate change has already shifted the policymaking terrain. Designation of the polar bear's critical habitat recognizes the ecological threat that climate change poses to the biosphere. The Arctic has been justifiably described as "the last great wilderness" (Kaye, 2006). Wilderness areas have long offered the promise of providing refuges "where the earth and its community of life are untrammeled by man, where man himself is a visitor who does not remain."[102] Despite their low levels of biodiversity, the Arctic and other cold and/or high-elevation locales may yet prove to be pivotal legal battlegrounds in the last-ditch effort to save the earth and its diverse forms of life from anthropogenically induced climate change. Although the law offers no conclusive answer to the question of "whether the [Endangered Species Act] is an effective or appropriate tool to address the threat of climate change,"[103] climate change and biodiversity conservation remain the most important things that the law can address.

Notes

1 Justin Smith Morrill Chair in Law, Michigan State University; Of Counsel, Technology Law Group of Washington, D.C. Charles R. McManis, Aman K. Gebru, and Andrew W. Torrance provided helpful comments. Marie Gordon provided exceptional administrative support. Special thanks to Heather Elaine Worland Chen.

2 *Missouri v. Holland*, 252 U.S. 416, 431, 435 (1920).

3 *Pierson v. Post*, 3 Cairns Rep. 175, 2 Am. Dec. 264 (N.Y. Sup. Ct. 1805); *Liesner v. Wanie*, 145 N.W. 374 (Wis. 1914); *Young v. Hichens*, 6 Q.B. 607, 611 (1844).

4 16 U.S.C. §§ 1533–44.

5 16 U.S.C. § 1538.

6 16 U.S.C. § 1532(18).

7 E.g., *United States v. McKittrick*, 142 F.3d 1170 (9th Cir. 1998), cert. denied, 525 U.S. 1072 (1999).

8 16 U.S.C. § 1532(6).

9 16 U.S.C. § 1538(a)(2)(B).

10 *Conservation Council for Hawaii v. Babbitt*, 2 F. Supp. 2d 1280, 1281 (D. Haw. 1998) (noting that critical habitat designations covered only 24 of approximately 700 plant species listed in 1998).

11 27 U.S.T. 1087 (1973), entered into force July 1, 1975.

12 Animal and Plant Health Inspection Act, 7 U.S.C. §§ 150aa-150jj; Plant Quarantine Act, 7 U.S.C. §§ 151–167; 7 C.F.R. parts 319, 340.

13 E.g., Availability of Determination of Nonregulated Status for Genetically Engineered Canola, 59 Fed. Reg. 55,250, 55,250–51 (November 4, 1994) (declining to restrict genetically engineered laurate canola varieties containing "sequences … derived from the plant pathogens *A. tumefaciens* and cauliflower mosaic virus").

14 42 U.S.C. §§ 4321-4370d.

15 *Thomas v. Peterson*, 753 F.2d 754, 764 (9th Cir. 1985).

16 *Hughes River Watershed Conservancy v. Glickman*, 81 F.3d 437, 445 (4th Cir. 1996).

17 *National Parks and Conservation Ass'n v. US Dep't of Transp.*, 222 F.3d 677, 680–81 and n.3 (9th Cir. 2000).
18 Art. 8(h), 31 I.L.M. 818 (1992).
19 *Gibbs v. Babbitt*, 214 F.3d 483, 494 (4th Cir. 2000), cert. denied, 531 U.S. 1145 (2001).
20 43 U.S.C. § 1701(a)(7).
21 43 U.S.C. § 1702(c).
22 *Public Lands Council v. Babbitt*, 529 U.S. 728, 732 (2000).
23 43 U.S.C. §§ 1701–85.
24 *Ibid.* § 1702(c).
25 *Ibid.* § 1702(h).
26 *Ibid.* § 1752(c).
27 *Grazing Administration – Exclusive of Alaska*, 60 Fed. Reg. 9894, 9900–06 (Feb. 22, 1995).
28 *Public Lands Council*, 529 U.S. at 741.
29 43 U.S.C. § 315b.
30 *Natural Resources Defense Council, Inc. v. Hodel*, 618 F. Supp. 848, 871 (E.D. Cal. 1985).
31 43 U.S.C. §§ 1901–08.
32 16 U.S.C. § 1131(c).
33 *Ibid.* § 1131(a).
34 *Ibid.*
35 16 U.S.C. § 1.
36 16 U.S.C. §§ 1531–44.
37 50 C.F.R. § 402.01.
38 16 U.S.C. § 1532(6).
39 *Ibid.* § 1532(20).
40 *In re Polar Bear Endangered Species Act Listing and Section 4(d) Rule Litig.*, 709 F.3d 1, 15 (D.C. Cir.), cert. denied, 134 S. Ct. 310 (2013).
41 *Oregon Natural Res. Council v. Daley*, 6 F. Supp. 2d 1139, 1151 (D. Or. 1998).
42 *Trout Unlimited v. Lohn*, 645 F. Supp. 2d 929, 954 n.18 (D. Or. 2007), aff'd, 559 F.3d 946 (9th Cir. 2009).
43 *Siskiyou Mountains Salamander (Plethodon stormi) and Scott Bar Salamander (Plethodon asupak)*, 73 Fed. Reg. 4380, 4381 (Jan. 24, 2008).
44 *Center for Biological Diversity v. Lubchenco*, 758 F. Supp. 2d 945 (N.D. Cal. 2010).
45 *Twelve-Month Finding on a Petition to List the American Pika as Threatened or Endangered*, 75 Fed. Reg. 6,438 (Feb. 9, 2010).
46 16 U.S.C. § 1533(a).
47 50 C.F.R. § 424.11(b).
48 16 U.S.C. § 1533(b).
49 *Building Indus. Ass'n v. Norton*, 247 F.3d 1241, 1246 (D.C. Cir. 2001).
50 *In re Polar Bear Endangered Species Act Listing and § 4(d) Rule Litig.*, 794 F. Supp. 2d 65, 106 (D.C. Cir. 2011), aff'd, 709 F.3d 1 (D.C. Cir.), cert. denied, 134 S. Ct. 310 (2013); *City of Las Vegas v. Lujan*, 891 F.2d 927, 933 (D.C. Cir. 1989).
51 16 U.S.C. § 1533(a)(3).
52 *Ibid.* § 1532(5)(A)(i).
53 *Ibid.* § 1532(5)(A)(ii).
54 *Ibid.* § 1533(b)(2).
55 *TVA v. Hill*, 437 U.S. 153, 184 (1978).
56 16 U.S.C. § 1533(f).
57 *Fund for Animals v. Babbitt*, 903 F. Supp. 96, 108 (D.D.C. 1995).
58 *Defenders of Wildlife v. Babbitt*, 130 F. Supp. 2d 121, 134 (D.D.C. 2001).
59 *Fund for Animals*, 903 F. Supp. at 111.
60 16 U.S.C. § 1536(a)(2).
61 *Sierra Club v. U.S. Fish and Wildlife Serv.*, 245 F.3d 434, 441 (5th Cir. 2001).
62 50 C.F.R. § 402.14(a); *Bennett v. Spear*, 520 U.S. 154, 158 (1997).
63 16 U.S.C. § 1536(a)(2); 50 C.F.R. § 402.14.
64 50 C.F.R. § 402.14(g); *Pacific Coast Fed'n of Fishermen's Ass'ns v. Gutierrez*, 606 F. Supp. 2d 1122, 1184 (E.D. Cal. 2008), *Natural Resources Defense Council v. Kempthorne*, 506 F. Supp. 2d 322, 374–76 (E.D. Cal. 2007).
65 *Colorado River Cutthroat Trout v. Salazar*, 898 F. Supp. 2d 191, 206–07 (D.D.C. 2012).
66 *Ibid.* at 207.

67 *Alliance for the Wild Rockies v. Lyder*, 728 F. Supp. 2d 1126, 1140 (D. Mont. 2010).
68 *Greater Yellowstone Coal, Inc. v. Servheen*, 665 F.3d 1015, 1026 (9th Cir. 2011).
69 *Kempthorne*, 506 F. Supp. 2d at 331–32 (E.D. Cal. 2007); Ruhl (2008), pp. 45–46.
70 *Marsh v. Oregon Natural Resources Council*, 490 U.S. 360, 377 (1989).
71 *In re Polar Bear Endangered Species Act Listing and Section 4(d) Rule Litig.*, 709 F.3d 1, 3 (D.C. Cir. 2013).
72 16 U.S.C. § 1538(a)(1).
73 16 U.S.C. § 1538(a)(1)(G).
74 16 U.S.C. § 1533(d).
75 50 C.F.R. § 17.31.
76 16 U.S.C. § 1532(19).
77 515 U.S. 687 (1995).
78 50 C.F.R. § 17.3.
79 16 U.S.C. § 1539(a).
80 *Ibid.* § 1536(a)(2); 50 C.F.R. § 402.01(b).
81 *Carson-Truckee Water Conservancy Dist. v. Clark*, 741 F.2d 257, 262 (9th Cir. 1984); *Florida Key Deer v. Stickney*, 864 F. Supp. 1222, 1237–38 (S.D. Fla. 1994).
82 16 U.S.C. § 1535 (authorizing cooperative species conservation agreements between states and the federal government).
83 Natural Communities Conservation Act, Cal. Fish and Game Code § 2800–2840; Tarlock (2002), p. 10,539.
84 515 U.S. at 709 (O'Connor, J., concurring).
85 *Palila v. Hawaii Dep't of Land and Natural Res.*, 852 F.2d 1106 (9th Cir. 1988).
86 *Center for Biological Diversity v. National Marine Fisheries Serv.*, 977 F. Supp. 2d 55 (D.P.R. 2013); 50 C.F.R. pts. 223, 226.
87 *American Lands Alliance v. Norton*, 242 F. Supp. 2d 1, 6 (D.D.C. 2003); 50 C.F.R. pt. 17.
88 *Bearded Seal*, 77 Fed. Reg. 76,740 (Dec. 28, 2012); *Ringed Seal*, 77 Fed. Reg. 76,706 (Dec. 28, 2012).
89 *In re Polar Bear Endangered Species Act Listing and Section 4(d) Rule Litig.*, 709 F.3d 1 (D.C. Cir. 2013), cert. denied, 134 S. Ct. 310 (2013).
90 *Alaska Oil and Gas Ass'n v. Salazar*, 916 F. Supp. 2d 974 (D. Alaska 2013).
91 *In re Polar Bear Endangered Species Act Listing and Section 4(d) Rule Litig.*, 818 F. Supp. 2d 214 (D.D.C. 2011).
92 *Determination of Threatened Status for the Polar Bear (Ursus maritimus) Throughout Its Range*, 73 Fed. Reg. 28,212 (2008).
93 Designation of Critical Habitat for the Polar Bear (*Ursus maritimus*) in the United States, 75 Fed. Reg. 76,085, 76,088 (Dec. 7, 2010).
94 *In re Polar Bear Endangered Species Act Listing and Section 4(d) Rule Litig.*, 709 F.3d 1, 6 (D.C. Cir.), cert. denied, 134 S. Ct. 310 (2013).
95 *Alaska Oil and Gas Ass'n v. Salazar*, 916 F. Supp. 2d 974, 1001–02 (D. Alaska 2013).
96 16 U.S.C. §§ 1373–74.
97 27 U.S.T. 1087 (1973).
98 *In re Polar Bear Endangered Species Act Listing and Section 4(d) Rule Litig.*, 818 F. Supp. 2d 214, 222–23 (D.D.C. 2011), aff'd, 709 F.3d 1 (D.C. Cir.), cert. denied, 134 S. Ct. 310 (2013).
99 16 U.S.C. § 1371(a)(5)(A).
100 *Center for Biological Diversity v. Salazar*, 695 F.3d 893 (9th Cir. 2012); *Center for Biological Diversity v. Kempthorne*, 588 F.3d 701 (9th Cir. 2009); *Incidental Take During Specified Activities*, 76 Fed. Reg. 47,010 (Aug. 3, 2011).
101 16 U.S.C. § 1371(a)(5)(A)(i)(I); *see Kempthorne*, 588 F. 3d at 710–11.
102 16 U.S.C. § 1131(c).
103 *In re Polar Bear Endangered Species Act Listing*, 818 F. Supp. 2d at 234.

References

Alexander, M.A., Kilbourne, K.H. and Nye, J.A. (2014) "Climate Variability During Warm and Cold Phases of the Atlantic Multidecadal Oscillation (AMO) 1871–2008", *Journal of Marine Systems*, Vol. 133, pp. 14–26.

Bean, M.J. (2002) "Overcoming Unintended Consequences of Endangered Species Regulation", *Idaho Law Review*, Vol. 38, pp. 409–20.

Blomquist, R.F. (2000) "Ratification Resisted: Understanding America's Response to the Convention on Biological Diversity, 1989–2002", *Golden Gate University Law Review*, Vol. 32, pp. 493–586.
Cassano, E.N., Cassano, J.J., Higgins, M.E. and Serreze, M.C. (2014) "Atmospheric Impacts of an Arctic Sea Ice Minimum as Seen in the Community Atmosphere Model", *International Journal of Climatology*, Vol. 34, pp. 766–79.
Channell, R. and Lomolino, M.V. (2000) "Dynamic Biogeography and Conservation of Endangered Species", *Nature*, Vol. 403, pp. 84–86.
Chapin III, F.S., Zavaleta, E.S., Eviner, V.T., Naylor, R.L. Vitousek, P.M., Reynolds, H.L., Hooper, D.U., Lavorel, S., Sala, O.E., Hobbie, S.E., Mack, M.C. and Díaz, S. (2000) "Consequences of Changing Biodiversity", *Nature*, Vol. 405, pp. 234–42.
Chen, J.M. (2015) "Αρκτούρος: Protecting Biodiversity Against the Effects of Climate Change Through the Endangered Species Act", *Washington University Journal of Law and Policy*, Vol. 47, pp. 11–27.
Cox, G.W. (1999) *Alien Species in North America and Hawaii: Impacts on Natural Ecosystems*, Island Press, Washington, DC.
Cokinos, C. (2000) *Hope Is the Thing with Feathers: A Personal Chronicle of Vanished Birds*, J.P. Tarcher, New York, NY.
Cummings, B. and Siegel, K.R. (2007) "Ursus maritimus: Polar Bears on Thin Ice", *Natural Resources and Environment*, Vol. 22, pp. 3–7.
Diamond, J.M. (1989) "Quaternary Megafaunal Extinctions: Variations on a Theme by Paganini", *Journal of Archaeological Science*, Vol. 16, pp. 167–75.
Dillard, A. (1974 [this edition 1999]) *Pilgrim at Tinker Creek*, Perennial Classics Edition, Perennial, New York.
Dodson, C.H. and Gentry, A.H. (1991) "Biological Extinction in Western Ecuador", *Annals of the Missouri Botanical Garden*, Vol. 78, pp. 273–95, St. Louis, MO.
Ehrlich, P.R. (1988) "The Loss of Diversity: Causes and Consequences", in E.O. Wilson (ed), *Biodiversity*, National Academy Press, Washington, DC, pp. 21–27.
Elton, C.S. (1958) *Ecology of Invasions by Animals and Plants*, The University of Chicago Press, Chicago, IL.
ESA (1981) *Oversight Hearings Before the Subcommission on Environmental Pollution of the Senate Environment and Public Works Commission.*, 97th Cong., 1st Sess.
Farber, D.A. (1997) "Taking Slippage Seriously", *Harvard Environmental Law Review*, Vol. 23, pp. 297–324.
Freeman, J. (2000) "The Contracting State", *Florida State University Law Review*, Vol. 28, pp. 155–213.
Glowka, L. (2000) "Bioprospecting, Alien Invasive Species, and Hydrothermal Vents: Three Emerging Legal Issues in the Conservation and Sustainable Use of Biodiversity", 13 *Tulane Environmental Law Journal* 329–60.
Goldschmidt, T. (1996) *Darwin's Dreampond: Drama in Lake Victoria*, The MIT Press, Cambridge, MA.
Hansen, E. (2000) *Orchid Fever: A Horticultural Tale of Love, Lust, and Lunacy*, Vintage, New York, NY.
Harper, D. (2016) Online Etymology Dictionary – Definitions of *arctic* and *Arcturus*, viewed November 11, 2016, from www.etymonline.com/index.php?term=arctic and www.etymonline.com/index.php?term=arcturus.
IPCC (2014) *Intergovernmental Panel on Climate Change, Climate Change 2013: The Physical Science Basis*, viewed November 11, 2016, from http://climatechange2013.org/images/uploads/WGIAR5_WGI-12Doc2b_FinalDraft_All.pdf.
Karkkainen, B.C. (1997) "Biodiversity and Land", *Cornell Law Review*, Vol. 83, pp. 1–104.
Karkkainen, B.C. (2002) "Collaborative Ecosystem Governance: Scale, Complexity, and Dynamism", *Virginia Environmental Law Journal*, Vol. 21, pp. 189–243.
Karkkainen, B.C. (2003) "Adaptive Ecosystem Management and Regulatory Penalty Defaults: Toward a Bounded Pragmatism", *Minnesota Law Review*, Vol. 87, pp. 943–98.
Kaye, R. (2006) *Last Great Wilderness: The Campaign to Establish the Arctic National Wildlife Refuge*, University of Alaska Press, Fairbanks, AK.
Kolbert, E. (2014) *The Sixth Extinction: An Unnatural History*, Henry Holt and Company, New York, NY.
Kunich, J.C. (2001) "Preserving the Womb of the Unknown Species with Hotspots Legislation", *Hastings Law Journal*, Vol. 52, pp. 1149–1253.
MacArthur, R. H. and Wilson, E. O. (1967) *The Theory of Island Biogeography (MPB-1)*, Princeton University Press, Princeton, NJ.
Matson, P.A., Parton, W.J., Power, A.G. and Swift, M.J. (1997) "Agricultural Intensification and Ecosystem Properties", *Science*, Vol. 275, pp. 504–09.
McCann, K.S. (2000) "The Diversity-Stability Debate", *Nature*, Vol. 405, pp. 228–33.
National Research Council (1995) *Science and the Endangered Species Act.*

Parmesan, C. (2006) "Ecological and Evolutionary Response to Recent Climate Change", *Annual Review of Ecology, Evolution and Systematics*, Vol. 37, pp. 637–69.
Parmesan, C. and Yohe, G. (2003) "A Globally Coherent Fingerprint of Climate Change Impacts Across Natural Systems", *Nature*, Vol. 42, pp. 37–42.
Rolston III, H. (1990) "Property Rights and Endangered Species", *University of Colorado Law Review*, Vol. 61, pp. 283–306.
Ruhl, J.B. (2008) "Climate Change and the Endangered Species Act: Building Bridges to the No-Analog Future", *Boston University Law Review*, Vol. 88, pp. 1–62.
Schorger, A.W. (1955) *The Passenger Pigeon: Its Natural History and Extinction*, University of Oklahoma Press, Norman, OK.
Sellars, R.W. (1997) *Preserving Nature in the National Parks: A History*, Yale University Press, New Haven, CT.
Simberloff, D. (1976) "Experimental Zoogeography of Islands: Effects of Island Size", *Ecology*, Vol. 57, pp. 629–48.
Tarlock, A.D. (2002) "Biodiversity Conservation in the United States: A Case Study in Incompleteness and Indirection", *The Environmental Law Reporter*, Vol. 32, pp. 10529–42.
Verissimo, D., MacMillan, D.C. and Smith, R.J. (2011) "Toward a Systematic Approach for Identifying Conservation Flagships", *Conservation Letters*, Vol. 4, pp. 1–8.
White, G.F. (1988) "Environmental Effects of the High Dam at Aswan", *Environment*, Vol. 30, no. 7, pp. 5–40.
White, R. (1991) *It's Your Misfortune and None of My Own: A History of the American West*, University of Oklahoma Press, Norman, OK.
Whitehead, D.R. and Jones, E.J. (1969) "Small Islands and the Equilibrium Theory of Island Biogeography", *Evolution*, Vol. 23, pp. 171–79.
Williamson, M. (1996) *Biological Invasions*, Chapman and Hall, London, UK.
Wilson, E.O. (1992) *The Diversity of Life*, Belknap Press, Cambridge, MA.
Wilson, E.O. (2002) *The Future of Life*, Vintage, New York, NY.
Wu, J., Huang, J., Han, X., Xie, Z. and Gao, X. (2003) "Three-Gorges Dam: Experiment in Habitat Fragmentation?", *Science*, Vol. 300, pp. 1239–40.

9
CHINA'S BIODIVERSITY LAW

John Copeland Nagle

China offers the best and the worst of biodiversity protection. China is a vast, varied nation that hosts an incredible range of ecosystems and species. "China's biodiversity ranks eighth in the world and first in the northern hemisphere" (China's Agenda 21, p. 1). Over 100,000 species of animals and nearly 33,000 plant species exist in 460 different types of ecosystems. Those ecosystems include forests, grasslands, deserts, wetlands, seas and coastal areas, and agricultural ecosystems. China hosts 212 different types of bamboo forests alone. China also has an unusual number of ancient and relic species because of its protection from historic geologic events, such as the movement of glaciers. Most famously, it is the only home of the giant panda, the symbol of many efforts to protect biodiversity throughout the world today. Such species and ecosystem diversity is complemented by an unsurpassed collection of genetic diversity. "The richness of China's cultivated plants and domestic animals [is] incomparable in the world. Not only did many plants and animals on which human survival depend originate in China, but it also retains large numbers of their wild prototypes and relatives" (Development Gateway). A 2005 report estimated that China's biodiversity is valued at nearly five hundred billion dollars (Maxey and Lutz, 1994).

China is also home to more than 1.35 billion people. Not surprisingly, it is difficult to generalize about China. Like *developing* countries, China is poor. Its per capita income remains in the bottom half of the world. Of the 1.37 billion people in the world who live on less than \$1.25 per day, 208 million live in China. Much of China's population lives as if it is a developing country. This is true both in the countryside, where the rural peasants often live in the same way that their ancestors did generations ago, and in the cities, where the unprecedented migration of people from the countryside to the cities in search of better economic opportunities has overwhelmed the ability of the cities to provide for them. China relies upon such evidence when it describes itself as "a low-income developing country" (The People's Republic of China, 2004, p. 1).

But like *developed* countries, China has one of the leading economies in the world. It is the world's leading producer of steel, producing four times as much as the United States. It produces nearly three times as much coal as the United States. It produces half of the world's cement and aluminum. It has the second largest gross domestic product in the world, trailing only the United States. It imports more oil than every country except the United States. These and other statistics are frequently cited in the many popular books reporting on China's ascent to economic powerhouse.

Threats to China's biodiversity

The rapid economic growth that China has experienced since 1980 strains the nation's ability to preserve ecosystems, species, and genetic resources. But the biodiversity of China has encountered countless threats for thousands of years, including the cultivation of more and more land for agriculture, and the consequences of numerous wars. During the Great Leap Forward of 1958 to 1960, Mao Zedong targeted the "Four Pests": rats, sparrows, flies, and mosquitoes. The attack on sparrows enlisted schoolchildren to knock down nests and to beat gongs so that the sparrows could not find a place to rest. Only after sparrows were virtually eliminated throughout China did the country's leaders recognize the value of the birds in controlling insects. China faces many of the same threats to biodiversity as in other countries, with the notable addition of the country's notorious air pollution. Habitat loss is the biggest threat to biodiversity in China. As in many other countries, rapid economic development and continued population growth exert relentless pressure on previously undeveloped areas of habitat to a diversity of wildlife and plants. The overgrazing of rangelands, erosion, and the adverse effects of tourism and mining further compromise the condition of ecosystems and species throughout China.

Forests have suffered an especially devastating toll throughout China. Mark Elvin describes "[t]he destruction of the old-growth forests that once covered the greater part of China" as "the longest story in China's environmental history" (Elvin, 2004, p. 23). The story unfolded because "the original core of classical Chinese culture was hostile to forests, and saw their removal as the precondition for the creation of a civilized world" (Elvin, 2004, p. 12). Trees were cut for fuel, to provide building materials, and to remove obstacles to farms and other human projects. But the disappearance of the forests caused other, albeit predictable, problems. Deforestation increased erosion, which resulted in huge amounts of sediment collecting along the coasts and the sides of lakes and rivers. Wood became scarce as early as 600 B.C. in some parts of the country. By the nineteenth century, a writer lamented that "[t]hese days, people have used their axes to deforest the mountains" (Elvin, 2004, p. 78). During the twentieth century, China encouraged the wholesale destruction of forests for their timber – which was the country's primary fuel until coal replaced it – or simply the removal of trees to facilitate agricultural crops. Trees were cut indiscriminately in a planned effort to generate revenue for local education, health, and infrastructure needs. As one villager remembered:

> When I was a child, there were jackals and foxes in the woods, but after the big trees were cut to fuel furnaces during the [Great Leap Forward], there wasn't even a rabbit. New trees grew, but then it was time to "learn from Dazhai." In fact, we didn't need terraces in our area, because the population was sparse. But our per-*mu* production was considered low. So we had to cut the trees. Whoever cut the most got the most political points, and the most grain.
>
> (Shapiro, 2001, p. 109)

Fires and pests further degraded forest ecosystems. The result was that forest cover in the lush provinces of southwest China declined from 30 per cent of the land in 1950 to 13 per cent by 1999, though it has since rebounded to almost 22 per cent. The loss of forests caused deadly flooding along the Yangtze River and devastated the natural ecosystems and the species within them. Tigers, for example, "stalk their prey from the cover and the shadows provided by forests. The relationship is pretty simple: no forests, no tigers" (Marks, 1998, p. 323). Forests continue to disappear at an alarming rate, with the remaining forests often broken into smaller, fragmented areas.

Other types of ecosystems confront similar threats. Overgrazing, farming, and plagues of rodents have caused the grassland steppes that account for one-third of China's total area to lose up to half of their grass yields in the past twenty years. Over seven million hectares of wetlands were reclaimed during the past thirty years. Once known as a "province of thousand lakes," Hubei Province now has only 326 lakes and rivers left. Lime mining and handicraft production by local residents have damaged 80 per cent of the coral reefs along the coast of Hainan Island. The overall result is that "continued destruction and deterioration of ecosystems has now become one of the most serious environmental problems in China" (Maxey and Lutz, 1994, p. 10). Furthermore, invasive species have begun to exact a heavy toll on China's biodiversity as well.

China's notorious pollution affects many of the country's ecosystems. China routinely has multiple cities in the lists of the world's most polluted cities, and air pollution damages croplands, fisheries, and other ecosystems. China's fisheries suffered $130 million in losses from 941 water pollution incidents in 2004 that affected 211,000 hectacres of freshwater ecosystems. A November 2005 factory explosion that polluted the Songhua River required the temporary termination of water supplies in the northwestern city of Harbin and had untold consequences for the freshwater ecosystem. The quantity of water is often a problem for biodiversity as well. Efforts to move freshwater to places where it is scarce, such as Beijing, include such controversial projects as the Three Gorges Dam in central China, which destroyed many of the nearby ecosystems. Further south, the damming of the Mekong River has wreaked ecological havoc there as well.

Biodiversity is also threatened by the direct exploitation of many species. China is the world's largest exporter, and a leading user of endangered species. Enforcement becomes even more difficult because of the huge demand for products derived from endangered species. Traditional Chinese medicine uses tiger bones (for arthritis and rheumatism), rhino horns (for fevers), and bear gall bladders. Nearly every tiger part is used as a tonic, an aphrodisiac, gourmet delicacies or some other purpose. Chinese pharmaceutical factories use 1,400 pounds of rhino horns annually, the product of about 650 rhinos. Panda pelts sell for as much as $10,000, tiger bones are priced at $500 per pound, and a rhino horn can earn as much as $45,000. Villagers can earn ten years' income from one tiger (Sharma, 2005, p. 215).

These pressures are evidenced in the placement of three native Chinese species among the World Wild Fund for Nature's list of the top ten most endangered species in the world. The giant panda is the most famous of those three species. Only one thousand pandas are left in the wild, and their numbers are still declining, albeit at a reduced rate. The threats to their survival include the loss of bamboo and habitat, a relatively small number of young pandas, genetic inbreeding, inability to survive in captivity, poaching, and the earthquake that devastated Sichuan Province in April 2008. The second species – the black rhinoceros – has suffered a 95 per cent drop in population since 1970, so that only two thousand are alive today. The third species – the Indo-Chinese tiger – is the most endangered. Perhaps only fifty tigers survive in the wild in China, and with two of the four native Chinese tiger species already extinct, many fear that this tiger could disappear by the end of the century. The disappearance of native species is obvious in other ways as well. The town of "Wild Yak Gully now has no wild yaks; Wild Horse Sands, no wild horses" (Maxey and Lutz, 1994, p. 15), and the town of Moose and the town of Gazelle have no moose or gazelles. Other notable Chinese species that are endangered include the Yangtze alligator, the crested ibis, and certain Mongolian horses.

China's nature reserves

China's primary response to the threat to its biodiversity has been the creation of nature reserves. The Dinghushan National Natural Reserve was the first such reserve, established in 1956 in Guangdong Province to protect the subtropical evergreen forests and accompanying rare plants and animals. By 2013, 2,697 reserves covered 14.8% of China's land (The Ministry of Environmental Protection of China, 2014, p. 1). More than a dozen of those reserves were for pandas, and the population of pandas in the wild increased from 1,114 in 2000 to 1,596 in 2005. Another reserve covers 45,000 square kilometers and protects sixty endangered animals and 300 rare plants. Forest ecosystems are well represented in the nature reserves. Wetland and coastal ecosystems have been included in reserves since the 1970s, while the creation of reserves for grassland and desert ecosystems is a new priority for the government. In Inner Mongolia, the creation of 184 reserves has been critical as countervailing pressures have seen an additional two million hectares of newly cultivated grasslands, a dramatic increase in coal production, and a doubling of livestock grazing on the grasslands (Ma, 2016, p. 1).

Nature reserves, however, do not solve all of the problems faced by China's biodiversity. Consider the Zhalong Nature Reserve in northeastern China's Heilonjiang Province, which is home to nine of the fifteen species of cranes in the world. In recent years it has suffered from a severe drought, extensive fires, and housing developments built within its borders, which now provides habitat for 60,000 people as well as for thousands of cranes. Droughts have reduced the wetlands from 36,000 hectares to less than 6,000 hectares, and the government worries that the area could become a "sea of sand" if conditions are not reversed. Another wetland reserve in northern China was seriously polluted by oil that leaked from a passenger airplane crash in 2004. Most reserves are simply no-hunting zones, rather than affirmative wildlife management areas. For example, over 15,000 people live in ninety villages within Xishuangbanna Nature Reserve in southwestern Yunnan Province, where "they engage in agriculture, forestry, animal production, fisheries, and small-scale retailing and commercial activities" (Tisdell, 1999, pp. 147–48). More generally,

> Some engineering projects go on even in the core areas of nature reserves. In other reserves or scenic spots, tourism is promoted to develop the local economy, and while tourism can assist conservation when it is carried out properly, the prospects for quick profits may lead to abuses of the natural systems and species which the reserves protect.
>
> (Maxey and Lutz, 1994, p. 21)

Additionally, "illegal hunting and poaching of endangered animal and plant species occurs frequently" in reserves (Maxey and Lutz, p. 21). There is no general law regulating the operation of nature reserves. Management difficulties and inadequate funding also threaten many reserves. Reserve administrators and employees are often untrained to protect the species in their care. Most reserves do not even possess a list of species that live there.

The Chinese government is aware of these shortcomings, though, and it has charted an ambitious program to improve the effectiveness of nature reserves in protecting the country's biodiversity. Proposed actions include restrictions on free access to sensitive reserves, better pay and living conditions for reserve personnel (including allowances for families to live in nearby cities), efforts to "improve relations with local people and find ways for them to make a living without depleting the natural resources," and the establishment of new nature reserves "in regions with urgent need of biodiversity conservation," such as the coral reefs of Dongshan Island and seven proposed reserves to conserve wild rice, soybeans, and other agricultural crops

(Maxey and Lutz, 1994, pp. 36–40). Likewise, in 2005, Sichuan Province "closed 78 mines and polluting companies in the giant panda's habitat to provide a better home for the endangered species" (Panda's Home Reducing Pollution, 2005).

The nature reserves are joined by zoos, botanical gardens, and scientific study institutes. China's twenty-eight zoological gardens and 143 zoological exhibition sites contain more than 600 species of animals. Over 13,000 species of plants are contained in more than 100 botanical gardens. Over one thousand scientists work together through the Chinese Research Network of Ecosystems to study and monitor ecosystem diversity. Genetic diversity is protected by "the world's largest resource bank of different varieties of crops, a number of gene and cell banks and 25 germ-plasm nurseries, which hold a total of 350 thousand specimens of germ-plasm for various species of trees and crops" (China's Agenda 21, 1994, p. 173).

China's environmental education campaigns

Educational campaigns serve as another primary feature of China's efforts to protect its biodiversity. China has traditionally relied on exhortational campaigns to change people's conduct. China's biodiversity conservation action plan begins with an emphasis on the need "[t]o enhance the nation's awareness of the critical importance of our biodiversity and its conservation is our urgent task of the highest priority" (China's Agenda 21, 1994 p. ii). Such an educational focus appears in China's Agenda 21 plan, which calls for media teaching about biodiversity, the promotion of public events such as Earth Day and Bird Loving Week, and the use of a traveling panda exhibition. China also held a National Program for Environmental Education and Publicity that drew upon the resources of such organizations as the government's environmental departments, the Ministry of Broadcasting and Television, and the Chinese Communist Youth League. One program to protect the 5,000 remaining *grus nigricollis* – a rare type of crane – was designed to "make the youth conscious of animal protection before they become poachers" (Nagle, 2007, p. 444). "Such efforts have helped convince 99% of the Chinese people that environmental pollution and ecological destruction are at least 'fairly serious' issues" (Nagle, 2007, p. 444). In particular, anyone who harms a panda must face "the censure of an angry public" (Nagle, 2007, p. 444).

Yet all agree that more environmental education needs to be done. The greatest problem exists in rural areas, where people ask why wild animals can no longer survive on their own, and where menus proclaiming "Rare Wild Animals Are Served" still appear in restaurants and hotels (Nagle, 2007, p. 445). Demand for the products of endangered species remains high. Years of teaching traditional Chinese medicine and delicacies is hard to reverse. How do you convince a billion people to take aspirin instead of rhino horn pills? "Many Chinese still believe that wildlife species are endowed with magical powers capable of curing a myriad of ills, and are angered by pressure from countries such as the United States to ban the sale of endangered species" (Chow, 1995, p. 299). Likewise, many still see tigers as pests, just as many ranchers fear the introduction of wolves and bears into the western United States. More generally, "[b]iodiversity conservation is a new technical term for many officials in the governments at all levels and for citizens who are lacking basic knowledge on biodiversity conservation" (Maxey and Lutz, 1994, p. 333).

The biodiversity conservation action plan reveals a keen understanding of the importance of gaining public support for the task at hand:

> In general, people want government policies that do not require them to change their lifestyles, provide material benefits and development, and provide benefits today that

> will be paid for later. Politics to conserve biodiversity would be the opposite, requiring fundamental changes in people's relationship with the environment, restricting access to resources, foregoing material benefits, and paying today for abstract future benefits. Unless the public is convinced of the value of conserving biodiversity, and the government changes its policies accordingly, the chance of saving biodiversity is small.
>
> (Maxey and Lutz, 1994, p. 360)

Thus the Chinese government seeks to help the media better publicize the importance of biodiversity conservation (Maxey and Lutz, 1994, p. 333), "[w]ork with local theater groups to write and perform plays with a biodiversity message" (Maxey and Lutz, 1994, p. 333), and teach students of all ages about biodiversity in the nation's schools (Maxey and Lutz, 1994, p. 333).

China's biodiversity regulation

Neither China's emphasis on nature reserves nor its use of educational campaigns actually regulates any conduct that threatens biodiversity. The development of Chinese wildlife law mirrors the development of Chinese environmental law (and indeed Chinese law) generally. Interest in the environment and interest in law both lagged until the 1970s, so not surprisingly, there was little Chinese environmental law. The People's Congress approved the Law on Environmental Protection – the first general Chinese environmental statute – in 1978. Article 15 of that law prohibits hunting and exploitation of rare wildlife. Then, in 1982, several provisions regarding environmental protection were added to China's constitution. Article 9 provides for state ownership of natural resources, ensures state protection of natural resources, and prohibits appropriation or damage of natural resources. Article 26 provides that "the State protects and improves the living environment and the ecological environment, prevents and remedies pollution and other public hazards." By 1994, China had enacted twelve national statutes, twenty national administrative regulations, over six hundred local laws and regulations, and three hundred other norms regulating the environment.

Chinese biodiversity law has developed in much the same fashion. To be sure, China's long history contains numerous examples of the law being used to protect the country's biodiversity. An edict issued in 336 A.D. stated that "[t]o take possession of the mountains, or to put the marshes under one's personal protection is tantamount to robbery with violence" (Elvin, 2004, p. 550). The *Respectfully Determined Laws and Precedents of the Great Qing* prescribed that anyone who "thievishly cuts down the trunks of trees, removes soil or stones, opens kilns for charcoal ... or starts fires to burn the mountains for short-term farming, *he shall be beheaded* as if he had stolen imperial vessels used for sacrifices to the gods" (Elvin, 2004, p. 294). Today, the Forestry Law prohibits the hunting of animals in protected areas. The Water Law provides that the government "shall protect water resources and adopt effective measures to preserve natural flora, plant trees and grow grass, conserve water sources, prevent and control soil erosion and improve the ecological environment." The Grassland Law directs the government to protect grassland ecosystems, vegetation, and rare plants, and it prohibits harmful reclamation and construction activities.

One law seeks to abate the transformation of once-fertile grassland ecosystems into lifeless deserts. Nomadic herders have lived in the grasslands of what is now Inner Mongolia for countless generations, but the 1950s brought a wave of Chinese immigrants adding more livestock and seeking to cultivate the naturally arid land bordering the Gobi Desert. Today, expanding desertification claims 2,500 square kilometers at a cost of $6.5 billion to China's economy each year. The effects of the dust have been seen as far away as Colorado, where particulate concentrations rose above permissible levels in April 2002 after the jet stream carried dust all the way from China. In March 2002, another dust storm dumped 30,000 tons of dirt on Beijing, even as billboards around the city trumpeted the "Green Olympics" to be held there in 2008. The resulting international publicity prompted local television newscasters to affirm the government's resolve to "outwit" the dust storms. The first law to try to match wits with the dust was enacted by the National People's Congress (NPC) in August 2001. The law against desertification:

- states that land occupants have a duty not only to prevent desertification but also to restore areas that have already become desert;
- promises unspecified preferential policies, tax breaks, subsidies, and technical support to offset the cost of this unfunded mandate;
- creates a new class of protected areas off-limits to development and calls for farmers and herders to be removed from those areas; and
- authorizes local governments to grant land-use rights of up to 70 years to desertified areas if the landholder promises to undertake restoration efforts.

As Qu Geping, the chair of the NPC Environment and Resources Committee, explained, the anti-desertification law was designed to prevent the frequent dust storms that have sounded "a warning bell from nature."

Endangered wildlife is also protected by Chinese law. The Ministry of Forestry established the first list of Rare and Precious Species of China in 1969. In 1988, the National People's Congress enacted the Wild Animal Conservation Act (WACA), which "charges the state to ensure the protection of wild animals and their habitats, organize regular field surveys of wildlife resources, and to improve ecological impact assessment for construction projects" (Sharma, 2005, pp. 226–28). Regulations promulgated pursuant to the WACA prohibit hunting, fishing, and collecting of key wild species.

The existence of such laws is one thing; their actual implementation is another. To be sure, there are examples of very stringent enforcement of wildlife laws in China. The government has imposed the death penalty for killing endangered pandas (Sharma, 2005, pp. 240–41). In 1995, nineteen hotels and restaurants on Hainan Island were closed and fined $34,000 for serving bear's paw, monkey brains, and other wildlife. China has promised to step up such efforts to punish those who kill endangered species for financial gain. China has also acted to prohibit patented medicines from containing ingredients taken from endangered species. A fishing ban on the Xiaolangdi Reservoir in central China soon resulted in the rediscovery of the copper cyprinid, a species that had been thought to be extinct. But the Chinese government admits to its failure to adequately enforce the existing laws protecting biodiversity.

> While many laws and regulations intended to protect biodiversity exist, in practice they are often not enforced or enforced strictly, or when the violators are apprehended, the court system treats them very leniently. As a result, illegal hunting and collection of endangered animal and plant species is very widespread, and disputes arise continuously between management of nature reserves and local residents, hindering biodiversity conservation efforts.
>
> (Maxey and Lutz, 1994, p. 32)

Alex Wang described the enforcement of China's environmental protection laws as "extremely weak" (Wang, 2007, p. 203). Wang Canfa, the director of the Center for Legal Assistance to

Pollution Victims (CLAPV), blames the failure to consider enforcement issues when legislation is drafted, the inability to promulgate regulations to implement statutes, the tendency of local governments to "pursue economic benefits while overlooking environmental protection," and the failure to consider public opinion (Canfa, 2007a, pp. 386–87). Jerome Cohen, the dean of America's Chinese law scholars, adds that "even in China, the central government's writ does not run very far. It doesn't have the financial resources because of an inadequate tax system" (Cohen, 2007, p. 402). Corruption is another major impediment to the implementation of the rule of law in China.

Non-governmental organizations (NGOs) have played a growing role in China's efforts to preserve biodiversity. The Nature Conservancy is active in Yunnan Province, which hosts abundant biodiversity along the borders with Vietnam, Myanmar, and Tibet. One of the organization's projects supports ecotourism, operates a community conservation development fund, and established a comprehensive fisheries management plan in the Lashi Lake watershed that serves as habitat for the endangered black-necked crane. Other projects target ecosystems that are home to snow leopards, the Yunnan golden monkey, Asiatic black bears, red pandas, and thousands of acres of forests and alpine ecosystems. "In collaboration with the State Environmental Protection Agency (SEPA), the State Forestry Administration (SFA), and the Chinese Academy of Sciences," The Nature Conservancy has been active in supporting biodiversity protection in Yunnan Province, and it is advising and assisting the Chinese government as it revises its national biodiversity conservation action plan. Even so, "China's leaders … have been careful to circumscribe both the number of NGOs and the scope of their activities, so the role that such groups will be able to play in preserving the country's biodiversity remains uncertain" (Economy, 2004, p. 130).

China's role in international biodiversity protection

A final part of China's biodiversity strategy is its active participation in international efforts to protect biodiversity. In 1980, China joined the Convention on International Trade in Endangered Species (CITES). In 1992, it signed the Ramsar Convention for the protection of wetlands. That year also saw China become one of the first nations to ratify the Convention on Biological Diversity that was negotiated in Rio de Janeiro. China then launched a "China Biodiversity Conservation Plan" in 1994, and it discussed the measures needed to protect biodiversity in its white paper documenting China's efforts to further its Agenda 21 environmental commitments. The Agenda 21 strategy states that "[t]he policy for biodiversity conservation in China is 'laying equal stress on both the development and utilization and the conservation and protection of natural resources' and 'he who develops, conserves; he who utilizes, compensates; he who destroys, restores'" (China's Agenda 21, 1994, pp. 171–72).

But critics question China's resolve to end its trade in endangered species. China resisted international calls for the destruction of existing rhino horn stocks. It declined to become a member of the Global Tiger Forum established by twelve Asian countries in 1994 to protect endangered tigers throughout Asia. It advanced a proposal that would create a farm to raise tigers in order to satisfy the demand for tiger parts, though that idea was withdrawn after environmentalists objected. China's limited efforts to stop that trade have subjected it to international criticism. For example, in 1993 the United States and other countries threatened to sanction China for failing to control the trade in tiger and rhino parts. That the United States decided not to penalize China was viewed as an exercise in diplomacy unrelated to China's actual progress in enforcing the treaty. China's efforts to protect its ecosystems suffer from similar limitations on resources and political will. As one observer writes, China's solid national biodiversity policy "has made very little difference to the peoples of southwest China, where many of the reserves lack staff, funds, infrastructure, or a management plan. The international conservation community has focused on the panda at the expense of other endangered species" (Studley, 1999, p. 30).

The ultimate success of these measures remains uncertain. China's State Council has admitted that "[t]he environmental situation remains extremely grim" (Wang, 2007, p. 201). Many scholars agree. Yet the attention that China receives, due both to its economic prowess and its remarkable biodiversity, ensures that China's natural heritage will not disappear quietly.

Conclusion

China understands the challenges that it faces, and it has stated ambitious official goals for meeting them. According to its 2014 report on its implementation of the Convention on Biological Diversity, China's short-term goal was to effectively contain "the trend of biodiversity decline in key regions" by 2015; its mid-term goal is for "biodiversity decline and loss [to] be basically controlled" by 2020; and its long-term is that "biodiversity will be effectively protected" by 2030. Those specific targets will enable the world to see whether China succeeds in preserving its rich biodiversity for future generations.

Acknowledgement

Much of this discussion expands and updates what I previously wrote in John Copeland Nagle, "The Effectiveness of Biodiversity Law", 24 *J. Land Use and Envtl. L.* 203 (2009); and John Copeland Nagle, J.B. Ruhl and Kalyani Robbins, *The Law of Biodiversity and Ecosystem Management* (3rd edn., 2012, Taylor & Francis).

Bibliography

Bergsten, C., Freeman, C., Lardy, N., and Mitchell, D.J. (2008), *China's Rise: Challenges and Opportunities*, Peter G. Institute for International Economics.

Canfa, W. (2007a) "Special Functions of Promoting Public Participation in Environmental Protection in Aiding Pollution Victims", *Vt. J. Envtl. L.*, Vol. 8, p. 379.

Canfa, W. (2007b) "Chinese Environmental Law Enforcement: Current Deficiencies and Suggested Reforms", *Vt. J. Envtl. L.*, Vol. 8, p. 159.

Chinese Government (1994) *China's Agenda 21: White Paper on China's Population, Environment, and Development in the 21st Century*, China Environmental Science Press.

Chow, D. (1995) "Recognizing the Environmental Costs of the Recognition Problem: The Advantages of Taiwan's Direct Participation in International Environmental Law Treaties", *Stan. Envtl. L.J.*, Vol. 14, p. 256.

Chunquan, Z., Taylor, R., and Guoqiang, F. (2004) *China's Wood Market, Trade and the Environment*, Science Press USA, Inc.,

Cohen, J. (2007) "An Introduction to Law in China", *Vt. J. Envtl. L.*, Vol. 8, p. 379.

Démurger, S., Fournier, M., and Shen, G. (2005) "Forest Conservation Policies and Rural Livelihood in North Sichuan Tibetan Areas" (unpublished manuscript available at http://papers.ssrn.com/sol3/papers.cfm?abstract_id=876870).

Development Gateway: *The Richness and Uniqueness of China's Biodiversity*, http://en.chinagate.cn/english/2029.htm.

"Draught Causes China's Wetland Nature Reserve to Shrink" (2005) Xinhua New Agency, www.redorbit.com/news/science/203155/drought_causes_chinas_wetland_nature_reserve_to_shrink/index.html.

Economy, E. (2004) *The River Runs Black: The Environmental Challenge to China's Future*, Cornell University Press.
Elvin, M. (2004) *The Retreat of the Elephants*, Yale University Press.
Fishman, T. (2006) *China, INC.: How the Rise of the Next Superpower Challenges America and the World*, Scribner.
Gifford, R. (2008) *A Journey into the Future of a Rising Power*, Random House.
Kynge, J. (2007) *China Shakes the World: A Titan's Rise and Troubled Future – And the Challenge for America*, Mariner Books.
Ma, W. (2016) Status of Nature Reserves in Inner Mongolia, *Sustainability*, Vol. 8.
Marks, R. (1998) *Tigers, Rice, Silk, and Silt: Environment and Economy in Late Imperial South China*, Cambridge University Press.
Maxey, C. and Lutz, J. (eds.) (1994) *Biodiversity Conservation Action Plan for China*, National Environmental Protection Agency.
Ming-hai, Z. and Shuang-ling, W. (2004) "Co-Management: Transformation of Community Affair Model in Chinese Nature Reserves", *J. Forestry Research*, Vol. 15, p. 313.
Nagle, J. (2007) "Why Chinese Wildlife Disappears as CITES Spreads", *Geo. Int'l Envtl. L. Rev.*, Vol. 9, p. 435.
Nagle, J. (2009a) "Discounting China's CDM Dams", *Loy. U. Chi. Int'l L. Rev.*, Vol. 7, p. 9.
Nagle, J. (2009b) "The Effectiveness of Biodiversity Law", *J. Land Use and Envtl. L.*, Vol. 24, p. 203.
Nagle, J., Ruhl, J.B., and Robbins, K. (2012) *The Law of Biodiversity and Ecosystem Management*, 3rd edn., Foundation Press, St. Paul.
NE China Reserve Sees Record Number of Migrating Cranes (2007) *People's Daily Online*, http://english.people.com.cn/90001/90782/6292613.html.
"Panda's Home Reducing Pollution" (2005) *Xinhua News Agency*, www.china.org.cn/english/2005/May/12855htm.
Quan, L. (2004) "Research on Spatial-Temporal Evolution of Wetland Water Resource in Zhalong Nature Reserve", *IEEE Int'l*, Vol. 7, p. 4686.
"Red-crested Crane Habitat Flame in NE China" (2005) *People's Daily Online*, http://english.peopledaily.com.cn/200503/24/eng20050324_17800html.
Shapiro, J. (2001) *Mao's War Against Nature: Politics and the Environment in Revolutionary China*, Cambridge University Press.
Sharma, S. (2005) "Chinese Endangered Species at the Brink of Extinction: A Critical Look at the Current Law and Policy in China", *Animal L.*, Vol. 11, p. 215.
Shenkar, O. (2006) *The Chinese Century: The Rising Chinese Economy and Its Impact on the Global Economy, The Balance of Power, And Your Job*, Wharton School Publishing.
Siberian Crane Wetland Project: Zhalong National Nature Reserve, www.scwp.info/china/zhalong.shtml (last visited May 17, 2009).
Studley, J. (1999) "Environmental Degradation in SW China", *P.R.C. Rev.*, Vol. 12, p. 28.
The Forest Law of the People's Republic of China (promulgated by the Standing Comm. Nat'l People's Cong., Sept. 20, 1984), art. 25, LawInfoChina (last visited May 17, 2009) (P.R.C.).
The Law on Combating Desertification Prevention in the People's Republic of China (2001), available at www.adb.org/Projects/PRC_GEF_Partnership/Desertification.pdf.
The Ministry of Environmental Protection of China (2014), China's Fifth National Report on the Implementation of the Convention on Biological Diversity.
The Ministry of Environmental Protection of China, China's Fifth National Report on the Implementation of the Convention on Biological Diversity (Mar. 2014).
The Nature Conservancy, China: How We Work, available at www.nature.org/wherewework/asiapacific/china/strategies/.
The People's Republic of China (2004) Initial National Communication on Climate Change: Executive Summary, www.ccchina.gov.cn/file/en_source/da/da200411090pdf.
Tisdell, C. (1999) *Biodiversity, Conservation and Sustainable Development: Principles and Practices with Asian Examples*, Edward Elgar Publishing.
Wang, A. (2007) "The Role of Law in Environmental Protection in China: Recent Developments", *Vt. J. Envtl. L.*, Vol. 8, p. 195.
Water Law of the People's Republic of China (promulgated by the Standing Comm. Nat'l People's Cong., Aug. 29, 2002, effective Oct. 1, 2002), art. 9, LawInfoChina (last visited May 17, 2009) (P.R.C.).

Wood, A. *et al.* (eds.) (2000) *The Root Causes of Biodiversity Loss*, Routledge.
Xu, H. *et al.* (2005), "Design of Nature Reserve System for Red-Crowned Crane in China", *Biodiversity and Conservation*, Vol. 14, p. 2275.
Zhao, Y. (2006) "The War Against Biotic Invasion – A New Challenge Of Biodiversity Conservation For China", *UCLA J. Envtl. L. and Pol'y*, Vol. 24, p. 459.
Ziegler, E. (2006) "China's Cities, Globalization, and Sustainable Development: Comparative Thoughts on Urban Planning, Energy, and Environmental Policy", *Wash. U. Global Studies L. Rev.*, Vol. 5, p. 295.

10

THE INTERNATIONAL TREATY ON PLANT GENETIC RESOURCES FOR FOOD AND AGRICULTURE

Toward the realization of farmers' rights as a means of protecting and enhancing crop genetic diversity

Regine Andersen

Introduction

Over the last ten millennia, farmers from all regions of the world have contributed to developing the enormous crop genetic diversity that is available today. This has been recognized in the International Treaty on Plant Genetic Resources for Food and Agriculture (ITPGRFA) as the basis for food and agriculture production throughout the world (Art. 9). During the last hundred years, the division of labour within the agricultural sector has increased, leading to the professionalization of plant breeding and the development of high-yielding varieties. These varieties have boosted agricultural production while simultaneously wiping out untold other varieties. Breeders' innovations have been protected and promoted with intellectual property rights, whereas the legal space for farmers to continue their contributions to the conservation and sustainable use of crop genetic resources has been reduced, and mechanisms to promote their contribution are lacking. The ITPGRFA was meant to balance this situation, with its multilateral system of access and benefit-sharing and its provisions on farmers' rights. However, the benefit-sharing mechanism is hardly functioning, and farmers' rights are only vaguely addressed in the Treaty. This reflects the great controversies that have surrounded these issues over the years of negotiation and implementation. In this chapter, I provide a historical overview of the process related to farmers' rights under the Treaty, and present a model for understanding these developments through a 'stewardship' and an 'ownership' approach. I suggest that a clear grasp of these approaches and their potential consequences is important to develop international norms and regulations that are needed to ensure the realization of farmers' rights.

The historical development of farmers' rights

The enormous diversity of food crops available today has developed through careful selection of seeds and propagating material and exchange over short and long distances, in close interaction with nature. At the core of this innovation are the farmers of the last ten millennia and more, the custodians of crop genetic diversity (Andersen, 2016). An estimated 7,000 species are used as crops worldwide (Wilson, 1992), with great diversity within species. The continuous growth of crop diversity was, however, brought to a halt in the last century, when modern plant breeding introduced genetically homogenous high-yielding varieties. Given the great value of crop genetic resources for food security, this caused international concern: plant genetic diversity has been argued to be more important for farming than any other environmental factor, because it enables farmers to adapt to changing environmental conditions, including climate change (Andersen, 2008; United Nations, 2009; Fujisaka *et al.*, 2011).

In response to the rapid erosion of crop genetic resources, the International Board for Plant Genetic Resources (IBPGR) was founded in 1974 under the auspices of the Consultative Group on International Agricultural Research (CGIAR).[1] Located at the FAO headquarters in Rome, it sent out collecting missions, and constructed and expanded gene banks at the national, regional, and international levels (FAO, 1998). Only 15% of the samples collected were designated for storage in developing countries, whereas 85% were stored in industrialized countries and in the gene banks of the International Agricultural Research Centres (IARCs) of the CGIAR (Fowler, 1994), most of which were then located in the developed world. The IBPGR and the IARCs did invaluable work in saving fast-eroding plant varieties from extinction – but in the process developing countries lost control over their own genetic resources. This led to the FAO Conference deciding in 1981 to draft the elements of a legal convention for the establishment of an international gene bank. This was reported back to the FAO Conference two years later (Fowler, 1994).

During the negotiations, a major conflict existed between those favouring plant breeders' rights over improved plant varieties and those in favour of unrestricted access to all varieties (Fowler, 1994, pp. 187–91). The USA and representatives of the seed industry were the leading proponents of the former stance, while developing countries represented the latter position. This point is worth noting since most developing countries were later to change their position on access in order to provide for control over their genetic resources and benefits from their use, which required a stricter regulation of access. This position was voiced a decade later under the Convention on Biological Diversity (CBD), in response to the emerging Agreement on Trade-Related Aspects of Intellectual Property Rights (TRIPS Agreement), which was then being negotiated in the Uruguay Round leading to the establishment of the World Trade Organization (WTO). This is an important background for understanding the access and benefit-sharing arrangements which eventually emerged under the CBD and its Nagoya Protocol.

When the International Undertaking on Plant Genetic Resources was adopted in 1983 by the 22nd session of the FAO Conference, it was adhered to by 113 countries. The adoption of the International Undertaking can be seen as a partial victory for developing countries because it was achieved despite the opposition of major industrialized countries led by the USA. The victory was only partial, however, because the new agreement ended up as a legally non-binding undertaking, without the adherence of industrialized countries that were important to the international management of plant genetic resources for food and agriculture (PGRFA).

The objectives of the International Undertaking (IU) were to ensure that PGRFA would be explored, preserved, evaluated and made available for plant breeding and scientific purposes.

The International Undertaking was based on 'the universally accepted principle that plant genetic resources are a common heritage of mankind and consequently should be available without restriction'. The two-pronged goal was clear: conservation and access.

Along with the International Undertaking, the Commission on Plant Genetic Resources (CPGR) was also established.[2] The CPGR was an intergovernmental body charged with ensuring the implementation of the International UndertakingIU and monitoring it, especially the operation of international arrangements for the management of PGRFA.

The main reason that developed countries did not adhere to the International Undertaking[3] was its statement that genetic resources should be available without restriction, which was seen to be in conflict with plant breeders' rights. Therefore, countries could only adhere to the International Undertaking if the text was modified in some way (Andersen, 2005). It was in this context that the concept of farmers' rights was taken up in the FAO for the first time. The first documented use of the concept was at a meeting of the working group in 1986 (FAO, 1986) and arose as a response to the increased demand for plant breeders' by drawing attention to the unremunerated innovations of farmers that were seen as the foundation of all modern plant breeding. The working group produced a report on how to deal with the reservations towards the International Undertaking and on how to attract greater adherence (FAO, 1986, para. 8), the third chapter of which is devoted to farmers' rights. It not only linked the issue to the question of access to genetic resources but also revealed substantial uncertainties as to the understanding of the concept, and called for further elucidation. At the second meeting of the working group in 1987, farmers' rights were hence addressed in greater detail, with particular attention to the need to reward farmers for their contribution to PGRFA. The rights holders were not to be single farmers or communities but, rather, entire peoples – that is, a form of a collective right. This concept can be regarded as the foundation for the stewardship approach to farmers' rights that is discussed later in this chapter.

The idea of developing farmers' and plant breeders' rights simultaneously in order to balance the two also emerged: 'The Working Group concurred that Breeders' Rights and Farmers' Rights were parallel and complementary rather than opposed and that the simultaneous recognition and international legitimization of both these rights could help to boost and speed up the development of the people of the world' (FAO, 1986, para. 12). At the second session of the CPGR in 1987, the contact group agreed that, '*while the so-called "farmers' rights" could not yet be given a precise definition, some sort of compensation for their most valuable contribution to the enrichment of the plant genetic resources of the world was well-founded and legitimate*' (FAO 1987). It was pointed out that one way of giving practical recognition to this right could be via a form of multifaceted international cooperation that included freer exchange of plant genetic resources, information and research findings, and training. Another way could be through monetary contributions for programmes furthering the objectives of the International Undertaking (FAO, 1986, Appendix G).

Thus, the contact group did not arrive at a definition of 'farmers' rights', but outlined some means of according them practical recognition within the framework of the International Understanding.

Nevertheless, deep controversies over these issues remained between the countries of the Organization for Economic Development and Co-operation (OECD), on the one hand, and the group of developing countries and their NGO supporters, on the other. These controversies were also fuelled by the Uruguay Round of the General Agreement on Tariffs and Trade (GATT), which ultimately led to the WTO, where intellectual property rights (IPR) were brought into the negotiations by the USA.[4] During the first years of the Uruguay Round, which started in 1986, an agreement on IPRs was strongly opposed by several developing countries. Indeed, by the 1988 mid-term review of the Round, it was determined that such an agreement would be impossible (Evans and Walsh, 1994). During 1989, however, those developing countries that were in opposition changed their positions and dropped their earlier resistance to an agreement on IPRs. This radical shift clearly resulted from their recognized need to make concessions within the negotiations, since a consensus on all of the agreements would be needed before the package could be adopted (Yusuf, 1998). Thus, the resulting Agreement on Trade-related Aspects of Intellectual Property Rights (TRIPS Agreement) excluded from patentability plants and animals (other than micro-organisms) and essential biological processes for the production of plants and animals (other than non-biological and microbiological processes), but it did oblige members to provide for the protection of plant varieties either by patents or an effective *sui generis* system, or a combination of these (Article 27.3.b). Even though several different *sui generis* systems are in operation, the term has most often been used with respect to the International Union for the Protection of New Varieties of Plant (UPOV) (Andersen, 2008, pp. 164–68).

The developments at the FAO Conference in 1989 should be seen in the light of the Uruguay Round, as what was sacrificed there was taken up again at the FAO in other ways. Two resolutions were adopted by this Conference: Resolution 4/89 on the Agreed Interpretation of the International Undertaking and Resolution 5/89 on Farmers' Rights (both Resolutions annexed to the International Undertaking). These Resolutions were adopted by consensus, but only arose as a result of tense negotiations: there had again been fierce resistance to the idea of plant breeders' rights among developing countries, and the interpretations that provided for the acceptance of these rights could only be adopted with the simultaneous recognition of farmers' rights (Andersen, 2005a).

Resolution 4/89 stated that '*Plant Breeders' Rights as provided for under UPOV... are not incompatible with the International Undertaking*' (para. 1) and that '*states adhering to the Undertaking recognize the enormous contribution that farmers of all regions have made to the conservation and development of plant genetic resources, which constitute the basis of plant production throughout the world, and which form the basis for the concept of Farmers' Rights*' (para. 3). Resolution 5/89 represented a milestone as the first recognition by the FAO Conference of farmers' contributions to the global pool of genetic diversity.

In 1991, a new annex to the International Undertaking was adopted as Resolution 3/91 (FAO, 1991). This time, the Conference stated that the concept of genetic resources as the heritage of mankind, as applied in the International Undertaking, was subject to the sovereignty of states.[5] This interpretation might be seen to have been heavily influenced by the ongoing negotiations for a Convention on Biological Diversity (CBD), which was adopted only six months later and which also incorporated the principle of national sovereignty in Article 3. As a result of the CBD negotiations (and in response to the emerging intellectual property regime), negotiators from developing countries demanded control over access to their genetic resources as well as the fair and equitable sharing of the benefits arising from their use. In many circles, this demand brought about a shift in thinking on genetic resources, from a perspective based on the common heritage of mankind to a bilateral approach to benefit-sharing, which was in turn a response to the IPR regime emerging from the Uruguay Round (Andersen, 2008). This shift can be seen as the beginning of the 'ownership approach' to farmers' rights, as set out below in this chapter.

After Resolution 3/91, FAO members stated that the conditions for access to plant genetic resources required further clarification (FAO, 1991, para. d). The original purpose of the International Undertaking – which was to ensure unrestricted access to genetic resources – was no longer clear, and the principles of 'the common heritage of mankind' that had controlled these resources were blurred.

The adoption of the Convention on Biological Diversity (CBD) in 1992 was a decisive event for the development of the IU regime. The CBD became the first legally binding international agreement to address the sustainable management of biological diversity worldwide[6] and was developed as a stand-alone convention as well as a framework convention (Andersen, 2008).

At the Conference for the Adoption of the Agreed Text of the Convention on Biological Diversity in May 1992, the Nairobi Final Act was adopted (UNEP, 1992), including a resolution on the inter-relationship between the CBD and the promotion of sustainable agriculture (Resolution 3). This resolution recommended that ways and means be explored to develop complementarity and cooperation between the CBD and the Global System for the Conservation and Sustainable Use of Plant Genetic Resources for Food and Sustainable Agriculture (*ibid.*, para. 2), which had been established under the FAO with the IU acting as a central component. Finally, the resolution recognized the need to seek solutions to two outstanding matters concerning PGRFA: (1) access to *ex-situ* collections that had not been acquired in accordance with the CBD, and (2) the question of farmers' rights.

At its 27th session in 1993, the FAO Conference accordingly requested the FAO director-general to provide a forum for negotiations for harmonizing the International Undertaking with the CBD (Resolution 7/93): this was the point of departure for the lengthy negotiations that finally resulted in the adoption of the ITPGRFA in 2001.

Revising the IU in harmony with the CBD was a challenging task. The specific features, uses and management needs of PGRFA had to be considered.[7] PGRFA constitute the basis of farming and are, except for their wild relatives, domesticated resources. Since access to PGRFA is a condition for the further domestication, and thus continued existence, of these resources, expeditious facilitation of access was a major concern to the negotiators. To ensure access, it was also important that transferred PGRFA should remain in the public domain and not be made subject to exclusive IPRs. A means of benefit-sharing other than that envisaged under the CBD had to be found, focussed on those who conserve and sustainably use the resources, rather than on the specific providers. This was because: (1) for most crops, it is difficult to identify the countries of origin (the countries entitled to provide access under the CBD; Andersen, 2001; Fowler, 2001); (2) all countries are interdependent on PGRFA so a complicated system of transfers between providers and recipients would hamper expeditious access (Palacios, 1998); (3) rewarding only the current providers of genetic resources would not be fair to farmers who maintain or develop genetic diversity that will benefit future generations.

Throughout the negotiations, farmers' rights were one of the most contested issues. Most developing countries, as well as some industrialized countries (e.g. Norway) had advocated comprehensive and internationally binding recognition of farmers' rights, a stance opposed by countries such as the USA and Australia. The controversies were complex, and a breakthrough seemed unlikely when, in 1999, negotiators from the North decided to meet some of the demands from the South – and this compromise led to the long-awaited breakthrough. What resulted was the final text of the ITPGRFA on farmers' rights as we know it today.[8]

When the ITPGRFA was finally adopted in November 2001, many observers had almost given up hope of ever reaching a consensus. Indeed, full consensus proved impossible and the Treaty had to be put to the vote: 116 countries voted in favour of the treaty and two countries abstained (Japan and the USA). The ITPGRFA was the first legally binding agreement to deal exclusively with PGRFA, and it was also incidentally the first international treaty of the new millennium.[9] Since then, both the USA and Japan have revised their policies and also ratified.[10]

The ITPGRFA entered into force on 29 June 2004,[11] and as of October 2016 it has been ratified by 141 states.

The objectives of the ITPGRFA are the conservation and sustainable use of PGRFA as well as the fair and equitable sharing of benefits arising from their use – in harmony with the CBD – for sustainable agriculture and food security (Article 1). The Treaty sets out that the contracting parties shall promote an integrated approach to the exploration, conservation and sustainable use of PGRFA (Article 5): suggested measures include improving *ex situ* conservation of plant varieties and wild crop species and providing farmers with support for on-farm management and conservation of PGRFA – the latter being particularly relevant for farmers' seed systems and farmers' rights.

The ITPGRFA stipulates that contracting parties shall develop and maintain appropriate policies and legal measures that promote sustainable use of PGRFA (Article 6). This provision is an obligation for all contracting parties. The ITPGRFA also sets out a Multilateral System of Access and Benefit-sharing (MLS) (in Articles 10–13[12]) which covers 35 food crops and 29 forage plants that are in the public domain and under the management and control of the contracting parties (the Annex I crops).[13]

In the preamble to the ITPGRFA, the contracting parties affirm that the past, present and future contributions of farmers in all regions of the world – particularly those in the centres of origin and diversity – in conserving, improving and making available these resources, constitute the basis of farmers' rights. They also affirm that the rights recognized in the ITPGRFA to save, use, exchange and sell farm-saved seed and other propagating material, to participate in relevant decision-making and to encourage fair and equitable benefit-sharing are fundamental to the realization of farmers' rights. Article 9 of the ITPGRFA recognizes the enormous contribution of farmers in the conservation and development of PGRFA and that this contribution constitutes the basis of food and agriculture production throughout the world. It explicitly states that responsibility for the implementation of farmers' rights, as they relate to the management of PGRFA, rests with national governments. Certain measures to protect and promote farmers' rights are suggested e.g. the protection of traditional knowledge, the right to participate in equitable benefit-sharing and the right to participate in decision-making at the national level. Also the rights that farmer have to save, use, exchange and sell farm-saved seeds and propagating materials are addressed, but without any particular direction. As these provisions are vague, contracting parties, in particular developing countries, have sought guidance and assistance for the implementation of Farmers' Rights since the entry into force of the Treaty, without much effect so far. There have however been consultation processes between the sessions of the Governing Body and negotiations related to the resolutions from the Governing Body that contribute to shaping a common ground of understanding of what it takes to realize farmers' rights. In order to make progress in this regard, it may be useful to analyse the negotiations and discussions along the lines of a 'stewardship' and an 'ownership' approach to realizing farmers' rights.

Two approaches to the realization of farmers' rights under the Treaty

As described above, farmers' rights constitute a cornerstone of the ITPGRFA.[14] Achieving the conservation and sustainable use of crop genetic resources as set out in Article 1 depends decisively on farmers and their ability to maintain these resources *in situ* on-farm, which in turn depends on farmers' rights. The provisions on access and benefit-sharing under the Treaty are vital to the realization of farmers' rights, but as the historical account above outlines, the topic of farmers' rights has been discussed in the contexts of different rationales, resulting in different perceptions on their main contents. Two basic ways of approaching the concept of farmers' rights have been advanced previously (Andersen 2006, 2016) and are further developed here: the 'stewardship approach', and the 'ownership approach'.

The 'stewardship approach'

The stewardship approach describes the idea that agrobiodiversity as a principle belongs to the common heritage of mankind and that it should be shared for the common good as part of the public domain. As such, the stewardship approach can be said to have been the dominant rationale throughout the history of agriculture until the advent of intellectual property rights. In terms of farmers' rights, a stewardship approach would refer to the rights that farmers must be granted collectively in order to enable them to continue as stewards and innovators of agro-biodiversity and reward them for this contribution. A core idea is to uphold and enhance the 'legal space' required for farmers to continue this role. Another core idea is that farmers involved in the maintenance of agro-biodiversity – on behalf of their generation and for the benefit of all mankind – should be rewarded and supported for their contributions, and that this principle should constitute the basis of a benefit-sharing system.

The 'ownership approach'

The ownership approach evolved when the interests in the commercial use of genetic resources increased along with the growing economic stakes of biotechnologies in the second half of the last century, followed by demands for intellectual property rights to protect and promote inventions. As intellectual property systems are costly institutions, the capacity of developing countries, rich in genetic resources, to develop and effectively use such systems was limited (Andersen, 2008). These emerging power asymmetries were met with much protest against intellectual property rights to genetic resources from the 'global South', along with the demands of securing control over their resources through systems regulating access on mutually agreed terms and prior informed consent between purported owners and users of these resources. There should be fair and equitable sharing of the benefits arising from the use of genetic resources between purported owners and users of these resources. This is the basis of the ownership approach, which describes the idea that establishing individual or collective ownership to genetic resources provide important incentives to promote breeding as well as the conservation and sustainable use of agrobiodiversity. Furthermore, it enables control over the genetic resources that are covered with ownership rights for the holders of such rights, the purported owners, and makes it possible to trade with them as well as to attract benefit-sharing. In terms of farmers' rights, an ownership approach would establish the right of farmers to be rewarded on an individual or collective basis for genetic material that has been obtained from their fields and used in commercial varieties and/or protected with intellectual property rights. The idea is that such a reward system is necessary to enable the equitable sharing of benefits arising from the use of agro-biodiversity and to establish an incentive structure for the continued maintenance of this diversity. Access and benefit-sharing legislation and farmers' intellectual property rights would be central instruments.

The distinctions between the two approaches are not clear-cut. An evolving ownership approach to the management of crop genetic resources will enable different actors to exclude each other from the access to, and use of, these vital resources, and thereby reduce the legal space for all to contribute to the conservation and sustainable use of crop genetic diversity (Andersen, 2008). A stewardship approach would maintain and enhance the legal space and possibilities to contribute to the conservation and sustainable use of crop genetic resources. The paradox is, however, that without sufficient measures to avoid it, the stewardship approach might result in genetic resources and information from the public domain being privatized and thus becoming subject to the ownership approach. Whereas the stewardship approach may result in misappropriation of crop genetic resources by third parties, the ownership approach may result in disincentives to share crop genetic resources among farmers and thus reduce the millennia-old traditions of seed exchange and distributions that have contributed to the agrobiodiversity we have today. It is important to understand the different rationales behind the two approaches, but also how they can be combined to achieve the conservation and sustainable use of genetic resources, the sharing of benefits arising from the use of these resources and the realization of farmers' rights.

The next section will examine in detail the four elements of farmers' rights – namely, protection of traditional knowledge, benefit-sharing, participation in decision-making and the rights to save, use, exchange and sell farm-saved seed – and how they can be interpreted under the stewardship and ownership approaches. It will also discuss how the two approaches can be combined to achieve the goals of the International Treaty.

Protecting farmers' traditional knowledge

Understanding traditional knowledge related to plant genetic resources for food and agriculture requires a holistic understanding of the dynamic nature of this knowledge, including factors such as livelihoods, cultures and landscapes. Traditional knowledge is vital to understanding the properties of plants, their uses and how they are cultivated. It includes knowledge on how to select seeds and propagating material, how to store them and how to use them for the next harvest. Thus, it also comprises the basic necessities for farmers to be able to maintain crop genetic diversity in the fields. Article 9.2(a) is the only provision on traditional knowledge in the ITPGRFA, and provides for 'the protection of traditional knowledge relevant to plant genetic resources for food and agriculture'. The Treaty provides no further guidance on how this article can be interpreted and operationalized. However, since the objectives of the ITPGRFA are to be implemented in harmony with the CBD (Article 1), Article 8j of the CBD is also relevant in this context. According to this article, each contracting party shall – as far as possible and as appropriate and pursuant to national legislation – respect, maintain and preserve traditional knowledge, innovation and practices and promote their wider application, with the approval of the holders of such knowledge, innovations and practices. The equitable sharing of benefits from its use should be encouraged.

Understanding the challenges related to the protection of traditional knowledge has significantly influenced current views about how Article 9.2(a) can be implemented. Examining the contents of this right from the stewardship and an ownership approaches suggests rather different possibilities:

1 *Protection against extinction* means ensuring that traditional knowledge is kept alive and can further develop among farmers. Under a stewardship approach, the best way to protect traditional knowledge from the threat of extinction is to share it – a widespread approach in the North – and, thus, the motto: 'protection by sharing'.
2 *Protection against misappropriation* is a different approach. It is based on the anticipation that farmers' varieties, and associated knowledge, could be appropriated by commercial actors without prior informed consent from the holders of this knowledge and benefit-sharing on mutually agreed terms. Thus, the sharing of knowledge should not take place unless measures are in place to avoid this. This view is often accompanied by a widespread regret that the fear of misappropriation has made it necessary to be cautious. An ownership approach to

> protecting traditional knowledge would mean providing farmers with the right to act against misappropriation of their knowledge and to decide about the use of this knowledge and related plant genetic resources.

In order to consider the two approaches it is important to assess the threat of misappropriation of crop genetic resources. To what extent is such misappropriation taking place? According to existing documentation, it seems that in developing new varieties, commercial plant breeders tend to use already improved varieties from their own stocks or from other plant breeders. Farmers' varieties are generally regarded as being difficult to work with due to their genetic heterogeneity. Only when particular traits are sought – those not found in their own stocks or other improved varieties – are farmers' varieties deemed necessary. When they are sought, they are normally obtained from gene banks and not from farmers' fields or markets. In gene banks, little traditional knowledge is typically included in the passport data. Thus, traditional knowledge related to crop genetic resources is still rarely used in commercial breeding. Generally, the genetic foundation for commercial plant breeding appears to be narrowing (Esquinas-Alcázar, 2005, p. 948). This situation, together with the effects of climate change, may well change demand for landraces and farmers' varieties – together with their associated knowledge – in the future (Esquinas-Alcázar, 2005).

In any case, based on the ownership approach, protection of traditional knowledge would mean offering ownership status to farmers with the right to act against misappropriation and to decide over the use of their knowledge and related plant genetic resources. In Norway, farmers stress that their traditional knowledge is about to disappear. Therefore, protection, as they understand it, must ensure that such knowledge does not die out (Andersen, 2011). To achieve this, knowledge must be shared in the broadest manner possible. Norwegian farmers are thus prone to a stewardship approach. They fear that an ownership approach to protection could provide disincentives to sharing knowledge between and among farmers. Proponents of the stewardship approach insist that ownership in this context has been an alien idea among farmers, and that it represents a profound break with traditional perceptions.

Ultimately, the measures that are chosen should reflect the situation. What is most important today, with the rapid erosion of traditional knowledge, is to protect traditional knowledge related to crop genetic resources from becoming extinct. Nevertheless, avoiding misappropriation is important, and considered a condition in many communities for sharing knowledge.

Ensuring equitable benefit-sharing

Article 9.2(b) of the ITPGRFA concerns a farmer's right to participate equitably in the sharing of benefits arising from the utilization of plant genetic resources for food and agriculture. To interpret this provision, some guidance can be found in Article 13 on benefit-sharing in the multilateral system. This Article lists the most important benefits as: (1) facilitated access to plant genetic resources for food and agriculture; (2) the exchange of information; (3) access to, and transfer of, technology; (4) capacity-building; and (5) the sharing of monetary and other benefits arising from commercialization. Moreover, it specifies that benefits arising from the use of plant genetic resources for food and agriculture that are shared under the multilateral system should flow primarily, directly and indirectly to farmers in all countries – especially in developing countries and countries with economies in transition – who conserve and sustainably utilize plant genetic resources for food and agriculture.

Whereas these provisions all relate to the multilateral system and not directly to the provisions on farmers' rights in the ITPGRFA, they reflect a line of thought on benefit-sharing that is relevant for interpreting Article 9.2(b) as a measure to protect and promote farmers' rights. Firstly, it is clear that there are many forms of benefit-sharing, of which monetary benefits comprise only one part. Secondly, the benefits are not only to be shared with those few farmers who happen to have plant varieties that are utilized by commercial breeding companies but also with farmers in all countries that are engaged in the conservation and sustainable use of agro-biodiversity.

Measures to ensure the equitable sharing of benefits arising from the use of genetic resources can be designed in many ways. Under an ownership approach, these measures would mandate the development of direct benefit-sharing in which the benefits would be shared directly between the purported 'owners' and 'buyers' of genetic resources – based on a prior informed consent on mutually agreed terms (as set out in the CBD).[15]

In the South, policies on benefit-sharing – if any – are normally present in the laws and regulations on access to biological resources, which are sometimes found in the national legislation on the protection of biological diversity. Countries with legislation on indigenous peoples' rights often include provisions on benefit-sharing in these laws, which then also cover indigenous farmers. Most of these regulations comprise forms of direct benefit-sharing between the 'owners' and the 'buyers' of genetic resources, often based upon prior informed consent on mutually agreed terms, as set out in the CBD. However, despite all of these efforts, so far there have hardly been any examples of direct monetary benefit-sharing between the providers and recipients of plant genetic resources for food and agriculture as a result of such legislation.

There are, however, other ways of sharing benefits, which are often referred to as indirect approaches to benefit-sharing. These approaches are in line with FAO's mandate in the early days of negotiations on farmers' rights, inspired by a stewardship approach. A basic principle was that benefits should be shared among 'entire peoples', the stewards of plant genetic resources in agriculture and society at large (FAO, 1987, Appendix F, section 8). This principle is based on the idea that it is farmers' legitimate right to be rewarded for their contributions to the global genetic pool from which we all benefit, and it is an obligation of the international community to ensure that such recognition and reward is provided.

Where should the funds come from to enable such benefit-sharing? First of all, as already noted, the benefit-sharing mechanism under the multilateral system specifies that the benefits from the system should flow primarily to farmers in all countries, especially in developing countries and countries with economies in transition, who conserve and sustainably use crop genetic resources (Article 13.3). The basic principles of the multilateral system is that the countries that are parties to the ITPGRFA include all the genetic material of their Annex I crops that are in the public domain and under their control in the multilateral system. This material is freely accessible upon signing a standard material transfer agreement. In order for this material to remain in the public domain, recipients are not allowed to seek intellectual property rights on the material in the form it is received. If recipients develop it further and then patent it, then a mandatory fixed payment is to be paid to the benefit-sharing fund under the multilateral system. If the developed material is commercialized, but without patenting, then a contribution is voluntary. Other voluntary payments may also be paid to the benefit-sharing fund, and most of the funds received so far belong to this latter category. However, it is uncertain how much funding can be generated by this mechanism and even whether this mechanism will be successful and make a substantial difference to the farmers it is supposed to be helping.

The funding strategy of the ITPGRFA (as set out in Article 18) is another important source in so far as it supports the implementation of conservation (Article 5), sustainable use (Article 6) and farmers' rights (Article 9), which would all greatly benefit diversity farmers. However, since there are to date no fixed mandatory contributions, it is uncertain how much money the fund can generate. Thus, for the time being, Article 7 on international cooperation and Article 8 on technical assistance are important provisions on benefit-sharing. In these Articles, the contracting parties agree to promote the provision of technical assistance to developing countries and countries with economies in transition, with the objective of facilitating the implementation of the ITPGRFA. The third source of benefit-sharing, and the most successful at the present time, is official development assistance (Brush, 2005; Andersen 2008). Official development assistance can be channelled through bilateral or multilateral cooperation or through NGOs.

A 2005 survey showed that – for many reasons – benefit-sharing is more promising when the primary target for funding is the farming community that actually contributes to the maintenance of plant genetic diversity rather than the providers of genetic resources to commercial plant breeders. Since then, many organizations have engaged in such forms of benefit-sharing, as documented in e.g. Andersen and Winge (2013) and Vernooy *et al.* (2015).

Still, the dominant view on benefit-sharing in many countries, particularly in the South, is the ownership approach, whereby direct benefit-sharing between purported 'owners' and 'buyers' is the preferred mode. While such an ownership approach might seem to be fair and equitable as a point of departure, there are many difficulties with it. These difficulties include the facts that:

- it is difficult to identify exactly who should be rewarded;
- the demand for farmers' varieties among commercial breeders is limited, so relatively few farmers would benefit and most of the contributors to the global pool of genetic resources would remain unrewarded;
- the approach could lead to disincentives to share seeds and propagating material among farmers because of the expectations of personal benefit or the benefit to a community;
- although several countries in the South have enacted legislation on direct benefit-sharing, few instances of such benefit-sharing have been reported so far with regard to agro-biodiversity; and
- in many countries, the transaction costs of establishing access and benefit-sharing legislation have been considerable.

Thus, the ownership approach has not proven to be especially promising so far, and these concerns must be taken into account when measures are designed to ensure benefit-sharing that is in line with the intentions of the ITPGRFA.[16]

There are many good examples of indirect forms of benefit-sharing, including a registry of farmers' varieties in the Philippines; community seed fairs in Zimbabwe; community gene banks and on-farm conservation in India; dynamic conservation and participatory plant breeding in France; participatory plant breeding in Nepal, which is adding value to farmers' varieties; capacity building for seed potato selection in Kenya; technical assistance to the Peruvian Potato Park in Cusco; and the reward for best farming practices with regard to crop genetic diversity in Norway (Andersen and Winge, 2008, 2013; Vernooy *et al.*, 2015). These are all examples of programs and developments that provide models for the further implementation of farmers' rights. The major challenge today is to find ways and means to scale up such activities – for example, through the national agricultural extension service systems and other ways of linking up with government policies, as exemplified in Nepal (Vernooy *et al.*, 2015). However, such initiatives are heavily dependent on political will, which is often lacking. In order to increase the political will, it is necessary to raise awareness in society in general on the vital importance of agro-biodiversity and farmers' rights.

Participation in decision-making

Article 9.2(c) deals with the right of farmers to participate in decision-making at the national level on matters related to the conservation and sustainable use of plant genetic resources for food and agriculture. However, no further guidance is provided in the ITPGRFA as to how such decision-making can be implemented in practice. To operationalize this measure, it will be necessary to specify the 'relevant matters' in which farmers can have the right to participate as well as the way in which they can participate.

The development of laws and regulations related to the management of plant genetic diversity in agriculture is clearly relevant for farmers' participation. At the current time, there are numerous examples of such laws and regulations, including seed acts, seed certification regulations, other regulations regarding seed distribution and trade, plant variety protection laws, patent laws, bio-prospecting laws or regulations, laws on the conservation and sustainable use of biodiversity in general or crop genetic resources in particular (as well as on several specific crops) and legislation on the rights of indigenous peoples and traditional knowledge. In addition, it is also important to consider any legislation that regulates mainstream agriculture since such legislation tends to produce incentive structures that are often detrimental to farmers' rights without providing any compensation. The extensive use of hearings at various stages in the decision process is an important measure to ensure participation. It is particularly important to ensure that farmers that are engaged in the management of plant genetic diversity are aware of the processes and are explicitly invited to participate through their organizations.

The implementation of laws and regulations is also relevant to farmers' participation. The way in which these regulations are interpreted and implemented often has an enormous influence on a farmer's management of these resources and also on his or her livelihood.

Ultimately, then, the implementation of farmers' rights requires farmers' participation. This is not only because of their unquestioned right in this regard, according to the ITPGRFA, but also because they are the ones who can best define the needs and priorities of farmers in the context of farmers' rights and they are also the central actors in the implementation process. Comprehensive consultative processes of various kinds are relevant – the better represented farmers are, the greater legitimacy the results will have, and the more likely it is that they will constitute effective measures for the realization of farmers' rights. In particular, it is important for farmers to actually be involved in the management of plant genetic diversity in order to participate in such processes since they constitute an important target group of the ITPGRFA.

In general, we find few examples of legislation on farmers' participation, although some countries in the South have extensive legislation on farmers' participation in decision-making (Andersen, 2005). All the same, the actual participation of farmers in decision-making processes seems marginal and is often limited to large-scale farmers who are normally not engaged in the maintenance of plant genetic diversity. In the North, the participation of farmers in decision-making processes is more common, even if diversity farmers are rarely represented, but such participation does not usually involve specific laws or policies. It should be noted that farmers in the North claim that their influence is decreasing, due to their countries' commitments to regional and international organizations and agreements such as the World Trade Organization (WTO) and the European Union (EU) (Andersen, 2005).

While the process of implementing participation has been slow, there have been a few success stories. The most comprehensive consultative process on the implementation of farmers' rights to date was carried out in Peru in 2008, involving 180 farmers from many different regions as well as numerous central decision-makers (Scurrah *et al.*, 2008).[17] Other success stories include capacity-building measures to prepare farmers for participating in decision-making in Malawi,

Zimbabwe, Philippines and Peru, and several successful advocacy campaigns regarding the implementation of elements of farmers' rights, where farmers have been directly involved, as for example in India, Norway and Nepal (Andersen and Winge, 2008, 2013).

Under both the stewardship and ownership approaches, participation in relevant decision-making is important but for different reasons. Under a stewardship approach, the most important objectives would be to ensure legal space for farmers to continue their practices as custodians and innovators of plant genetic resources and to establish reward mechanisms for farmers' contributions to the global genetic pool. Under the ownership approach, the goals would be to ensure appropriate legislation on access and benefit-sharing as well as to safeguard farmers' intellectual property rights to the genetic resources in their fields and related knowledge. It is clear that these two sets of objectives could be conflicting. However, the overall objectives of the ITPGRFA to conserve, sustainably use and share benefits from crop genetic resources for sustainable agriculture and food security may serve as guiding principles. Measures that limit a farmer's ability to take part in these activities would go against the intentions of the treaty.

Farmers' rights to save, use, exchange and sell farm-saved seed

Farmers' customary use of propagating material – to save, use, exchange and sell farm-saved seed and propagating material – is a pivotal element of farmers' rights and rooted as a 10,000-year-old tradition that enabled mankind to develop today's rich agro-biodiversity. However, the ITPGRFA is vague on farmers' rights to save, use, exchange and sell farm-saved seed. Section 9.3 of the treaty states that nothing in the relevant article (Article 9 on farmers' rights) 'shall be interpreted to limit any rights that farmers have to save, use, exchange and sell farm-saved seed, subject to national law and as appropriate', but this article does not really offer much direction, except for labelling these practices as 'rights'. Despite this lack of precision, the general line of thought would seem clear. It is important to grant their rights to save, use, exchange and sell farm-saved seed, but individual countries are free to define the legal space that they deem to be sufficient.

The freedom to define such legal space for farmers is restricted by other international commitments. Most countries in the world are members of the WTO and are thus obliged to implement the WTO Agreement on Trade Related Aspects of Intellectual Property Rights (TRIPS Agreement).[18] According to the TRIPS Agreement, all WTO member countries must protect plant varieties either by patents, by an effective *sui generis* system (a system of its own kind), or a combination of both (Article 27.3.b). The limits to a *sui generis* system and the meaning of an 'effective' *sui generis* system are not explicitly defined in the text. In other words, countries have to introduce some sort of plant breeders' rights.

The Union for the Protection of New Varieties of Plants (UPOV) explains that the most effective way to comply with the provision concerning an effective *sui generis* system is to follow the model of the International Convention for the Protection of New Varieties of Plants (UPOV Convention).[19] There are several versions of the UPOV model. The most recent (the 1991 Act of the UPOV Convention) provides that plant breeders are to be granted comprehensive rights – to the detriment of farmers' customary rights to save, re-use, exchange and sell seeds. It is possible to make exceptions for small-scale farmers but only within strict limits. Exchange and sales of seeds among farmers are prohibited. It should be noted, however, that these regulations apply only to seeds protected by plant breeders' rights, and not to traditional or other varieties.

The UPOV model has met resistance from some countries and many organizations that fear that ratification of the Convention would be detrimental to the rights of farmers to save and share propagating material. The TRIPS Agreement provides only minimum standards, leaving enough scope for the development of other solutions that are more compatible with the demand for farmers' rights. The challenge in the context of the ITPGRFA is thus for WTO member countries to meet their TRIPS obligations regarding plant breeders' rights, while also maintaining the necessary legal space to realize farmers' rights to propagating material.

A further constraint to farmers' rights in many countries is the introduction of seed laws that affect all propagating material, whether it is protected with intellectual property rights or not. The most important factor is that these laws also affect traditional varieties and farmers' varieties. They require that all varieties be officially approved for release and that seed and propagating material be certified before they are offered on the market. The original reason for these regulations was to ensure plant health and seed quality. However, in many countries, the regulations have gone so far that they now hinder the maintenance of crop genetic resources in the fields in two ways. First, since traditional varieties are normally not genetically homogeneous enough to meet the requirements for approval and certification, these varieties are excluded from the market and gradually disappear from active use when those farmers who currently use them begin to give them up. Second, many seed laws also stipulate that only authorized seed shops are allowed to sell seeds, and they prohibit all other seed marketing (with rare exceptions).

This is the case in most of Europe. The EU has tried to solve these hurdles with a specific directive on conservation varieties. However, EC Directive 62/2008 (EU Conservation Varieties Directive) is not adequate with regard to farmers' rights, because (1) seed exchange and sale are still prohibited among farmers; (2) only varieties deemed interesting by certain authorities can be covered by the system, which limits diversity; (3) the variety release and certification criteria are still too strict to allow for the release of many traditional and farmers' varieties; (4) the marketing and use of the varieties are limited to the regions of origin; (5) only limited quantities may be used; and (6) the conservation varieties may not be further developed by farmers. A comprehensive evaluation was carried out by the EU Commission to provide a foundation for revisions of the EU directives on seeds and propagating material. The evaluation led to a proposal to simplify the whole structure of relevant directives and solve many of the constraints highlighted above. The proposal was approved by the EU Parliament, but eventually turned down by the EU Commission, thus further prolonging these issues.

When combined, these two processes – restrictions on plant variety release and seed marketing laws – may constitute serious obstacles to the implementation of the ITPGRFA in terms of *in situ* on-farm conservation and sustainable use, as well as to farmers' rights. It is a paradox that rules originally intended to protect plant health have, in fact, contributed to removing the very basis for ensuring plant health in future – namely, the diversity of genetic resources.

What possibilities are there to make such laws more compatible with the customary rights of farmers, which are so crucial to the maintenance of agro-biodiversity for food security, today and in the future?

Under a stewardship approach, the goal would be to grant the rights to save, use, exchange and sell farm-saved seed, whether from protected or non-protected varieties. Due to the present constrains of existing legislation, however, the challenge seems rather to uphold or re-establish sufficient legal space for farmers to continue their crucial role as custodians and innovators within the existing legal framework on plant breeders' rights, variety release and seed distribution.

Under an ownership approach, on the other hand, the goal would be to provide farmers with intellectual property rights on the varieties in their fields on equal footing with breeders' rights. Arguments related to this objective have been discussed earlier in this chapter.[20]

There are several pertinent stories on how legal space for farmers' rights can be established and maintained in order to allow farmers to maintain their traditional practices and innovation in agriculture (see, for example, Andersen and Winge, 2008, 2013). These include India's 2001 Protection of Plant Varieties and Farmers' Rights Act;[21] Norway's 'no' to stricter plant breeders' rights in order to maintain the balance with farmers' rights and the ways in which farmers are circumventing the law in the Basque Country in Spain. Nevertheless, establishing and maintaining legal space for farmers' rights to save, use, exchange and sell farm-saved seed constitutes the main barrier to implementing the ITPGRFA in terms of the conservation and sustainable use of crop genetic diversity and of the realization of farmers' rights. Solutions are urgently needed.

Undoubtedly, there are other means of combining the stewardship and ownership approaches in order to realize farmers' rights to seed and propagating material. What matters in this context is that the approach that is chosen must not conflict with the principles of the stewardship approach, which has been the primary goal of the FAO since the issue was first taken up as well as the rationale behind the ITPGRFA.

Future directions: how can farmers' rights be realized?

Whereas the implementation of farmers' rights under the ITPGRFA is a national responsibility, the Governing Body of the ITPGRFA is responsible for promoting the full implementation of the Treaty, including the provision of policy direction and guidance, and monitoring of implementation (Art. 19). According to Article 21, the Governing Body is to ensure compliance with all provisions of the ITPGRFA, and the Preamble of the Treaty highlights the necessity of promoting farmers' rights at national as well as international levels. In this final section, I will consider how the Governing Body has carried out its responsibilities, with a view to national implementation.

In the first session of the Governing Body in 2006, the issue of farmers' rights was on the working agenda. Since then the topic has been discussed at each session of the Governing Body, resulting resolutions with the decisions made (see Resolutions 2/2007, 6/2009, 6/2011, 8/2013 and 5/2015 of the Governing Body of the International Treaty). There has been extensive consultation processes prior to each of the sessions of the Governing Body, most notably the Informal International Consultation on Farmers' Rights in Lusaka, Zambia, in 2007; the Global Consultation on Farmers' Rights in Addis Ababa, Ethiopia, in 2010; and the Global Consultation on Farmers' Rights in Bali, Indonesia, in 2016. Each of the first consultations resulted in comprehensive reports and summarizing input papers which were presented at the Governing Body at its sessions in 2007 (by Norway and Zambia) and 2011 (by Ethiopia).

The results of these consultations are, at first sight, promising, and some of the decisions can be regarded as a breakthrough for the negotiations. Since the entry into force of the Treaty, much has been achieved in terms of establishing a joint understanding of important issues related to the realization of farmers' rights. However, little is happening from the side of Contracting Parties or of the Secretariat in terms of implementation. Many of the provisions are made subject to the availability of funding, which is mostly scarce. Some NGOs and IGOs are doing substantial work to realize farmers' rights in many countries, but national efforts are lagging behind. To strengthen the work on farmers' rights, much more attention to the topic and its pivotal importance for the implementation of the treaty is required.

Since 2007, developing countries, along with some developed ones, have demanded that voluntary guidelines be prepared to guide and assist countries in the implementation of farmers' rights. There was strong resistance against that from several developed countries. Nevertheless, the demand is being repeated with greater strength at each session of the Governing Body. It is demanded that the guidelines be developed through a participative, inclusive and transparent manner. Not only would such guidelines provide necessary guidance and assistance for contracting parties and other stakeholders, it would also provide an important arena for establishing a common ground of understanding with regard to why farmers' rights are important for the implementation of the treaty and what it takes to realize these rights.

Conclusions

The International Treaty on Plant Genetic Resources for Food and Agriculture is the single most important international instrument that currently exists to ensure the sustainable management of crop genetic resources. After ten years of implementation, it is evident that progress so far has been slow. Developing countries are demanding a functioning benefit-sharing mechanism and greater emphasis on the realization of farmers' rights in order to support the Treaty, e.g. by placing their genetic resources in the multilateral system. Action in this regard is urgently required. By understanding the different rationales behind the discussion in the Governing Body, i.e. the stewardship and the ownership approaches, it might become clearer how they affect the conservation and sustainable use of crop genetic resources for food and agriculture. The consequences of the ownership approach might be detrimental to the on-farm conservation and sustainable use of crop genetic resources, as shown in this chapter. The stewardship approach could, seen in isolation, provide a solid basis for the on-farm conservation and sustainable use of crop genetic resources. The paradox, however, is that the resources from the public domain can be made subject to private ownership and thus be turned into a part of the ownership approach. Thus, the stewardship approach could not be the sole approach under the treaty, but may need to be complemented by elements of the ownership approach.

Much has been achieved with regard to developing a joint understanding of farmers' rights, their importance and the steps required for their realization – and there are many success stories, mainly at a local level. Much still remains to be done to ensure that these rights are realized on a scale that is required to enable farmers to continue to maintain and further develop the crop genetic diversity. This is a contribution to ensuring the basis of local and global food security, and to recognize and reward these farmers for their contributions to the global genetic pool. Awareness of the challenges, political priority, and international cooperation are required to make farmers' rights a reality.

Acknowledgement

This chapter is a shortened and modified version of the following:

Andersen, R. (2017) 'Stewardship or Ownership? How to Realise Farmers' Rights'. In Hunter, D., Guarino, L., Spillane, C. and McKeown, P. (eds) *Routledge Handbook of Agricultural Biodiversity*. Routledge, London and New York.

Notes

1 In 1974, the International Board for Plant Genetic Resources (IBPGR) was transformed into the IPGRI, which is now Bioversity International, a part of the Consultative Group on International Agricultural Research (CGIAR). The CGIAR was founded in 1971 on the initiative of the Ford and Rockefeller Foundations to unite privately funded international agricultural research centres (IARCs)

into one network. As an informal association of public and private donors that support the IARCs, it is a donor-led group that has provided a forum for discussion of research priorities and coordination of funding (FAO, 1998, p. 248). As divisions of the network, the IARCs have their own governing bodies. The United Nations Environment Programme (UNEP), the FAO, the United Nations Development Programme and the World Bank co-sponsor the system, and the CGIAR is headquartered at the premises of the World Bank in Washington, DC (FAO, 1998).

2 It was established by FAO Conference Resolution 9/83. It was later renamed the Commission on Genetic Resources for Food and Agriculture (CGRFA), as its mandate was broadened (as discussed below).

3 At that time, there were still only 74 signatories.

4 General Agreement on Tariffs and Trade, 30 October 1947, 55 UNTS 194.

5 This principle was first voiced at the 1972 United Nations Conference on the Human Environment in Stockholm in the form that states have sovereign rights to exploit their natural resources in accordance with their own environmental priorities (Stockholm Declaration on the Human Environment, 16 June 1972, 11 ILM 1416 (1972), Principle 21).

6 According to the Treaty Reference Guide of the United Nations Office of Legal Affairs, the term 'agreement' can be used for legally binding as well as non-binding agreements (see http://untreaty.un.org/ola-internet/Assistance/Guide.htm#agreements (last accessed 15 June 2012)).

7 This section is based on Andersen *et al.* (2010).

8 A thorough analysis of the recognition of farmers' rights in the ITPGRFA is found in Batta Bjørnstad (2004). Further analyses of the ITPGRFA provisions on farmers' rights are provided by the Farmers' Rights Project, www.farmersrights.org; see also Moore and Tymowski (2005).

9 The Cartagena Protocol on Biosafety to the Convention on Biological Diversity, 29 January 2000, http://sedac.ciesin.org/pidb/texts-menu.html. It is not dealt with in this chapter, but, as a protocol to the CBD, it is a part of an already established regime.

10 The United States has also signed the CBD but has not ratified it.

11 An interesting analysis of the contents and prospects of the ITPGRFA is found in Fowler (2004). Explanations on the background and contents of the ITPGRFA are presented in Moore and Tymowski (2005).

12 This section is based on Andersen (2008) and Andersen *et al.* (2010).

13 For example, rice, wheat, maize, rye, potatoes, beans, cassava and bananas. Not included are other important crops, including soybeans, tomatoes, cotton, sugarcane, cocoa, groundnuts, as well as many vegetables and important tropical forage plants.

14 This chapter is based on the results of the Farmers' Rights Project of the Fridtjof Nansen Institute, an international project designed to support the implementation of farmers' rights, as they are addressed in the ITPGRFA. Started in 2005, it has been a long-term project with many different components, comprising research and surveys as well as policy guidance, facilitation of consultations, information and capacity building. For an overview of the research reports and activities, see www.farmersrights.org. International Treaty on Plant Genetic Resources for Food and Agriculture, 29 June 2004, www.planttreaty.org/texts_en.htm [ITPGRFA].

15 Convention on Biological Diversity, 31 ILM 818 (1992).

16 An agreement on access to teff genetic resources in Ethiopia, and the fair and equitable sharing of benefits derived from their use has been hailed as one of the most advanced of its time. A thorough study of this agreement between a Dutch company and Ethiopian authorities shows however that the implementation failed. As a result of several circumstances, Ethiopia was left with fewer possibilities for generating and sharing the benefits from the use of teff genetic resources than before (Andersen and Winge, 2012).

17 Progress is slow, however, due to a lack of resources and political attention.

18 Agreement on Trade Related Aspects of Intellectual Property Rights, Annex 1C of the Marrakech Agreement Establishing the World Trade Organization, 15 April 1994, 33 ILM 15 (1994).

19 International Convention for the Protection of New Varieties of Plants, 2 December 1961, www.upov.int/en/publications/conventions/index.html.

20 Protection of Plant Varieties and Farmers' Rights Act, http://agricoop.nic.in/seeds/farmersact2001.htm.

21 Protection of Plant Varieties and Farmers' Rights Act, 2001, http://agricoop.nic.in/seeds/farmersact2001.htm.

References

Andersen, R. (2001) 'Conceptualizing the Convention on Biological Diversity: Why is it difficult to determine the "country of origin" of agricultural plant varieties?' *FNI Report no. 7/2001*, Fridtjof Nansen Institute (FNI), Lysaker, Norway.

Andersen, Regine (2005a) 'The Farmers' Rights Project – Background Study 1: The History of Farmers' Rights: A Guide to Central Documents and Literature', *FNI Report no. 8/2005*, Fridtjof Nansen Institute, Lysaker.

Andersen, R. (2005b) 'The Farmers' Rights Project – Background Study 2: Results from an international stakeholder survey on Farmers' Rights', *FNI Report no. 9/2005*, FNI, Lysaker, Norway.

Andersen, R. (2006) 'Realising farmers' rights under the International Treaty on Plant Genetic Resources for Food and Agriculture, summary of findings from the Farmers' Rights Project (Phase 1), *FNI Report no. 11/2006*, FNI, Lysaker, Norway.

Andersen, R. (2008) *Governing Agrobiodiversity: Plant Genetics and Developing Countries*, Ashgate, Aldershot, UK.

Andersen, R. (2009) Information Paper on Farmers' Rights Submitted by the Fridtjof Nansen Institute, Norway, Based on the Farmers' Rights Project, input paper submitted to the Secretariat of the ITPGRFA, 19 May 2009, Doc. IT/GB-3/09/Inf, Rome, Italy.

Andersen, R. (2011) 'Farmers' Rights in Norway: a case study', *FNI Report no. 11/2011* (Norwegian edition) and *FNI Report no. 17/2012* (English edition), FNI, Lysaker, Norway.

Andersen, R. (2016) 'Realising farmers rights to crop genetic resources', chapter 7 in M. Halewood (ed.) *Farmers' Crop Varieties and Farmers' Rights: Challenges in Taxonomy and Law*, Earthscan for Routledge, Abingdon, UK.

Andersen, R. and Winge, T. (2008) 'The Farmers' Rights Project – background study 7: success stories from the realization of farmers' rights related to plant genetic resources for food and agriculture', *FNI Report 4/2008*, FNI, Lysaker, Norway.

Andersen, R. and Winge, T. (2012) 'The Access and Benefit-sharing Agreement on Teff Genetic Resources: Facts and Lessons', *FNI Report 6/2012*, FNI, Lysaker, Norway.

Andersen, R. and Winge, T. (2013) *Realising Farmers' Rights to Crop Genetic Resources: Success Stories and Best Practices*, Routledge, Abingdon, UK.

Andersen, R., Tvedt, M.W., Fauchald, O.K., Winge, T., Rosendal, K. and Schei, P.J. (2010) 'International Agreements and Processes Affecting an International Regime on Access and Benefit-sharing under the Convention on Biological Diversity: Implications for its Scope and Possibilities of a Sectoral Approach', *FNI Report 3/2010*, FNI, Lysaker, Norway, 47.

Bjørnstad, B.S-I. (2004) 'Breakthrough for 'the South'? *FNI Report 13/2004*, FNI, Lysaker, Norway.

Brush, S.B. (2005) 'Protecting traditional agricultural knowledge', *Washington University Journal of Law and Policy*, Vol. 17, pp. 59–109.

Centro Internacional de la Papa and Federación Departemental de Comunidades Campesinas (2006) *Catálogo de Variedades de Papa Nativa de Huancavelica, Peru*, Centro Internacional de la Papa and Federación Departemental de Comunidades Campesinas, Lima, Peru.

Esquinas-Alcázar, J. (2005) 'Protecting crop genetic diversity for food security: political, ethical and technical challenges', *Nature Reviews Genetics*, Vol. 6, pp. 946–53.

Evans, P. and Walsh, J. (1994) *The EIU Guide to the New GATT.* Inter-American Institute for Cooperation on Agriculture, San José, Costa Rica.

Food and Agriculture Organization of the United Nations (FAO) (1986) Report of the Working Group of the FAO Commission on Plant Genetic Resources, 2-3 June 1986, CPGR/87/3, October 1986, FAO, Rome, Italy.

FAO (1987) *Report of the Second Session of the Commission on Plant Genetic Resources*, Doc. CL 91/14, FAO, Rome, Italy.

FAO (1989) FAO C 1989/REP, Conference resolution 5/89, Farmer's Rights, *Report of the Conference of FAO*, 25th session, Rome, Italy.

FAO (1991) FAO C 1991/REP, Conference resolution 3/91, *Report of the Conference of FAO*, 26th session, Rome, Italy.

FAO (1998) *State of the World's Plant Genetic Resources for Food and Agriculture*, FAO, Rome.

Fowler, C. (1994) *Unnatural Selection: Technology, Politics and Plant Evolution*, Gordon and Breach, Yverdon, Switzerland.

Fowler, C. (2001) 'Protecting famer innovation: the Convention on Biological Diversity and the question of origin', *Jurimetrics*, Vol. 41, pp. 477–88.

Fowler, C. (2004) 'Diversity and protectionism – use of genebanks: trends and interpretations', pp. 49–51 in *Food Security and Biodiversity: Sharing the Benefits of Plant Genetic Resources*, Basel, Switzerland.

Fujisaka, S., Williams, D. and Halewood, M. (eds) (2011) 'The Impact of Climate Change on Countries' Interdependence on Genetic Resources for Food and Agriculture', *Background Paper No. 48*, Rome, Italy: Food and Agriculture Organization of the United Nations (FAO).

Global Alliance for the Future of Food (2016) *The Future of Food: Seeds of Resilience; A Compendium of Perspectives on Agricultural Biodiversity from around the World*, http://futureoffood.org/wp-content/uploads/2016/09/FoF_Print_Compendium_82616_FA.pdf.

Moore, G. and Tymowski, W. (2005) 'Explanatory Guide to the International Treaty on Plant Genetic Resources for Food and Agriculture', *IUCN Environmental Policy and Law Paper no. 57*, International Union on the Conservation of Nature, Gland, Switzerland and Cambridge, UK.

Palacios, X. F. (1998) 'Contribution to the estimation of countries' interdependence in the area of plant genetic resources', *Background Study Paper no. 7*, Commission on Genetic Resources for Food and Agriculture, FAO, Rome, Italy.

Scurrah, M., Andersen, R. and Winge, T. (2008) 'Farmers' Rights in Peru: Farmers' perspectives', *FNI Report no. 16/2008*, FNI, Lysaker, Norway.

Scurrah, M., de Haan, S. and Winge, T. (2013) 'Cataloguing Potato Varieties and Traditional Knowledge from the Andean Highlands of Huancavelica, Peru', in Andersen, R. and Winge, T. (eds), *Realising Farmers' Rights to Crop Genetic Resources: Success Stories and Best Practices*. Routledge, Abingdon (UK), pp. 64–79.

United Nations (2009) *Seed Policies and the Right to Food: Enhancing Agrobiodiversity and Encouraging Innovation*, Interim report of the UN Special Rapporteur on the Right to Food, transmitted by the Secretary General to the members of the General Assembly of the UN for its sixty-fourth session, UN, New York, NY, USA.

United Nations Environment Programme (UNEP) (1992) *Nairobi Final Act of the Conference for the Adoption of the Agreed Text of the Convention on Biological Diversity*, UNEP, Nairobi, Kenya.

Vernooy, R., Shrestha, P. and Sthapit, B. (eds) (2015) *Community Seed Banks: Origins, Evolution and Prospects*, Earthscan for Routledge, Abingdon, UK.

Wilson, E.O. (1992) *The Diversity of Life*, Penguin, London, UK.

Yusuf, A.A. (1998) 'TRIPS: background, principles and general provisions', pp 3–21 in C.M. Correa and A.A. Yusuf (eds), *Intellectual Property and International Trade: TRIPS Agreement*, 2nd edn, Kluwer Law International, Alphen aan den Rij, the Netherlands.

PART III

Access and benefit-sharing

11

ACCESS TO AND BENEFIT-SHARING OF MARINE GENETIC RESOURCES BEYOND NATIONAL JURISDICTION

Developing a new legally binding instrument

Carlos M. Correa

This chapter examines the underlying principles and main elements of binding and non-binding international instruments relating to access and benefit-sharing of genetic resources. It discusses the gap left by the United Nations Convention on the Law of the Sea and some of the elements that might be considered in developing a new legally binding instrument on access and benefit-sharing, derived from the exploitation of marine genetic resources found in areas beyond national jurisdiction, as mandated by the UN General Assembly. The chapter further elaborates on some of the problems faced in the implementation of the Convention on Biological Diversity and other international instruments on genetic resources and the extent to which the principles and operative mechanisms of those instruments may provide guidance for the development and adoption of that new instrument.

Introduction

An international legal framework relating to the conservation, sustainable use, access to and sharing of the benefits arising from the exploitation of genetic resources was established more than twenty years ago. The Convention on Biological Diversity (CBD),[1] adopted in 1992, set out the main principles and mechanisms to that end. The CBD's objectives are the conservation of biological diversity, the sustainable use of its components and the fair and equitable sharing of the benefits arising out of the utilization of genetic resources (Article 1). With 196 contracting parties,[2] the CBD has become the main framework for action by the international community against the loss of biodiversity. Pursuant to the adoption of the Aichi Biodiversity Targets,[3] a set of strategic goals has been defined under the Strategic Plan for Biodiversity 2011–2020.[4]

Further steps were taken by the international community to apply such principles to plant genetic resources for food and agriculture with the adoption of the FAO Treaty on Plant Genetic Resources for Food and Agriculture (ITPGRFA),[5] and to clarify and strengthen the CBD provisions on benefit-sharing through the Nagoya Protocol on Access to Genetic Resources and the Fair and Equitable Sharing of Benefits Arising from their Utilization ('the Nagoya Protocol').[6] Developing countries, as proponents of this Protocol, sought to strengthen the mechanism for benefit-sharing, in view of the limited income generated by such mechanism under the CBD (Ruiz Muller, 2015), among other things, due to the limitations of the CBD's regime regarding the commercial exploitation of derivatives from genetic resources (Nijar, 2011).

In addition to these binding international instruments,[7] under the auspices of the World Health Organization (WHO), a Pandemic Influenza Preparedness (PIP) Framework was adopted in 2011 to implement a global approach to pandemic influenza preparedness and response. A central objective of this Framework was to improve the sharing of information about influenza viruses with human pandemic potential and to set up a regime for benefit-sharing, while facilitating access by developing countries to vaccines and other pandemic-related supplies.[8]

The CBD and the Nagoya Protocol[9] are wide in scope. They cover all kinds of genetic resources,[10] including those of marine origin. The ITPGRFA, in contrast, is a specialized agreement that only covers, as noted, plant genetic resources for food and agriculture. The specific characteristics of these resources (generated through breeding practices), the difficulty to determine the commercial worth of individual accessions, the monitoring costs of tracking geneflows (if possible at all to identify the respective countries of origin), the interdependence among countries in their use and the importance of such resources for food security explained the need for a separate instrument (Laliberté *et al.*, 2000), which includes a Multilateral System under which resources are available without the need to enter into bilateral transactions, as is the case under the CBD and the Nagoya Protocol (Frison *et al.*, 2011).

The regulatory gap

One pillar of the legal framework established by the three binding international treaties referred to above is the recognition of States' sovereign rights on the genetic resources that reside in their territories.[11] They apply to genetic resources that may be found and accessed *within national jurisdictions*. The Nagoya Protocol, however, also requires the contracting parties to consider the establishment of a 'global multilateral benefit-sharing mechanism'. Article 10 of the Protocol stipulates that

> Parties shall consider the need for and modalities of a global multilateral benefit sharing mechanism to address the fair and equitable sharing of benefits derived from the utilization of genetic resources and traditional knowledge associated with genetic resources that occur in transboundary situations or for which it is not possible to grant or obtain prior informed consent. The benefits shared by users of genetic resources and traditional knowledge associated with genetic resources through this mechanism shall be used to support the conservation of biological diversity and the sustainable use of its components globally.

The wording 'or for which it is not possible to grant or obtain prior informed consent' has been interpreted as implying that the global multilateral mechanism might be applicable to marine genetic resources in *areas beyond national jurisdiction* (ABNJs) (Ruiz Müller, 2015). However, the Preamble of the Protocol reaffirms 'the sovereign rights of States over their natural resources and according to the provisions of the Convention'. Hence, it would seem difficult to consider that the Protocol is outside the boundaries set out by the CBD, which is firmly rooted in national jurisdictions (Fedder, 2013, p. 51).

The geographical coverage of the CBD and the Nagoya Protocol leaves a noticeable gap in the international regime regarding marine genetic resources in ABNJs. It seems an obvious observation that policies for the conservation and sustainable use of these resources are not less important than those relating to resources under national jurisdiction. Many of such resources exist both in maritime areas subject to costal states' jurisdiction – as defined in accordance with the United Nations Convention on the Law of the Sea (UNCLOS) – and outside such jurisdiction, while some may only exist beyond those national maritime areas.[12]

Several studies and reports have noted the immense[13] and still vastly untapped value of marine genetic resources in the deep seabed and the high seas for a range of industries including pharmaceuticals, food and beverage, cosmetics, agriculture and industrial biotechnology (Wynberg, 2015, p. 397).[14] Those resources are considered today to be more valuable than the mineral resources found in the seabed or terrestrial genetic resources (Centre for International Law, 2016, p. 10). It has been observed that deep seabed mining remains uncompetitive compared to land-based mining (Scovazzi, 2010, p. 6), and that the 'deepest of ironies' in the context of the UNCLOS is that while the exploitation of mineral resources in the seabed has not yet became economically attractive, bioprospecting and the commercial use of marine genetic resources already opens promising opportunities (Glowka, 1995). It has also been argued that

> there is a strong difference between the yields of marine biodiversity versus terrestrial biodiversity. If a marine species is isolated, there is a greater chance to derive a biotechnological application from it than it if was a terrestrial species.
>
> (Centre for International Law, 2016, p. 37)

Marine bioprospecting in areas within and beyond national jurisdiction is taking place. There is particular interest in marine species that live in 'extreme environments, such as hydrothermal vents, seamounts, cold seeps and submarine trenches ("extremophiles"), which trigger organisms to adopt new biosynthetic pathways that generate interesting compounds' (Wynberg, 2015, p. 398). However, there is a deep asymmetry in the capacity of developed and developing countries to undertake those activities, given the financial resources and complex technologies and equipment needed to explore the deep seabed. Thus, the daily cost of renting an oceanography vessel for commercial research may range from US$40,000 to US$100,000. This order of investment is affordable to only a few governments and private companies (Centre for International Law, 2016, p. 49), particularly in view of the high risks of sampling in the deep sea, the low chances of success and the regulatory hurdles for product marketing approval.[15] A workshop report noted in this regard that

> Five countries including Japan, U.S.A., France, Russia and China possess manned research submersibles able to dive under 4,000 meter below sea level in 2012. And United Kingdom, Germany and Korea aggressively conduct marine survey. [The] [c]urrent situation indicates that limited countries can access to BBNJ [Biodiversity Beyond National Jurisdiction]. This causes [*sic*] that a first-come-first-served rule under the principle of freedom in the high sea is criticized by other countries.
>
> (National Institute of Genetics, 2015, p. 29)

But while most marine bioprospecting has been undertaken by developed countries, the sampling is often conducted in developing, tropical countries (Wynberg, 2015, p. 399). Moreover, 'there are at present few if any restrictions on access to marine genetic resources in areas beyond national jurisdiction or substantive controls as to how the acquisition of marine genetic resources is to be undertaken' (Directorate General for Internal Policies, 2014, p. 47).

The scientific and economic value of marine genetic resources has been widely recognized by the international community. Thus, the United Nations General Assembly (UNGA) noted in Resolution 65/37 ('Oceans and the law of the sea', 2010) 'the abundance and diversity of marine genetic resources and their value in terms of the benefits, goods and services they can provide' (para. 168) as well as 'the importance of research on marine genetic resources for the purpose of enhancing the scientific understanding, potential use and application, and enhanced management of marine ecosystems' (para. 169).[16] The Rio+20 Conference outcome document, *The Future We Want*,[17] further recognized 'the importance of the conservation and sustainable use of marine biodiversity beyond areas of national jurisdiction' and made a commitment 'to address, on an urgent basis, the issue of the conservation and sustainable use of marine biological diversity of areas beyond national jurisdiction, including by taking a decision on the development of an international instrument under the Convention on the Law of the Sea' (para. 162).

Soon after the Rio+20 Conference, an important decision was made by UNGA on the subject. On 19 June 2015 it agreed to develop 'an international legally-binding instrument under the Convention [on the Law of the Sea] to deal with the conservation and sustainable use of marine biological diversity of areas beyond national jurisdiction'.[18] This resolution set in motion one of the most important negotiating processes currently underway within the UN system.

The negotiations on the proposed 'international legally-binding instrument' should address the topics identified in the so-called BBNJ Working Group's '2011 package', namely the conservation and sustainable use of marine biological diversity in the ABNJ, including questions regarding the sharing of the benefits arising from the exploitation of such marine genetic resources.

Designing a legally binding instrument

Importantly, the UNGA resolution 69/292 mandated UN members to negotiate 'under the Convention [on the Law of the Sea]'. This means that the resulting agreement, if adopted, should be consistent with UNCLOS and eventually fill its gaps in relation to genetic resources found in ABNJs. Therefore, a starting point of the negotiating exercise will necessarily be the interpretation of the UNCLOS in relation to the issues that need to be addressed.

A key interpretative issue in this regard is which regime under UNCLOS is applicable to marine genetic resources in the ABNJ. The positions held around this issue reflect the asymmetry mentioned above regarding the capacity to explore and benefit from such resources and the products derived therefrom.

In accordance with one view, marine genetic resources and their derivatives should be deemed to be subject to the principle of Freedom of the High Seas. If this principle were accepted, there would be free access to genetic resources in the ABNJ, freedom to undertake not only scientific research but also bioprospecting activities for commercial purposes, without any obligation to ensure the public availability of the information obtained nor to share benefits with States not involved in those activities (Centre for International Law, 2016, p. 5). While fishing is a freedom of the High Seas recognized under UNCLOS (Article 87.1(e)), there is no reference in this Convention to bioprospecting. This is understandable, as UNCLOS did not allude to genetic resources as such. Although Article 87.1 does not contain an exhaustive list of 'freedoms' (as the chapeau refers to '*inter alia*'), this is not sufficient to support the interpretation that bioprospecting of genetic resources is subject to the same regime as fishing. Bioprospecting, unlike fishing, involves obtaining samples of biological materials – new compounds can now be identified from mere micrograms of source material[19] – that contain genetic information of actual or potential value, rather than capturing large volumes of specimens. Despite the small quantities needed, sampling may have adverse effects on the environments where it is conducted; however, so far the International Seabed Authority (ISA) has only issued 'Regulations on prospecting and exploration for polymetallic nodules in the Area' (ISA/6/A/18), while the Legal and Technical Commission adopted 'Recommendations for the guidance of the contractors for the assessment of the possible environmental impacts arising from exploration for polymetallic nodules in the Area' (ISA/7/LTC/1/Rev.1, 10 July 2002).

If Freedom of the High Seas were the accepted premise of the forthcoming negotiations pursuant to UNGA Resolution 69/292, the asymmetry generated by the differences in capacity to undertake bioprospecting would remain unsolved. That principle, however, is not absolute (Scovazzi, 2010, p. 1). The adoption of a new binding instrument regarding bioprospecting of genetic resources would have to limit, hence, some of the currently existing rights in the High Seas by providing for some exceptions to the 'freedoms' recognized under the UNCLOS[20] and introducing obligations on bioprospectors, notably regarding benefit-sharing.

A different view is that all the resources of the ABNJ, including marine genetic resources, should be deemed to be the 'common heritage of the mankind' (Article 136, UNCLOS) and, consequently, all activities (including bioprospecting) should 'be carried out for the benefit of mankind as a whole' (Article 140.1, UNCLOS) subject to 'the equitable sharing of financial and other economic benefits' (Article 140.2, UNCLOS). These and other provisions in Part XI of UNCLOS aim at ensuring that all States have open and free access to such resources, and that a principle of equitable sharing of financial and other economic benefits is applied, taking particularly into account the interests and needs of developing States (Article 140.1). The fundamental character of the principle of 'common heritage of mankind' in the context of the UNCLOS was confirmed by its Article 311.6, according to which

> States Parties agree that there shall be no amendments to the basic principle relating to the common heritage of mankind set forth in Article 136 and that they shall not be party to any agreement in derogation thereof.

An important limitation to this view is that Article 133(a) of UNCLOS stipulates that for the purposes of Part XI, '"resources" means all solid, liquid or gaseous mineral resources *in situ* in the Area at or beneath the seabed, including polymetallic nodules'. This means that Part XI applies to mineral resources of the seabed.[21]

The legal implications of the absence of any reference in Part XI to genetic resources is debatable. When UNCLOS was negotiated it was thought that, in the absence of light at the bottom of the sea, no living organisms could be found there; today it is known that a wide variety of organisms live at or near the seafloor of the Area (Centre for International Law, 2016, p. 8). In addition, Resolution 25/2749 (1970)[22] of the UNGA, which declared that the Area and its resources were the common heritage of mankind and provided the basis for Part XI of the UNCLOS, was not limited to minerals.[23]

Whatever the reason for the exclusion, a reading of the UNCLOS in accordance with the interpretative principles codified in Article 31 of the Vienna Convention on the Law of the Treaties, would make it difficult to assert that marine genetic resources are effectively covered under Part XI of UNCLOS. Nothing, however, would prevent an extension of its provisions, through a new binding instrument to be developed pursuant to the UNGA mandate, to marine genetic resources in the ABNJs. The fact that marine genetic resources are renewable, while the mineral resources are not (Centre for International Law, 2016, p. 8) is clearly insufficient to support the view that the principle of Freedom of the High Seas, rather than Common Heritage of Mankind, must apply to marine genetic resources in the ABJN. Similarly, the argument that the concept of common heritage of mankind was intended to protect the Area against any attempt at appropriation by a State (Centre for International Law, 2016, p. 41) does not support the conclusion that it is not applicable to such resources. In fact, that same concept was incorporated in the International Undertaking on Plant Genetic Resources for Food and Agriculture adopted by the Food and Agriculture Organization (FAO) in 1981 in order to counter the privatization of genetic resources through intellectual property rights (GRAIN, 2001). The fact that the applicability of this concept in the area of plant genetic resources was later revisited with the adoption of the bilateral model of access and benefit-sharing under the CBD and of the Multilateral System created by the ITPGRFA (in relation to a number of listed crops) does not mean that it is not applicable to marine genetic resources in the ABNJs. The major difference is, of course that, as mentioned, the CBD (as well as the Nagoya Protocol) and the ITPGRFA are grounded on the principle of national sovereignty.

Under the UNCLOS, a distinction is made between 'scientific research' and other activities that may eventually involve commercial purposes. This distinction has clear operative implications, as it determines the applicability of different legal rules. Given that the future legally binding instrument on access and benefit-sharing for marine genetic resources in the ABNJs should be framed under the UNCLOS, the distinction made by the Convention in relation to different activities needs to be properly taken into account.

Scientific research is one of the freedoms that States enjoy under UNCLOS in the High Seas (Article 87.1(f)). This freedom is not absolute, however, since any activity must be undertaken with due regard to the interests of other States (Article 87.2) and for peaceful purposes only (Article 88). In addition, in accordance with Article 256 of UNCLOS, '[a]ll States, irrespective of their geographical location, and competent international organizations have the right, in conformity with the provisions of Part XI, to conduct marine scientific research in the Area'.

Part XIII of UNCLOS introduces a number of obligations in relation to the conduct of marine scientific research, including:

(1) States and competent international organizations shall, in accordance with the principle of respect for sovereignty and jurisdiction and on the basis of mutual benefit, promote international cooperation in marine scientific research for peaceful purposes (Article 242.1 UNCLOS).
(2) Information on proposed major marine scientific research programs and their objectives shall be made available by publication and dissemination through appropriate channels (Article 244.1 UNCLOS).
(3) The knowledge resulting from marine scientific research shall also be made available by publication and dissemination through appropriate channels (Article 244.1 UNCLOS).
(4) States, both individually and in cooperation with other States and with competent international organizations, shall actively promote the flow of scientific data and information and the transfer of knowledge resulting from marine scientific research, especially to developing States (Article 244.2 UNCLOS).
(5) States, both individually and in cooperation with other States and with competent international organizations, shall actively promote the strengthening of the autonomous marine scientific research capabilities of developing States through, *inter alia*, programmes to provide adequate education and training of their technical and scientific personnel (Article 244.2).

In addition, the provisions of Part XI of the UNCLOS apply to marine scientific research undertaken in the Area (Article 256). This means, in particular, that such research, as mentioned, must be carried out 'exclusively for peaceful purposes and for the benefit of mankind as a whole, in accordance with Part XIII' (Article 143.1) and that international cooperation in marine scientific research in the Area must be promoted (Article 143.3).

One important question is whether the described framework for scientific research, which requires active cooperation and dissemination (sharing) of the outcomes of scientific research, is applicable to marine genetic resources. Importantly, under Article 244 of UNCLOS, States 'shall actively promote the flow of scientific data and information and the transfer of knowledge resulting from marine scientific research…'.

The UNCLOS provisions on scientific research are broadly drafted. They are not circumscribed to specific subjects of research. No reference is made in Articles 143.1 and 143.3 to the 'resources' in the Area, a reference that would have otherwise limited (given the definition of 'resources' mentioned above) the applicability of those provisions to mineral resources. In the absence of such a reference, those provisions should be deemed equally applicable to scientific research conducted on marine genetic resources (Greiber, 2011, p. 14).[24]

Another relevant question is whether 'bioprospecting' may be deemed to be part of 'scientific research'. A study prepared in the context of the CBD suggested that the former inherently encompasses a commercial intent, and suggested that the UNCLOS provisions on scientific research were not applicable to bioprospecting (UNEP/CBD/SBSTTA, 2003, 11).[25] It has been argued, however, that despite the absence of any definition of 'marine scientific research' this concept covers projects 'of direct significance for the exploration and exploitation of natural resources, whether living or non-living' (UNCLOS, 246. 5(a)) and hence, it should be interpreted as encompassing bioprospecting (Scovazzi, 2010, pp. 12–13). In any case, nothing would prevent the contracting parties to clarify the scope of 'marine scientific research' or agree on the extension of such provisions to bioprospecting in the ABNJs.

Lessons from the current international regime

In negotiating the new proposed legally binding instrument, it would be important to consider a number of difficulties that arose in the development and implementation of the international regime for genetic resources under national jurisdiction. The new process put in motion by UNGA resolution 62/292 opens an opportunity for overcoming some of those difficulties and developing a regime that provides sufficient legal certainty to all interested parties while ensuring an equitable distribution of benefits between those who invest in bioprospecting and the international community.

Definitions

Some of the difficulties referred to emerged from definitional problems. Thus, the distinction between 'genetic' and 'biological' resources under the CBD has led to different interpretations: while for some they are synonymous (Cabrera Medaglia *et al.*, 2007, p. 29), for others the former can only be deemed a subset of the latter (Fedder, 2013, p. 34). The lack of a clear distinction between these two concepts has brought about confusion in the national implementation of the CBD provisions.

Further, 'genetic resources' are defined under that Convention as 'materials'. The CBD and the other instruments on genetic resources referred to above have been conceived to deal with access to *materials* that consist of or contain genetic information. This characterization overlooks the *intangible* nature of the information, which is the object of actual interest for biotechnological developments (Ruiz Muller, 2015, p. 4). Developments in genetic science have allowed researchers to obtain workable information from data bases. The *dematerialization* of genetic resources through 'digital bioprospecting' means that access to physical samples is not required to obtain complete and usable information.

The dematerialization of genetic information poses a new challenge to the access and benefit-sharing system built up on the basis of the international instruments referred to above, particularly if the relevant information is freely accessible through genome databases.[26] It is indeed crucial to determine whether the use of genetic information would constitute an 'access' to material or not and the implications of the utilization of synthetic biology (National Institute of Genetics, p. 37).[27] The design of a new instrument on marine genetic resources in the ABNJs opens the possibility of exploring a system that clearly encompasses both the utilization of materials as well as genetic information obtainable without getting access to a physical sample of the relevant biological material.

Another problem with the CBD has been the lack of a definition of what is meant by the 'utilization' of genetic resources for the purposes of the application of access and benefit-sharing legislation. Gray areas have included, for instance, taxonomic studies (Cabrera Medaglia *et al.*, 2007, p. 30) where no commercial utilization of the genetic resources or their components is made.

The Nagoya Protocol incorporated a definition of 'utilization of genetic resources' that has helped to clarify the scope of the obligations of the contracting parties:

> 'Utilization of genetic resources' means to conduct research and development on the genetic and/or biochemical composition of genetic resources, including through the application of biotechnology as defined in Article 2 of the Convention (Article 2(c)).

This definition confirms that research and development on the 'biochemical composition' of genetic resources is a form of utilization of genetic resources. This clarification was needed and is useful. Although the concept of 'research and development' is widely used[28] and generally well understood (Greiber *et al.*, 2010, p. 65), some differences in interpretation may remain regarding which activities are effectively covered. An illustrative list of such activities would help to give more certainty to providers and users of genetic resources. In this regard, the Group of Legal and Technical Experts on Concepts, Terms, Working Definitions and Sectoral Approaches, established under the CBD, have suggested a non-exhaustive list of uses of genetic resources:

- genetic modification;
- biosynthesis (use of genetic material as a 'factory' to produce organic compounds);
- breeding and selection;
- propagation and cultivation in the form received;
- conservation;
- characterization and evaluation; and
- production of compounds naturally occurring in genetic material (i.e., extraction of metabolites, synthesis of DNA segments, and production of copies).[29]

In addressing benefit-sharing, a differentiation between commercial and non-commercial activities may be made under the CBD. This differentiation is specifically articulated in the Nagoya

Protocol,[30] but is difficult to operationalize. Activities that may be planned and initiated as non-commercial, may end up with the development of commercial products. In relation to the PharmaSea project it has been noted, for instance, that

> The PharmaSea Project is non-commercial work at the beginning but is clear commercial work and there is also a lot of third-party transfer. The PharmaSea project is therefore a very complicated example for considering ABS. We really have to track every movement of samples very carefully and have to make a lot of agreements not only in terms of access, PIC and MAT, but also in terms of material transfer between academic and commercial partners.
>
> (National Institute of Genetics, 2015, p. 33)

A fragmented set of rules based on the commercial or non-commercial purpose of research and development may, given their blurred contours, be problematic and deepen the asymmetries between countries with and without financial and other resources to undertake research on marine genetic resources in the ABNJs, and create inequalities in sharing the benefits arising from their exploitation.

Benefits

Under the CBD and the other international treaties on genetic resources mentioned above, benefits to be shared arising from the exploitation of such resources may be monetary or non-monetary. Even though the CBD entered into force in December 1993, the royalties and other payments directly obtained by providing countries acting under the Convention seem to have been quite limited (Ruiz Muller, 2015, p. 24). They are certainly disappointing in view of the expectations generated by the Convention upon its adoption as a source of new income for countries providing genetic resources, particularly those rich in biological diversity. It has been noted in this regard that

> Global experience since the CBD in 1992 has proved that benefit-sharing, both as an incentive for conservation as well as royalties for access to traditional knowledge, is a proposition akin to fetching water in a sieve. Hardly any successful model of ABS that is a sustainable source of supplementary income for the rural communities is known.
>
> (Prathapan *et al.*, 2011, p. 290)[31]

The bilateral approach adopted by the CBD has not been successful in promoting responsible bioprospecting. According to one commentator,

> Over a period of more than two decades, relatively few bioprospecting projects exist where PIC [prior informed consent] and MAT [mutually agreed terms] correspond to the contours envisioned by the CBD and national ABS [access and benefit-sharing] frameworks.
>
> (Ruiz Muller, 2015, p. 33)

Arguably, one of the problems that has contributed to the poor performance of access and benefit-sharing legislation adopted pursuant to the CBD framework has been the apparent limitation of the CBD to benefits arising from the exploitation of the genetic resources *as such*, that is, biological materials that contain functional units of heredity,[32] thereby excluding (or at least leaving in a nebulous situation) those obtainable from *derivatives* of such resources. This gap was specifically addressed by the Nagoya Protocol, which defines 'biotechnology' as 'any technological application that uses biological systems, living organisms, or derivatives thereof, to make or modify products or processes for specific use', and 'derivative' as 'a naturally occurring biochemical compound resulting from the genetic expression or metabolism of biological or genetic resources, even if it does not contain functional units of heredity' (Article 2(d) and (e)).

The extent to which the Nagoya Protocol and, in particular, the introduction of the concept of 'derivatives' may effectively enhance the implementation of the CBD bilateral approach is an open question. It is, at present, too early to assess the outcomes of the Protocol, which also includes an improved compliance regime (Articles 15 and 16).

The effectiveness of the ITPGRFA in generating monetary benefits has also been questioned, particularly by developing countries (López Noriega, Wambugu and Mejías, 2013). The lack of monetary flows generated under the Multilateral System established under the ITPGRFA has triggered proposals for amendment of the treaty and/or the Standard Material Agreement adopted by the Governing Body of the Treaty. Discussions are ongoing in the context of FAO to develop a 'subscription system', with the aim of increasing the income available for benefit-sharing through user payments calculated on the sales of one or more crops (Tsioumani, 2015). Monetary payments are only provided for under the current Multilateral System in situations where a recipient who commercializes a product that is a plant genetic resource for food and agriculture and that incorporates material accessed from the Multilateral System, imposes a restriction on others from engaging in further research, breeding or training (Article 13.2(d)(ii)). In reality, it was not reasonable to expect that, subject to these conditions, the system could have generated a significant flow of income.

Benefit-sharing may also be realized on the basis of non-monetary benefits, such as the transfer of know-how, increased numbers/quality of scientific publications (including through co-publication), capacity building and infrastructure development, and biodiversity conservation. Some of these benefits often reflect practices which are common within research communities (Ruiz Muller, 2015, p. 24). In some cases, such as the agreements for benefit-sharing entered into in Brazil, Colombia and the Philippines, the benefits obtained were, mainly or entirely, non-monetary, as no royalties were paid in exchange for the access granted to genetic resources (Prathapan *et al.*, 2011, p. 291). This may also be the case in many cases of bioprospecting in the ABNJs (National Institute of Genetics, 2015, p. 36), particularly when research takes a long time and there are low chances of commercial success (Wynberg, 2015, p. 403).

The 'Bonn Guidelines on Access to Genetic Resources and Fair and Equitable Sharing of the Benefits Arising out of their Utilization' have listed possible non-monetary benefits to be derived from the exploitation of genetic resources.[33] The relative importance of such benefits as compared to monetary benefits would logically vary case by case and would strongly depend on the research entities' policies and the behavior of the researchers involved. In implementing the referred to PharmaSea project, for instance, it has been found that

> it is extremely difficult to motivate the scientists on board to provide the information on where they are sampling, when they are going to sample, who they are going to send the samples to, and whether they have a material transfer agreement or not. It will take some more years before it will become common practice in the scientific community to take ABS seriously as one of the things scientists just do as part of their work.
>
> (National Institute of Genetics, 2015, p. 33)

In the case of the ITPGRFA, the Multilateral System provides for a range of non-monetary benefits, including exchange of information, technology transfer and capacity building.[34] In addition, it is considered that the free access to germplasm under the Multilateral System, is *per se* a benefit that contracting parties can obtain through mere participation in the system.[35]

With regard to marine genetic resources in the ABNJs, both monetary and non-monetary benefits may be contemplated. It has been argued that the most important benefits may be non-monetary, and that most of the research done will be conducted on samples that come from national government-run cruises (Centre for International Law, 2016, p. 49). However, the potential of obtaining significant monetary benefits has also been mentioned. Thus, it has been noted that

> there are seven drugs from the marine environment in the clinic right now. The total benefits from that, if royalties were granted to the country of origin, would be between US$10 and US$50 million per annum. With a blockbuster drug grossing US$1 billion per year, royalties would likely amount to no more than 3 per cent, which would total US$30 million per year. If ten drugs were being developed, royalties may amount to hundreds of millions of dollars...
>
> (Centre for International Law, 2016, p. 49)[36]

The new legally binding instrument to be developed in relation to genetic resources in the ABNJs should carefully define and balance the monetary and non-monetary benefits obtainable from bioprospecting in the ABNJs. While the experiences under the CBD and the ITPGRFA may be taken into account, they should be understood in the particular normative contexts in which they were developed. Neither of the sovereignty-based bilateral or multilateral mechanisms adopted under those treaties would be applicable; hence, a true global mechanism would need to be created that facilitates dissemination of the new information and knowledge obtained and that, while providing adequate incentives for publicly and privately funded research and development, ensures a fair sharing between all countries of the monetary and non-monetary benefits that might be generated.[37] Such a mechanism should also include effective monitoring and compliance measures. While clearance for marine scientific research may have been tightened in some jurisdictions, including requests on environmental impact and lists of equipment to be used (National Institute of Genetics, 2015, p. 28), non-compliance with access and benefit-sharing regulations may be significant. It has been observed in this regard that

> It is highly probable that most cases of access and use to these [marine genetic] resources may have not been granted by national authorities when falling within the EEZ [exclusive economic zone], thus probably leaving the patent applicant as the sole collector of benefits.
>
> (UNCTAD, 2014, p. 12)

Intellectual property

Finally, an important question is whether the new international regime to be developed would allow for the appropriation of innovations developed from such marine genetic resources in the ABNJs via patents or other intellectual property rights.[38]

While patents on genetic resources as found in nature may not be granted in some countries, other countries with more liberal approaches regarding patentability may permit isolated genes and other substances to be subject to patent protection. For instance, under the European

Directive 98/44/EC on the legal protection of biotechnological inventions (1998), 'biological material which is isolated from its natural environment or produced by means of a technical process may be the subject of an invention even if it previously occurred in nature' (Article 3.2).

A study revealed that 677 international claims regarding marine genetic resources were deposited between 1991 and 2009, covering 8,648 sequences from 520 species.[39] The number of patent applications relating to such resources is likely to be growing exponentially (Arnaud-Haond *et al.*, 2011). Reflecting the asymmetries in national capacities to conduct marine bio-prospecting, the study found that 'claims associated with marine genes originate from only 31 of the 194 countries in the world. Patent holders from ten countries own 90% of the patents deposited with marine genes, with 70% belonging to the top three' (Arnaud-Haond *et al.*, 2011).[40] Several deep seabed organisms have already been patented and used for commercial applications, such as anti-cancer, anti-allergy and anti-coagulant agents and skin protection products (Smith *et al.*, 2015).

Currently patent applicants are not subject in most countries to an obligation to disclose the geographical origin or source of the claimed genetic materials (Vivas-Eugui, 2012). It is often unknown, hence, whether the patented marine genetic resources were obtained within or beyond national jurisdictions (Chiarolla, 2013, p. 39). This might be solved if an international obligation to provide that information were incorporated into the Agreement on Trade Related Aspects of Intellectual Property Rights ('TRIPS Agreement'), as formally requested by a large group of developing countries (Vivas Eugui, 2012, p. 17). Such an obligation might be reinforced, in the case of marine genetic resources, by a *juris tantum* presumption that they have been found in the ABJN in case that no or wrong information is provided by the patent applicant (Millicay, 2007).

Since the negotiation of the CBD, the possible impact of intellectual property rights on the implementation of the Convention has been a controversial issue. Article 16.5 of the CBD only establishes an obligation to cooperate and spells out as an objective of contracting parties' cooperation to ensure that intellectual property rights 'are supportive of and do not run counter' to the CBD objectives.[41] While developing countries have generally expressed their fear that such rights could undermine or impede the implementation of the CBD, developed countries have argued that, on the contrary, they may support the realization of the Convention's objectives (Correa, 2015, p. 19).

In view that the TRIPS Agreement allows for, or requires (as is the case for rmicroorganisms and plant varieties), the protection of genetic materials, the Council for TRIPS was mandated by the WTO's Doha Ministerial Declaration (paragraphs 12 and 19) and by the Hong Kong Ministerial Declaration (paragraph 39), to examine the compatibility between the TRIPS Agreement and the CBD. One key concern of developing countries has been that said Agreement does not contain any provisions to ensure that patent applicants have obtained prior informed consent from the countries of origin of the genetic resources and complied with national regimes on access and benefit-sharing. These countries submitted proposals to the Council of TRIPS for an amendment to the TRIPS Agreement aiming at incorporating an obligation to disclose the source of materials claimed in patent applications (Correa, 2015, p. 20). However, no action has been taken so far on these proposals.

In the context of the ITPGRFA, the issue of appropriation of genetic resources obtained from the Multilateral System was specifically addressed. In accordance with Article 12.3(d):

> Recipients shall not claim any intellectual property or other rights that limit the facilitated access to the plant genetic resources for food and agriculture, or their genetic parts or components, in the form received from the Multilateral System.

While the expression 'in the form received' may be interpreted as allowing for the patenting of modified materials, it sets an interesting precedent for the negotiation of the legally binding instrument to be developed under the UNCLOS.

The implications of the dematerialization referred to above for the protection of genetic resources under intellectual property rights are currently discussed in the context of the CBD.[42] One of the consequences of that development is that, once publicly available in a database, genetic sequences' information would not comply with the novelty requirement necessary to obtain a patent. However, such information may be redesigned and escape an obstacle of lack of novelty. As noted by an expert,

> Gene information is everywhere. Everybody deposits genes online very quickly, and a marine genetic resource or its genetic information can be readily obtained and redesigned. The DNA can be redesigned so thoroughly that it would be very difficult to trace it back to its roots. Those genes could then be recombined from different sources into an organism, which could then be grown, resulting in a product in the end. It would be very hard to track that product to the MGR it came from.
>
> (Centre for International Law, 2016, p. 51)[43]

Importantly, Article 137.2 of UNCLOS provides that

> All rights in the resources of the Area are vested in mankind as a whole, on whose behalf the Authority shall act. These resources are not subject to alienation. The minerals recovered from the Area, however, may only be alienated in accordance with this Part and the rules, regulations and procedures of the Authority.

As noted above, the definition of 'resources' is limited under the UNCLOS. However, it would be difficult to argue that a treatment different from that accorded to minerals – as a common heritage of mankind – should be applied to marine genetic resources in the ABNJs. Economic value can equally be derived from both types of resources. A commentator has argued in this regard that

> Indeed, should MGRs fall under the common heritage of mankind as many developing states maintain, equitable benefit-sharing would apply and, consequently, products derived from these resources should also be the common heritage of mankind... Should bioprospecting be considered marine scientific research, it would appear that granting patents protecting inventions derived from such activities would be contrary to UNCLOS provisions regarding scientific research that prohibit the use of such activities as the legal basis for any claim to any part of the marine environment or its resources. Furthermore, many states are concerned about the use of IP rights as a means to privatize genetic resources without sharing benefits.
>
> (Heafey, 2014, pp. 508–09)

Whether or not marine genetic resources are characterized as part of the common heritage of mankind, it would seem necessary to seek a regulation that prohibits the appropriation of such resources, as found in the ABNJs, via intellectual property rights. An unrestricted access to discoveries as well as upstream research results relating to marine genetic resources may be critical for the advancement of science (Chiarolla, 2013, p. 40). Such resources are international public goods that, due to their very nature, should not be subject to sovereign rights. Nor should they be appropriated under private rights conferred by national governments. The development of the new legally binding instrument may offer a unique opportunity to explore new models to promote innovation that do not rely on exclusive rights (such as patents) but on cooperative actions and accessibility to the outcomes of innovation.[44] Such models should be based on a public domain approach; common pools of samples and related data could be shared through an international network of biorepositories and databases (National Institute of Genetics, 2015, p. 36). A pool of research funds may be envisaged to support, in particular, research in developing countries, without discrimination as to its commercial or non-commercial nature. A research program may be agreed upon for the use of such funds, subject to the condition that any research results are contributed to the common pool (Department for Environment Food and Rural Affairs, 2014, pp. 13–14).

Conclusions

The development of a new legally binding instrument on access and benefit-sharing derived from the exploitation of marine genetic resources found in the ABNJs offers the opportunity to fill an important gap in the UNCLOS. Such resources may give rise to new scientific discoveries, technologies and products, the benefits of which should be equitably shared within the international community. Some of the UNCLOS principles – notably the concept of Common Heritage of Mankind – and UNCLOS obligations regarding the sharing of and cooperation in scientific research, may be further developed in order to regulate access to and utilization of those resources as international public goods. The experience in the application of the CBD and other international instruments relating to genetic resources may be useful in designing this new instrument. However, in doing so, the problems discussed above in relation to these pre-existing instruments need to be adequately addressed and, as far as possible, avoided. New models to encourage research and development without relying on the granting of legal monopolies may be constructively explored.

Notes

1 Available at www.cbd.int/convention/text/.
2 See www.cbd.int/information/parties.shtml.
3 See www.cbd.int/sp/targets/.
4 **Strategic Goal A**: Address the underlying causes of biodiversity loss by mainstreaming biodiversity across government and society;
Strategic Goal B: Reduce the direct pressures on biodiversity and promote sustainable use;
Strategic Goal C: To improve the status of biodiversity by safeguarding ecosystems, species and genetic diversity;
Strategic Goal D: Enhance the benefits to all from biodiversity and ecosystem services;
Strategic Goal E: Enhance implementation through participatory planning, knowledge management and capacity building.
See www.cbd.int/sp/elements/.
5 Adopted in 2001 by Thirty-First Session of the Conference of the Food and Agriculture Organization of the United Nations, www.fao.org/plant-treaty/overview/texts-treaty/en/.
6 The Nagoya Protocol is a supplementary agreement to the CBD. It was adopted in 2010 and entered into force on 12 October 2014.
7 It should also be noted that a bracketed text in a draft developed within the Intergovernmental Committee on Genetic Resources, Traditional Knowledge and Folklore (IGC) of WIPO proposes to exclude marine biodiversity from areas beyond national jurisdiction from a regime to be adopted on genetic resources. See www.wipo.int/tk/en/genetic/.
8 See www.who.int/influenza/resources/pip_framework/en/.

9 The Nagoya Protocol, however, recognized that 'where a specialised international access and benefit-sharing instrument applies that is consistent with, and does not run counter to the objectives of the convention and this protocol, this protocol does not apply for the party or parties to the specialised instrument in respect of the specific genetic resource covered by and for the purpose of the specialised instrument' (Art. 4.4).
10 With the exclusion of human genetic resources.
11 See Art. 3 of the CBD; Art. 6.1 of the Nagoya Orotocol; Preamble of the ITPGRFA. The PIP Framework also recognizes 'the sovereign right of States over their biological resources and the importance of collective action to mitigate public health risks' (Preamble, para. 11).
12 In accordance with one source, 'There has been virtually no example of the industrial application of MGRs [marine genetic resources] in the ABNJ. Since two-thirds of the world oceans are under the ABNJ, ABS issues could emerge when development of science and technology reaches to the ABNJ in the future' (National Institute of Genetics, 2015, p. 41).
13 In accordance with one source, the current sales volume of medicinal drugs originated from marine biological resources would reach about US$2.4 billion (National Institute of Genetics, 2015, p. 29). Another source estimates that, although only a small number of products derived from marine organisms have reached the commercialization phase, the global market for marine biotechnology products may reach US$4.9 billion by 2018 (Wynberg, 2015, p. 397).
14 The PharmaSea project, for instance, financed by the European Union, aims at collecting samples from some of the hottest, deepest and coldest places that will be screened to discover marine microbes and new bioactive compounds to evaluate their potential as novel drug leads, antibiotics or ingredients for nutrition or cosmetic applications (see www.pharma-sea.eu/).
15 Wynberg (2015, pp. 398/399) notes that 'it took more than three decades for Prialt®, the first pharmaceutical based on a marine source – the poison released by a tropical marine cone snail to paralyse its prey – to be approved in the United States as a treatment for chronic pain'.
16 Available at https://documents-dds-ny.un.org/doc/UNDOC/GEN/N10/514/76/PDF/N1051476.pdf?OpenElement.
17 UNGA Resolution A/RES/66/288, 27 July 2012, www.un.org/ga/search/view_doc.asp?symbol=A/RES/66/288&Lang=E.
18 General Assembly Resolution 69/292: Development of an international legally binding instrument under the United Nations Convention on the Law of the Sea on the conservation and sustainable use of marine biological diversity of areas beyond national jurisdiction, www.un.org/depts/los/biodiversity/prepcom.htm.
19 As a result of advances in analytical technologies, such as nuclear magnetic spectroscopy and mass spectrometry (Wynberg, 2015, p. 400).
20 They include (a) freedom of navigation; (b) freedom of overflight; (c) freedom to lay submarine cables and pipelines, subject to Part VI; (d) freedom to construct artificial islands and other installations permitted under international law, subject to Part VI; (e) freedom of fishing, subject to the conditions laid down in section 2; (f) freedom of scientific research, subject to Parts VI and XIII (Art. 87.1 UNCLOS).
21 In 1994 an Agreement relating to the implementation of Part XI was adopted to address certain concerns about the seabed mining provisions contained therein. The Agreement entered into force in 1996. See www.un.org/depts/los/convention_agreements/convention_overview_part_xi.htm.
22 Available at www.un-documents.net/a25r2749.htm.
23 The Resolution stated that 'The sea-bed and ocean floor, and the subsoil thereof, beyond the limits of national jurisdiction (hereinafter referred to as the area), as well as the resources of the area, are the common heritage of mankind' (para. 1), and that 'The exploration of the area and the exploitation of its resources shall be carried out for the benefit of mankind as a whole, irrespective of the geographical location of States, whether land-locked or coastal, and taking into particular consideration the interests and needs of the developing countries' (para. 7).
24 In fact, the ISA seems to have recognized that marine biodiversity is one of the areas of scientific research as defined under the UNCLOS. See Millicay (2007).
25 The study noted that 'bioprospecting' has been identified as 'the exploration of biodiversity for commercially valuable genetic and biochemical resources' and further defined it as 'the process of gathering information from the biosphere on the molecular composition of genetic resources for the development of new commercial products' (para. 49).
26 This issue has been addressed in the context of PIP Advisory Group's on the sharing of influenza genetic sequence data. See, e.g., Pandemic Influenza Preparedness (PIP) Framework Advisory Group Technical

Working Group (TWG) on the sharing of influenza genetic sequence data, Optimal Characteristics of an Influenza Genetic Sequence Data Sharing System Under The PIP Framework, 6 April 2016, www.who.int/influenza/pip/advisory_group/gsd/en/. It has also been considered in the CBD Conference of the Parties held in Cancun on 2–17 December 2017. See 'Digital sequence information on genetic resources', CBD/COP/DEC/XIII/16, www.cbd.int/conferences/2016/cop-13/documents.

27 In the case of the referred to drug Prialt®, the polypeptide was 'obtained through standard pharmaceutical manufacturing processes and exploitation of the natural source of this compound was never considered for supplying sufficient material for development' (Wynberg, 2015, p. 399).

28 The Oxford Dictionary's definition of research is 'the systematic investigation into and study of materials and sources in order to establish facts and reach new conclusions'.

29 Group of Legal and Technical Experts on Concepts, Terms, Working Definitions and Sectoral Approaches, UNEP/CBD/WG-ABS/7/2, 12 December 2008, www.cbd.int/doc/meetings/abs/abswg-07/official/abswg-07-02-en.pdf.

30 In accordance with of the Protocol, in 'the development and implementation of its access and benefit-sharing legislation or regulatory requirements, each Party shall:

> (a) Create conditions to promote and encourage research which contributes to the conservation and sustainable use of biological diversity, particularly in developing countries, including through simplified measures on access for non-commercial research purposes, taking into account the need to address a change of intent for such research' (Art. 8(a)).

31 This mismatch in expectations may also occur in relation to marine bioprospecting. See, e.g., National Institute of Genetics (2015), p. 38.

32 See Art. 2 of the CBD.

33 See Bonn Guidelines, www.cbd.int/abs/bonn/.

34 See Art. 13.2(a) (b) and (c) of the ITPGRFA.

35 Art. 13.1 of the ITPGRFA reads as follows: 'The Contracting Parties recognize that facilitated access to plant genetic resources for food and agriculture which are included in the Multilateral System constitutes itself a major benefit of the Multilateral System…'.

36 An example of a successful drug development is 'Halaven', an anti-cancer drug obtained from Halichondrin molecules which has obtained marketing approval in more than 50 countries, including Japan, the United States and in Europe. Although it was discovered in Japan, sponges supplied from New Zealand permitted further preclinical testing and subsequent development of a synthetic analogue (about 300 mg of Halichondrin B were isolated from 1 tonne of collected sponges). The product is currently synthesized but essentially derived from the natural marine product (National Institute of Genetics, 2015, pp. 12 and 17). See also www.eisai.com/news/news201351.html, accessed 27 February 2017.

37 For possible options for the way forward, see e.g., Greiber, 2011, pp. 44–51.

38 Trade secret protection would be excluded if an obligation to effectively disseminate the outcomes of research were established.

39 See http://concarneau.mnhn.fr/sites/concarneau.mnhn.fr/files/upload/arnaudhaond_rvcc_2015.pdf, accessed December 21, 2016.

40 Another study of 460 patents suggested 'the predominance of the pharmaceutical and bio-technology industries for a range of applications from anti-cancer drugs through to applications in the fields of ecotoxicology, bioremediation, anti-fouling and biofuel production. A significant focus includes micro-algae for large-scale production of polyunsaturated fatty acids for inclusion in dietary supplements; a range of biomedical applications such as fluorescent proteins, bone fillers and bioceramic coatings; polysaccharides derived from algae for use in the food and health sectors; and use of "extrem-ozymes" in industrial applications' (Wynberg, 2015, p. 399). See also Wright *et al.* (2016).

41 Art. 16.5: 'The Contracting Parties, recognizing that patents and other intellectual property rights may have an influence on the implementation of this Convention, shall cooperate in this regard subject to national legislation and international law in order to ensure that such rights are supportive of and do not run counter to its objectives'.

42 See the Decision by the CBD Conference of the Parties XIII/16 'Digital sequence information on genetic resources', CBD/COP/DEC/XIII/16 16, December 2016, www.cbd.int/doc/decisions/cop-13/cop-13-dec-16-en.pdf.

43 Based on a statement by Professor Marcel Jaspars (University of Aberdeen).

44 An interesting discussion and set of proposals in this regard have been made in relation to medical technologies. See, e.g., World Health Organization (2012).

References

Arnaud-Haond, S., Arrieta, J. and Duarte, C. (2011), 'Marine Biodiversity and Gene Patents', *Science*, AAAS, March 24, www.sciencemag.org/.

Cabrera Medaglia, J. and López Silva, C. (2007), 'Addressing the Problems of Access: Protecting Sources, While Giving Users Certainty', *IUCN, Environmental Policy and Law paper No. 67/1.*

Centre for International Law (2016), *Conservation and Sustainable Use of Marine Biological Diversity of Areas beyond National Jurisdiction: Preparing for the PrepCom*, Report of the BBNJ Workshop of the Centre for International Law, National University of Singapore, http://cil.nus.edu.sg/wp/wp-content/uploads/2015/10/CIL-report-of-BBNJ-workshop-21-March-2016-final-2.pdf.

Chiarolla, C. (2013), 'Intellectual Property Rights', *IUCN Information Papers for the Intersessional Workshop on Marine Genetic Resources 2–3 May 2013,* Bonn, www.un.org/depts/los/biodiversityworkinggroup/documents/IUCN%20Information%20Papers%20for%20BBNJ%20Intersessional%20Workshop%20on%20MGR.pdf.

Correa, C. (with contributions from Shashikant, S. and Meienberg, F.) (2015), 'Plant Variety Protection in Developing Countries. A Tool for Designing a Sui Generis Plant Variety Protection System: An Alternative to UPOV 1991', APBREBES, Berne Declaration, Development Fund, SEARICE and TWN, www.apbrebes.org/news/new-publication-plant-variety-protection-developing-countries-tool-designing-sui-generis-plant.

Department for Environment Food and Rural Affairs (2014), 'Valuing the Deep: Marine Genetic Resources in Areas Beyond National Jurisdiction', http://randd.defra.gov.uk/Default.aspx?Menu=Menu&Module=More&Location=None&Completed=0&ProjectID=18977.

Directorate General for Internal Policies (2014), *Towards a Possible International Agreement on Marine Biodiversity in Areas Beyond National Jurisdiction*, Study for the ENVI Committee, European Parliament, Brussels, www.europarl.europa.eu/RegData/etudes/STUD/2014/536292/IPOL_STU(2014)536292_EN.pdf.

Fedder, B. (2013), *Marine Genetic Resources, Access and Benefit-sharing. Legal and Biological Perspectives*, Earthscan, London.

Frison, C., López, F. and Esquinas-Alcazar, J. (2011), *Plant Genetic Resources and Food Security: Stakeholder Perspectives on the International Treaty on Plant Genetic Resources for Food and Agriculture*, Routledge, London.

Glowka, L. (1995), *The Deepest of Ironies: Genetic Resources, Marine Scientific Research and the Area*, IUCN, Gland.

GRAIN (2001), 'International Undertaking on Plant Genetic Resources: The Final Stretch', www.grain.org/Article/entries/90-international-undertaking-on-plant-genetic-resources-the-final-stretch.

Greiber, T. (2011), *Access and Benefit-sharing in Relation to Marine Genetic Resources from Areas Beyond National Jurisdiction. A Possible Way Forward*, Federal Agency for Nature Conservation, Bonn, www.bfn.de/fileadmin/MDB/documents/service/Skript_301.pdf.

Greiber, T., Peña Moreno, S., Åhrén, M., Nieto Carrasco, J., Chege Kamau, E. Cabrera Medaglia, J., Oliva, M.J. and Perron-Welch, F. in cooperation with Ali, N. and Williams, C. (2010), *An Explanatory Guide to the Nagoya Protocol on Access and Benefit-sharing*, IUCN, http://cmsdata.iucn.org/downloads/an_explanatory_guide_to_the_nagoya_protocol.pdf.

Heafey, E. (2014), 'Access and Benefit-sharing of Marine Genetic Resources from Areas beyond National Jurisdiction: Intellectual Property – Friend, Not Foe', *Chicago Journal of International Law* Vol. 14, No. 2, http://chicagounbound.uchicago.edu/cgi/viewcontent.cgi?article=1390&context=cjil.

Laliberté, B., Engels, J. and Fowler, C. (2000), 'The International Undertaking on Plant Genetic Resources: Its Relevance to Botanic Gardens', *BGCI*, Vol. 3 No. 5, www.bgci.org/resources/Article/0156/.

López Noriega, I., Wambugu, P. and Mejías, A. (2013), 'Assessment of Progress to Make the Multilateral System Functional. Incentives and challenges at the country level', in M. Halewood, I. López Noriega, S. Louafi (eds), *Crop Genetic Resources as a Global Commons: Challenges in International Law and Governance*, Earthscan, Oxon.

Millicay, F. (2007), 'A Legal regime for the Biodiversity of the Area', in Nordquist, Long, Heider and Moore (eds), *Law, Science and Ocean Management*, Brill, Leiden.

National Institute of Genetics (2015), *Access to Marine Genetic Resources and Benefit-sharing from Their Academic Use, Report of MGR Workshop in Japan* (ed. Hajimu Morioka), Tokyo, http://nig-chizai.sakura.ne.jp/abs_tft/wp-content/uploads/2016/08/Marine-genetic-resources.pdf.

Nijar, G. (2011), *The Nagoya Protocol on Access and Benefit-sharing of Genetic Resources: Analysis and Implementation Options for Developing Countries*, Research Paper 36, South Centre, Geneva.

Prathapan, K. and Dharma Rajan, P. (2011), 'Biodiversity access and benefit-sharing: weaving a rope of sand', www.researchgate.net/publication/233764492_Biodiversity_access_and_benefit-sharing_Weaving_a_rope_of_sand.

Ruiz Muller, M. (2015), *Genetic Resources as Natural Information: Implications for the Convention on Biological Diversity and Nagoya Protocol* (Routledge Studies in Law and Sustainable Development), Earthscan, London.

Scovazzi, T. (2010), 'The Conservation and Sustainable Use of Marine Biodiversity, Including Genetic Resources, in Areas Beyond National Jurisdiction: A Legal Perspective', www.un.org/Depts/los/consultative_process/ICP12_Presentations/Scovazzi_Presentation.pdf.

Smith, H., Suárez de Vivero, J. and Agardy, T. (eds) (2015), *Routledge Handbook of Ocean Resources and Management*, Routledge, London.

Tsioumani, E. (2015), 'Intersessional Working Group Proposes Subscription System for ABS under the ITPGRFA MLS', IISD, 5 October, http://sdg.iisd.org/news/intersessional-working-group-proposes-subscription-system-for-abs-under-the-itpgrfa-mls/.

UNCTAD (2014), *The Oceans Economy: Opportunities and challenges for small island developing states*, UNCTAD/ DITC/TED/2014/5, New York and Geneva, http://unctad.org/en/PublicationsLibrary/ditcted2014d5_en.pdf.

UNEP/CBD/SBSTTA (2003), *Study of the relationship between the Convention on Biological Diversity and the United Nations Convention on the Law of the Sea with regard to the conservation and sustainable use of genetic resources on the deep seabed (Decision II/10 of the Conference of the Parties to the Convention on Biological Diversity*, UNEP/CBD/SBSTTA/8/INF/3/REV1, www.cbd.int/kb/record/meetingDocument/4761?Event=SBSTTA-08.

Vivas-Eugui, D. (2012), *Bridging the Gap on Intellectual Property Intergovernmental Committee (IGC)*, ICTSD, Geneva, www.ictsd.org/sites/default/files/event/2012/02/bridging-the-gap-on-intellectual-property-and-genetic-resources-in-wipos-intergovernmental-committee-igc.pdf.

World Health Organization (2012), 'Research and Development to Meet Health Needs in Developing Countries: Strengthening Global Financing and Coordination. Report of the Consultative Expert Working Group on Research and Development: Financing and Coordination', www.who.int/phi/cewg_report/en/.

Wright, G., Rochette, J., Druel E. and Gjerde, K. (2016), *The Long and Winding Road Continues: Towards a New Agreement on High Seas Governance, IDDRI*, www.iddri.org/Publications/Collections/Analyses/ST0116_GW%20et%20al._high%20seas.pdf.

Wynberg, R. (2015), 'Marine Genetic Resources and Bioprospecting in the Western Indian Ocean' in *Regional State of the Coast Report Western Indian Ocean*, UNEP, http://wedocs.unep.org/bitstream/handle/20.500.11822/11349/rsocr_printedition.compressed_Part30.pdf?sequence=31andisAllowed=y.

12

THE IMPACT OF NATURAL PRODUCTS DISCOVERY PROGRAMS ON OUR KNOWLEDGE OF THE FLORA OF MADAGASCAR

James S. Miller and Porter P. Lowry II

When the Convention on Biological Diversity entered into force in 1993 (Glowka *et al.*, 1994), it established new expectations that benefits would accrue to countries from products developed using their biological sources. The premise was that the development and commercialization of drugs or other products derived from natural sources, including plants, could potentially provide monetary and/or capacity-building support in a fair and equitable manner. Much of the discussion subsequent to the Convention entering into force has focused on the potential for monetary benefits and for building research capacity through direct contributions and/or training (e.g. Miller, 2015), but another important benefit from natural products discovery programs has received less attention. In almost all cases, sampling for these programs was led and conducted by institutions whose mission is to document and understand the diversity of some major group of living organisms. The funding provided by natural products discovery programs to support field collection of samples for screening thus also resulted in valuable and important natural history collections that have helped advance biological knowledge and in many cases have helped promote conservation. This review will focus on two of those efforts, both conducted by the Missouri Botanical Garden (MBG) in Madagascar, and will outline how these programs contributed to advancing the biological research and conservation goals of MBG, its local partners, and the Malagasy government.

In 1986, the Natural Products Branch of the Developmental Therapeutics Program of the United States National Cancer Institute (NCI) issued a call for proposals for contracts to collect and supply plant samples to be extracted and assayed at NCI against a panel of cancer cell lines. Three contracts were issued, one each for the three major tropical regions of the world, i.e., the Neotropics, tropical Asia, and Africa and Madagascar. The Missouri Botanical Garden was issued the contract for the third of these regions. Collecting programs were initially established in several African countries, and in 1989 the program was expanded to the island nation of Madagascar through a collaboration with the *Centre National d'Application et des Recherches*

Pharmaceutique (CNARP). An agreement between CNARP and the NCI was signed in 1990, the first ever that ensured a share of benefits for a source country in exchange for access to its biodiversity (reviewed in Miller *et al*., 2007). This program remained active until 2006 and resulted in a significant array of benefits for Madagascar (Miller, 2007).

Nearly a decade later, a highly innovative natural products discovery effort was launched via the Fogarty International Center of the United States National Institutes of Health (NIH). The International Cooperative Biodiversity Groups (ICBG), initially funded jointly by NIH, the National Science Foundation (NSF), and the US Agency for International Development (AID), were launched in 1993 with the complementary goals of drug discovery, building research capacity in low- and middle-income countries, and conserving the biological diversity that was targeted as the basis for drug development. During the 20+ years the program has operated, it has evolved considerably and the agencies funding this effort have undergone changes; AID no longer supports the program, but the Department of Energy has joined NIH and NSF. The ICBG program initially focused on several biodiversity-rich parts of the world, including Peru, Mexico, and Vietnam, but perhaps the most emblematic target country was Madagascar, where a series of activities were carried out between 1998 and 2013. The benefits that accrued to Madagascar through this program have been presented elsewhere (Miller, 2015), but in this review we focus specifically on the impact the ICBG program has had on our understanding of the flora of Madagascar.

One of the original ICBG groups funded in 1993 focused on Suriname and involved a partnership between Virginia Polytechnic Institute and State University (VPI), Conservation International (CI), Bristol-Myers Squibb, Bedrijf Geneesmiddelen Voorziening Suriname (BGVS), a pharmaceutical research group that operated in Suriname, and MBG. During the first five years of the program, numerous problems were encountered, especially with regard to obtaining permission to operate in country, and when a renewal application was submitted in 1998 for a second ICBG grant, the decision was made to move the program to Madagascar and to work with CNARP, building on the collaboration that had been established during the NCI program.

The Missouri Botanical Garden has been active in Madagascar since the early 1970s and has had full-time in-country presence there since the mid-1980s. MBG's program initially focused on surveying the island's immensely diverse flora, but soon expanded to include local training and capacity-building, as well as support of national research and education centers, and subsequently, in the early 2000s, it diversified further to include a broad array of conservation activities. The ICBG's commitment to promoting the understanding and conservation of biological diversity and building local research capacity was thus very well aligned with the goals of MBG's Madagascar program.

The natural products discovery and development activities of the ICBG project in Madagascar were conducted by VPI, Bristol-Myers Squibb, and CNARP, while MBG's primary role was to work with CNARP to coordinate and conduct the field work that supplied the bulk plant samples analyzed in the search for novel bioactive compounds. Field collecting during the first contract, from 1998 to 2003, was focused on the humid, mid-elevation forests in and around Zahamena National Park in east-central Madagascar, and from 2003 to 2008 the focus shifted to dry, low-elevation forests in northeastern Madagascar. A fourth five-year grant extended the ICBG program in Madagascar, but its focus shifted to marine invertebrates and soil microbial organisms, and thus while support for collection and research continued, plants were no longer a target.

The appeal of Madagascar as a site for bioprospecting activities stemmed from the country's amazing botanical diversity. With perhaps as many as 14,000 native species of plants, nearly 90%

of which are found nowhere else on earth, combined with the fact that less than 7% of the original forest cover remains intact and much of that is still very poorly explored (Phillipson *et al.*, 2006; Callmander *et al.*, 2011), Madagascar is widely recognized as an exceptionally important and highly threatened biodiversity hotspot (Myers *et al.*, 2000; Ganzhorn *et al.*, 2001). It also yields species new to science at a rate unequalled by just about any other place in the world. Moreover, since many natural compounds are produced by just one or a few species, Madagascar's unique flora offers a rich and unique source of material for natural products discovery.

The NCI and ICBG programs have made impressive contributions to the scientific community's efforts to document, understand and protect the flora of Madagascar. From the outset, MBG conceived and designed the inventory and collection component of both programs to provide maximum benefit for botanical research, while at the same time ensuring that the samples used for screening were of high quality and of known provenance.

In both programs, carefully prepared voucher specimens were collected in association with all plant samples, providing a permanent reference collection of known geographic origin to ensure accurate identification of source materials and to facilitate re-location and re-collection of material as needed. Complete sets of duplicate specimens were deposited at one or both of Madagascar's national herbaria and at MBG, and duplicates of most collections were also deposited in CNARP's herbarium and sent to the *Muséum National d'Histoire Naturelle* in Paris, which houses the largest reference collection of plant specimens from Madagascar. Comprehensive data associated with each collection were entered into MBG's TROPICOS database (www.tropicos.org), which is freely accessible on-line. While it is difficult to know precisely how many of these collections represent species new to science, because many families and genera are still undergoing study and others remain to be revised, there are surely dozens of novelties collected during the NCI and ICBG programs, one of which was described and named *Ludia craggiana* in honor of Gordon Cragg, director of NCI's Natural Products Branch (Rogers *et al.*, 2006).

The NCI and ICBG programs supported inventory work in many botanically rich areas that had previously been poorly explored. This made a significant contribution to MBG's overall effort to document the flora of key areas, which has covered hundreds of sites throughout the island over the last four decades. For example, almost 2,700 collections were made in the area of Zahamena National Park, representing nearly three-quarters of the records in TROPICOS from this protected area and nearby forests. In all, the NCI program generated nearly 5,550 collections, and almost 6,850 were made through the ICBG program, which together represent just under 21% of the ca. 60,000 collections made by MBG staff, students and collaborators in Madagascar since the early 1980s.

The information generated from this inventory work also contributed to MBG's effort in the mid-2000s to identify priority areas for plant conservation in Madagascar. In all, more than 70 areas were identified whose protection would be needed to ensure the conservation of all of the island's botanical diversity. This information was directly integrated into Madagascar's national effort to increase, by more than a factor of three, the total area managed for conservation, an effort that has led to the establishment of 55 new, officially designated protected areas, including the Ankeniheny-Zahamena Corridor to the south of Zahamena National Park and the Oronjia Reserve in the far north. The argument for establishing both of these new reserves was based largely on richness of their biodiversity, documented mostly through collections made during the NCI and ICBG projects.

Another positive benefit of the NCI and ICBG projects was the training in botanical field methods and plant identification techniques provided to nearly a dozen staff members of both CNARP and MBG, along with as many Malagasy students. This constitutes one of the core benefits explicitly targeted by the Convention on Biological Diversity, and the training made possible through the NCI and ICBG projects contributed significantly to MBG's overall efforts to develop and strengthen in-country capacity and expertise. This in turn has directly contributed to the development of a strong community of botanists in Madagascar whose members have the knowledge and skills to conduct botanical inventories, identifying the plants collected, and publishing the results.

One of the explicit goals of the ICBG program was to promote conservation, and in this regard the effort in Madagascar was very successful. The specimens generated contributed to documenting plant diversity at each site, a critical step in arguing for protected status. However, ensuring long-term, sustainable conservation of such areas in an economically disadvantaged country such as Madagascar, where the rural population is very poor and highly dependent on access to natural resources to meet their basic needs, required an integrated effort that also focused on addressing the needs of local communities. Addressing the twin objectives of conservation and improving livelihoods in a fully participative manner was essential to making these efforts sustainable. The ICBG program provided a broad range of support, including for health care and the establishment of medical facilities, improving education and upgrading infrastructure to improve water quality.

The second five-year ICBG grant supported collecting in northeastern Madagascar, an area that was badly under-represented in the country's network of protected areas. Using the information generated by the ICBG project about this area's flora, the Oronjia forest, located just to the east of Antsiranana, was identified as a priority area for plant conservation, which prompted MBG to select it as a target for a community-based conservation program. In 2015, Orangea was officially designated by the Government of Madagascar as a new protected area, one of twelve managed by MBG in collaboration with the local community, an outcome that would never have been achieved without the initial involvement of the ICBG project.

For two decades, the NCI and ICBG natural products discovery programs supported the collection of more than one-fifth of the plant specimens collected by botanists at the Missouri Botanical Garden and its Malagasy collaborators. This provided an essential basis for documenting plant diversity at key areas of high importance for biodiversity and for securing their status as legally protected areas. More broadly, the discovery efforts of the NCI and ICBG programs made a significant and lasting contribution to the overall documentation of Madagascar's plant diversity, especially in areas that had been poorly explored, and to the country's efforts to achieve sustainable conservation of its newly expanded network of protected areas.

References

Callmander, M.W., P.B. Phillipson, G.E. Schatz, S. Andriambololonera, M. Rabarimanarivo, N. Rakotonirina, J. Raharimampionona, C. Chatelain, L. Gautier and P.P. Lowry II. (2011). 'The endemic and non-endemic vascular flora of Madagascar updated'. *Plant Ecology and Evolution* 144: 121–25, http://dx.doi.org/10.5091/plecevo.2011.513.

Ganzhorn, J.U., P.P. Lowry II, G.E. Schatz and S. Sommer. (2001). 'The biodiversity of Madagascar: one of the world's hottest hotspots on its way out'. *Oryx* 35: 346–48.

Glowka, L., F. Burhenne-Guilmin, and H. Synge. (1994). 'A guide to the Convention on Biological Diversity'. *Environmental Policy and Law Paper No. 30*, IUCN Environmental Law Centre.

Myers N., R.A. Mittermeier, C.G. Mittermeier, G.A.B. da Fonseca and J. Kent. (2000). 'Biodiversity hotspots for conservation priorities'. *Nature* 403: 853–58, http://dx.doi.org/10.1038/35002501.

Miller, J.S. (2007). 'Impact of the Convention on Biological Diversity: The lesson of ten years of experience with models for equitable sharing of benefits'. pp. 58–67 in C.R. McManis (ed.), *Biodiversity and the Law: Intellectual Property, Biotechnology, and Traditional Knowledge*, Earthscan, London and Sterling, VA.

Miller, J.S. (2015). 'The realized benefits from bioprospecting in the wake of the Convention on Biological Diversity'. *Washington University Journal of Law and Policy* 47: 51–60.
Miller, J.S., R. Andriantsiferana, G.M. Cragg, and P.P. Lowry II. (2007). 'Appendix: History of a landmark collecting agreement: The origin of the National Cancer Institute's Letter of Intent, a precursor to modern bioprospecting agreements'. pp. 68–70 in C.R. McManis (ed.), *Biodiversity and the Law: Intellectual Property, Biotechnology, and Traditional Knowledge*, Earthscan, London and Sterling, VA.
Phillipson, P.B., G.E. Schatz, P.P. Lowry II and J.-N. Labat. (2006). 'A catalogue of the vascular plants of Madagascar'. pp. 613-27 in S.A. Ghazanfar and H.J. Beentje (eds.), *Taxonomy and Ecology of African Plants, their Conservation and Sustainable Use: Proceedings of the 17th AETFAT Congress, Addis Ababa, Ethiopia*, Kew Publishing, London.
Rogers, Z.S., A. Randrianasolo and J.S. Miller. (2006). 'A new Species of *Ludia* (Salicaceae) from Madagascar's Eastern Littoral Forest'. *Novon* 16: 409–12.

13

REGULATORY MEASURES ON ACCESS AND BENEFIT-SHARING FOR BIOLOGICAL AND GENETIC RESOURCES

National and regional perspectives from the Philippines, Singapore and ASEAN

Lye Lin-Heng and Rose-Liza Eisma-Osorio

Introduction

The Convention on Biological Diversity was signed by many nations at the UN Conference on Environment and Development (the Earth Summit) in Rio de Janeiro in June, 1992. It entered into force on 29 December 1993, 90 days after the 30th ratification. It has three main objectives, namely:

1. The conservation of biological diversity.
2. The sustainable use of the components of biological diversity.
3. The fair and equitable sharing of the benefits arising out of the utilization of genetic resources.

The CBD's Article 15 recognizes that States have sovereign rights over their natural resources, and therefore the authority to determine access to genetic resources rests with the national governments and is subject to national legislation. Article 15 provides that each party "shall endeavour to create conditions to facilitate access to genetic resources for environmentally sound uses by other contracting parties and not to impose restrictions that run counter to the objectives" of the Convention. In conjunction with this, succeeding paragraphs elaborate on conditions for access to genetic resources. Paragraphs 4 and 5 state that such access shall be given "upon mutually agreed upon terms" and "subject to prior informed consent of the party providing the resources". Further, contracting parties shall fully participate in developing and carrying out scientific research on genetic resources with the benefits of biodiversity shared in a "fair and equitable way" (Art. 15, paras 6 and 7). Such distribution of benefits shall be done through financial mechanisms established by Articles 20 and 21 of the CBD.

The Nagoya Protocol on Access to Genetic Resources and the Fair and Equitable Sharing of Benefits Arising from their Utilization was adopted at the tenth meeting of the CBD's Conference of the Parties on 29 October 2010, in Nagoya, Japan. It seeks to provide a transparent legal framework for the effective implementation of one of the three objectives of the CBD: the fair and equitable sharing of benefits arising out of the utilization of genetic resources. It came into effect on 12 October 2014. The Philippines acceded to the Protocol on 29 September 2015, but Singapore is not yet a party.

This chapter now seeks to examine the position firstly, in the Philippines and then in Singapore, in the context of the laws and procedures regulating access and benefit-sharing of genetic resources in these two countries in Southeast Asia, as well as attempts by the Association of South East Asian Nations (ASEAN) to address these issues at a regional level.

The Philippines

The Philippines is an archipelagic country composed of about 7,600 islands with 300,000 square kilometers of land area.[1] As a state party to the Convention on Biological Diversity, the Philippines has established national legislative and policy frameworks in regard to access to genetic resources and equitable sharing of benefits (hereinafter "access and benefit-sharing" or "ABS"). The Philippines is known to have some of the most advanced legislation regulating ABS (CISDL, 2012). Executive Order 247, enacted in 1995, aims to regulate the prospecting of biological and genetic resources so that these resources are protected and conserved, are developed and put to the sustainable use and benefit of the national interest. This was prompted by the Manila Declaration on the Peaceful Settlement of International Disputes in 1992, which sought to promote the ethical use of bioresources to protect the rights of communities and researchers in developing countries.[2]

Regulatory mechanisms that are now in place include the establishment of competent national authorities, procedures for commercial and academic research agreements, prior informed consent processes, and benefit-sharing measures. An overarching legislation, the 1987 Philippine Constitution, has facilitated these trail-blazing policy developments for ABS. Under the Constitution, natural resources in the country are owned by the State (Article XII, section 2). Hence, any activity relating to its exploration, development and utilization is primarily controlled by the government subject to certain exceptions.

The national government is given the authority to allow bioprospecting of natural resources subject to strict regulations. Bioprospecting is defined as "the research, collection and utilization of biological and genetic resources for purposes of applying the knowledge derived therefrom solely for commercial purposes" (section 5(a), Republic Act 9147). This activity is regulated by national agencies in the light of treaty obligations under the CBD and in accordance with the Nagoya Protocol to which the Philippines acceded in 2015. Despite the adoption of national laws and implementing guidelines on bioprospecting, the Philippines continues to face challenges in the implementation of these laws and administrative measures. These will now be examined.

National legal and policy framework on access and benefit-sharing

1987 Constitution and the Regalian Doctrine

Article XII, s. 2 of the Constitution states that "All lands of the public domain, waters, minerals, coal, petroleum, and other mineral oils, all forces of potential energy, fisheries, forests or timber, wildlife, flora and fauna, and other natural resources are owned by the State". It is clear from the Philippine Constitution that the State has full control of the utilization of such natural resources. The Supreme Court has emphasized the importance of these resources which constitute the exclusive heritage of the Filipino nation, thus, the exploration, development and utilization thereof are matters vital to the public interest and general welfare of the people.[3] In essence, natural resources can be developed by the State in cooperation with qualified individuals or corporations by virtue of co-production, joint venture or production sharing agreements. Development of natural resources is limited to Filipino citizens or corporations of which 60 per cent of the capital must be owned by Filipino citizens (Article XII, section 2, 1987 Constitution).

A specific exception is, however, expressly provided for in the Constitution with regard to technical or financial assistance for large-scale exploration, development and utilization of minerals, petroleum, and other mineral oils according to the general terms and conditions provided by law, based on real contributions to the economic growth and general welfare of the country. What this means is that the President may enter into agreements with foreign-owned corporations subject to certain conditions. The Supreme Court has held that the State, in the exercise of its police power, may not be precluded from altering, modifying and amending such mining leases or agreements (Bernas, 1997).

Special considerations arise when parties seek access to biological and genetic resources found on ancestral lands and/or domains of indigenous communities. The Constitution provides that "the State recognizes and promotes the rights of indigenous cultural communities within the framework of national unity and development" (Article II, section 22, 1987 Constitution). This engenders the protection of the culture, traditions and institutions of indigenous peoples. Hence, the State protects the rights of indigenous cultural communities to their ancestral lands, including their rights to genetic resources, traditional medicines, medicinal plants, and other indigenous knowledge, systems and practices. The Indigenous Peoples Rights Act (IPRA) of 1997 (Republic Act 8371) was passed by Congress to govern the applicability of customary laws to their property rights or relations with respect to ownership and extent of ancestral domains. Ancestral domain is an all-embracing concept which refers to lands which are considered a part of a cultural region, including lands not yet occupied, such as deep forests, while ancestral lands are the occupied portions of the ancestral domain (Bernas, 1997).

The foregoing principles and policies on full control by the state over natural resources must be understood in the light of the state policy on advancing the right of the people to a healthy environment (Article II, section 15, 1987 Constitution). Consistent with this, the Supreme Court has expanded this policy by recognizing the right to a balanced and healthful ecology as a basic human right, in the landmark *Oposa* ruling. It was held that this human right was concerned with the right of preservation and self-perpetuation, the advancement of which predated all governments, having assumed their existence from the inception of mankind. This right implies the judicious management and conservation of the country's forests, hence, the duty to refrain from impairing the environment, for the benefit of the present and future generations.[4]

Post-Convention on Biological Diversity

The Philippines is recognized as one of 17 countries worldwide with the highest biodiversity (Mittermeier, Gil and Mittermeier, 1997). However, its biodiversity is under serious threat. To address the ongoing challenges, the Philippine Biodiversity Strategy and Action Plan (PBSAP) 2015–2028 seeks to achieve its vision by 2028 to have biodiversity "restored and rehabilitated, valued, effectively managed and secured, maintaining ecosystem services to sustain healthy, resilient Filipino communities and delivering benefits to all". The PBSAP was crafted as the country's roadmap to conserve its biodiversity. This was formally adopted by virtue of DENR Administrative Order 2016-12, 13 June 2016.

Bio-prospecting under Executive Order 247 (1995)

Executive Order 247 (EO 247), enacted in 1995, established the first regulatory framework in the Philippines for the prospecting of biological and genetic resources for scientific and commercial purposes. Specific guidelines under Administrative Order No. 96-20 (DAO 96-20) were issued by the Department of Environment and Natural Resources (DENR) in 1996.

EO 247 sought to meet two general objectives. First, to regulate the prospecting of biological and genetic resources so that these resources are protected, conserved, developed and put to the sustainable use and benefit of the national interest; and secondly, to promote the development of local capacity in science and technology to achieve technological self-reliance in selected areas (section 1). Further, the law emphasized the need for prior informed consent of indigenous cultural communities before bioprospecting can be conducted in ancestral lands and domains. Such consent is to be obtained in accordance with the customary laws of such communities. A similar provision also requires the prior informed consent of local communities before bioprospecting is carried out in other areas (section 2).

Prospecting for biological and genetic resources is to be allowed in cases when a recommendation is given by the Inter-Agency Committee on Biological and Genetic Resources (IACBGR). The IACBGR is an agency of the Department of Environment and Natural Resources (DENR) and is composed of representatives from DENR, Department of Science and Technology, Department of Agriculture, Department of Health, Department of Foreign Affairs, the National Museum, academic experts on biodiversity, biotechnology, genetics, natural products chemistry or similar disciplines, non-government organizations active in biodiversity protection and people's organizations. Specific roles and functions of each member agency are set out in DAO 96-20.

Two kinds of agreements for bioprospecting are provided in EO 247. If the research and collection of biological and genetic resources is intended, directly or indirectly, for commercial purposes, the agreement must be a Commercial Research Agreement (CRA). On the other hand, if the prospecting of biological and genetic materials is intended primarily for academic purposes, the agreement shall be an Academic Research Agreement (ARA). The Academic Collector may, however, apply for a CRA when it becomes clear that the research and collection being done has commercial prospects. Only duly-recognized Philippine universities and academic institutions, domestic governmental entities and intergovernmental entities may apply for an ARA.

There are minimum terms that must be included in both CRAs and ARAs. Some important terms include: (1) a limit on the samples obtained and exported, (2) an obligation to deposit a complete set of all specimens collected with the designated governmental entity, (3) an obligation to grant access to collected specimens and relevant data to all Filipino citizens and Philippine governmental entities whenever these specimens are deposited in depositories abroad, (4) a duty to inform the Philippine Government as well as the affected local and indigenous cultural communities of all discoveries from the activity conducted in the Philippines, if a commercial product is derived from such activity, and (5) a provision for payment of royalties or other forms of negotiated compensation to the National Government, local or indigenous cultural community and individual person or designated beneficiary in case commercial use is derived from the biological and genetic resources taken.

Moreover, if the commercial collector or its principal is a foreign person or entity, Filipino scientists must be actively involved in the research and collection process and, where applicable and appropriate, in the technological development of a product derived from the biological and/or genetic resources taken from any area in the Philippines. However, any commercial collector and/or its principal shall be encouraged to use the services of Philippine universities and academic institutions. A fixed fee must be paid to the DENR in accordance with a schedule of fees formulated by the IACBGR in order for such research to proceed.

Under EO 247, the CRA is differentiated from an ARA in terms of duration as well as coverage. While a CRA is given for three years, the ARA is for a term of five years. However, both are subject to renewal upon review by the IACBGR. Further, the ARA may be more comprehensive in scope and include as many research areas as desired by the applicant. It may stipulate that all scientists and researchers affiliated with a duly-recognized university, academic institution, governmental and intergovernmental entity need not apply for a different research agreement but may conduct research and collection activities in accordance with an existing agreement so long as all conditions are complied with by the affiliated scientist or researcher; affected communities must have given their prior informed consent to the activities to be undertaken.

Wildlife Resources Conservation and Protection Act of 2001

The Wildlife Resources Conservation and Protection Act of 2001 (Republic Act 9147) updated EO 247 and essentially changed some of the bioprospecting procedures under Administrative Order 96-20. This sought to correct the deficiencies found in EO 247 which was viewed as highly restrictive of research activities without providing the necessary support and incentives.[5]

Under the Wildlife Act, bioprospecting shall be allowed upon execution of an undertaking by any proponent, stipulating its compliance with and commitments to reasonable terms and conditions imposed by the Secretary of DENR or Department of Agriculture, as the case may be (section 14). The Bioprospecting Undertaking (BU) replaced the commercial and academic research agreements that were previously enforced under EO247. The BU, however, contains similar requirements, for instance, prior informed consent certificate from indigenous cultural communities, protected area management board, local communities or private individual or entity. Full disclosure is also expected from the applicant in terms of the intent and scope of the bioprospecting activity.

Under the new regulations, licensing conditions for an applicant who is a foreign entity or individual are more stringent. The active involvement of local institutions is required especially in terms of the research, collection and appropriate technological development of products derived from biological and genetic resources of the country. The former regulation, in contrast, only required a letter of acceptance from Filipino counterparts authorized by or representing the host institutions, to cooperate in bioprospecting activities in the Philippines, where applicable (DAO 96-20, section 6.1.2a).

Aside from this amended provision, the regulatory framework on bioprospecting under the Wildlife Act has been supplemented by other laws. For instance, section 35 of the Indigenous Peoples' Rights Act of 1997 (IPRA) (Republic Act No. 8371), in relation to the National Integrated Protected Areas System of 1992 (NIPAS) Act (Republic Act No. 7586), provides that access to biological and genetic resources and to indigenous knowledge related to the conservation, utilization and enhancement of these resources, shall be allowed within ancestral lands and domains of the ICCs/IPs or within protected areas therein, only after free and prior informed consent of such communities has been obtained in accordance with the customary laws of the concerned community. In addition, the Local Government Code of 1991 (Republic Act No.

7160), the Strategic Environmental Plan for Palawan Act of 1992 (Republic Act No. 7611), the Fisheries Code (Republic Act No. 8550), the Seed Industry Development Act (Republic Act No. 7308) and the Traditional and Alternative Medicine Act of 1997 (Republic Act No. 8423) are complementary laws that regulate certain contexts in which bioprospecting may take place as well as related activities. One example is the regulation of medicinal plants for traditional or alternative medical use, which is primarily governed by the Traditional and Alternative Medicine Act.

To further ensure full implementation, joint implementing guidelines specific for bioprospecting have been issued in accordance with Joint DENR- DAPC (Department of Agriculture-Palawan Council) Sustainable Development Implementing Rules and Regulations (IRR) pursuant to Republic Act No. 9147 (Rule 14.1) in 2005.

2005 Joint Guidelines for Bioprospecting Activities

The Joint Guidelines for Bioprospecting were issued in 2005 by the Department of Environment and Natural Resources (DENR), Department of Agriculture (DA), Palawan Council for Sustainable Development (PCSD) and the National Commission on Indigenous Peoples (NCIP) (JAO 1-2005). Seeking to ensure fair and equitable sharing of benefits, the Guidelines apply to all bioprospecting activities conducted by any resource user (including government agencies) in all areas, covering protected areas under NIPAS and on private lands, as well as ancestral domains and ancestral lands, consistent with Indigenous Peoples Rights Act (IPRA). Specific types of bioprospecting activities are identified in the new regulation. These include bioprospecting of any biological resource found in the Philippines including wildlife, micro-organisms, domesticated or propagated species and exotic species. It is also applicable to all *ex-situ* collections of biological resources sourced from the Philippines, except for collections currently accessed under international agreements where the Philippines is a party. Any bioprospecting activity involving species listed under CITES and the IUCN Red List, whenever allowed by law, shall be governed by the Guidelines in addition to specific regulations on the conservation of these species. Any violation of the Guidelines, such as non-compliance with the provisions of the BU, is considered a violation of the Wildlife Act and is subject to administrative and criminal sanctions. Any person who conducts bioprospecting without a BU will be subject to penalties for collection without a permit.

To address some of the practical issues arising from the earlier regulations on bioprospecting, these new Guidelines added more activities that are exempt from compliance apart from traditional uses by indigenous community in accordance with written or unwritten rules, usage, customs and practices. These exemptions include the use of biological resources for subsistence consumption, conventional commercial consumption for direct use such as logging or fishing, and scientific research on wildlife and agrobiodiversity for purely academic purposes. These are exempted under the Guidelines JAO 1-2005. Nevertheless, if after scientific research, there is a subsequent transfer of biological resources and use of research findings for commercial purposes, these activities will no longer be exempt (see section 3).

These Joint Guidelines also detail the processes for securing a BU in accordance with the Wildlife Act. Dispensing with the Inter-Agency Committee (IACBGR) under EO 247, separate jurisdictions are now recognized for the Department of Agriculture and DENR Secretaries in the execution of the undertaking. When the bioprospecting activity is to be conducted in the Province of Palawan, the PCSD is a co-signatory to the BU. The respective Secretary shall also evaluate the BU through separate or joint technical committees, especially with respect to requirements for PIC and benefit-sharing negotiations with concerned resource providers.

Salient access and benefit-sharing provisions

Bioprospecting Undertaking

The Bioprospecting Undertaking (BU) refers to the "undertaking" or "permit" that allows the resource user access to biological resources for bioprospecting purposes, subject to the conditions stated therein. It contains, in addition to the negotiated terms of benefit-sharing, standard terms and conditions relating to compliance with complementary regulations and other basic contractual terms (section 9.1, JAO 1-2005). It also indicates the specimen and the amounts of the samples to be collected. Specified amounts of samples to be taken by the resource user are set out under Annex III of the guidelines, unless the collection of samples in excess of the limits can be justified based on an appropriate resource inventory and considering the conservation of the resource (section 10.1). Nevertheless, when a BU is awarded, it does not imply automatic access to traditional knowledge associated with these resources. A separate research proposal is required to access such associated traditional knowledge.

Benefit-sharing arrangements are set out in more detail under new joint guidelines. There must be an express agreement regarding payment of monetary and non-monetary benefits between the resource user and providers. The negotiated benefits arising from the use of biological resources shall be given by the resource user, in the amounts and periods agreed upon, directly to the national government and resource providers. The bioprospecting fee accrues to the national government, while up-front payments accrue to the resource providers. In practice, after the applicant has paid the standard application fee of 500 Philippine pesos (approx. US$10) and posted a rehabilitation or performance bond, in the form of a surety bond (equivalent to 25% of the project cost), a bioprospecting fee will be assessed by the appropriate government agency – DENR, Department of Agriculture and/or Palawan Council for Sustainable Development. A minimum bioprospecting fee of US$3,000 for each BU and US$1,000 for each collection site is payable. These amounts may be increased but may not be greater than three times the minimum fee. The bioprospecting fee is payable to the applicable agency. Such fees are earmarked for a special fund (the Wildlife Management Fund or the Protected Area Fund, as appropriate). Otherwise, the fees shall be divided equally among the three agencies.

Moreover, royalties are to be shared between the national government and the resource providers; while local governments will share in the amounts received by the national government, consistent with the provisions of the Local Government Code (RA 7160). This is in contrast to the previous guidelines, where no outright benefits were shared with local governments; the previous guidelines merely provided that for research agreements in general, all immediate, medium- and long-term benefits, resulting directly or indirectly from the bioprospecting activities conducted, were to be shared equitably and upon mutual consent among the Philippine government, the communities concerned and the principal researcher.

Another distinct feature in the new guidelines is an express provision regarding bioprospecting fees. While a minimum amount for each BU is fixed at US$3,000 and US$1,000 per collection site, a new amendment has been a welcome development for Filipino researchers. Lower fees are now imposed for Filipino resource users with no foreign collaborators or investors at 10% of the assessed amount ($3,000 or $1,000). In the case of Filipino students who are fulfilling academic requirements in a local institution and who are undertaking the research with no foreign collaborator or investors, their fees for bioprospecting are at 3% of the assessed amount.

Despite the lowered bioprospecting fee provisions for local researchers, there have not been any BUs issued by the DENR through its implementing agency, the Biodiversity Management Bureau. According to the Biodiversity Management Bureau of the DENR – what has been issued so far are permits for scientific researchers. Presumably, these researchers refer to the scientific wildlife researchers as provided under section 15 of the Wildlife Act, as differentiated from the bioprospecting provision under section 14. Under section 15, if the collection of biological resources is for scientific purposes and not for commercial purposes, a gratuitous permit shall be issued by the Secretary of the DENR or authorized representative. This opens up a new collection process which is not covered under the Bioprospecting Guidelines. However, there is no clear delineation between the activities that fall within section 14 and section 15. What adds to the confusion is the absence of clear guidelines for the scientific research collection under section 14.

Prior informed consent

Specific provisions on prior informed consent and benefit-sharing are found in the Philippine bioprospecting guidelines (JAO 1-2005). In general, the guidelines require that a prior informed consent (PIC) certificate must be obtained from indigenous cultural communities/indigenous peoples (IPs), local communities, protected area management boards and private land owners. Any violation in the procurement of PIC under the BU is subject to an administrative investigation. Formal complaints on alleged violations of the terms and conditions of the BUs, particularly the procurement of PIC and the collection of materials, may be lodged with any of the implementing agencies. The concerned agency will undertake a fact-finding mission upon finding a prima facie violation. Non-compliance with the agency's instructions will result in the automatic cancellation/revocation of the agreement and confiscation of collected materials in favour of the government, the forfeiture of bonds and the imposition of a perpetual ban on access to biological resources in the Philippines upon the violator. Such misconduct is also considered a violation of the Wildlife Act and is subject to criminal prosecution and penalties amounting to a maximum of two years' imprisonment and a maximum fine of 300,000 Philippine pesos (about US$6,000).

Under the previous regulation DAO 96-20, a PIC certificate signed by the recognized head of the indigenous community, Municipal or City Mayor of the Local Government Unit, Protected Area Management Board (PAMB) or Private Land Owner concerned, together with proof of public notification and/or sectoral consultation must be submitted by the Principal/Collector. However, the academic research agreement (ARA) may be executed even without the PIC certificate, provided that such consent can be secured prior to commencement of actual bioprospecting activity. This did not apply to a commercial research agreement (CRA). This apparent loophole in ARAs was somewhat addressed in the latest regulation (JAO 1-2005), where the Bioprospecting Undertaking ensures that notification and consultation are essential prerequisites before any activity is undertaken. This means that the ARA cannot be executed until the PIC Certificate is issued, either by the head of the community (village chairperson) or management board, a private landowner or other concerned agencies within 30 days after they have been consulted.

For free prior informed consent (FPIC), secured from Indigenous Peoples, the provisions of the Indigenous Peoples' Rights Act (Republic 8371) shall prevail over the provisions of the joint guidelines. While this legislation refers to FPIC, this is no different from the PIC referred to in succeeding regulations. The addition of the term "free" indicates that consent was given by the community without external pressure but according to their free will. Access to traditional knowledge is explicitly set out in the application for FPIC and reflected in the FPIC Certificate.

The State explicitly recognizes the rights of indigenous groups. This is in line with national legislation and jurisprudence. Aside from acknowledging the existence of the indigenous cultural communities or indigenous peoples (ICCs/IPs) as a distinct sector in Philippine society, the 1997 Indigenous Peoples' Rights Act (RA 8371), or IPRA, grants them the ownership and possession of their ancestral domains and ancestral lands, and defines the extent of these lands and domains. The ownership given is the indigenous concept of ownership under customary law which traces its origin to native title (*Cruz v. Secretary*). The landmark decision of the Supreme Court in *Cariño v. Insular* recognized the long occupancy of land by an indigenous member of the cultural communities as one of private ownership, which, as a legal concept, is termed "native title" (7 Phil. 132 [1906]), which is evidenced by a Certificate of Ancestral Domain Title (CADT). Land held by this title is presumed to "never have been public land" (*Cruz v. Secretary*).

While the abrogation of the Filipinos' ancestral rights in land and the introduction of the concept of public domain were the most immediate fundamental results of Spanish colonial theory and law (*Cruz v. Secretary*), this was essentially corrected in the 1987 Constitution. The 1987 Constitution expressly guarantees the rights of tribal Filipinos to their ancestral domains and ancestral lands. By recognizing their right to their ancestral lands and domains, the State has effectively upheld their right to utilize the natural resources therein. Furthermore, indigenous communities are considered as the natural stewards of the world's biological resources but often forgotten partners in biodiversity conservation (Sobrevila, 2008). However, these communities are more likely to be highly marginalized and often require capacity building to improve awareness of their rights and responsibilities especially with respect to biodiversity conservation. In the case of the Aytas (*Magbukün* tribe) in the Bataan Natural Park in the Philippines, the empowerment of the indigenous communities as stewards of biological resources was a necessary element of the bioprospecting process for research into natural products conducted by a local academic institution and its foreign collaborators.[6] When their capacities are strengthened, the rights and benefits of indigenous communities are ensured.

Comments

While the bioprospecting regulations specifically ensure that the rights of indigenous cultural communities, indigenous peoples and local communities are protected, the right of control over the exploration, utilization and development of natural resources ultimately remain with the state. It is clear then that state control is based on state ownership of natural resources which is enshrined in the Constitution. Hence, despite the rights given to indigenous communities/peoples to utilize natural resources within ancestral domains, the authority to regulate bioprospecting of biological and genetic resources therein is still retained by the state under the bioprospecting law and regulations. FPIC requirements are merely one of the ways to ensure that indigenous communities are consulted, their consent is given without undue pressure, and they share in the benefits derived from the exploitation of biological and genetic resources. In the end, such benefit-sharing provisions demonstrate a healthy compromise between the national government and indigenous communities.

Singapore

Singapore is a tiny city-state of some 720 square kilometers, with a population of 5.28 million.[7] It has received many accolades for being a "clean and green" city, managing to move from third world to first in some forty years (Lee, 2000) through good governance and management, cleaning and greening its environment in the process. While much of the original vegetation was destroyed first by its colonial masters and then by its post-independence government in the quest to industrialize and urbanize, Singapore has made valiant efforts not only to reclaim its biodiversity, but to be a leader in greening the urban environment. Singapore signed the CBD at the Earth Summit in Rio on 12 June 1992, and ratified on 21 December 1995. In October 2010, Singapore presented its Cities Biodiversity Index at Nagoya, at the 10th Conference on the Parties on the CBD. This is a self-assessment tool for cities to evaluate and monitor the progress of their biodiversity conservation efforts against their own individual baselines. Unlike the Philippines, Singapore is not party to the Nagoya Protocol on Access to Genetic Resources and the Fair and Equitable Sharing of Benefits Arising from their Utilization.

Again, unlike the Philippines, there is no provision in the Singapore Constitution that natural resources including water, timber, wildlife, forests, and flora and fauna are owned by the State. The issue of ownership of natural resources is thus determined under the principles of English common law, received via the Application of English Law Act, 1993 (Act 35 of 1993, Cap. 7A), as well as under Singapore's statutory laws. The former provides that "The common law of England (including the principles and rules of equity), shall apply in Singapore so far as it was part of the law of Singapore immediately before 12th November 1993" and "so far as it is applicable to the circumstances of Singapore and its inhabitants and subject to such modifications as those circumstances may require". Section 4 of Land Titles Act provides that "land ... includes any estate or interest therein and all vegetation growing thereon and structures affixed thereto" (Act 27 of 1993, Cap. 157, Rev. Ed. 2004).

Singapore is also the repository of a vast store of biological and genetic resources obtained by its former colonial masters. The Singapore Botanic Gardens Herbarium (international acronym SING) houses a main collection of about 750,000 herbarium specimens, as well as a supporting spirit collection. The Herbarium, founded in 1955, serves as a documentation of the rich plant resources in Southeast Asia, with more 25,000 specimens in its holding. Parts of the reptile and amphibian collection are available online and searchable on Herpnet, along with two moss databases, East Asian and Malesian.[8] The Herbarium collections mainly include materials from the Malesian region (Peninsular Thailand, Malaysia, Singapore, Brunei, Indonesia, the Philippines and New Guinea) and adjacent areas (East Asia, mainland SE Asia, the Southwest Pacific), with the most extensive collections from Singapore and Peninsular Malaysia dating from the 1880s. The SING Herbarium collection is the national herbarium of Singapore.

The Raffles Museum of Biodiversity Research (RMBR), now known as the Lee Kong Chian Natural History Museum, or LKCNHM, located at the National University of Singapore (NUS), has its origins in the Raffles Museum which was founded in 1849 as a result of an idea mooted by Sir Stamford Raffles, an eminent naturalist and founder of Singapore. The LKCMNH now houses one of the largest collection of Southeast Asian fauna specimens in the region. The Zoological Reference Collection (ZRC) is internationally renowned, containing at least 500,000 catalogued specimens.

Interesting questions arise in relation to ownership of these *ex-situ* resources – given that most of these genetic resources were collected pre-CBD. Post-CBD, Singapore has not enacted any laws to address the ABS issues associated with such *ex-situ* collections of genetic resources.

Protection of flora and fauna

In the context of wildlife, the common law provides that property in wildlife only passes to the landowner when it is caught (Gray and Gray, 2009, 1.2.84). In Singapore, the first wildlife protection law (the Wild Birds Protection Ordinance) was passed in 1884. Today, all wild animals and birds with the exception of six birds are protected in Singapore under the Wild Animals and Birds Act (Act 5 of 1965, Cap. 351) and its subsidiary laws which include the Wild Animals (Licensing) Order and the Birds Sanctuaries Order (Lye, 1991).

Nature reserves and parks

Singapore's nature reserves date back to 1908 when land was designated as "reserve forests" to protect "forests and forest produce" (Chun, 2008, p. 208). Today, Singapore has four nature reserves, a marine park and over 300 parks. Together, the four nature reserves cover more than 4.5% of Singapore's land area. A network of green space, parks and park connectors, comprising an additional 4.5% of the land area, supports and buffers these Nature Reserves. All come under the control of the National Parks Board (NParks) and are governed by the Parks and Trees Act (Act 4 of 2005, Cap. 216) and its subsidiary laws which include the Parks and Trees Regulations (S.519/2005) and the Preservation of Trees Order.

National biodiversity strategy and action plan

In 2009, NParks developed and launched Singapore's National Biodiversity Strategy and Action Plan (NBSAP), entitled "Conserving Our Biodiversity". This provides a framework to guide biodiversity conservation efforts in Singapore. It states at the forefront that, while seeking to promote biodiversity conservation, it is mindful of Singapore's limited land space, and thus adopts a pragmatic approach to conservation, requiring unique solutions to be developed for the challenges ahead. It aims to "create an urban biodiversity conservation model that champions environmental sustainability in an urban setting with well-endowed natural heritage". The following principles guide its implementation:

- The biodiversity resources of Singapore are our natural heritage and should be conserved for future generations.
- Considerations on biodiversity and ecosystems are factored into the national planning process.
- A balanced view is adopted of national priorities and international and regional obligations.

NParks is currently developing more specific targets based on the CBD Strategic Plan and the Aichi Targets for 2011–20. This would enable better tracking of Singapore's biodiversity conservation efforts, which would also support the achievement of global targets.

National Biodiversity Centre

The National Biodiversity Strategy and Action Plan (NBSAP) is administered by the National Biodiversity Centre (NBC) which was established on 22 May 2006 to serve as Singapore's "one-stop hub for biodiversity-related information and activities".[9] The Centre consolidates and manages all available resources on biodiversity in Singapore that are generated, stored and updated by different organizations and individuals, by linking them in a single meta-database called the Biodiversity and Environment Database System (BIOME). The Centre represents NParks in its role as the government's Scientific Authority on Nature Conservation. In addition, it represents Singapore in various biodiversity-related international/regional conventions and fora, including the Convention on Biological Diversity and the ASEAN Centre for Biodiversity. The Centre's functions include the formulation, implementation and coordination of strategies, policies and guidelines on biodiversity conservation; as well as the formulation, coordination and monitoring of programs on biodiversity, maintaining reliable biodiversity databases for informed decision-making; providing relevant biodiversity data in support of research and development; functioning as an information portal to enable access to local biodiversity information; establishing the Clearing-house Mechanism (CHM) to promote exchange of biodiversity information as well as the programmes of the CBD and operating as a one-stop centre for research permit application. The Centre is also Singapore's focal point for the ASEAN Working Group for Nature Conservation and Biodiversity (AWGNCB) and the CBD.

Permits for research

Singapore encourages research for non-commercial purposes. Its Research Permit Application Form from the National Biodiversity Centre (NBC) states that the NBC "is always keen to learn of the outcomes of research performed within the permit framework. As such, we would like to invite researchers who wish to share their findings to give a talk to officers within NParks, so that any potentially useful outcomes can be explored from a management perspective.

Research permits are issued by the National Parks Board (NParks) under the authority of the National Parks Board Act (Cap. 198A), the Parks and Trees Act 2005 (Act 4 of 2005) and the Parks and Trees Regulations 2005. Terms and conditions appear on the website.[10] A research permit is required for anyone intending to carry out taxonomic or ecological field studies or to collect any material for research in a national park, nature reserve or parkland under the management and maintenance of NParks. The research permits are specific to the area identified in the permit application. NParks' officers may confiscate any unauthorized samples or specimens collected, and they have the right to check the field collection(s) anytime at the site. Each permit granted has a validity period, which is usually a period of one year or less.

Samples that need to be taken out of Singapore require a Letter of Authorization (LOA) and/or an Export Permit from the Agri-Food and Veterinary Authority of Singapore (AVA). The AVA administers the Wild Animals and Birds Act (WABA), the Animals and Birds Act (ABA) and the Endangered Species (Import and Export) Act, which implements the Convention on International Trade in Endangered Species of Wild Flora and Fauna (CITES).

LOAs are issued by AVA for scientific collections of specimens of species (whether endangered or non-endangered, whether alive, dead or preserved) found in areas that are not within the jurisdiction of NParks (i.e. not nature reserves, national parks or public parks). They are issued for exports of specimens collected in Singapore that do not fall under the definition of "animals" under the ABA. Examples include invertebrates such as spiders, ants, insects. LOAs are issued after consultation with NParks to ascertain whether the collection of such species would threaten wild populations. Export permits are issued under ABA for animals belonging to higher taxonomy such as mammals, e.g. bats, and also birds and fish (defined to include crustaceans, aquatic molluscs, turtles, marine sponges and any form of aquatic life), and their tissue samples. Following an amendment to the definition of animals in 2015, export permits may also be required for collected reptiles and amphibians. If the collected specimens belong to CITES species, CITES export permits would also be required.

Flora specimens (including live, dead or preserved plants, aquatic plants and ornamental plants) can be exported with a phytosanitary certificate, or other pre-shipment treatment or lab certification as required by the importing country, from the AVA. The export of CITES specimens of flora requires the exporter to obtain an additional permit (CITES export permit) from the AVA. Similar requirements apply to persons who wish to import any biological specimens for research purposes – they must apply for a Letter of Authorization and/or Import Permits from the AVA.

The applicant is required to specify in the research permit application form whether the research is for commercial or non-commercial purposes. If it is for commercial purposes, the NBC will consult with other agencies in Singapore and will need to negotiate a benefit-sharing agreement. If the research is initially non-commercial in nature, but may potentially have commercial spin-offs subsequently, the researcher will then need to engage in a separate negotiation process at that time.

It is still unclear what terms are to be contained in the case of research for commercial purposes. In response to our query, NParks has replied that they do not have standard terms and conditions relating to access to genetic resources for commercial/bioprospecting requests. Only one bioprospecting agreement has been signed with a company some years ago, a mutually negotiated contract which has since expired. These issues are currently being re-examined by NParks, reviewing, in particular, whether they are the most appropriate agency to enter into such bioprospecting agreements on behalf of the government.

The Singapore Botanic Gardens (SBG) has three collections of plant genetic material: (i) herbarium specimens, (ii) living plants and (iii) cryogenically preserved orchid seeds. It is also planning to start up a small seed bank for native species. The SBG allows free access to the SING herbarium collection for research and educational purposes by bona fide researchers and research institutions. Loans of herbarium specimens are regulated by the terms and conditions stipulated by the SING Herbarium.

For SBG's living plant collection, its Living Collection Policy (Version 2.0, 2015) on Access, states that "The SBG will encourage the use of its living collections to the greatest extent possible". Physical access to the living collections in the Gardens is unrestricted during normal hours of operation. Detailed information and records relating to the collections are "freely available" to persons with approved research permits. Propagation, DNA and herbarium material are only available to qualified individuals who have obtained permission from relevant officers. Access and use will be regulated by the terms and conditions of any Material Transfer Agreement (MTA) signed by the SBG with parties seeking to obtain samples of the plant in question. For SBG to exchange/transfer interested living plant materials (sapling, cuttings, seeds) of which SBG has "ownership" (i.e. native species or SBG created hybrids), or plants that are freely available from its nurseries, SBG will require the requesting institution/researcher to sign an MTA. The general practice is that SBG will not transfer living plant materials to private individuals and private commercial entities. All this is set out in the SBG's Living Collection Policy, which ends with a note that "a small but historically important role of the Botanic Gardens has been in decline since 1992", with the emergence of the CBD. This relates to the development of commercially valuable plant collections for the horticultural trade, utilizing plant diversity in the plant collections and the skills of the staff in the plant breeding and selection process. These processes must now comply with the CBD and must also be commercially viable to offset the costs of research and breeding activities. Some examples include new orchid cultivars, hybrids, ginger selections and native species with ornamental potential.

The Lee Kong Chian Museum of Natural History (LKCMNH) serves as a nexus for biodiversity research, particularly in the area of wildlife taxonomy. Its Zoological Reference Collection (ZRC) is a valuable scientific resource for which many researchers come to Singapore to study. The museum maintains an "open-door policy" with regard to the use of its Collection for research purposes, and requests to examine the material in the Collection are usually accommodated. The Museum also routinely sends out material on loan for study by scientists overseas. Requests for biological material other than whole specimens (i.e. tissues for DNA sequencing) are also routinely granted. The loan forms state clearly that "These are for research and educational purposes, and NOT FOR COMMERCIAL USE". Loans of specimens are usually made for a period of six months to two years. The LKCMNH also makes specimen donations to other institutions for research purposes, via a Deed of Donation that again clearly states that the items donated are meant for research and educational purposes, and not for commercial use.

It would appear that Singapore is not ready to allow access and benefit-sharing of its biological resources for commercial purposes. Permits currently issued by NParks, the Singapore Botanic Gardens and the Lee Kong Chian Museum of Natural History only relate to research for non-commercial purposes. However, important issues remain unresolved, as it is unclear how researchers should proceed should they subsequently discover that these biological resources they have accessed have potential commercial use, NParks has indicated that this is an issue which they are currently working on.

ASEAN Centre for Biodiversity

The ASEAN Centre for Biodiversity (ACB) was established in 2005. ACB's beginnings date back to 1998, when the ten ASEAN Member States established the ASEAN Regional Centre for Biodiversity (ARCBC) Project with funding support from the European Union. The project was implemented with two key objectives: (i) intensify biodiversity conservation, and (ii) promote institutional networking among ASEAN countries and between ASEAN and European Union partner organizations. When this project came to an end in 2005, the environment ministers of ASEAN agreed to form the ACB to continue with the work of the ARCBC. ACB's Vision is "An ASEAN region whose biodiversity is conserved, sustainably managed and used, and equitably shared for the well-being of its peoples". Its mission is "To effectively facilitate regional cooperation and deliver capacity building services to the AMS in conserving biodiversity". The ACB seeks to:

- Facilitate coordination and cooperation on regional biodiversity conservation programs.
- Deliver capacity-building services.
- Develop regional mechanisms to sustainably manage and protect biodiversity and ecosystems and strengthen ASEAN regional positions in negotiations and in compliance with relevant multilateral environmental agreements.
- Impart and deliver to ASEAN Member States the knowledge and tools on managing biodiversity.

The ASEAN Member States have dabbled with at least two international legal frameworks to regulate access and benefit-sharing practices in relation to the region's biological diversity, neither of which has got off the ground. These will be outlined in brief below.

Draft ASEAN Framework Agreement on Access to Biological and Genetic Resources, 2000

In February 2000, an ASEAN Framework Agreement on Access to Biological and Genetic Resources was drafted and presented in June 2000 to the ASEAN Working Group for Nature Conservation and Biodiversity.[11] This was a detailed agreement aimed at setting minimum standards in regulating access to biodiversity, and ensuring uniform and consistent regulations within ASEAN region on ABS.

Article 4 (scope and coverage) states that:

- access to biological and genetic resources does not automatically mean access to the traditional knowledge associated with the resource; such must be specifically stated in the application for access;
- *ex-situ* materials from the ASEAN region collected prior to the CBD shall be considered as "held in trust for the benefit of humankind where the application of intellectual property rights shall not be allowed";
- member states "shall not allow the patenting of plants, animals, micro-organisms or any parts thereof, and traditional and indigenous knowledge" – these are "regarded as a sacred heritage for all mankind";
- the prospecting and application of intellectual property rights on genetic materials of human origin is disallowed; and member states call for the establishment of a process to regulate access, use and commercialization of human genetic material;
- this agreement shall not apply to traditional uses of biological and genetic resources by indigenous and local communities in accordance with their customary practices and traditions.

Article 10 on Prior Informed Consent (PIC) requires that PIC must be obtained before access is allowed. Further,

- Member states must establish legal procedures for access, but must involve "indigenous people and local communities embodying traditional lifestyles".
- PIC must comply with all customary laws, practices and protocols of indigenous peoples and local communities.
- Each application for PIC must contain full disclosure of certain information in Article 10 (a) to (g).

Article 11 on fair and equitable sharing of benefits provides that:

- All resource providers, especially indigenous peoples must be included in negotiations of benefits on the basis of full disclosure.
- At a minimum, the conditions must include:
  - the participation of nationals in research activities;
  - the sharing of research results, including all discoveries;
  - a complete set of all voucher specimens left in national institutions;
  - access by nationals to all national specimens deposited in international *ex situ*-collections;
  - the receipt by resource providers, without payment of a royalty, of all technologies developed from research on provided materials;
  - fees, royalties and financial benefits; and
  - the donation to national institutions of equipment used as part of research.

Article 12, Common Fund for Biodiversity Conservation:

- Contributions to this Fund come from "a share in the revenues derived from commercialization of the use of common and shared resources among Member States".
- ASEAN Working Group on Nature Conservation and Biodiversity (AWGNCB) shall make recommendations on the implementation mechanisms for the Fund.

Article 13, Environmental and Social Impact and Biosafety Concerns:

- Must conform with all guidelines on biosafety but AWGNCB may develop separate and distinct protocol.
- Framework Agreement must take into account the various environmental and social impacts of access to genetic resources in conformity with national, regional and international guidelines.

The ACB is designated as the clearing house, whose role is to maintain a database on biological diversity and biological and genetic resources, access agreements and resources, provide technical and legal support to national authorities, report the status of implementation to the Standing Committee and adopt a warning system to warn other states of applications that have been denied by a member state and the reasons given for such denials.

The draft Framework Agreement has not been signed and does not appear to have received a single signature from any of the ASEAN Member States to date.

Bonn Guidelines on Access and Benefit-sharing, 2002

The Bonn Guidelines on Access and Benefit-sharing (ABS) were approved by Decision VI/24 of COP VI, at the Hague in 2002. They were meant to be voluntary and intended to promote transparency, covering all genetic resources except the human genome. However, some countries were unhappy with these Guidelines, and a Group of Like-Minded Mega-Diverse Countries (together holding 70 per cent of the world's biodiversity) comprising 17 developing states including Indonesia, Malaysia and the Philippines was formed in Cancun, Mexico. At the 7th COP in Malaysia (2004), they pushed for an international regime on ABS, arguing that national laws were insufficient. They demanded provisions for (i) the mandatory disclosure of the country of origin and traditional knowledge in applications for intellectual property; (ii) undertakings that domestic laws have been respected; and (iii) mandatory specific consequences for non-compliance.

The Bonn Guidelines remain voluntary and, at its eighth meeting, in decision VIII/4 B, the Conference of the Parties urged Parties to continue implementing the Bonn Guidelines and to share experiences and lessons learned in their implementation as well as in the development and implementation of national and sub-national measures.[12]

Conclusion

It is clear that Singapore and the Philippines have very different regimes for access and benefit-sharing, given their differences in socio-economic conditions and geographic landscapes. The Philippines is party to the Nagoya Protocol and has detailed laws, procedures and guidelines regulating ABS in relation to wildlife research permits and bioprospecting, prior informed consent process and benefit-sharing measures. The country's bioprospecting laws recognize the rights of indigenous and local communities over natural resources and require the prior informed consent of these communities. Singapore is an urban city-state with no indigenous population. It has comprehensive laws protecting nature reserves and wildlife and is an *ex-situ* repository of biological and genetic materials sourced from countries in the region. It has instituted a system of permits to facilitate ABS only for purposes of non-commercial research.

At the regional level, the ASEAN Centre for Biodiversity was established in 2005 and is hosted by the Philippines. The 2000 ASEAN Framework Agreement on ABS has not been signed, let alone ratified, and appears to have been abandoned. It would appear that member states are not comfortable with its provisions and have rejected them. The ASEAN states, particularly the mega-diverse states of Indonesia, Malaysia and the Philippines, have also expressed dissatisfaction with the Bonn Guidelines. Singapore has yet to implement the Nagoya Protocol. The region, however, continues to seek to build capacity in biodiversity conservation. In 2008, NParks in Singapore initiated Regional and Sub-Regional Capacity-Building Workshops on Implementing NBSAPs (National Biodiversity Strategies and Action Plans) and Mainstreaming Biodiversity for South, Southeast and East Asia.[13] Efforts at capacity building in access and benefit-sharing within the region continue, supported by ACB. Thus, in September 2012, an ASEAN-India Capacity Building Workshop on Access and Benefit-sharing and Traditional Knowledge was held in New Delhi, India, funded by India. The goals of the CBD, the Nagoya Protocol and the Aichi Targets continue to be supported by the ASEAN members. At the 28th and 29th ASEAN Summits in Vientiane, Laos PDR (6 September 2016); an ASEAN Joint Statement to the 13th CBD Conference of the Parties was made. Key excerpts are provided below in Box 13.1 below. Most recently, at the 30th ASEAN meeting hosted by the Philippines on 29 April 2017, the Chairman reaffirmed ASEAN's commitment to the conservation and sustainable management of biodiversity and natural resources in the ASEAN region (para. 101). However, it would appear that, at least in the context of regulating access to and benefit-sharing of biological resources, this is still a work in progress at the regional level for ASEAN, and at the national level for most of the ASEAN states, with the possible exception of the Philippines. It is hoped that capacity-building in this field will continue to be a priority for the governments of the ASEAN Member States and will further regional efforts to achieve ASEAN Environmental Legal Integration (Koh *et al.*, 2016).[14]

Box 13.1

> "**REAFFIRMING** our commitment to implement the Strategic Plan for Biodiversity 2011–2020 and achieve the Aichi Biodiversity Targets; and to the 2030 Agenda for Sustainable Development and its sustainable development goals (SDGs), in particular Goals 14 and 15 on the conservation and sustainable use of coastal and marine resources, and terrestrial ecosystems respectively; ..."

ASEAN member states agreed to:

> "**Accelerate** efforts to effectively implement the Strategic Plan for Biodiversity 2011–2020 and contribute to the achievement of the Aichi Biodiversity Targets by 2020; and to increase the efficiency and effectiveness of support from the Global Environmental Facility and other possible sources of funding to implement the ASEAN Member States' National Biodiversity Strategies and Action Plans;
>
> **Promote** the effective implementation of legislative, administrative or policy measures on access and benefit-sharing to support the third objective of the Convention, as appropriate;
>
> **Promote** cooperation on species conservation and wildlife management, ecosystem restoration and where relevant, agricultural biodiversity, health and biodiversity and urban biodiversity to further support mainstreaming of biodiversity activities in the region..."

Acknowledgements

The writers would like to thank Prof. Peter Ng, Dr. Jose Christopher E. Mendoza and Mr. N Sivasothy, Lee Kong Chian Museum of Natural History, National University of Singapore; and Dr. Lena Chan and Ms. Wendy Yap, National Parks Board, Singapore; as well as Dr. Theresa Mundita S. Lim, Biodiversity Management Bureau, DENR, Philippines, for their kind assistance.

Notes

1 See http://web0.psa.gov.ph/sites/default/files/2014%20PIF.pdf.
2 Cruz, L., "Bioprospecting in the Ancestral Domain of Aytas in Bataas". Presentation at the ASEAN Centre for Biodiversity in 2010, https://searca.org/phocadownload/ADSS 2010/UPLB 2010 Bioprospecting in BNP.pdf.
3 *Miners Association of the Philippines, Inc. v. Factoran et al.*, 240 SCRA 100, January 16, 1995.
4 *Minors Oposa et al. v. Factoran, Jr. et al.*, G.R. No. 101083, July 30, 1993.
5 Cruz, L., "Bioprospecting in the Ancestral Domain of Aytas in Bataas". Presentation at the ASEAN Centre for Biodiversity in 2010, https://searca.org/phocadownload/ADSS 2010/UPLB 2010 Bioprospecting in BNP.pdf.
6 *Ibid.*
7 See www.singstat.gov.sg/statistics/latest-data.
8 See http://lkcnhm.nus.edu.sg/dna/places/details/64.
9 See www.nparks.gov.sg/biodiversity/national-biodiversity-centre.
10 See www.nparks.gov.sg/biodiversity/resources-and-research-permits.
11 See www.mabs.jp/countries/others/pdf/321e.pdf.
12 Information provided to the Secretariat related to the implementation of the Bonn Guidelines has been compiled in information documents UNEP/CBD/WG-ABS/5/INF/2 and UNEP/CBD/WG-ABS/5/INF/2/Add.1.
13 See www.cbd.int/nbsap/workshops/south-southeast-east-asia.shtml.
14 See Koh, K.-L., N.A. Robinson and L. Lin-Heng (2016). "ASEAN Environmental Legal integration – Sustainable Goals?", in J.H.H. Weiler and Tan H.-L. (eds.) *Integration Through Law – The Role of Law and the Rule of Law in ASEAN Integration,* Cambridge University Press.

References

Bernas, J.G. (1997). *The 1987 Philippine Constitution, A Reviewer-Primer.* 3rd edn. Rex Book Store, Quezon City, Philippines.

Centre for International Sustainable Development Law. (2012) *Overview of national and regional measures on access to genetic resources and benefit-sharing: Challenges and Opportunities in Implementing the Nagoya Protocol.* 2nd edn.

Chun, J. (2008). "Wildlife Law in Singapore: Protecting Wildlife in the 'Garden City'", chapter 5 in *Wildlife Law - A Global Perspective,* R. Panjwani ed., American Bar Association.

Gray, K. and Gray, S.F (2009). *Elements of Land Law,* 5th edn. Oxford University Press.

Jenks, D.T. (1995). "The Convention on Biological Diversity – An Efficient Framework for the Preservation of Life on Earth." 15 *Nw. J. Int'l L. and Bus.* 636 (1994–1995).

Koh, K.L., Robinson, N. and Lye, L.H. (2016). *ASEAN Environmental Legal integration – Sustainable Goals?,* Cambridge University Press.

Lee, K.Y. (2000). *From Third World to First: The Singapore Story,* 1965–2000, Marshall Cavendish.

Lye, L.-H. (1991). "Wildlife Protection Laws in Singapore", *SJLS* 287–319.

Mittermeier, R.A., Robles-Gil, P. and Mittermeier, C.G. (eds) (1997). *Megadiversity. Earth's Biologically Wealthiest Nations.* CEMEX/Agrupaciaon Sierra Madre, Mexico City.

Peña, N. (1997). *Philippine Law on Natural Resources.* Rex Printing Company, Inc. Quezon City, Philippines.

Sobrevila, C. (2008). *The Role of IPs in Biodiversity Conservation.* IBRD and WB, Washington, USA.

14
ONE STEP FORWARD, TWO STEPS BACK?
Implementing access and benefit-sharing legislation in South Africa

Rachel Wynberg[1]

Introduction and context

South Africa has been at the global forefront of regulating access and benefit-sharing (ABS) and is also one of the few African countries to have a comprehensive ABS legal and policy framework in place (Crouch *et al.*, 2008; Lowman, 2012; Taylor and Wynberg, 2008). Indeed, an ABS policy process was initiated in South Africa as early as 1996, following ratification of the Convention on Biological Diversity (CBD)[2] in 1995, and amid public controversies and concern that the country's rich natural and cultural heritage was being "sold for a song", without proper controls and oversight.

Coinciding with the democratic elections of 1994, this heralded a new era for the country. Over the next ten years, conservation and social justice became integrally intertwined in a new set of biodiversity and bioprospecting laws and policies that entrenched equity and benefit-sharing (Kepe *et al.*, 2005; Wynberg, 2002). After decades of often unscrupulous exploitation, companies and researchers wishing to use the country's biological resources – or traditional knowledge associated with these resources – were now required to demonstrate that they had both received the prior informed consent of communities who were resource- or knowledge-owners, and negotiated a benefit-sharing agreement based on mutually agreed terms (Taylor and Wynberg, 2008). Without a so-called ABS permit, issued by the Department of Environmental Affairs (DEA) as required by South Africa's National Environmental Management Biodiversity Act (NEMBA), Act 10 of 2004 (hereafter referred to as the Biodiversity Act) and its 2008 Bioprospecting, Access and Benefit-sharing (BABS) Regulations, those found to be non-compliant faced the risk of a hefty fine or even imprisonment.[3]

The eight years of implementation since promulgation of the BABS Regulations have witnessed the negotiation of a suite of benefit-sharing agreements. These include the well-known example of *Hoodia* spp., with benefit-sharing agreements developed between indigenous San peoples and the South African-based Council for Scientific and Industrial Research (Wynberg *et al.*, 2009) based on the properties of this anti-obesity succulent plant; those between the South African San Council[4] and HGH pharmaceuticals focused on *Sceletium tortuosum* (Chennells, 2013), a succulent plant developed as an anti-depressant and mood enhancer; and an agreement between the national Khoi-San Council and Cape Kingdom Nutraceuticals focused on the commercial development of the indigenous buchu shrub (*Agathosma betulina*) for medicinal purposes.

At the same time, there have been considerable challenges in implementing the laws, despite ongoing stakeholder consultations and several legal amendments (Crouch *et al.*, 2008; Lowman, 2012). This has been due in part to the complexity of the issues under consideration, but mostly due to concerns about the cumbersome and ambiguous nature of the regulatory framework and permit approval process. Moreover, policies have struggled to keep up with profound scientific, technological and market changes over the past twenty years (Laird and Wynberg, 2016) as well as major legal developments due to the ratification by South Africa of the Nagoya Protocol on Access to Genetic Resources and the Fair and Equitable Sharing of Benefits Arising from their Utilization and its coming into force.[5] While many of these factors are not unique to South Africa, the country's decades-long experience of regulating ABS, combined with its strategic role in natural product development, provides an important platform from which to learn lessons. This chapter aims to provide a historical and contemporary review of ABS policy and implementation in South Africa, with the objective of working towards an improved ABS system that better accommodates social justice, economic development and biodiversity conservation. The analysis draws on an appraisal of archival policy records since 1994; semi-structured interviews with stakeholders representing different industry sectors, government bodies, NGOs and community groups; and a substantial archive of personal observations and notes gathered over the past twenty years. Following this introduction, the chapter sets out the history of ABS regulation in South Africa. This is followed by a review of the range of implementation issues which have confronted both regulators and those attempting to comply with the law. The chapter closes with a set of recommendations that have application both in South Africa and across the globe.

Legal frameworks for access and benefit-sharing in South Africa – a brief history

Developing a Biodiversity Act

The history of ABS regulation in South Africa in many ways has laid the foundation for current approaches. Due to its apartheid policies and exclusion from the United Nations, South Africa was still a pariah in the early 1990s, when the CBD was negotiated and signed. Just a handful of academics, religious leaders and journalists from the country participated in the United Nations Conference on Environment and Development (the Rio "Earth Summit"). At this meeting, questions of biopiracy came to the fore, amidst claims that developed countries had exploited the biodiversity and traditional knowledge of developing countries to build their pharmaceutical and agricultural empires (Shiva, 1992). Returning to South Africa, a small group of researchers began work to explore the extent to which bioprospecting was taking place in the country. This took place as part of a broader post-apartheid law reform initiative, which included a consultative process of developing a biodiversity policy that represented the interest of all citizens.

A report was commissioned on the status of bioprospecting in South Africa by the Land and Agriculture Policy Centre, a policy think tank of the ruling party at the time, the African National Congress (Laird and Wynberg, 1996). For the first time an overview was obtained about the scale of bioprospecting operations, the nature of the partnerships being developed and the key policy issues requiring resolution. Through a series of engagements with stakeholders, comments were incorporated into ABS policy in a gazetted Biodiversity White Paper in 1997 (Department of Environmental Affairs and Tourism, 1997).

A further consultation process was initiated for the development of a new Biodiversity Act. Again, a research process commenced to review legislation elsewhere in the world and advise on appropriate ABS models. A strategic review was also conducted of national bioprospecting initiatives and the actors involved. Numerous drafts of the legislation were developed for discussion and consultation, with the firm intent to create framework legislation that was easy to implement and that encouraged research and commercialization without compromising community rights and economic development. Commercial interests were represented to some extent, largely through the research institutions involved in bioprospecting at the time.

Unfortunately, the period between finalizing the draft and submitting it to the state legal advisor coincided with a change in leadership at the national Department of Environmental Affairs (DEA), and a consequent decline in ABS capacity and understanding. At some point the final version was thus transformed,[6] creating a legal instrument that was almost unworkable from an ABS perspective. This was due in large part to its very wide scope, which now regulated all indigenous biological resources, as opposed to only genetic resources,[7] and all phases of research, development and commercialization. A cumbersome administrative process was established to accompany all applications, proving near impossible to implement in its earliest renditions. A trust whereby funds could be deposited in situations where knowledge was shared among several communities, or where it was difficult to identify traditional knowledge owners, was abandoned. Instead of creating an instrument to accommodate these ambiguities, a purely mechanical and non-discretionary "bank account" was established, to be used simply to disburse funds. In many respects, therefore, the considerable amount of research and consultation that had taken place to inform the Act was effectively annulled.

Regulations for bioprospecting, access and benefit-sharing

In 2004, the National Environmental Management: Biodiversity Act (10 of 2004) was promulgated, Chapter 6 of which was specifically focused on ABS. The three objectives of the Act mirrored those of the CBD, providing for[8] the management and conservation of biodiversity; the sustainable use of indigenous biological resources; and the fair and equitable sharing among stakeholders of benefits arising from bioprospecting involving indigenous biological resources. The Biodiversity Act provided only a broad framework for ABS, however, leaving the detail to be dealt with in subordinate national legislation (Box 14.1). Thus a process commenced in 2005 to develop detailed ABS regulations through wide consultation with various stakeholders.

Box 14.1 Key provisions of the National Environmental Management: Biodiversity Act (10 of 2004) and its 2008 Bioprospecting, Access and Benefit-Sharing Regulations

The Biodiversity Act and BABS Regulations require anyone carrying out bioprospecting that involves indigenous biological resources and, if applicable, associated traditional use or knowledge, to obtain a permit.

A permit is also required for anyone exporting indigenous biological resources for bioprospecting or other research, and export must be in the public interest.

Foreign individuals or companies must apply jointly with South African individuals or companies for bioprospecting or export permits.

A permit will only be issued if there has been material disclosure to stakeholders, if their prior informed consent to the bioprospecting has been obtained, and if the Minister (of Environmental

Affairs) is satisfied that certain conditions as set out in the legislation have been met. Consent must be reflected in a benefit-sharing agreement, which could include both monetary and non-monetary benefits.

The Act includes two categories of stakeholders whose prior informed consent to a bioprospecting project must be obtained.

They are:

- those who give access to the indigenous biological resources (e.g. a landowner or a gene bank); and/or
- indigenous communities whose knowledge or traditional use of indigenous biological resources has contributed to, or may contribute to, the bioprospecting.

Benefit-sharing agreements (BSAs) must be entered into with both categories of stakeholders and, in addition, a material transfer agreement (MTA) must be entered into with stakeholders who give access to the indigenous biological resources.

Importantly, the law makes a distinction between the "discovery phase" of a bioprospecting project and the "commercialization phase". In the discovery phase, researchers attempt to find out if an indigenous biological resource has any commercial potential. In the commercialization phase, that potential has already been identified. Those doing discovery phase research need to notify the minister about what they are doing, and do not require a bioprospecting permit. A bioprospecting permit is needed only for the commercialization phase.

"Bioprospecting", in relation to indigenous biological resources, means any research on, or development or application of, indigenous biological resources for commercial or industrial exploitation, and includes:

- the systematic search, collection or gathering of such resources or making extractions from such resources for purposes of such research, development or application;
- the utilization for purposes of such research or development of any information regarding any traditional uses of indigenous biological resources by indigenous communities; or
- research on, or the application, development or modification of, any such traditional uses, for commercial or industrial exploitation.

"Indigenous biological resources" includes:

- any indigenous biological resources whether gathered from the wild or accessed from any other source, including any animals, plants or other organisms of an indigenous species cultivated, bred or kept in captivity or cultivated or altered in any way by means of biotechnology;
- any cultivar, variety, strain, derivative, hybrid or fertile version of any indigenous species or of any animals, plants or other organisms referred to above; and
- any exotic animals, plants or other organisms, whether gathered from the wild or accessed from any other source which, through the use of biotechnology, have been altered with any genetic material or chemical compound found in any indigenous species or any animals, plants or other organisms referred.

The Act also establishes a Bioprospecting Trust Fund, into which all money arising from benefit-sharing agreements must be paid, and from which all payments to stakeholders will be made.

The Minister of Water and Environmental Affairs is the issuing authority for bioprospecting permits. She is advised by a Bioprospecting Committee that comprises national and provincial government departments as well as parastatal organizations such as the South African National Biodiversity

Institute and the South African National Parks. Foreign individuals or companies must apply jointly with South African individuals or companies for bioprospecting or export permits. Export of indigenous biological resources must be for a purpose that is in the public interest. A benefit-sharing agreement template is included in the BABS Regulations. This lists possible monetary and non-monetary benefits that may be appropriate.

Since promulgation, several amendments have been made to the Biodiversity Act and BABS Regulations. In 2009 the Biodiversity Act was amended to, among others, insert definitions for commercialization[a] in relation to indigenous biological resources; distinguish between the commercialization phase[b] and discovery phase of bioprospecting;[c] introduce notification requirements in the discovery phase of a bioprospecting project; and to take into consideration knowledge of specific individuals when issuing bioprospecting permits. The net effect of this was to remove any requirements for benefit-sharing from the discovery phase and to entitle individuals to engage as beneficiaries in benefit-sharing agreements.

A second set of amendments to the Biodiversity Act, published late 2013, inter alia, extended the definitions of "bioprospecting" and "commercialization" to include "the trading in and exporting of indigenous biological resources to produce products";[d] and included "commercial exploitation"[e] as a form of commercialization. Further, the purpose of Chapter 6 was amended to include genetic resources as well as biological resources, and an additional purpose of utilizing indigenous genetic and biological resources sustainably, while promoting social and economic development, particularly in areas where resources and knowledge are accessed was inserted. Prior to this amendment there was some ambiguity about whether biotrade was included within the ambit of the Biodiversity Act but the amendment clarified that biotrade fell squarely within the regulatory framework, despite concerns that this approach was inappropriate and contrary to the intent of the Nagoya Protocol.

At the time of writing, further amendments to the Regulations have been tabled, along with a redrafting of the Biodiversity Act, anticipated for completion in 2018.

Notes

[a] **"Commercialization"**, in relation to indigenous biological resources, includes the following activities:

(a) the filing of any complete intellectual property application, whether in South Africa or elsewhere;
(b) obtaining or transferring any intellectual property rights or other rights;
(c) commencing clinical trials and product development, including the conducting of market research and seeking pre-market approval for the sale of resulting products; or

the multiplication of indigenous biological resources through cultivation, propagation, cloning or other means to develop and produce products, such as drugs, industrial enzymes, food flavours, fragrances, cosmetics, emulsifiers, oleoresins, colours and extracts.

[b] **"Commercialization phase of bioprospecting"** means any research on, or development or application of, indigenous biological resources where the nature and extent of any actual or potential commercial or industrial exploitation in relation to the project is sufficiently established to begin the process of commercialization.

[c] **"Discovery phase of bioprospecting"** means any research on, or development or application of, indigenous biological resources where the nature and extent of any actual or potential commercial or industrial exploitation in relation to the project is not sufficiently clear or known to begin the process of commercialization.

[d] Such as: drugs, industrial enzymes, food flavours, fragrances, cosmetics, emulsifiers, oleoresins, colours, extracts, and essential oils.

[e] Meaning engaging in any bioprospecting activity with the intention of making a profit.

In 2008, the BABS Regulations came into effect following a protracted period of public engagement (Taylor and Wynberg, 2008). No fewer than 14 drafts of the regulations were produced prior to their promulgation, indicating the complexity of the issues being dealt with in the regulations (Taylor and Wynberg, 2008). Drafts were prepared by an independent legal consultant, who was guided by a technical ABS expert (the author) – both from outside of government. Prior to their publication, drafts were vetted by a State legal advisor who made significant modifications, in some instances unfortunately transforming carefully drafted text into a published version that did not always accord with the initial intent. The final BABS Regulations thus represented a far from perfect outcome and were marred to an even greater extent by their necessity to stay within the ambit of a flawed Biodiversity Act (Crouch *et al.*, 2008; Taylor and Wynberg, 2008; Wynberg and Taylor, 2009). To some extent, therefore, the legislation was doomed to present problems even before implementation began.[9]

Figure 14.1 below gives an overview of the current regulatory process, whereas Figures 14.2 and 14.3 illustrate the different requirements for discovery and commercial bioprospecting.

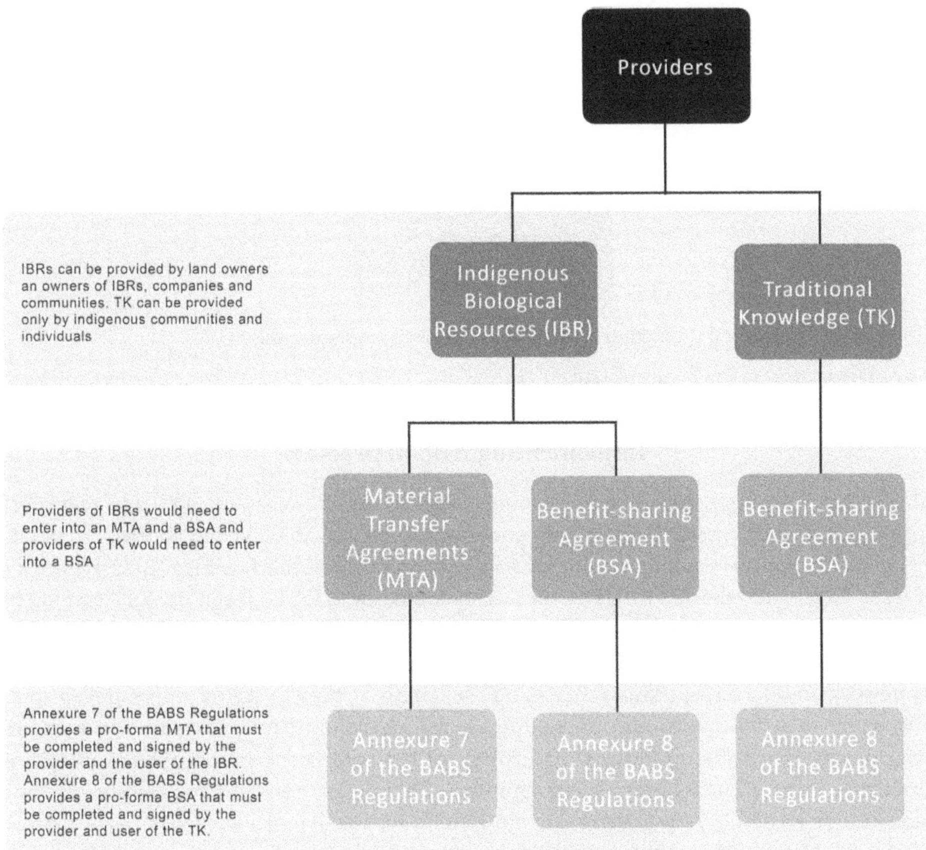

Figure 14.1 The ABS regulatory process in South Africa

Source: Department of Environmental Affairs (2012).

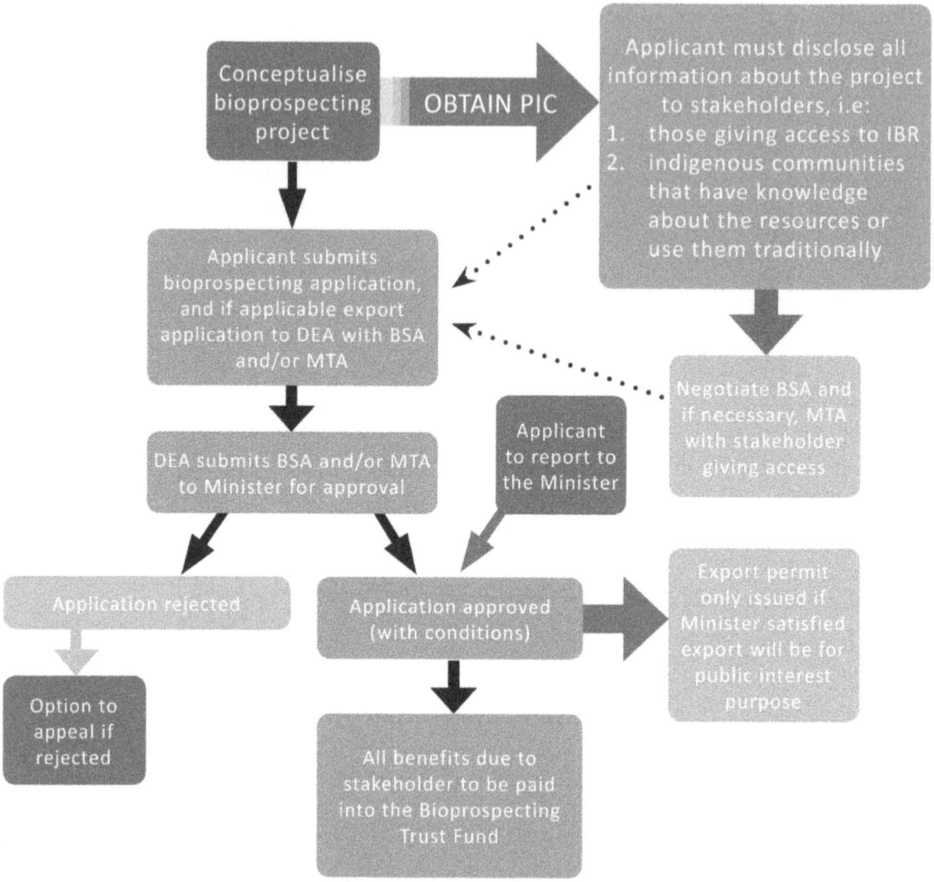

Figure 14.2 The general application process for commercial bioprospecting

Source: Department of Environmental Affairs (2012).

Implementing ABS in practice

Almost ten years later, and despite ongoing consultation and legal amendments, regulatory approaches for ABS in South Africa have continued to be very difficult and complex to implement and the general sentiment is one of negativity and frustration (Lowman, 2012; University of Cape Town and others, 2014). The approach adopted by DEA has been one of "learning by doing" which, while bold in nature, has unsurprisingly also had negative ramifications.

Substantive concerns have been expressed about the cumbersome nature of the regulatory framework and permit approval process and the length of time to secure a permit – in some cases more than two years. The ambiguities of the legislation have presented serious stumbling blocks, especially regarding the inclusion of biotrade (or commodity trade in natural products) as part of the bioprospecting definition. Implementation has also been stymied by a lack of understanding from government about industry needs and a lack of business acumen within the National Focal Point. This has been exacerbated by the substantial challenges of getting prior informed consent and finding communities with whom to negotiate benefit-sharing agreements. Combined with questions about the actual benefits that communities are realizing from agreements, and a neglect of attention to benefits arising in the research and

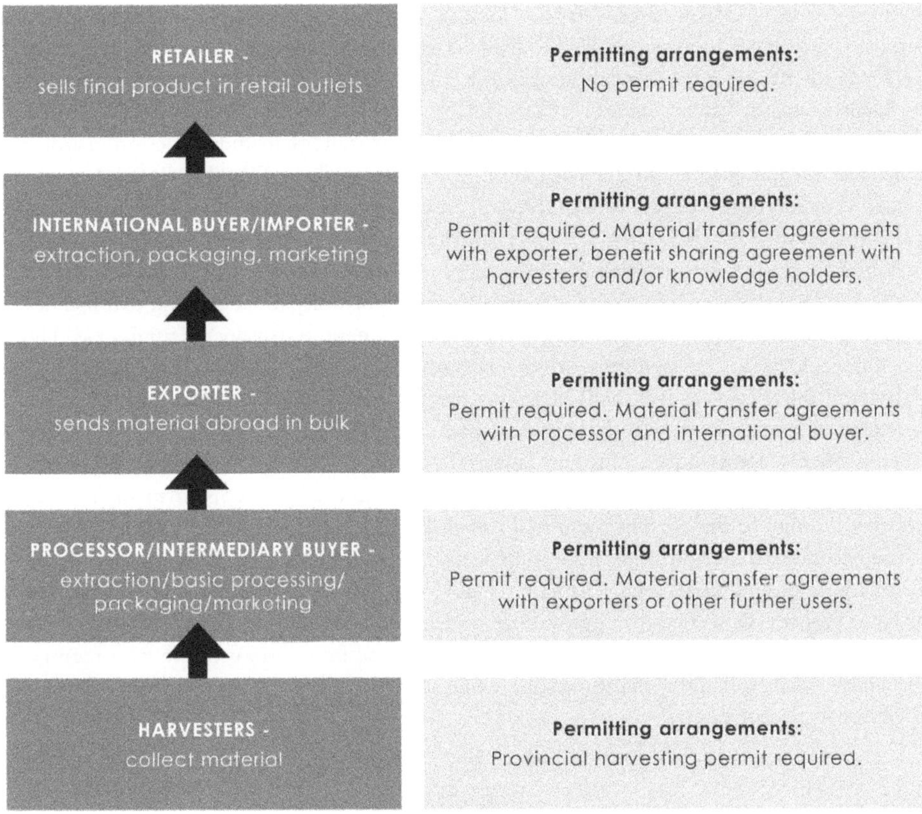

Figure 14.3 Permits required along the value chain

Source: Department of Environmental Affairs (2012).

development process, many have raised questions about whether the intent of the law has been realized.

Each of these concerns is dealt with in further detail below.

Multiple laws and institutions

A multiplicity of laws, policies and initiatives has unfolded in South Africa over the past twenty years with relevance for ABS. These have been ignited by government interest in the commercial potential of the country's biological diversity, as well as a strong national interest in biotechnology. An equally diverse set of institutions, often with divergent objectives, has emerged to manage these laws and policies (see Department of Environmental Affairs, 2012, Appendix 5). In addition to the Biodiversity Act described above, these range from laws focused on intellectual property protection; requirements for disclosure of origin on patent applications; the protection of traditional knowledge; plant variety protection; and biotechnology, among others. They are administered by a range of government bodies, including the Department of Environmental Affairs; the Department of Science and Technology, the Department of Trade and Industry; and the Department of Agriculture.

Some of these are framed by specific strategies and policies, including the Bio-economy Strategy of the Department of Science and Technology which aims to "provide an enabling environment for all stakeholders" and to "identify areas where public policy can remove barriers and encourage innovation" (DST, 2014). Three key economic sectors are identified as the focus of the strategy – agriculture, health and industry. In 2015, almost simultaneously, the Department of Environmental Affairs (DEA) developed a Biodiversity-Based Economic Development Strategy (DEA 2015). Key elements of the strategy include (1) coordination and leadership; (2) sustainable commercialization; (3) transformation of the biodiversity economy sector; (4) establishing a facility to promote and enhance use of biodiversity compounds; and (5) leveraging funding for the strategy and its implementation.[10]

At the heart of these initiatives is a conviction that the commercialization of South Africa's biodiversity and traditional knowledge represents an opportunity for addressing national development priorities of poverty alleviation and unemployment, especially in rural areas. This is best encapsulated in the government-led programme dubbed "Biopanza", modelled on the Malaysian "Big, Fast Results" methodology, aiming to transform the biodiversity sector into an engine of economic growth. While this remains a common objective, its interpretation varies considerably across institutions and laws. Traditional knowledge is an especially disputed domain – from new legislation envisaging its protection as a *sui generis* form of intellectual property, through to the establishment of a national database (known as the National Recordal System) to protect and promote traditional knowledge (DST, 2014). There are also different sets of commitments to ABS. For example, South Africa has yet to sign the International Treaty on Plant Genetic Resources for Agriculture, and has not formally recognized farmers' rights. There also remains little understanding about the practical implications of the ABS regime for agriculture. As a result, the agricultural sphere of ABS remains ambiguous and poorly developed (Wynberg *et al.*, 2012).

Scope

A central concern is the very wide scope of the Biodiversity Act, which, in contrast to the narrow definition of genetic resources embraced by the CBD and Nagoya Protocol, defines "indigenous biological resources" broadly in relation to bioprospecting to include "any living or dead … organism of an indigenous species; any genetic material or derivatives of such organisms, or any chemical compounds and products obtained through use of biotechnology".[11] The term "bioprospecting" is also broadly defined to include "any research on, or development or application of, indigenous biological resources for commercial or industrial exploitation".[12] Biotrade – commodity trade in biological resources such as the indigenous rooibos tea (*Aspalathus lineraris*) – South Africa's most successful natural product industry, is thus included within the remit of the BABS Regulations. This view has been subsequently entrenched in a 2013 amendment to the Biodiversity Act.

In practice, this reflects a wider international confusion and definitional challenge about the distinctions between biological and genetic resources. Biological resources include genetic resources and are usually traded as raw materials for use in low to medium technology industries. They are typically used as whole organisms in industrial processing and manufacture for food and beverages, cosmetics or botanical medicines. Genetic resources, in contrast, include genetic material of actual or potential value, with functional units of heredity. They may be accessed physically, but increasingly this occurs through digital databases that include genetic sequence information. Genetic resources are typically used to produce new products, processes or ingredients in research-intensive and high technology industries such as pharmaceuticals, biotechnology and agriculture. However, there are also grey definitional areas, especially where species have multiple uses in more than one sector. For example, research and development for foods, beverages, cosmetics and botanical medicines of species used in biotrade might include original research on genetic resources and traditional knowledge, as companies look for species with particular uses, and guides to safety and efficacy. At this stage, under the CBD, these activities would be characterized as bioprospecting, or genetic resource use. After companies have investigated new properties or traditional knowledge, demand shifts into the biological resource trade, or biotrade, involving biological resources that are part of slower-moving, lower technology industries (Laird and Wynberg, 2012).

The regulation of biotrade seems to stem from a concern that material traded as a commodity may subsequently be transferred to third parties and enter a new research and development cycle. A controversial case involving research and development on extracts of rooibos and another indigenous tea, honeybush (*Cyclopia* spp.), brought these concerns to the fore (Berne Declaration and Natural Justice, 2010). In this case, the food giant Nestlé secured rooibos and honeybush plant material from a local South African processor, did research on extracts and filed patents, but without the requisite agreements in place. Although the material was obtained from a local processor, it could equally have been purchased off the shelves of any European supermarket, raising questions about the challenges of regulating research and development on commodities such as rooibos tea that are already commercially available.

The wide scope of the Biodiversity Act has significant implications, in that it regulates an extensive range of activities, which is also contrary to ABS approaches in neighbouring countries. This has led to conflicting regulatory approaches for species that occur across political boundaries in southern Africa and are widely traded and commercialized. Examples include devil's claw (*Harpagophytum prucumbens* and *H. zeyherii*), incorporated into a multitude of botanical and pharmaceutical products to treat arthritis and rheumatism; baobab (*Adansonia digitata*), used widely as a cosmetic and recently approved as a novel food in Europe; and *Hoodia gordonii*, under development as an appetite suppressant. Traditional knowledge about the plants is similarly widely shared in the region – and differently regulated – creating an inordinately complex and inharmonious array of approaches. In this regard, the call for a multilateral mechanism, as proposed by Article 10 of the Nagoya Protocol, is especially timely, despite concerns about the practicalities of implementation (Winands-Kalkuhl and Holm-Müller, 2015).

Regulating biotrade is important when the volumes are large and where resource overexploitation is a concern (Laird *et al.*, 2010). Regulation can also be an important tool to promote value-adding and beneficiation and to bring the equity concerns of ABS to the commodity raw material trade for herbal medicines, cosmetics and food products. However, addressing these concerns requires measures quite different from those called for in bioprospecting and the utilization of genetic resources, which, as defined by the Nagoya Protocol, means "to conduct research and development on the genetic and/or biochemical composition of genetic resources, including through … biotechnology". Biotrade per se should not require benefit-sharing agreements and prior informed consent, a fact also recognized by the limited scope of the Nagoya Protocol.

The South African experience of conflating the two in a single regulatory system has led to negative impacts on harvester communities, on traditional knowledge holders and on the industries creating economic opportunities (Wynberg *et al.*, 2015). As examples, benefit-sharing agreements required to trade the medicinal plant *Pelargonium sidoides* saw ruling elites come to the fore, capturing benefits intended for harvesters but who were in no position to navigate the complex legal requirements associated with ABS agreements (Morris, 2016; van Niekerk and Wynberg, 2012). In this industry, legal requirements also enabled "ABS monopolies" for those savvy enough to secure bioprospecting and trade permits (van Niekerk and Wynberg, 2012). In other southern African natural product industries, such as those for devil's claw (*Harpagophytum* spp.) and baobab (*Adansonia digitata*), the inappropriate regulation of biotrade through ABS has effectively reduced trade in species that have been widely traded for decades (PhytoTrade, 2015). This has potential impacts on harvester communities who rely on these trades for their livelihoods. Associated industries and research institutions have publicly stated that they would "go elsewhere" to find the same resources, which occur widely in the region, or find alternatives.

Impractical permitting

The seemingly inefficient permitting process has also emerged as a central concern. As of 2014, only 15 bioprospecting permits had been issued, including 11 Integrated Export and Bioprospecting Permits and four Bioprospecting Permits. By mid-2017 this had increased to 53 bioprospecting permits (29 Integrated Export and/or Biotrade and Bioprospecting Permits, 20 Bioprospecting Permits and four Biotrade Permits),[13] suggesting increased efficiencies in the system. While the total number of applications is unknown, the multiple reports of declined or pending permits suggest that only a fraction of applications are receiving approval. In part, this low proportion is due not only to government inefficiencies, but also to the poor quality of permit application documentation (DEA, pers. comm., January 2014). Over time, numerous administrative issues have emerged, focused in particular on the lengthy time period from lodging an application through to receipt of the permit; an apparent ineptitude within government, characterized by the loss of applications and inability to make decisions; the lack of guidance from DEA; and the cost of applications – both with regard to the time involved and the actual cost of the permit which is the same regardless of the size of the company and/or the nature of the application. Inadvertent impacts have also arisen. Onerous permit requirements, for example, have led to the creation of monopolies in some instances (for those companies who do receive permits) with negative impacts on communities and other companies (who do not receive permits).

Table 14.1 illustrates a snapshot of the types of permits issued and the applicants involved. The DEA distinguishes between the following sectors: cosmetic, pharmaceutical, complementary medicine, agricultural and nutraceutical. Applications are processed by a Bioprospecting Committee and are reviewed against their compliance with the Act and Regulations. Mandatory procedural requirements form the initial screening criteria for any application (DEA Guidelines, 2012), including requirements such as the need for a company to partner with a local institution, with factors such as company reputation considered secondarily. Nine of these permits are Internationally Recognised Certificates of Compliance under the Nagoya Protocol, serving as evidence that the genetic resource which it covers has been accessed in accordance with prior informed consent and that mutually agreed terms have been established. The certificate contains information that can assist in monitoring the utilization of genetic resources by users throughout the value chain (Article 17; see https://absch.cbd.int).

While the Act and its Regulations are open to legal interpretation with regard to the permits that are required at different steps of the value chain, the DEA has taken a conservative position, requiring each harvester, trader, exporter and processor to be in possession of a bioprospecting permit (Figure 14.3). These permits are in addition to licences required for the harvesting of the resource, which are typically issued by provincial or national conservation agencies. Multiple permitting requirements at each step of the chain are widely considered to be arduous, unrealistic and not rooted in reality. For example, hundreds of ingredient suppliers and brand owners are now required to show evidence of a bioprospecting permit for every ingredient supplied from South Africa. These ingredients will typically constitute just a few of several dozen other ingredients purchased by a single company. The same is likely for any new product development using an existing extract. The concern – borne out by current practices – is that in all

Table 14.1 Permits granted to date by DEA as at January, 2017

Permit holder	*Type of permit*
HGH Pharmaceuticals	Integrated Export and Bioprospecting permit
The Regents of the University of California jointly with the University of Free State	Integrated Export and Bioprospecting permit
Dennis Noel de Villiers	Integrated Export and Bioprospecting permit
Edakeni Muthi Futhi Trust	Bioprospecting permit
Gowar Enterprises	Integrated Export and Bioprospecting permit
Council for Scientific and Industrial Research (CSIR)	Bioprospecting permit
Essential Amathole	Integrated Export and Bioprospecting permit
African Aloe (Pty) Ltd	Integrated Export and Bioprospecting permit
Organic Aloe	Integrated Export and Bioprospecting permit
Parceval (Pty) Ltd	Integrated Export and Bioprospecting permit
Council for Scientific and Industrial Research (CSIR)	Bioprospecting permit
Skimmelberg Fynbos Oils (Pty) Ltd	Integrated Export and Bioprospecting permit
Incosmetics (Pty) Ltd	Integrated Export and Bioprospecting permit
Ecoproducts	Integrated Export and Bioprospecting permit
Schibuna Aloe cc	Integrated Export and Bioprospecting permit
Ottos's Aloe	Integrated Export and Bioprospecting
Aloe Ferox Trust	Integrated Export and Bioprospecting permit
Cape Kingdom Neutraceuticals (Pty) Ltd)	Integrated Export and Bioprospecting permit
University of Western Cape	Bioprospecting permit
Council for Scientific and Industrial Research (CSIR)	Bioprospecting permit
Afriplex (Pty) Ltd	Integrated Export and Bioprospecting permit
The Esse Trust	Integrated Export and Bioprospecting permit
KP Botanicals	Integrated Export and Bioprospecting permit
Gert Olivier	Integrated Export and Bioprospecting permit
Council for Scientific and Industrial Research (CSIR)	Bioprospecting permit
Cosmetic Ingredients (Pty) Ltd	Bioprospecting permit
Nestle South Africa (Pty) Ltd	Bioprospecting permit
Rain Africa Innovations CC	Integrated Export and Bioprospecting permit
University of Pretoria	Bioprospecting permit
University of Pretoria	Bioprospecting permit
Parceval (Pty) Ltd jointly with Provital Group, Spain	Integrated Biotrade and Bioprospecting permit
Parceval (Pty) Ltd jointly with Naturamus GmbH, Germany	Integrated Biotrade and Bioprospecting permit
University of Cape Town	Integrated Biotrade and Bioprospecting permit
University of Cape Town	Integrated Export and Bioprospecting
Deon Williams	Bioprospecting permit
Parceval (Pty) Ltd jointly with Schwabe Extracta	Integrated Biotrade and Bioprospecting permit
Quintessence Collectins CC	Bioprospecting permit
Parceval (Pty) Ltd jointly with Heel Biologische Heilmittel GmBH	Integrated Biotrade and Bioprospecting permit
Parceval (Pty) Ltd jointly with Mast Jagermeister SE	Integrated Biotrade and Bioprospecting permit

(*continued*)

Table 14.1 (cont.)

Permit holder	*Type of permit*
University of Pretoria	Bioprospecting permit
University of Pretoria	Bioprospecting permit
Council for Scientific and Industrial Research (CSIR)	Bioprospecting permit
DeVilliers Aloe Exporters	Biotrade Permit
Totally Wild (Pty) Ltd	Bioprospecting permit
University of Pretoria	Bioprospecting permit
University of Pretoria	Bioprospecting permit
University of Pretoria	Bioprospecting permit
Heights Tea (Pty) Ltd	Integrated Biotrade and Bioprospecting
Aloe Trust	Integrated Biotrade and Bioprospecting
Tartan Timber	Integrated Biotrade and Bioprospecting
Pasillo Seed Oils	Integrated Biotrade and Bioprospecting
Mokuti International Herbs	Biotrade permit
Croc cure (Pty) Ltd	Integrated Biotrade and Bioprospecting
Eco-Products cc	Biotrade permit

Source: www.environment.gov.za/projectsprogrammes/bioprospectingaccess_benefitsharing_babs_clearinghouse.

likelihood, additional permit requirements will simply turn companies away from South Africa and/or South African species and towards other countries with the same resource, or replacement ingredients, with less bureaucracy. One producer remarked: "I'm not telling buyers about ABS requirements otherwise they will avoid me and find the resource in another country", while another stated: "The more I am in it and try to comply the more difficult it gets". South Africa's regulatory actions thus clearly have implications elsewhere in the region, underscoring the importance of regional ABS coherence.

In many ways, these frustrations are the result of both ambiguous legislation and, in all likelihood, a Committee developing its approaches "on the hoof" and apprehensive of making decisions about potentially controversial cases. Applications are thus treated conservatively, with government erring on the side of caution, even if alternative legal opinion suggests otherwise. A central concern is the control of third-party transfers of material for new types of utilization, yet many have voiced concerns that requiring permits each step of the way is unlikely to resolve this issue, and that tracking will be near impossible. Although these concerns are well recognized, with several proposals tabled for a simplified and integrated permitting system as part of the redrafted Biodiversity Act, it may well require independent legal challenge to clarify some of these decisions and achieve legal certainty.

Lack of awareness and understanding of science, markets and industry

A central problem, and one that is not necessarily limited to South Africa, is the lack of awareness amongst regulators about the science, the markets, the different industries involved and the realities on the ground. Bioprospecting is associated with different models that are dynamic in nature and vary across different sectors and resources. For example, much of South Africa's current bioprospecting activity is focused on microbes. Current ABS policies and strategies, however, are largely directed at the management of plant materials, overlooking microbes and the publication of sequenced genetic information and its subsequent use. This is evident in the government's "Biopanza" programme, which identifies 25 plant species as the foundation for its success, as well as in its biodiversity-based economy strategy (DEA, 2015). Moreover, much of the framing of the Biodiversity Act assumes that any indigenous biological resources are physically obtained,[14] with no mention of digital or synthetic sources. This oversight is a global phenomenon which is receiving attention by the CBD, but, in the interim, regulation remains all but absent (Third World Network, 2016).

Marine organisms raise similar issues. Pharmaceutical development is very different from producing essential oils, or new foods and cosmetics. Final chemical analogues that end up being used as drugs may be significantly different from the naturally occurring molecules and will be, ultimately, synthetically synthesized and independent of the natural resource (Laird, 2013). At the same time company models are hugely divergent – from small, family-based enterprises through to multinational corporations. Regulating all within the same template, as is currently done, is highly problematic, a principle that is well established internationally (Laird and Wynberg, 2012).

As an example of the policy implications of such capacity gaps, one senior government official recently made the following comment in a public forum: "They are looting us, and the biggest culprits are the large pharmaceutical companies who are developing drugs. It is uncontrollable right now." This erroneous statement, which may have been true 20 years ago but nowadays no longer applies due to the dwindling of interest in natural products by pharmaceutical companies, reflects the extent to which government remains poorly informed about the state of play of science and development in bioprospecting. The reality today is far different, especially as attention shifts towards microbes, biotechnology, bioinformatics and the use of private collections of genetic resources (Ruiz Perez Muller, 2015). Decision-making, it seems, has been operating largely in a research and data-deficient vacuum, and high turnovers of staff within South African government departments do not help to build institutional understanding and capacity.

Company size and scope is a factor

Amongst commercial entities who utilize South Africa's biodiversity, the size and nature of companies has definitely determined their ability to comply with the South African ABS legislation. Small local companies working with multiple ingredients sourced from multiple areas, for example, have found it impossible to comply with the legislation (let alone to understand it) and the threat of a fine is enough to force them to work on non-indigenous species with well-developed science and uncontroversial profiles.[15] Small companies working on a single species, in contained communities with whom they have established relationships, have found it easier to comply with legislation, although questions remain regarding the permit requirements for passing on material to third parties. Large consortiums with well-established IP offices, large budgets and good links to government departments have also found it easier to jump through regulatory hurdles.

The complexities of the legal system have had negative impacts on smaller biotrade companies. One industry representative remarked: "They are trying to catch tuna with a shadecloth", implying that local, small companies have borne the cost of inappropriate regulation. Moreover, the government has not distinguished between small and large traders: "All must have benefit-sharing agreements and bioprospecting permits and it baffles everyone", remarked one company.

Identifying traditional knowledge holders and implementing benefit-sharing

The difficulties of identifying traditional knowledge holders and finding representative communities with whom to negotiate benefit-sharing agreements have been especially challenging, particularly where shared knowledge is involved. Companies have been concerned about their capacity to engage in negotiating benefit-sharing agreements – both in terms of the resources required and their own ability and expertise. At the same time, they have faced pressures to comply with legislation, in order to secure the necessary permits. Negotiations have thus commenced either with traditional authorities, or with groups who may not necessarily represent all knowledge holders, but who are sufficiently organized to engage and negotiate with industry. The concern, as reported by Wynberg *et al.* (2015), Morris (2016) and others, is that this can lead to elite capture, especially by traditional authorities who are not democratically elected and may not be accepted. Remarked one industry representative: "The system of giving money to chiefs is a disaster waiting to happen; now that DEA have begun to roll out this approach it will be impossible to go back" (pers. comm., 2016). Industry is partly to blame for this situation due to their tendency to look for "easy" groups with whom to partner and sign benefit-sharing agreements but who may not necessarily be representative of knowledge or resource owners. The government, however, has also tended towards rather naïve solutions, often accepting the legitimacy of traditional authorities and other groups without asking deeper questions about what is at stake.

The question of priority – or "who was first" – has been especially controversial, largely due to claims by representatives of indigenous San and Khoi groups over traditional knowledge of all Southern African biodiversity (Chennells Albertyn, 2010). In the case of rooibos tea, for example, demands have been made by the South African San Council and the National Khoisan Council[16] for a stake of the industry's benefits, based on their claims to indigenous knowledge about the plant. A subsequent report commissioned by the DEA suggests that "any individual or organisation involved in bioprospecting or biotrade using rooibos or honeybush species [is urged to] engage with the Khoi or San communities or people to negotiate a benefit-sharing agreement in terms of NEMBA and the BABS regulations" (DEA, 2015). The industry position, although changing, has been that there is no convincing evidence of San traditional knowledge of rooibos tea prior to the commencement of the industry, and thus that it would not be open to entering into a benefit-sharing agreement. Questions have also been raised about the long chain of rural communities, individuals, researchers and companies that have contributed in different ways towards the success of the rooibos industry (Wynberg, 2016). Recent developments in the rooibos industry point towards a resolution, but negotiations around traditional knowledge claims have all but overwhelmed other ABS issues such as the large number of patents held on rooibos by foreign companies, and the environmental destruction of the resource itself and the habitat in which it occurs (Wynberg, 2016).

Protecting national interests and strengthening benefits from research and technology

It is fair to say that, to date, most of the efforts of the South African ABS regulatory system have focused on regulating biotrade rather than on protecting national interests and strategically strengthening the research and technology benefits during bioprospecting. However, global experiences of ABS have revealed that some of the most significant benefits of bioprospecting lie in the research and development phase (Miller, 2007; Gámez 2007). These include research collaborations, technology transfer, joint publications, capacity development and contributions towards biodiversity knowledge. Recognition of these benefits led to the initial inclusion of the so-called discovery phase in the Biodiversity Act and its regulations. In response, universities and research institutions expressed serious concerns that requirements for PIC and benefit-sharing at this early stage would stymie research. This ultimately led to a change in the regulations, with the result that researchers involved in the discovery phase were no longer required to secure a bioprospecting permit and conclude a benefit-sharing agreement. Instead, they simply needed to notify DEA of their activities in the discovery phase and of any intent to commercialize (Crouch *et al.*, 2008).

The extent to which this regulatory environment has influenced biodiversity-based research, development and commercialization is variable. In general, the removal of a permit requirement in the discovery phase has meant that the impact of regulation on the research community has been minimal. Existing national and regional collaborative platforms for bioprospecting have continued to thrive, examples being national consortia comprising major South African scientific research institutions and universities, focused on the discovery of drugs for malaria and tuberculosis (TB) from indigenous plants (Naidoo *et al.*, 2010; Chibale *et al.*, 2012).

Some have interpreted the lack of regulation in the discovery phase as a missed opportunity for benefit-sharing. While it is clearly not possible to predict the financial benefits of potential commercialization during the discovery phase, the requirement for benefit-sharing does provide leverage to optimize non-monetary benefits. This might include student exchanges, co-publications, technology transfer or capacity development. *De facto*, the exclusion of the discovery phase from permit requirements has meant that many of the conventional bioprospecting projects involving research and development on genetic resources and, sometimes, associated traditional knowledge, are now rather bizarrely excluded from any requirements for benefit-sharing.

Towards a revitalized and transformed regulatory framework for ABS

Despite the challenges noted, there appears overall to be a positive attitude to ABS and a general willingness on the part of companies to comply with ABS laws. ABS legal compliance is a reality and gives legal certainty to users and providers. Companies want to do the "right" thing, and see this as a marketing advantage. Noted one company: "ABS is positive as it is an endorsement; you get to show you got material in a legally and morally correct way. It can give you a competitive edge in the market." In some cases, ABS requirements have triggered new community projects. "Softer" benefits have also emerged. Companies have spoken of how increased interactions with communities have "enriched" them personally, leading to ongoing social commitments.

The government has also noted a strengthening of long-term relationships between different actors along various supply and bioprospecting chains and increased dialogue and understanding. Efforts made by government to establish a Bioprospecting Forum to improve dialogue and trust between different groups have paid off. Industry is undoubtedly much more aware of the need for equity and sustainability than in the past. There has also been increased openness and transparency from industry and, with some variability across departments, from government. At a political level, there is increased recognition among South African policy-makers of the need to ensure that trade based on indigenous biodiversity is not characterized by the export of raw materials to global markets for value adding and beneficiation outside of the country. While this recognition has yet to bear real fruit through policy change, it represents an important step in the right direction. The DEA is also driving a consultative process to completely overhaul the ABS system to address these concerns, with planned promulgation of a revised Biodiversity Act in 2018/19.

Learning from these experiences is perhaps the biggest challenge of all, in the context of government systems and approaches that are typically highly managerial, siloed and territorial of their individual mandates. A persistent backdrop is the politically charged nature of bioprospecting and the oft-competing directives of neo-liberal governments to commoditize biodiversity and traditional knowledge. The concern is that governments react to these complexities and to the imperative of implementing the Nagoya Protocol by entangling themselves in unworkable laws and by adopting a somewhat mechanical "box-ticking approach" that aims to ensure regulatory compliance rather than creating a climate that is conducive to ensuring social justice, and stimulating economic development.

A number of key lessons have emerged over the past ten years. From these, recommendations can be drawn to advise the further development and implementation of ABS legislation, both in South Africa and elsewhere in the world.

First, having informed regulators is vital. It is critical that government has the capacity to understand the technicalities of ABS and to regulate and implement appropriately and decisively. Ongoing efforts to develop capacity among senior and junior officials alike should be strongly supported.

It is important to regulate both the discovery and commercialization phases of bioprospecting activity, albeit to different extents. Not regulating the discovery phase in any way is a missed opportunity for benefit-sharing. Careful solutions need to be crafted to leverage benefits during this phase, without impeding research through unnecessary permit requirements. Strategic consideration needs to be given to ways in which benefits can be enhanced during this phase, in consultation with the research community.

Regulating biotrade is necessary (1) when the volumes are large, (2) in cases where resource over-exploitation is a concern, (3) to promote local value adding, as opposed to the export of raw materials, and (4) to prevent exploitation of harvesters. However, addressing these concerns requires quite different measures to ABS. The two should not be conflated in a single regulatory system.

It is important to include explicit measures concerning third party transfer of biological resources for R&D or utilization in another application. This is often the rationale for regulating biotrade. It is much more effective to regulate this specific activity rather than the entire trade.

Bioprospecting is associated with different models that are dynamic in nature and vary across different sectors. ABS legislation needs to veer away from standard provisions to recognize this diversity.

It is important that foreign researchers and companies are required under the ABS framework to partner with local researchers or companies so that benefits arising from such collaborations can be optimized.

Conservation and sustainable use as a benefit is not gaining ground in ABS agreements. This is despite the fact that ABS is intended to create economic incentives for biodiversity conservation, slowing biodiversity loss driven by the need for income by providing sustainable alternatives that embrace rather than destroy biodiversity (World Wide Foundation for Nature, 2007; Pavoni, 2013). Although support for conservation is noted explicitly as a potential benefit in Annexure 8 of the BABS Regulations, it has been difficult to operationalize this intent, due perhaps to the public nature of conservation benefits in a bilateral agreement. This is an important issue to build into agreements and legislation more concretely.

Organized constituencies, alongside informed legal support, have helped companies engage with PIC and benefit-sharing provisions, to realize that ABS is possible – and can even work to their advantage. Legal support for communities needs to be developed and implemented in a more concerted and creative manner.

"Learning by doing" is an important and useful approach, but incurs significant costs. An improved approach would build in reflection and research to ensure that the learning is institutionalized and acted upon. While this is not necessarily a proposal for legal reform, a more reflexive approach could be effected by making permanent the existing Bioprospecting Forum, and providing it with a steering committee, a budget and an external facilitator to enable its effective functioning.

Many of these lessons could apply well beyond South Africa. Simple and effective permitting, equitable agreements, approvals in good time, reduced bureaucracy, improved capacity and a vision that brings different actors together in a common quest for environmental sustainability, economic development and social justice is a sure formula for realizing the goals of the CBD and the Nagoya Protocol.

Acknowledgements

This work in based on research supported by the South African Research Chairs Initiative of the Department of Science and Technology and National Research Foundation of South Africa. Any opinion, finding and conclusion or recommendation expressed in this material is that of the author and the NRF does not accept any liability in this regard.

Notes

1 Associate Professor, DST/NRF South African Research Chair: Social and Environmental Dimensions of the Bio-Economy, Department of Environmental and Geographical Science, University of Cape Town, South Africa.

2 Convention on Biological Diversity (1992). UNEP, Montreal.

3 Biodiversity Act, section 98(2).

4 The South African San Council was established in 2001 as part of the Working Group of Indigenous Minorities in Southern Africa (WIMSA). WIMSA is charged with uniting and representing San communities from Botswana, Namibia and South Africa. As Chennells *et al.* (2009) explain, the South African San Council represents the modern form of San leadership, aiming to represent different San communities in South Africa democratically. Although the council is not the only body that claims to represent San communities, it is the largest, and has been a central actor in negotiating benefit-sharing agreements based on traditional knowledge claims.

5 Nagoya Protocol on Access to Genetic Resources and the Fair and Equitable Sharing of Benefits Arising from their Utilization to the Convention on Biological Diversity (2010). The Secretariat of the Convention on Biological Diversity, Toronto.

6 The reason for this change was unknown, even to those in the Ministry at the time who were leading the process. The common belief is that there was a lack of understanding and capacity among state legal drafters about the nuances and implications of language changes.

7 The broad definition of "bioprospecting" in the Act means "any research on, or development or application of, indigenous biological resources for commercial or industrial exploitation". This definition arguably goes beyond research involving genetic material or biochemical material and the utilization and development of that material for commercial purposes as defined by the CBD and Nagoya Protocol. The implication is that wider trade in biological resources (biotrade) is included within the ambit of the Act, whether or not it has a research and development component. This significantly increases the scope of the Act to include all species traded commercially in South Africa. Since South Africa has a thriving and long-established biotrade industry, based upon dozens of species, this has considerable implications for existing enterprises.

8 Biodiversity Act, section 2.

9 A number of problems are associated with the Biodiversity Act, summarized in Wynberg and Taylor (2009). These include the fact that the Act does not vest ownership of genetic resources in the state. This has the effect that the state has no right to benefit in bioprospecting unless the collection of the resources occurs on state land. It also entrenches existing inequalities in land ownership in South Africa, distorted through decades of colonialism and apartheid, and thus goes against the principle of equitable benefit-sharing. In addition, as described earlier, the wide scope of the Act with regard to its definition of indigenous biological resources created a regulatory infrastructure that was almost impossible to implement, alongside considerable legal confusion. Biotrade and bioprospecting were effectively treated in the same way, without giving recognition to the significant differences in the industries involved and the research and development process. The requirement for applicants to identify relevant "indigenous communities" also created difficulties, and, as described later, has led to community conflict in many instances. Conflating requirements for "discovery" and "commercial" research in earlier drafts of the regulations added to the legal confusion. Many of these concerns are now well recognized, and form the basis for the redrafting of the Biodiversity Act.

10 Presentation to the Biodiversity Economy Indaba, Polokwane, 11 November, 2013.
11 Biodiversity Act, section 1.
12 Biodiversity Act, section 1.
13 www.environment.gov.za/projectsprogrammes/bioprospectingaccess_benefitsharing_babs_clearinghouse.
14 For example, section 2(a)(i) of the Biodiversity Act notes that "indigenous biological resource" includes those gathered from the wild or accessed from any other source, including any animals, plants or other organisms of an indigenous species cultivated, bred or kept in captivity or cultivated or altered in any way by means of biotechnology. No mention is made of digital or synthetic sources.
15 An example is a local South African company that has made a decision to reduce its investment in indigenous plants and rather focus on "low risk" species such as Lavender and Rosemary, which are not perceived to have an ABS profile.
16 Although absent from *Hoodia* negotiations, the National Khoisan Council, established by former President Nelson Mandela in 1999 to accommodate Khoisan historical leadership within South Africa's constitutional framework, has increasingly become a partner to various benefit-sharing agreements, in collaboration with the South African San Council. The Khoisan historically comprise five main groupings, namely San, Griqua, Nama, Koranna and Cape Khoi.

References

Berne Declaration and Natural Justice (2010) "Dirty business for clean skin: Nestlé's Rooibos robbery in South Africa", Briefing paper, www.cbd.int/abs/side-events/resumed-abs-9/id2114-berne-policy-brief.pdf.

Chennells Albertyn. Letter to the Director General, Department Water and Environmental Affairs, 11 October 2010.

Chennells, R. (2013) "Traditional knowledge and benefit-sharing after the Nagoya Protocol: three cases from South Africa". *Law, Environment and Development Journal*, Vol. 9, no. 1, p. 163.

Chennells, R., Haraseb, V., and Ngakaeaja, M., (2009) "Speaking for the San: Challenges for Representative Institutions", in Wynberg, R., Schroeder, D. and Chennells, R. (eds), Indigenous Peoples, Consent and Benefit-sharing. Learning from the San-Hoodia Case, Springer-Verlag, Berlin, pp. 165–92.

Chibale, K., Davies-Coleman, M., and Masimirembwa, C. (2012) *Drug Discovery in Africa: Impacts of genomics, natural products, traditional medicines, insights into medicinal chemistry, and technology platforms in pursuit of new drugs*, Springer-Verlag, Berlin, Heidelberg.

Crouch, N.R., Douwes, E., Wolfson, M.M., Smith, G.F., and Edwards, T.J. (2008) "South Africa's bioprospecting, access and benefit-sharing legislation: current realities, future complications, and a proposed alternative". *South African Journal of Science*, Vol. 104, no. 9–10, pp. 355–66.

Department of Environmental Affairs (2012) "South Africa's Bioprospecting, Access and Benefit-Sharing Regulatory Framework. Guidelines for Providers, Users and Regulators". www.environment.gov.za.

Department of Environmental Affairs (2014) *Traditional Knowledge Associated with Rooibos and Honeybush Species in South Africa*. Siyanda Samahlubi Consulting for Department of Environmental Affairs, Pretoria www.environment.gov.za/sites/default/files/reports/traditionalknowledge_rooibosandhoneybushspecies_report.pdf.

Department of Environmental Affairs (2015) "National Environmental Management: Biodiversity Act, 2004 (Act No. 10 of 2004): Biodiversity Economy Strategy". www.environment.gov.za/sites/default/files/gazetted_notices/nemba10of2004_biodiversityeconomystrategy_gg39268.pdf.

Department of Environmental Affairs and Tourism (1997) White Paper on the Conservation and Sustainable Use of South Africa's Biological Diversity, *Government Gazette* Notice 1095 of 1997, Vol. 385, no. 18163, www.polity.org.za/govdocs/white_papers/diversity.html.

Department of Science and Technology (2014) "The Bio-Economy Strategy". www.dst.gov.za.

Department of Science and Technology (2014) "Indigenous Knowledge Systems: Overview of DST Activities", Presentation to the Parliamentary Portfolio Committee, 20 August 2014, http://pmg-assets.s3-website-eu-west-1.amazonaws.com/140720pcscience.pdf.

Gámez, R. (2007) "Chapter 7. The link between biodiversity and sustainable development: Lessons from INBio's bioprospecting programme in Costa Rica", in C. McManis (ed.) *Biodiversity and the Law. Intellectual property, biotechnology and traditional knowledge*, Earthscan, London.

Kepe, T., Wynberg, R., and Ellis, W. (2005) "Land reform and biodiversity conservation in South Africa: complementary or in conflict?" *International Journal of Biodiversity Science and Management*, Vol. 1, pp. 3–16.

Laird, S. (2013) *Bioscience at a Crossroads. Implementing the Nagoya Protocol on access and benefit-sharing in a time of scientific, technological and industry change. The pharmaceutical sector.* Secretariat of the Convention on Biological Diversity, Montreal.
Laird, S. and Wynberg, R. (1996) *Biodiversity Prospecting in South Africa: Towards the development of equitable partnerships.* The Land and Agriculture Policy Centre, Johannesburg.
Laird, S. and Wynberg, R. (2008) *Access and Benefit-Sharing in Practice: Trends in partnerships across sectors, Volumes I, II and III. CBD Technical Series 38.* Secretariat of the Convention on Biological Diversity, Montreal.
Laird, S. and Wynberg, R. (2012) *Bioscience at a Crossroads. Implementing the Nagoya Protocol on access and benefit-sharing in a time of scientific, technological and industry change. Overview brief.* Secretariat of the Convention on Biological Diversity, Montreal.
Laird, S.A. and Wynberg, R.P. (2016) "Locating Responsible Research and Innovation Within Access and Benefit-sharing Spaces of the Convention on Biological Diversity: the Challenge of Emerging Technologies". *NanoEthics*, Vol. 10, no. 1, pp. 1–12.
Laird, S., Wynberg, R., and McLain, R. (2010) "The state of NTFP policy and law", in Laird, S., McLain, R., and Wynberg, R. (eds) *Wild Product Governance: Finding policies that work for non-timber forest products*, Earthscan, London.
Lowman, M. (2012) "Implementation of access and benefit-sharing legislation in South Africa", MPhil dissertation, University of Cape Town at Cape Town, South Africa.
Miller, J. (2007) "Chapter 5. Impact of the Convention on Biological Diversity: The Lessons of Ten Years of Experience with Models for Equitable Sharing of Benefits," in C. McManis (ed.) *Biodiversity and the Law. Intellectual property, biotechnology and traditional knowledge*, Earthscan, London.
Morris, C. (2016) "Royal pharmaceuticals. Bioprospecting, rights and traditional authority in South Africa". *American Ethnologist*, Vol. 43, no. 3, pp. 525–39.
Naidoo, D., Maharaj, V.J., Moodley, P., Sewnarain, P., and Ndlebe, V.J. (2010) "Accelerated approach of discovering plant derived drug leads for treatment of TB". http://researchspace.csir.co.za/dspace/handle/10204/4124.
Pavoni, R. (2013) "Channelling investment into biodiversity conservation: ABS and PES schemes", in Dupuy, P.-M. and Viñuales, J.E. (eds), *Harnessing Foreign Investment to Promote Environmental Protection: Incentives and safeguards.* Cambridge University Press, Cambridge.
PhytoTrade Africa (2015) *Access and Benefit-sharing in Southern Africa: Developing policy and implementing best practice.* PhytoTrade Africa, London.
Ruiz Perez Muller, M. (ed) (2015) *Genetic Resources as Natural Information: Implications for the Convention on Biological Diversity and Nagoya Protocol.* Routledge, London.
Shiva, V. (1992) *Biodiversity: A Third World Perspective.* Third World Network, Penang.
Taylor, M. and Wynberg, R. (2008) "Regulating access to South Africa's biodiversity and ensuring the fair sharing of benefits from its use". *South African Journal of Environmental Law and Policy*, Vol. 15, no 2, pp. 217–43.
Third World Network (2016) "Biodiversity Convention considers genetic sequence data and benefit-sharing". www.twn.my.
University of Cape Town and 16 others (2014) "Memorandum to the Ministers of Environmental Affairs, Science and Technology, Trade and Industry and Western Cape Ministry of Agriculture, Economic Development and Tourism", 1 December 2014.
Van Niekerk, J. and Wynberg, R. (2012) "The trade in *Pelargonium sidoides*. Rural livelihood relief or bounty for the bio-buccaneers?" *Development Southern Africa*, Vol. 29, no 4, pp. 530–47.
Winands-Kalkuhl, S. and Holm-Müller, K. (2015) "Bilateral vs Multilateral? On the economics and politics of a global mechanism for genetic resource use". *Journal of Natural Resources Policy Research*, Vol. 7, no. 4.
World Wide Foundation for Nature (2007) ABS: What's in it for conservation? An inquiry aimed at stimulating debate on the role of conservation organizations in ABS. Prepared by WWF for discussion with experts at the Working Group on Access and Benefit-sharing, Montreal, 8 October.
Wynberg, R.P. (2002) "A decade of biodiversity conservation and use in South Africa: tracking progress from the Rio Earth Summit to the Johannesburg World Summit on Sustainable Development". *South African Journal of Science*, Vol. 98, May/June, pp. 233–43.
Wynberg, R., van Niekerk, J., Williams, R., and Mkhaliphi, L. (2012) *Policy Brief. Securing Farmers' Rights and Seed Sovereignty in South Africa.* Biowatch South Africa and the Environmental Evaluation Unit, University of Cape Town, South Africa.
Wynberg, R., Schroeder, D., and Chennells, R. (eds) (2009) *Indigenous Peoples, Consent and Benefit-sharing. Learning from the San-Hoodia case.* Springer, Berlin.

Wynberg, R. and Taylor, M. (2009) "Finding a path through the ABS maze: challenges of regulating access and ensuring fair benefit-sharing in South Africa," in E.C. Kamau and G. Winter (eds) *Genetic Resources, Traditional Knowledge and the Law: Solutions for access and benefit-sharing,* Earthscan, London, pp. 203–24.
Wynberg, R., Laird, S., van Niekerk, J. and Kozanayi, W. (2015) "Formalization of the natural product trade in Southern Africa: Unintended consequences and policy blurring in biotrade and bioprospecting". *Society and Natural Resources: An International Journal,* Vol. 28, no. 5, pp. 559–74.
Wynberg, R. (2016) "Making Sense of Access and Benefit-sharing in the Rooibos Industry: Towards a Holistic, Just and Sustainable Framing". *South African Journal of Botany*, in press.

15

DE-MATERIALIZING GENETIC RESOURCES

Synthetic biology, intellectual property and the ABS bypass

Margo A. Bagley[1]

A dagger with a rhino horn handle is quite the status symbol in some parts of the globe, but it will cost you: rhino horn is more expensive by weight than gold, ivory, or cocaine, commanding up to $60,000/kg (Roberts, 2017). Unfortunately, it costs the rhino even more: its life, which is why the trade in rhino products is illegal in most countries. However, demand for rhino horn, for aesthetic as well as traditional medicine purposes, is fueling a poaching epidemic in South Africa that, uncurbed, could result in the extinction of rhinos in as few as ten years. Pembient, Inc. is trying to change that with its genetically engineered, 3-D printed, synthetic rhino horn that is genetically identical to the real thing but involves no loss of life. However, the synthetic biology-based product was developed using rhino DNA, and Pembient is currently negotiating a benefit-sharing arrangement with the South African government, recognizing that country's sovereign ownership of the genetic resources upon which Pembient's invention is based. An article discussing Pembient's efforts claimed that such negotiations were not necessary because Pembient's founder, Matthew Markus, "could have taken what he needed without the government's blessing or used a sample from a rhino in a U.S. zoo, but he was wary of Pembient being seen as a 'biopirate' looting South Africa for its natural resources" (Roberts, 2017).

The idea that researchers can take what they want in relation to genetic resources is an attitude that has indeed led to companies and researchers being labelled "biopirates" (Hammond, 2013), and potentially being subject to serious repercussions under domestic legislation implementing the Convention on Biological Diversity (CBD) and its Nagoya Protocol. These agreements, along with the United Nations Food and Agriculture Organization's International Treaty on Plant Genetic Resources for Food and Agriculture (Plant Treaty) were designed to address the utilization of genetic material, and provide for the sharing of benefits to the mutual advantage of users and providers of the resources. However, advances in DNA sequencing and synthetic biology more broadly are resulting in the "de-materialization" of genetic material, with the potential to enable researchers and corporate entities to bypass the obligations to obtain permission to access and to share benefits of use with providers of genetic resources.

This chapter considers current challenges associated with this de-materialization, in the context of, among other things, intellectual property protection for synthetic biology outputs, implementation issues concerning the 2010 Nagoya Protocol on Access and Benefit-sharing to the Convention on Biological Diversity, and emerging concerns regarding synthetic biology and digital genetic information misappropriation in relation to commercial and non-commercial research projects (Bagley, 2015).

Synthetic biology in brief

Synthetic biology is based on the idea that any biological system can be viewed as a combination of functional elements or parts which can be organized in new ways to modify living organisms (Danchin and di Lorenzo, 2008; European Commission, 2014).[2] Researchers engaged in "fundamental" or "bottom-up" synthetic biology seek to identify and establish design principles for biologic parts and modules in order to build living systems from raw, non-living components (Fussenegger and Weber, 2009). For example, as of this writing, multiple teams of researchers from across the globe are working to develop a synthetic yeast genome. Such an organism could, in theory, be tailored (and thus more useful) for designer research and commercial applications (Richardson *et al.*, 2017). Also, in 2014, researchers at the start-up Synthorx, Inc. reported their creation of a bacterium with an expanded six-letter genetic alphabet, adding new bases X and Y to the standard G, A, T, and C bases (Alexander, 2015). It is theorized that the addition of more bases could eventually enable the engineering of bacteria or other organisms to produce completely new therapeutic proteins containing unnatural amino acids.

Simultaneously, "translational" or "top-down" synthetic biology researchers seek solutions to health, agricultural, environmental, and other challenges by redesigning gene sequences in existing organisms to achieve new or improved functionality. For example, researchers are exploring the use of a variety of genome engineering and synthetic biology tools such as recombineering, clustered regularly interspaced short palindromic repeats (CRISPR), and bacterial cell-cell signalling mechanisms for pathogen targeting, to combat the emerging antimicrobial resistance health crisis involving multidrug resistant pathogenic bacteria (Krishnamurthy *et al.*, 2016). Researchers also designed and produced a synthetic copy of thebaine, the opiate morphine precursor harvested from poppies for millennia, using yeast embedded with genetic sequence information from several plant species, a bacterium, and a rodent (Service, 2015).

While public awareness of synthetic biology remains low, developments in synthetic biology are advancing rapidly, sometimes amid controversy (Woodrow Wilson Center, 2014), especially in relation to gene drives, also known as genetic extinction technologies (Civil Society Working Group on Gene Drives, 2016) and synbio substitutes for naturally occurring compounds such as saffron, vanillin, and stevia (Dressel, 2016; Kishore *et al.*, 2017). Furthermore, a growing cadre of companies is marketing synthetic biology-based products in areas including biofuels, specialty chemicals, agriculture, and therapeutics.

A 2014 Wilson Center report shows synthetic biology research is continuing to expand, taking place in approximately 30 countries and among 565 unique entities in universities, government laboratories, private companies, and community laboratory space (Woodrow Wilson Center, 2014). Although the US dominates, countries such as Denmark, Canada, China, South Korea, Brazil, Japan, Peru, South Africa, and Singapore all have scientists engaging in synthetic biology research and many are sharing information across borders (Oldham, Hall, and Burton, 2012). Moreover, commercial activity is not limited to small start-ups. Large, multinational corporations such as Johnson & Johnson, Merck, Cargill, Monsanto, GlaxoSmithKline, and Goodyear are increasingly incorporating synthetic biology projects and partnerships into their portfolios. As the cost of DNA synthesis decreases and the ease of making genetic modifications and genetic information digitally available increases, new questions and potential obligations may be arising regarding synthetic biology and the access and benefit sharing (ABS) regimes of the CBD/Nagoya Protocol, and Plant Treaty.

The potential benefits and risks of synthetic biology are the subject of considerable disagreement across the globe, and discussions and actions are in varying stages of maturity in different fora. Surprisingly, discussions in the CBD and Nagoya Protocol still are not focused on the temporal scope issues concerning synthetic biology and genetic resources which were left unresolved at the end of negotiations on the Protocol itself in 2010 (Bagley and Rai, 2013), and are just beginning to include breadth of coverage issues in the context of digital sequence, or natural, information.[3] Parties also are grappling with the more fundamental issues of the nature of synthetic biology research and the risks associated with the release of synthetic biology-based products.

In October 2014 at the Twelfth Meeting of the Conference of the Parties (COP) to the CBD, which also was the First Meeting of the Parties to the Nagoya Protocol (COP12-MOP1) in Pyeongchang, South Korea, some member states viewed synthetic biology as a simple extension of genetic engineering. Others characterized work in the field as sufficiently new and different to justify application of the precautionary approach (Conference of the Parties to the Convention on Biological Diversity, 2014; Nordhaus, 1969; Raffensperger and Tickner, 1999).[4] Ultimately, the COP concluded that there was insufficient information available to classify synthetic biology as a new and emerging biodiversity issue. However, it did agree to urge Parties to employ the precautionary approach and to approve organisms resulting from synthetic biology techniques for field trials only after appropriate risk assessments had been carried out (Conference of the Parties to the Convention on Biological Diversity, 2014).

During the Thirteenth Meeting of the COP/Second Meeting of the Parties to the Nagoya Protocol (COP13-MOP2) in December 2016 in Cancun, Mexico, the parties began discussing the controversial cross-cutting issue of application of the CBD/NP to digital sequence information from genetic resources, but ultimately punted the issue to COP14-MOP3. The COP requested an Ad Hoc Technical Expert Group (AHTEG) on synthetic biology to continue and expand its work; specifically, to report on technological developments in synthetic biology, and digital sequence information on genetic resources, and the implications of both for the objectives of the CBD/NP (Conference of the Parties to the Convention on Biological Diversity, 2016a and 2016b). It seems unlikely that the COP-MOP will soon resolve scope issues in relation to synthetic biology, genetic material, and information. Indeed, it may ultimately choose to leave the decisions for member states to continue to address at the national level, which would entrench the lack of harmonization and legal uncertainty for researchers regarding which genetic materials are subject to Nagoya Protocol obligations.

Intellectual property protection often is sought for the fruits of inventive activity, including synthetic biology products (Nature Biotechnology, 2016). Patents in particular offer an exclusivity that provides a connection to the access and benefit-sharing concerns that animate the Nagoya Protocol (Bagley *et al.*, 2014). However, as discussed in the following section, recent developments have both cast doubt on the patent eligibility of some synthetic biology outputs, and shed light on the potential availability of another form of intellectual property protection in the future.

Synthetic biology and intellectual property protection

Researchers are not of one mind when it comes to intellectual property (IP) protection for advances in synthetic biology. Two philosophical camps have emerged in the field: an open source community focused on disclosure, sharing, and free accessibility of synthetic biology engineered parts and information (Ledford, 2013),[5] and those with a more traditional patent-protection-as-an-incentive-for-disclosure-and-investment philosophy (Burk and Lemley, 2003; Fisher, 2001; Kieff, 2001).[6] Proponents of open source believe that the free availability of new knowledge will lead to more rapid discoveries that will benefit humanity. Those in favour of patenting synthetic biology products counter that research requires a financial investment and patents allow for the recoupment of returns on that investment. These two approaches, while clearly in tension, are playing out in tandem in the rapidly evolving synthetic biology space, with some researchers both freely distributing certain novel sequences while seeking patent protection for other, more commercially viable outputs. The focus on open source options is also creating interest in another type of IP protection: copyright. Patents and copyrights raise interesting intellectual property protection issues; the full eligibility of synthetic biology subject matter for coverage under either regime is not free from doubt at the time of this writing.

Patents have been, and likely will continue to be, the primary form of IP protection for synthetic biology. Numerous patents have been issued on synthetic biology products and processes, ranging from methods of artemisinin production for treating malaria (Keasling *et al.*, 2007; Wallart and Bouwmeester, 2009), to fuels made using a modified microorganism (Renninger, Newman and Reiling, 2010), and pharmaceutical compositions containing synthetic peptide amides for treating pain and inflammation (Schteingart *et al.*, 2016). However, recent judicial decisions on gene patent subject matter eligibility in the United States and Australia have eliminated patent protection for some synthetic biology inventions. In particular, the United States Supreme Court's decision in *Association for Molecular Pathology v. Myriad Genetics, Inc.* (2013) eliminated patent protection for isolated genomic DNA (gDNA) and other products of nature that do not qualify as machines, compositions of matter, articles of manufacture, or processes made by man.[7]

The Supreme Court did distinguish between gDNA and synthesized complementary DNA (cDNA), such as would be involved in synthetic biology research, holding that most cDNA claims would pass the patent eligibility hurdle. However, the Court cautioned that short cDNA sequences might be unpatentable if indistinguishable from "natural" DNA (McFarlane, Sharp, and Aquino, 2014). Moreover, even longer synthetic sequences could face problems if they are not "markedly different" from what exists in nature.[8] Claims already have been rejected for cDNA sequences on that basis in at least one synthetic biology-based patent application (Parida *et al.*, 2014). Several district court and appellate decisions implementing the *Myriad* holding also have invalidated patent claims deemed not markedly different to what exists in nature, such as the cloned animal claims in the *In re Roslin Institute* (2015) application.[9] In addition, the United States Patent and Trademark Office's initial interpretation of *Myriad* and other Supreme Court patent eligibility decisions drew severe criticism and consternation from the biotech industry and patent attorneys, as the Office's guidelines for examination appeared to go significantly beyond the Court's decisions in restricting patentability. However, more recent guidance from the Office indicates a less stringent approach to inventions based on products of nature (United States Patent and Trademark Office, 2014; 2015).

Patent law is territorial in nature and patents only take effect within the national/regional borders of the countries that grant them. Thus, the US *Myriad* decision only has effect in the United States; researchers still may be able to obtain patent protection in other countries, such as EU member states, as the European Union Biotechnology Directive explicitly allows for patents on gene sequences that would fail US patent eligibility requirements (Council Directive 1998; Bagley 2014).[10]

In *D'Arcy v. Myriad Genetics* (2015), the High Court of Australia invalidated both gDNA and cDNA[11] claims as not comprising a "manner of manufacture." It applied a two-part test, assessing the substance of the claim and whether the substance was "made," and concluded that the substance of both types of claims was genetic information that was "discerned" and not made. Nevertheless, some DSI inventions remain eligible for patent protection in Australia after the decision. For example, in Arrowhead Research Corporation (2016), the Australian Patent

Office concluded that synthetic double-stranded RNA (dsRNA) was a patent-eligible product, as the "substance" of the claim was more than just genetic information.

Some scholars and researchers see copyright protection as a preferable alternative to patents as it may produce a more "socially desirable balance" of permitted versus restricted uses of DNA sequences (Torrance, 2011). Copyright protection lasts longer than patents – life of the author plus 70 years (or 95 years from publication, for works for hire) versus 20 years from filing – but the protection is not as strong. Unlike for patents, independent creation is a defense to copyright infringement, there are limits on damages for innocent infringement, and the copyright fair use defense might reasonably allow many uses of protected sequences, such as for experimentation and instruction, not allowed by the strict liability patent law system (Torrance, 2011). Moreover, copyright protection is seen by some as better able than patents to foster an open source biology regime (Torrance, 2011).

Copyright protects original works of authorship fixed in tangible mediums of expression such as literary works, musical works, architectural designs, and even computer programs. Several commentators, making an analogy to computer software, have suggested that copyright may be appropriate for synthetic biology, noting that some synthetic DNA sequences meet the originality and fixation requirements and may require expressive choices (Holman, 2011; Karjala, 2011; Torrance and Kahl, 2014; Murray, 2014; Torrance, 2010; Torrance, 2011). Moreover, for open source proponents, the exclusivity provided by copyright law possibly could be used to impose sharing requirements on users, an approach that some in the free/open software movement have used effectively with "copyleft" licenses (Rai and Boyle, 2007).

However, the possibility of copyright protection for synthetic biology is far less certain than that for patents. Detractors argue that copyright is a poor fit for synthetic biology, as sequences are generally dictated by the desired function they are to perform, leaving little room for an author's expressive choices. In view of the challenges posed by patent and copyright law, some commentators have suggested that a sui generis IP regime for synthetic biology might be most appropriate (Samuelson, 2013; Rai and Boyle, 2007). In the United States, where the lion's share of synthetic biology research is performed, the copyright office has indicated that DNA sequences are not copyright-eligible subject matter. Consequently, copyright protection is not currently available for DNA sequences (Holman, Gustafsson, and Torrance, 2016; Ledford, 2013). However, as with computer software, which the copyright office originally rejected as ineligible for copyright protection, that position may change over time (Holman, 2011).

Synthetic biology, biodiversity, and the Nagoya Protocol

The Conference of the Parties (COP) is the governing body of the CBD and makes decisions at periodic meetings to advance implementation of the Convention. One such decision was the adoption of the Nagoya Protocol on Access and Benefit-sharing to the Convention on Biological Diversity at its Tenth meeting in 2010 in Nagoya, Japan. The Protocol was considered necessary because, while the CBD obligated Parties to facilitate access to their genetic resources, and to fairly and equitably share benefits arising from the utilization of such resources with provider countries, it provided almost no detail on how ABS, prior informed consent (PIC), and mutually agreed terms (MAT) should be accomplished in practice. Consequently, provider countries have either not yet implemented ABS provisions or implemented widely varying legislation, resulting in legal uncertainty for users faced with often burdensome rules for ABS/PIC that vary, sometimes significantly, by country.

The Protocol, which came into effect on October 12, 2014, was designed to reduce uncertainty and provide increased transparency for both users and providers of genetic resources and associated traditional knowledge. It is a binding agreement and is "*the* instrument for implementation of the access and benefit-sharing provisions of the [CBD]" (Greiber *et al.*, 2012). While the CBD and Protocol may be amenable to interpretations that exclude synthetic biology from their purview, it is more likely that researchers in this area will be subject to the provisions of these agreements as they are implemented in national legislation in provider and user countries. As such, it is worth considering whether the Protocol contains provisions that act as "ceilings" and/or "floors" for national legislation. "Ceilings" are upper limits on the kinds of obligations member states can impose on users in relation to ABS/PIC/MAT. Conversely, "floors" are minimum standards that leave countries free to impose more stringent requirements and sanctions.

The Nagoya Protocol contains several important "floors," minimum standards provisions aimed at achieving the objective of the fair and equitable sharing of benefits from the utilization of genetic resources and associated traditional knowledge (GRAATK). For example, it requires Parties that choose to impose prior informed consent (PIC) for access to GRAATK to take necessary legislative, administrative, or policy measures to, *inter alia*, provide fair and non-arbitrary genetic resource access rules, and information on how to apply for PIC (Nagoya Protocol 2010, Art. 6b–c). Also, it obligates each member to take appropriate steps to "provide that genetic resources utilized within its jurisdiction" have been accessed in accordance with the domestic ABS/PIC/MAT requirements of another Party. Furthermore, members must cooperate, as far as possible and as appropriate, in cases where another Party's domestic ABS legislation has been violated (Convention on Biological Diversity, 2016b).[12] However, the mandatory "shall" language in such Protocol provisions is weakened by the insertion of vague, broad phrases such as "effective and appropriate legislative, administrative or policy measures," "as far as possible," and "as appropriate." In fact, the Protocol has been called "a masterpiece in creative ambiguity" on a variety of topics (Oliva, 2011).

A close look at all of the Nagoya Protocol provisions shows only floors – minimum obligations – and not ceilings – upper limits on the kinds of ABS laws and penalties a country can impose. In fact, the only explicit limitation appears to be Article 12(4), which obligates countries to avoid restricting customary uses of GRAATK among indigenous groups and local communities (Nagoya Protocol, 2010). Thus, while the Protocol's requirements should improve transparency and increase certainty overall, researchers will still face a panoply of differing, un-harmonized PIC/ABS laws that may vary significantly in scope, obligations, and penalties. Moreover, while the Parties adopted a recommended compliance process at the 2014 COP12-MOP1 in South Korea pursuant to Article 30, the procedure is designed to be "non-adversarial, cooperative, simple, expeditious, advisory, facilitative, flexible and cost-effective in nature" (Conference of the Parties to the Convention on Biological Diversity, 2014).[13] It thus will lack the teeth of, for example, the WTO TRIPS dispute settlement mechanism with its trade-based sanctions regime (Agreement on Trade Related Aspects of Intellectual Property 1994; World Trade Organization, 2015).

Regrettably, the speed with which fifty countries deposited the necessary instruments of ratification for the Nagoya Protocol to come into effect is unlikely to be replicated in the Protocol's national implementation phase. The Protocol is complex, and while many countries had some type of ABS measure prior to the Protocol going into effect, only six countries and the EU had notified Protocol implementing legislation to the CBD prior to the Protocol's October 12, 2014 effective date (Medaglia, Perron-Welch and Phillips, 2014). As of this writing, thirty-one countries have submitted ABS legislative or policy instruments to the new ABS Clearinghouse but many of those documents (other than the EU member states' submissions of the EU implementing directive) are pre-Nagoya Protocol laws (Convention on Biological Diversity, 2016a).

Moreover, promptness in drafting implementing legislation has been no guarantee of a smooth operationalization process. For example, the EU adopted a Directive to implement certain aspects of the Protocol in April of 2014, but the Directive has already been the subject of two legal challenges (Maler, 2015). Also, French Polynesia rapidly developed and passed Nagoya Protocol Implementing legislation in 2012, but the legislation currently is not fully in operation. Moreover, after various changes in political leadership at multiple levels, the government is revisiting whether the law as drafted will meet the needs of the many constituents affected by it domestically and abroad. An intensive analysis began in the second half of 2015, across ministries and in consultation with indigenous communities and local stakeholders, to evaluate what, if any, changes may be needed to operationalize a Protocol-compliant and effective ABS framework.[14]

It seems likely that many other countries will face a similar raft of challenges in implementing the agreement, such that it could be many years before all member states are in compliance with their Protocol obligations.[15] Nevertheless, progress is being made; a total of ninety-five countries have now ratified the Protocol and as of this writing, fifty-two International Certificates of Compliance (ICC) have been issued by a total of five countries (Convention on Biological Diversity, 2016a; Mohan, 2015). Under the Protocol, such an ICC is clear evidence that a researcher has complied with the PIC/ABS/MAT requirements of a country providing genetic resources.

A 2014 study of ABS laws pre- and post- the Protocol reveals a variety of national approaches existing and emerging on this topic (Medaglia, Perron-Welch, and Phillips, 2014). PIC/ABS/MAT legislation differs on numerous issues, such as: whether research projects by foreigners require domestic scientist involvement to enhance capacity building, the types and number of compliance checkpoints, the nature and extent of indigenous and local community involvement, and whether users must use due diligence to determine if there has been compliance with a provider country's laws.

Whether domestic Protocol implementing legislation will be deemed to apply to synthetic biology researchers may be influenced by the nature of their research. Natural compounds historically have played an important role in the development of products in a number of industries including cosmetics, pharmaceuticals, and agricultural products. The role may have peaked for pharmaceuticals in the early 1990s, when approximately 80% of marketed drugs were natural products or their analogs; however, natural products are still being developed, and interest in natural products and ingredients is on the rise. In 2009, more than 100 natural product-based drugs were in clinical studies, and 13 natural product-derived drugs received FDA approval between 2005 and 2007 (Li and Vederas, 2009). Natural products and traditional knowledge can be particularly important for advances in the cosmetics industry where stories of traditional uses of cosmetic ingredients can enhance the exotic "experience" for consumers using certain products, and there is an increasing consumer demand for environmentally friendly and socially responsible offerings (Laird and Wynberg, 2013). In agriculture, the identification of bacteria that facilitate plant growth, pest resistance, and drought tolerance, and the recognition that there are potentially multiplied thousands of other such bacteria, are also fuelling natural products research (Vrieze, 2015).

The vast amounts of documented and unexplored biodiversity contribute to natural compounds being a significant potential source of new product ingredients. Fewer than 15% of higher plant species are believed to have been examined for bioactivity, fewer than 1% of micro-organisms are easily cultured (Li and Vederas, 2009), although synthetic biology advances are aiding progress in their characterization and analysis. And much remains to be learned from animal genetic information as well. Researchers recently identified the production of multiple copies of p53, a gene known to inhibit cancer growth, in African elephants, a finding that helps to explain why, despite their size, life span, and significantly higher number of total cells, elephants are much less likely to develop cancer than humans (ABC News, 2015). Additionally, researchers are studying a mechanism found in fruit flies that controls their production of L-2HG molecules, an oncometabolite found in brain and kidney cancer patients, which causes tumor formation and growth. In turn these findings could lead to better cancer treatments for humans (Genetic Engineering and Biotechnology News, 2017). Interestingly, developments in synthetic biology may increase the importance and viability of natural compounds in all of these areas.

The ability to synthesize near copies of natural compounds or genes with designer modifications could both significantly advance knowledge and prove extremely profitable in eliminating some of the regulatory and transactional barriers to new product development (Fussenegger and Weber, 2009; Laird and Wynberg, 2013; Zerbe *et al.*, 2014). However, these developments may have significant negative implications for biodiversity and economic development in low and middle income countries. Some synthetic biology companies are developing products for the lucrative flavours and fragrances market that have the potential to displace natural products (Meienberg *et al.*, 2015; Dressel, 2016). While this might appear to be beneficial for biodiversity, the benefits may be countered by economic losses to developing country farmers whose livelihoods depend on obtaining high prices for their crops, such as vanilla, saffron, and stevia (Dressel, 2016). To the extent they are driven to farm different crops that may be less suitable to the local geography, or abandon farming altogether, the long-term effects on people and place may be profound. In addition, such synthetic biology products, many of which are produced using genetically modified yeast, require vast amounts of sugar as a critical production input. Increasing the amount of land being planted to sugarcane can have a deleterious effect on biodiversity as well.

Because much synthetic biology work focuses on mimicking natural products and processes and then making changes to them, significant amounts of both tangible genetic resources and genetic resource information have been and are being used in synthetic biology research. This has raised questions regarding whether synthetic biology products and developers are subject to the Nagoya Protocol. Also, the increasing amount of DNA sequence information being made freely available in online databases, combined with the dramatic reduction in cost and difficulty of DNA sequencing from commercial labs, is raising digital misappropriation concerns among NGOs and developing country officials. However, the ability to use intangible genomic information in synthetic biology research makes assuring compliance by such researchers far from simple.

Use and misuse of digital genetic information

One of the driving forces behind both the CBD and the Nagoya Protocol has been a desire to prevent the misappropriation and/or misuse of GRAATK, a practice broadly characterized as "biopiracy." "Biopiracy" is a rather pejorative, multi-dimensional term that is often used as a shorthand label for some form of egregious misappropriation. By one definition, it is "the patenting of (often spurious) inventions based on biological resources and/or traditional knowledge that are extracted without adequate authorization and benefit-sharing from other (usually developing) countries, indigenous or local communities" (Robinson, 2010). This creates a clear connection to the word "piracy" which is defined as "the unauthorized use of another's production, invention, or conception especially in infringement of a copyright" (*Merriam Webster Dictionary*, 2015). As explained by one commentator: "What developing tropical nations are saying is that if the West cries foul over piracy of intellectual property, [such as] computer software, then biopiracy in Western labs of jungle extracts should also be considered a high economic crime" (Faiola, 1999). Pressure to address biopiracy has led to the use of the term by governmental entities and corporate titans alike: Peru has a National Anti-Biopiracy Commission that tracks global efforts to patent a variety of native Peruvian species, and the Sanofi Group has published a factsheet on biodiversity and biopiracy (Moeller IP Advisors, 2014; Sanofi, 2013).

The biopiracy label is controversial, and is itself subject to criticism, as it has been applied to a wide range of activity and used in vague, imprecise, and politicized ways (Chen, 2006; Heald, 2003; Robinson, 2010). Unauthorized uses of tangible genetic material and intangible genetic information can involve quite different levels of culpability in terms of knowledge and intent. For example, finding and using a DNA sequence of interest in a public database through "digital bioprospecting" seems far removed from intentionally removing genetic material from an *in situ* location without permission. Would both be deemed biopiracy? It probably depends on who you ask, with certain NGO and developing country personnel perhaps more likely to use the term than developed country researchers and representatives. Would both be deemed violations of national law? Perhaps yes, if national legislation covers intangible genetic information; but whether in fact that would be the case will vary from country to country and will only be known as Protocol implementing legislation is developed, operationalized, and enforced, as discussed below.

The CBD and Nagoya Protocol were both drafted primarily with tangible genetic resources in mind,[16] and do not facially address the rather different set of "virtual" concerns implicated by recent developments in the field of synthetic biology involving the digitization of genetic material. As the NGOs ETC Group and Friends of the Earth note:

> While "traditional" biopiracy involves the physical removal of material from a community to private hands, synthetic biology enables "digital biopiracy" where the DNA of an organism is sequenced *in situ*, uploaded to the internet as information, and then transferred digitally to a DNA synthesizer to be copied and re-built elsewhere. This digital transfer of DNA sequences does not even require a Material Transfer Agreement (MTA) since no physical material is transferred. Yet, the technology allows corporations, governments and individuals to freely take genetic material for private use in new synthetic organisms, which can then be patented as inventions.
>
> (SynBioWatch, 2013)

As discussed above, the term "biopiracy" is controversial in this context, but it is a catchphrase that instantly conveys the nature of the activities and fears at issue. Such fears seem justifiable in light of the wide availability of genome information and tools that can be used to construct modified or fully novel gene sequences that can be emailed or uploaded to commercial foundries and synthesized to specification.[17] For example, in a report advocating a "broad and dynamic" interpretation of the Protocol definition of genetic resources to include digital information, the Fridtjof Nansen Institute noted that, "new knowledge and technologies [such as synthetic biology] may create new and inventive uses of genetic resources with a future potential for ABS" (Fridjtof Nansen Institute, 2010).

Concerns regarding the potential for "digital biopiracy" may create reluctance on the part of some provider countries to enter into non-commercial research agreements. This is perhaps due to fear that DNA sequence information obtained from analyzing genetic material under a PIC/ABS agreement may be uploaded to publicly accessible databases and then used by synthetic biology researchers to develop lucrative, ABS-free, modified organisms and products for commercial applications (Pottage, 2006). This is particularly ironic and problematic as Article 8 of the Protocol explicitly encourages countries to ease access requirements for non-commercial research projects, such as the Moorea Biocode project in the South Pacific and the Smithsonian-based DNA Barcoding project, both of which are likely to generate voluminous amounts of digital sequence information.[18]

These developments have the potential to influence how the Nagoya Protocol and other treaties may be modified and interpreted in the future. Moreover, the increasing likelihood of such issues led the International Civil Working Group on Synthetic Biology to make the following recommendation regarding the Protocol:

> The Conference of the Parties [COP] should further invite the parties to the Nagoya Protocol on Access and Benefit-sharing to consider extending agreements on access and benefit-sharing to cover digital genetic sequences and products derived from natural sequences using synthetic biology tools such as directed evolution techniques.
> (International Civil Society Working Group on Synthetic Biology, 2011)[19]

Even if the Protocol is not construed to cover digital information and products derived therefrom, countries may still incorporate such coverage into their national legislation explicitly or by interpretation.[20] However, ensuring compliance for such digital information is likely to be significantly more challenging than for tangible genetic material. It is possible to watermark a DNA sequence without interfering with gene coding by inserting a watermark in a non-coding region of the DNA (Yamamoto *et al.*, 2014). J. Craig Venter used such a technique when developing "Synthia," the first cell controlled by a synthetic genome (Saenz, 2010). However, such a process may not be economically feasible or efficient for large quantities of DNA sequences. Moreover, watermarks may be susceptible to degradation through, for example, mutation. Finally, it may be possible for third parties to identify the watermark (e.g., if it contains a start signal) and remove it from the DNA sequence. Moreover, many sequences, or portions of sequences, can be found in multiple natural sources. Thus, identifying the source of genetic information used in creating an invention and downstream user compliance with PIC/ABS conditions may be impossible tasks.

In addition, researchers may not even realize that the use of such digital information is objectionable. A provider country may consider obtaining sequence data on a genetic resource from an internet database to be the functional equivalent of receiving a physical sample from the researcher who uploaded it (in which case compliance with Nagoya Protocol implementing legislation would be required). Yet, a researcher may view obtaining sequence data on the internet as no different to obtaining information/data from a publicly available scientific publication.

This difference in views regarding the free availability and/or use of digital sequence information has the potential to create additional controversies between users and providers of genetic resources. In addition, imposing too many barriers to accessing digital data, or generating uncertainty regarding the legality of using digital data, could negatively impact the development of new products and information.

This difference in views is also reminiscent of controversies in other areas where advances in digital copying and an increase in easily accessible digital information have created "piracy" concerns, for example music and movie file-sharing and 3-D printing. Considering that all three phenomena are the result of the increasing digitization of information (Brynjolfsson and McAfee, 2014), this is not surprising. For more than a decade, the Recording Industry Association of America and the Motion Picture Association of America have been fighting the facilitators and users of unauthorized downloads of copyrighted works from the internet, labelling such actions as "theft" and obtaining monetary and, in some cases, criminal penalties against the perpetrators – individuals and organizations alike (Palmedo, 2015). Likewise, 3-D printing is enabling individuals and companies to cheaply copy patent and copyright protected articles from internet-accessible digital files (Depoorter, 2014; Desai and Magliocca, 2014; Holbrook and Osborn, 2015; Osborn, 2014).

If the 3-D printing controversy is in its infancy, the digital misappropriation controversy created by de-materialization is, at best, embryonic; not many people are even aware of the issue. However, there is an important difference between music, movie file-sharing, and 3-D printing on the one hand, and genomic DNA sequence information on the other. In particular, the former content is protected by traditional intellectual property rights (copyrights and patents) and is seen as creative; the latter, generally speaking, is not.[21] Moreover, the IP-holding music and movie stakeholders are well-organized and have deep pockets to pay for lobbying, lawsuits, and the media and educational outreach efforts of their sustained campaigns. The indigenous people groups, local communities, developing country governments most likely to be negatively impacted by digital misappropriation are neither well-organized nor well-funded on this issue, and the origins of the sequence information may be difficult to ascertain.[22]

If government officials come to believe that data obtained in non-commercial research projects are being converted to commercial purposes without concomitant benefit-sharing, they may eliminate simplified access procedures for such research, put stringent limits on the public sharing of such research results, or even stop such research projects altogether, as contemplated by the Indonesia government in 2014 (Hakam, 2014). And if synthetic biology researchers believe they may be subject to various sanctions, have to engage in high transaction costs negotiations, and/or be labelled as "biopirates" as a consequence of using sequences from publicly accessible databases, they may not respond proactively, like Matt Markus of Pembient, who sought a win-win scenario for the user and provider. Rather, they may be inhibited from using such data, which could limit the beneficial discoveries they would otherwise make.

To avoid such scenarios, the CBD/NP COP should begin exploring ways to ameliorate the digital misappropriation concern. Article 10 of the Nagoya Protocol explicitly mentions the potential need for a multilateral benefit-sharing fund in the context of transboundary genetic resources and associated traditional knowledge, or for situations in which it is not possible to obtain prior informed consent. Some third-party commercial uses of genetic resource information from publicly accessible databases can be analogized to a transboundary situation or a situation where it is not feasible to obtain consent. This is because synthetic biology researchers may be using fragments of DNA sequences from many different species in designing new biosynthesis pathways to generate new, or enhanced compounds. Providing for PIC/ABS obligations to be met for such uses through payments into a multilateral benefit-sharing fund may be a viable way to avoid the twin spectres of provider countries limiting access for non-commercial endeavours and the deterrence of researchers from using digital genetic resource information.

One of the Decisions from the 2016 COP13-MOP2 is on "the need for and modalities of a global multilateral benefit-sharing mechanism," under Article 10. It invites the submission and compilation of information on "practical experiences, if any, on situations in which it is not possible to grant or obtain prior informed consent in relation to *in situ* or *ex situ* genetic resources and associated traditional knowledge" (COP-MOP 2016). The Decision also refers to developments in the Plant Treaty, which has a multilateral benefit-sharing mechanism (Plant Treaty, 2015). Under the Plant Treaty, Parties that commercialize products from material accessed through the Treaty's multilateral system agree to pay a fixed percentage of sales into a benefit-sharing fund that provides grants to crop diversity enhancing projects across the globe (Governing Body of the Treaty, 2006).

Unfortunately, the benefit-sharing mechanism has not been a success to-date; virtually all contributions to the fund have come from countries, not commercial enterprises, and the fund has collected millions of dollars less than had been forecast. As a result, the funding and benefit-sharing mechanism of the Plant Treaty system is currently under review. A variety of plans are being considered, including proportional contributions by countries based on domestic seed sales, government collection of fees from the seed industry for forwarding to the fund, and national subscription systems that would also feature payments into the fund by governments (Food and Agriculture Organization of the United Nations, 2017). Lessons learned and approaches adopted by the Plant Treaty could be instructive for the CBD/NP COP in exploring solutions to intangible genetic information benefit-sharing challenges.

Conclusion

The de-materialization of genetic resources into useable digital information is proceeding rapidly, and far outpacing legal and policy efforts to grapple with its implications. On the positive side, synthetic biology has the potential to yield important breakthroughs in a wide range of fields, including medicine and agriculture. However, it also seems likely to allow entities benefiting from such developments to bypass treaty-based and domestic ABS obligations.

New benefit-sharing approaches will be required to ensure that access to the vast wealth of untapped biodiversity in provider countries is not stymied by legitimate ABS fairness concerns in relation to genomic information. At the same time, domestic policymakers need to be aware of the increasing complexity of genomic sequence combinations and the challenges of tracking and reporting geographical origin information when a multiplicity of sequences, from publicly available sources, are altered and/or combined in creating an invention. Otherwise, the risk that researchers will unintentionally run afoul of national laws implementing and interpreting a range of treaties is bound to increase, resulting in unintended consequences for genetic resource users and providers that may be difficult to mitigate.

Notes

1 This chapter borrows significantly from my report "Digital DNA: The Nagoya Protocol, Intellectual Property Treaties, and Synthetic Biology" (2015), produced for, and with generous support from, the Woodrow Wilson International Center for Scholars.

2 A strand of synthetic biology research has also focused on *de novo* organism research. However, in an effort to create a controlled terminology, a 2014 European Commission Scientific Committee report adopted a definition of synthetic biology that begins with a living organism, relegating pre-life *de novo* research to the field of chemistry. Whether this new terminology will be universally adopted remains to be seen.

3 "Temporal scope" refers to the question of what genetic resources are covered by the Nagoya Protocol from a time perspective. Specifically, whether the Protocol applies only to genetic resources physically accessed (i.e. crossed a border) after the Protocol comes into force, or to genetic resources which are utilized after the Protocol comes into force, but were/are physically accessed at any time after the CBD came into force, or to genetic resources physically accessed at any time and utilized after the Protocol came into force. "Breadth of coverage" refers to the issue of whether the definitions of genetic resources and genetic material, should or will be interpreted broadly enough to include digital genetic information such as is used in synthetic biology research. Both of these issues are discussed in more detail in Margo A. Bagley and Arti K. Rai, *The Nagoya Protocol and Synthetic Biology: A Look at the Potential Issues*, Wilson Center: Synthetic Biology Project (November 2013), www.wilsoncenter.org/publication/the-nagoya-protocol-and-synthetic-biology-research-look-the-potential-impacts.

4 In the environmental context, the precautionary principle reflects the idea that even if there is not conclusive scientific proof of the risk associated with a product or activity, precautions should be taken to protect against the risk or else the product should not be used or the activity should not be undertaken.
5 See, e.g., www.biobricks.org/about-foundation; www.igem.org/Main_Page. It should be noted that some primarily open source advocates still seek patents on specific outputs on a strategic basis.
6 As the Supreme Court stated in *Bonito Boats, Inc. v. Thunder Craft Boats, Inc.*, 489 U.S. 141, 150–51 (1989) (internal quotations and citations omitted): "The applicant ...who is willing to reveal to the public the substance of his discovery and the best mode ... of carrying out his invention is granted the right to exclude others from making, using, or selling the invention throughout the United States for a period of [20] years. The federal patent system thus embodies a carefully crafted bargain ... [The inventor] may keep his invention secret and reap its fruits indefinitely. In consideration of its disclosure and the consequent benefit to the community, the patent is granted."
7 *Association for Molecular Pathology v. Myriad Genetics, Inc.* (2013) 133 S. Ct. 2107. In addition, *Mayo Collaborative Services v. Prometheus Laboratories, Inc.*, 132 S. Ct. 1289 (2012) ("Mayo") in conjunction with *CLS Bank v. Alice*, 134 S. Ct. 2347 (2014) imposed an "inventive concept" requirement for claimed inventions involving a law of nature or abstract idea; thus both cases are particularly relevant to biotech diagnostic or therapeutic method claims.
8 *Diamond v. Chakrabarty* (1980) 447 U.S. 303.
9 *In Re Roslin Institute* (2015) 750 F.3d 1333. A related patent claimed the cloning methods used to produce Dolly the sheep. Those method claims were not subject to challenge under the *Myriad* reasoning.
10 Directive 98/44/EC of the European Parliament and of the Council of 6 July 1998 on the legal protection of biotechnological inventions, Art. 5(2), states: "An element isolated from the human body or otherwise produced by means of a technical process, including the sequence or partial sequence of a gene, may constitute a patentable invention, even if the structure of that element is identical to that of a natural element."
11 *D'Arcy v. Myriad Genetics, Inc.* (2015) HCA 35. The Court noted that the isolated nucleic acid claims "[embrace] a nucleic acid sequence or protein removed from its naturally occurring environment and includes recombinant or cloned DNA isolates and chemically synthesized analogs or analogs biologically synthesized by heterologous systems."
12 Denmark's draft legislation provides one approach to complying with this obligation, as it prohibits the utilization of GRs when the use is based on GRs acquired in violation of GR access regulations in the country where the GRs were accessed. Violations are punishable by fines, or up to two years in prison if willful or grossly negligent, and foreign states and persons appear to have standing to bring relevant actions in Danish courts.
13 Art. 27 of the CBD does contain a dispute settlement procedure; however, it has never been used. Art. 30 of the NP requires Members at the first meeting of the parties, to develop and approve cooperative mechanisms to promote compliance and to address cases of non-compliance.
14 Communication with personnel in the French Polynesia Ministry for the Environment.
15 Interviews with French Polynesia Ministry of Environment and Ministry of Research personnel.
16 Although there were some discussions of synthetic biology issues in the Nagoya Protocol negotiations, they did not gain sufficient traction to warrant an explicit reference in the final agreement.
17 One such tool is the free "Gene Designer" software available from www.DNA2.0.com. As a press release from the J. Craig Venter Institute explains: "Genomic science ... is enabling researchers to 'read' the genetic code of organisms from all branches of life ... Sequencing genomes has now become routine, giving rise to thousands of genomes in the public databases. In essence, scientists are digitizing biology by converting the A, C, T, and G's of the chemical makeup of DNA into 1's and 0's in a computer."
18 See www.mooreabiocode.org; www.barcodeoflife.org/content/about/what-dna-barcoding. Art. 8 of the Nagoya Protocol, Art. 8 states in part:

"In the development and implementation of its access and benefit-sharing legislation or regulatory requirements, each Party shall:

(a) Create conditions to promote and encourage research which contributes to the conservation and sustainable use of biological diversity, particularly in developing countries, including through simplified measures on access for non-commercial research purposes, taking into account the need to address a change of intent for such research."

19 The previously mentioned NGOs ETC Group and Friends of the Earth are also members of the Working Group.

20 See, e.g., Brazilian Provisional Act, No. 2,186-16, Title II, Art. 7, ¶ 1 August 23, 2001, which broadly defines genetic heritage as "*information* of genetic origin, contained in samples of all or part of a plant, fungal, microbial or animal species, in the form or molecules or substances originating in the metabolism of these living beings, and in extracts obtained from *in situ* conditions, ..." (emphasis added). See also, Art. 1 of the Andean Community Decision 391 which defines "access" as "[t]he obtaining and use of genetic resources conserved *in situ* and *ex situ*, of their by-products and, if applicable, of their *intangible components*, for purposes of research, biological prospecting, conservation, industrial application and commercial use, among other things." Decision Number 391 Establishing the Common Regime on Access to Genetic Resources, Bol.-Colom.-Ecuador-Peru, July 17, 1996, 213 Cartagena Agreement Official Gazette (emphasis added), www.wipo.int/wipolex/en/details.jsp?id=9446.

21 While isolated genomic DNA sequences may be eligible for patent protection outside of the United States, not surprisingly, there appears to be no major effort underway to patent, at considerable cost, the large quantities of DNA sequence information obtained during non-commercial research expeditions.

22 Due in part to transboundary issues where species are common to more than one territorial jurisdiction.

References

ABC Science News 2015, "Secret to Why Elephants Rarely Get Cancer Is in Their Genes," www.abc.net.au/news/2015-10-09/genes-explain-why-elephants-rarely-get-cancer/6841096.

Ad Hoc Open-Ended Working Group on Access and Benefit-sharing 2010, *Report of the First Part of the Ninth Meeting of the Ad Hoc Open-Ended Working Group on Access and Benefit-sharing*, 26 April 2010, UNEP/CBD/WG-ABS/9/3.

Agreement on Trade Related Aspects of Intellectual Property 1994, 1869 UNTS 299, opened for signature 15 April 1994, entered into force 1 January 1995.

Alexander, B. 2015, "Synthetic Life Seeks Work," *MIT Technology Review*, www.technologyreview.com/s/540701/synthetic-life-seeks-work/.

Arrowhead Research Corporation 2016, APO 70 (13 October 2016), http://www.austlii.edu.au/cgi-bin/viewdoc/au/cases/cth/APO/2016/70.html?context=1;query=Arrowhead.

Bagley, M. 2015, "Digital DNA: The Nagoya Protocol, Intellectual Property Treaties, and Synthetic Biology," Wilson Center: Synthetic Biology Project, www.synbioproject.org/site/assets/files/1395/digital_dna_final.pdf.

Bagley, M. 2014, "Patent Barbarians at the Gate: The Who, What, When, Where, Why and How of U.S Patent Subject Matter Eligibility Challenges," in R. Okediji and M. Bagley (eds), *Patent Law in Global Perspective*, Oxford University Press, New York, NY, p. 165.

Bagley, M., Okediji, R., and Erstling, J. 2013, *International Patent Law and Policy*, West Academic Publishing, St. Paul, MN, pp.154–58.

Bagley, M. and Rai, A. 2013, "The Nagoya Protocol and Synthetic Biology: A Look at the Potential Issues," Wilson Center: Synthetic Biology Project, www.wilsoncenter.org/publication/the-nagoya-protocol-and-synthetic-biology-research-look-the-potential-impacts.

Brynjolfsson, E. and McAfee, A. 2014, *The Second Machine Age: Work, Progress, and Prosperity in a Time of Brilliant Technologies*, W.W. Norton and Company, New York, NY, pp. 57–69.

Burk, D., and Lemley, M. 2003, "Policy Levers in Patent Law," *Virginia Law Review*, Vol. 89, pp. 1597–99.

Chen, J. 2006, "There's No Such Thing as Biopiracy... And It's a Good Thing Too," *McGeorge Law Review*, Vol. 37, p. 1.

Civil Society Working Group on Gene Drives 2016, "The Case for a Global Moratorium on Genetically-Engineered Gene Drives," www.etcgroup.org/sites/www.etcgroup.org/files/files/cbd_cop_13_gene_drive_moratorium_briefing.pdf.

Conference of the Parties to the Convention on Biological Diversity 2014, *Report of the Twelfth Meeting of the Conference of the Parties to the Convention on Biological Diversity*, 17 October 2014, UNEP/CBD/COP/12/29, www.cbd.int/doc/meetings/cop/cop-12/official/cop-12-29-en.pdf, p. 178.

Conference of the Parties to the Convention on Biological Diversity 2016a, *Thirteenth Meeting: Digital Sequence Information on Genetic Resources*, 16 December 2016, UNEP/CBD/COP/13/L.29.

Conference of the Parties to the Convention on Biological Diversity 2016b, *Thirteenth Meeting: Synthetic Biology*, 16 December 2016, UNEP/CBD/COP/13/L.34.

Conference of the Parties to the Convention on Biological Diversity Serving as the Meeting of the Parties to the Nagoya Protocol on Access to Genetic Resources and the Fair and Equitable Sharing of Benefits

Arising from Their Utilization 2016 (COP-MOP 2016), *Decision Adopted by the Parties to the Nagoya Protocol on Access and Benefit-Sharing*, 17 December 2016, CBD/NP/MOP/DEC/2/10.

Convention on Biological Diversity 2016a, *The Access and Benefit-sharing Clearinghouse*, last viewed 4 April 2017, https://absch.cbd.int.

Convention on Biological Diversity 2016b, *The Access and Benefit-sharing Clearinghouse: Legislative, Administrative, or Policy Measures – Denmark*, https://absch.cbd.int/search/nationalRecords?schema=measure.

Council Directive 98/44/EC of the European Parliament and of the Council of 6 July 1998 on the Legal Protection of Biotechnological Inventions, *Official Journal of the European Communities* L 213/13, 30.7.1998. Article 5(2).

Danchin, A. and di Lorenzo, V. 2008, "Synthetic Biology: Discovering New Worlds and New Words: The New and Not So New Aspects of this Emerging Research Field," *Emboreports*, Vol. 9, pp. 822–27.

Depoorter, B. 2014, "Intellectual Property Infringements and 3D Printing: Decentralized Piracy," *Hastings Law Journal*, Vol. 65, pp. 1483–504.

Desai, D. and Magliocca, G. 2014, "Patents, Meet Napster: 3D Printing and the Digitization of Things," *Georgetown Law Review*, Vol. 102, p. 1691.

Dressel, H. (ed.) 2016, "Synthetic Biology, Biodiversity, and Farmers: Case Studies Exploring the Impact of Synthetic Biology on Natural Products, Livelihoods, and Sustainable Use of Biodiversity," *ETC Group*, pp. 32, 54, 59.

European Commission 2012, *Proposal for a Regulation of the European Parliament and of the Council on Access to Genetic Resources and the Fair and Equitable Sharing of Benefits Arising from Their Utilization in the Union*, COM (2012) 576 Brussels: Commission of the European Communities.

European Commission Health and Food Safety Scientific Committees 2014, *Final Opinion on Synthetic Biology*, 23 September 2014, https://ec.europa.eu/health/sites/health/files/scientific_committees/emerging/docs/scenihr_o_044.pdf.

European Patent Convention 1973, 29 I.L.M *1417*, opened for signature 5 October 1973, entered into force 7 October 1977.

Faiola, A. 1999, "Amazon Cash Crop," *The Washington Post*, 9 July 1999, www.washingtonpost.com/archive/politics/1999/07/09/amazon-cash-crop/eea4b835-db25-49ef-b995-e62a499a0bc3/?utm_term=.59c92176133e.

Fisher, W. 2001, "Theories of Intellectual Property," in *New Essays in the Legal and Political Theory of Property*, Cambridge University Press.

Food and Agriculture Organization of the United Nations 2015, *Fourth Meeting of the Ad-Hoc Open-Ended Working Group to Enhance the Functioning of the Multilateral System: Exploring a Proposal to Develop a Mechanism of Contributions by Contracting Parties to the Benefit-Sharing Fund*, September 2015, IT/OWG-EFMLS-4/15/Inf.5.

Food and Agriculture Organization of the United Nations 2017, *Eighth Meeting of the Ad-Hoc Advisory Committee on the Funding Strategy: Exploring a Proposal to Develop a Mechanism of Contributions by Contracting Parties to the Benefit-Sharing Fund*, February 2017, IT/ACFS-8/17/Inf.4.

Fridtjof Nansen Institute 2010, *The Concept of "Genetic Resources" in the Convention on Biological Diversity and How It Relates to a Functional International Regime on Access and Benefit-Sharing*, 19 March 2010, UNEP/CBD/WG-ABS/9/INF/1.

Fussenegger, M. and Weber, W. 2009, "The Impact of Synthetic Biology on Drug Discovery," *Drug Discovery Today*, Vol. 14, no. 19–20, pp. 956–63.

Greiber, T., Morena, S., Ahren, M., Carrasco, J., Kamau, E., Medaglia, J., Oliva, M., and Perron-Welch, F. 2012, "An Explanatory Guide to the Nagoya Protocol on Access and Benefit-Sharing," *International Union for Conservation of Nature Environmental Policy and Law Paper*, no. 83.

Genetic Engineering and Biotechnology News 2017, "Fruit Fly Finding May Lead to New Cancer Therapies," 25 January 2017, www.genengnews.com/gen-news-highlights/fruit-fly-finding-may-lead-to-new-cancer-therapies/81253768?utm_medium=newsletter&utm_source=GEN+News+Highlights+of+the+Week&utm_content=02&utm_campaign=GEN+News+Highlights+of+the+Week_20170128.

Governing Body of the Treaty 2006, "The Standard Material Transfer Agreement for the Multilateral System," www.fao.org/3/a-bc083e.pdf.

Hakam, J. 2014, "Indonesia to Put Moratorium on Foreign Research to Protect Its Biodiversity," *Eukuatorial*, 21 October 2014. www.ekuatorial.com/2014/10/indonesia-to-put-moratorium-on-foreign-researches-to-protect-its-biodiversity/#!/story=post-9230&loc=-12.897489183755892,132.36328125,4.

Hammond, E. 2013, "Biopiracy Watch: A Compilation of Some Recent Cases," *Third World Network*, Vol. 1.

Hart Research Associates 2014, "Perceptions of Synthetic Biology and Neural Engineering," *Woodrow Wilson International Center for Scholars Synthetic Biology Project*, 18 April 2014, www.synbioproject.org/site/assets/files/1073/focusgroup_2014.pdf.

Heald, P. 2003, "The Rhetoric of Biopiracy," *Cardoza Journal International & Comparative Law*, Vol. 11, p. 519.

Holbrook, T. and Osborn, L. 2015, "Digital Patent Infringement in an Era of 3D Printing," *U.C. Davis Law Review*, Vol. 48, p. 319.

Holman, C. 2011, "Copyright for Engineered DNA: An Idea Whose Time Has Come?" *West Virginia Law Review*, Vol. 113, p. 699.

Holman, C., Gustafsson C., and Torrance, A. 2016, "Are Engineered Genetic Sequences Copyrightable? The U.S. Copyright Office Addresses a Matter of First Impression," *Biotechnology Law Report*, Vol. 35, pp. 103–11.

International Civil Society Working Group on Synthetic Biology 2011, *A Submission to the Convention on Biological Diversity's Subsidiary Body on Scientific, Technical and Technological Advice (SBSTTA) on the Potential Impacts of Synthetic Biology on the Conservation and Sustainable Use of Biodiversity*, 17 October 2011, www.cbd.int/doc/emerging-issues/Int-Civil-Soc-WG-Synthetic-Biology-2011-013-en.pdf.

J. Craig Venter Institute 2010, *First Self-Replicating Synthetic Bacterial Cell*, press release, 20 May 2010, www.jcvi.org/cms/research/projects/first-self-replicating-synthetic-bacterial-cell/overview/#sthash.2TuEjP2I.dpuf.

Jasanoff, S. 2005, *Designs on Nature: Science and Democracy in Europe and the United States*, Princeton University Press, Princeton, NJ.

Karjala, D. 2011, "Protecting Innovation in Computer Software, Biotechnology and Nanotechnology," *Virginia Journal of Law and Technology*, Vol. 16, pp. 55–61.

Keasling, J., Martin, V., Pitera, D., Withers, S., and Newman, J. 2007, Biosynthesis of amorpha-4,11-diene, U.S. Patent No. 7,192,751.

Kieff, F. S. 2001, "Property Rights and Property Rules for Commercializing Inventions," *Minnesota Law Review*, Vol. 85, p. 697.

Kishore, G., Motion, M., Hicks, P., Hansen, J., Houghton-Larsen, J., Hansen, E., Mikkelsen, M., Tavares, S., and Blom, C., 2017, Production of Steviol Glycosides in Microorganisms, U.S. Patent No. 9,562,251.

Krishnamurthy, M., Moore. R., Rajamani, S., and Panchal, R. 2016, "Bacterial Genome Engineering and Synthetic Biology: Combating Pathogens," *BMC Microbiology*, https://bmcmicrobiol.biomedcentral.com/articles/10.1186/s12866-016-0876-3.

Laird, S. and Wynberg, R. 2013, "Bioscience at a Crossroads: Access and Benefit-sharing in a Time of Scientific, Technological and Industry Change: The Cosmetics Sector," *Secretariat of the Convention on Biological Diversity*, www.cbd.int/abs/policy-brief/default.shtml/.

Ledford H. 2013, "Bioengineers Look Beyond Patents," *Nature*, Vol. 433, 3 July 2013, www.nature.com/news/bioengineers-look-beyond-patents-1.13320.

Li, J. and Vederas, J. 2009, "Drug Discovery and Natural Products: End of an Era or an Endless Frontier?" *Science*, Vol. 325, no. 5937. pp.161–65.

Maler, T. 2015, "EU Must Deliver Legal Certainty for Seed and Plant Sectors," *Agra Europe*, www.agra-net.net/agra/agra-europe/analysis/interview-eu-must-deliver-legal-certainty-for-seed-and-plant-sectors-458540.html.

Marrakesh Agreement Establishing the World Trade Organization 1994, 1867 U.N.T.S. 154, signed 15 April 1994.

McFarlane, M., Sharp, T., and Aquino, J. 2014, "Stopped at the Threshold: The Practical Impact of the Supreme Court's *Mayo* and *Myriad* Decisions on Biotechnology Patent Practices," *Bloomberg BNA Life Sciences Law and Industry Report*.

Medaglia, J., Perron-Welch, F., and Phillips, F. 2014, "Overview of National and Regional Measures on Access and Benefit-sharing: Challenges and Opportunities in Implementing the Nagoya Protocol," *CISDL Biodiversity and Biosafety Law Research Programme*, 3rd edn.

Meienberg, F, Sommer, L. Lebrecht, T., Lovera, M., Gonzalez, S., Luig, B., von Bremen, V., Steiner, K., Glauser, M., and Kienle, U. 2015, "The Bittersweet Taste of Stevia: Commercialisation of Stevia-derived sweeteners by violating the rights of indigenous peoples, misleading marketing and controversial SynBio production," www.publiceye.ch/fileadmin/files/documents/Biodiversitaet/BD_STEVIA_REPORT_EN.pdf.

Merriam Webster Dictionary 2015, www.merriam-webster.com/dictionary/piracy.

Moeller IP Advisors 2014, "Peru Accuses Chinese Companies of Biopiracy Involving Maca Root," www.moellerip.com/peru-accuses-chinese-companiesof-biopiracy-involving-maca-root/.

Mohan, V. 2015, "India Allows UK Researcher to Study Medicinal Knowledge of Ethnic Group in Gujarat," *The Times of India*, http://timesofindia.indiatimes.com/india/India-allows-UK-researcher-to-study-medicinal-knowledge-of-ethnic-group-in-Gujarat/articleshow/49489130.cms.

Murray, M. 2014, "Post-Myriad Genetics Copyright of Synthetic Biology and Living Media," *Oklahoma Journal of Law and Technology*, Vol. 10, pp. 106–11.

Nature Biotechnology 2016, "Recent Patents in Synthetic Biology," Vol. 34, no. 8, p. 822.

Nayoga Protocol 2010, *Nagoya Protocol on Access to Genetic Resources and the Fair and Equitable Sharing of Benefits Arising from Their Utilization to the Convention on Biological Diversity*, C.N.782.TREATIES-1, entered into force 29 October 2010.

Nordhaus, W.D. 1969, *Invention, Growth, and Welfare: A Theoretical Treatment of Technological Change*, MIT Press, Cambridge.

Oldham, P. 2011, "Tracking and Monitoring Genetic Resources in the Patent System," *ESRC Centre for Economic and Social Aspects of Genomics and UNU*, www.abs-initiative.info/fileadmin/user_upload/Activities/2011/Addis_IP_09-2011/Documents/2_-_Tracking___Monitoring__GRs_in_the_Patent_System___Paul_Oldham__ESRC_Cesagen___UNU_.pdf.

Oldham, P., Hall, S., and Burton, G. 2012, "Synthetic Biology: Mapping the Scientific Landscape," *Plos One*, Vol. 7, no. 4, e34368, http://journals.plos.org/plosone/Article?id=10.1371/journal.pone.0034368.

Oliva M. 2011, "Sharing the Benefits of Biodiversity: A New International Protocol and Its Implications for Research and Development," *Planta Medica*, Vol.77, pp.1221–23.

Open Source Seeds 2017, www.opensourceseeds.org/en.

Osborn, L. 2014, "Regulating Three-Dimensional Printing: The Converging Worlds of Bits and Atoms," *San Diego Law Review*, Vol. 51, p. 553.

Palmedo, M. 2015, "Over 200,000 John Does Sued for File Sharing in the U.S. Since 2010," *InfoJustice.Org*, http://infojustice.org/archives/4724.

Parida, M., Shukla, J., Sannarangiah, S., Sharma, S., Rao, V., and Vijayaraghavan, R. 2014, Oligonucleotides and Process for Detection of Swine Flu Virus, U.S. Patent Application Publication No. 2012/0231445 A1.

Plant Treaty 2015, *Sixth Session of the Governing Body of the International Treaty on Plant Genetic Resources for Food and Agriculture*, www.planttreaty.org/news/sixth-session-governing-body-international-treaty-plant-genetic-resources-food-and-agriculture.

Pottage, A. 2006, "Too Much Ownership: Bio-Prospecting in the Age of Synthetic Biology," *Biosocieties*, Vol.1, no. 2, pp. 137–58.

Presidential Commission for the Study of Bioethical Issues 2010, *New Directions: The Ethics of Synthetic Biology and Emerging Technologies*, pp. 43–46.

Raffensperger C. and Tickner J. (eds) 1999, "Protecting Health and the Environment: Implementing the Precautionary Principle," pp. 353–54.

Rai, A. and Boyle, J. 2007, "Synthetic Biology: Caught between Property Rights, the Public Domain, and the Commons," *Plos Biology*, Vol. 5(3), no. e58, http://journals.plos.org/plosbiology/Article?id=10.1371/journal.pbio.0050058.

Renninger, N., Newman, J., and Reiling, J. 2010, Fuel Components, Fuel Compositions and Methods of Making and Using Same, U.S. Patent No. 7,854,774.

Richardson, S., Mitchell, L., Stracquadanio, G., Yang, K., Dimond, J., DiCarlo, J., Lee D., Huang, C., Chandrasegaran, S., Cai, Y., Boeke, J., and Bader, J. 2017, "Design of a Synthetic Yeast Genome," *Science*, Vol. 355, no. 6329. pp. 1040–44.

Roberts, J. 2017, "Can Biotech Save the Rhino: Or Will It Speed the Animal's Demise?" *Chemical Heritage Foundation*, www.chemheritage.org/distillations/magazine/can-biotech-save-the-rhino.

Robinson, D. 2010, *Confronting Biopiracy: Challenges, Cases and International Debates*, 6th edn, Earthscan, London.

Saenz, A. 2010, "Secret Messages Coded into DNA of Venter Synthetic Bacteria," 24 May 2010, http://singularityhub.com/2010/05/24/venters-newest-synthetic-bacteria-has-secret-messages-coded-in-its-dna/.

Samuelson, P. 2013, "Are Gardens, Synthetic DNA, Yoga Sequences, and Fashions Copyrightable?" http://law.scu.edu/wp-content/uploads/Samuelson-Gardens-DNA-Yoga-Fashion-Abstract.pdf.

Sanofi 2013, "Sanofi Position on Biodiversity and Biopiracy," http://en.sanofi.com/csr/download_center/download_center.aspx.

Schteingart, C., Menzaghi, F., Jiang, G., Alexander, R., Sueiras-Diaz, J., Spencer, R., Chalmers, D., Luo, R. 2016, Synthetic Peptide Amides, U.S. Patent No. 9,359,399.

Service, R. 2015, "Modified Yeast Produce Opiates from Sugar," *Science*, Vol. 349, no. 6249, p. 677.
SynBioWatch 2013, "101 Fact Sheets from Friends of the Earth," www.synbiowatch.org/2013/05/synthetic-biology-101-some-technical-details-from-friends-of-the-earth/.
Tai, W., Wang, C., Sheu, P., Tsai, J. 2013, "Data Hiding in DNA for Authentication of Plant Variety Rights," *Journal of Electronic Science and Technology*, Vol. 11, no.1, pp. 38–42.
Torrance A. 2010, "Synthesizing Law for Synthetic Biology," *Minnesota Journal of Law Science and Technology*, Vol. 11, pp. 642–48.
Torrance, A. 2011, "DNA Copyright," *Valapraiso University Law Review*, Vol. 46, pp.1–41.
Torrance A. and Kahl L. 2014, "Bringing Standards to Life: Synthetic Biology Standards and Intellectual Property," *Santa Clara High Technology Law Journal*, Vol. 30, pp. 226–27.
United States Patent and Trademark Office 2014, *2014 Interim Guidance on Patent Subject Matter Eligibility*, www.uspto.gov/patent/laws-and-regulations/examination-policy/2014-interim-guidance-subject-matter-eligibility-0.
United States Patent and Trademark Office 2015, *July 2015 Update: Subject Matter Eligibility*, www.uspto.gov/sites/default/files/documents/ieg-july-2015-update.pdf.
Villalobos, A., Ness, J., Gustafsson, C., Minshull, J., and Govindarajan, S. 2006, "Gene Designer: A Synthetic Biology Tool for Constructing Artificial DNA Segments," *BMC Bioinformatics*, www.ncbi.nlm.nih.gov/pmc/Articles/PMC1523223/.
Vrieze, J. 2015, "The Littlest Farmhands," *SCIENCE*, Vol. 349, no. 6249, pp. 680–83.
Wallart T. and Bouwmeester, H. 2009, Transgenic amorpha-4, 11-diene synthesis, U.S. Patent No. 7,541,172.
Woodrow Wilson Center 2014, "Synthetic Biology Still in Uncharted Waters of Public Opinion," press release, 14 May 2014, www.synbioproject.org/news/project/6683/.
Woodrow Wilson International Center for Scholars 2010, "Trends in Synthetic Biology Research Funding in the United States and Europe," *Synthetic Biology Project*, June 2010.
World Trade Organization 2015, *Dispute Settlement*, www.wto.org/english/tratop_e/dispu_e/dispu_e.htm.
Yamamoto, N., Kajiura, H., Takeno, S., Suzuki, N., and Nakazawa, Y. 2014, "A Watermarking System for Labeling Genomic DNA", *Plant Biotechnology*. K. Aoki (ed.), www.jstage.jst.go.jp/article/plantbiotechnology/advpub/0/advpub_14.0609b/_pdf.
Zerbe, P., Chiang A., Dullat H. O'Neil- Johnson M., Starks C., Hamberger B., and Bohlmann J. 2014, "Diterpene Synthases of the Biosynthetic System of Medicinally Active Diterpenoids in Marrubium Vulgare," *The Plant Journal*, Vol. 79, pp. 914–27.

PART IV

Traditional knowledge protection

16
TRADITIONAL KNOWLEDGE
Lessons from the past, lessons for the future

Michael J. Balick

The nature of traditional knowledge and its devolution

Traditional knowledge, here considered as a body of information and set of skills developed by a group of people over time, is in a constant state of change. As each generation matures, skills perceived as immediately useful are maintained, others with a lesser perception of immediate value may be lost, and new information, skills and techniques may be incorporated into the body of traditional knowledge (Vandebroek and Balick, 2012). Thus the body of traditional knowledge is never static but rather dynamic in its shape and substance. It is this aspect of traditional knowledge that renders it susceptible to various kinds of intellectual property protection (McManis, 2004). In order to consider the 'preservation' of traditional knowledge, perhaps it would be useful first to explore the nature of this system, how it evolves over time, and identify some of the forces involved in its disintegration. This part of the current volume addresses the question of the composition of traditional knowledge and whether and how it might be protected.

In preparing this chapter, I have chosen to look at several site-specific examples, based on the notion that a study of the past may provide perspectives for the future.

In looking at the loss of information considered as traditional knowledge, Wolff and Medin (2001) suggested that 'with modernization, it may be that knowledge about living things has decreased, or as we say here, *devolved*'. The concept of devolution was derived from their study of undergraduate students at Northwestern University who were provided a list of 80 trees and asked to circle the species they 'had *heard* of before, regardless of whether they knew anything about them'. One result of the survey was that fewer than 50 per cent of the students recognized a group of trees that were frequently found in the area of their university, including alder, buckeye, catalpa, hawthorn, larch and others. The results were suggested to support the devolution hypothesis that linked modernization directly with loss of knowledge about living things. They suggested that this could be offset via 'cultural support', or 'sufficient amounts of indirect experience with the natural world, through a culture's media, talk and values … the degree to which a society promotes a particular area of knowledge'. For example, adults can teach children about living things, and thus help offset devolution as it relates to knowledge about the natural world and its components.

The extinction of language is an excellent example of devolution, as related to cultural knowledge.

> The last speakers of probably half the world's languages are alive today. As they grow old and die, their voices will fall silent. Their children and grandchildren – by overwhelming majority – will either choose not to learn or will be deprived of the opportunity to learn the ancestral language. Most of the world's languages have never been written down anywhere or scientifically described… .An immense edifice of human knowledge, painstakingly assembled over millennia by counless minds, is eroding, vanishing into oblivion.
>
> (Harrison, 2007, p.3)

Concern for the loss of traditional knowledge is the driving force behind many of the ethnobotanical and culture-related projects now underway throughout the world. Workers in the ethno-sciences are collecting data, specimens and craft objects, and using modern technology to catalogue and study this information (Salick, Konchar and Nesbitt, 2014). In the realm of ethnobotany, field studies are sometimes referred to as 'salvage ethnobotany', along the lines of the 'salvage botany' efforts that have been carried out for many years in endangered habitats of the tropics. These projects, for the most part, employ scientifically or technologically based approaches to recording information as their primary vehicle for the preservation of information. As I will discuss later in this chapter, the scientific paradigm may be effective in documentation of information and data collection, but not as useful with regard to long-term preservation of the actual knowledge.

The Micronesia Ethnobotany Project

In an attempt to quantify the rate of loss (or change) of information about traditional activities on Pohnpei, an island in the Federated States of Micronesia, Lee *et al.* (2001) studied what they referred to as 'cultural dynamism and change'. For example, with the trade in different species and varieties of food plants between islands in Micronesia, *Alocasia macrorrhiza* (L.) G. Don. var *macrorrhizos*, an edible taro that was once a pre-eminent food source, has been replaced by other taro species such as *Colocasia esculenta* (L.) Schott and *Cyrtosperma merkusii* (Hassk.) Schott. These latter introductions are considered more palatable, and thus more desirable, and as a consequence of their adoption as a major food crop, it is likely that an important part of the cultural knowledge associated with *Alocasia macrorrhiza* var. *macrorrhizos* has diminished, even become extinct.

As part of the effort known as the Micronesia Ethnobotany Project, a great deal of formal and informal dialogue on the loss of cultural knowledge was held with traditional leaders and ordinary people on Pohnpei. Some of the results will be discussed further in this chapter. Lee *et al.* (2001) reported that:

> The traditional leaders we spoke to in Micronesia were concerned with a related, but qualitatively and quantitatively different phenomenon. Instead of their culture changing and evolving at a relatively slow 'background' rate, over the last two generations a large percentage of traditions and skills specific to Micronesia have not been passed on, and will become extinct if an active program is not put into place to keep them an active part of local life.
>
> (Lee *et al.* 2001, p. 9)

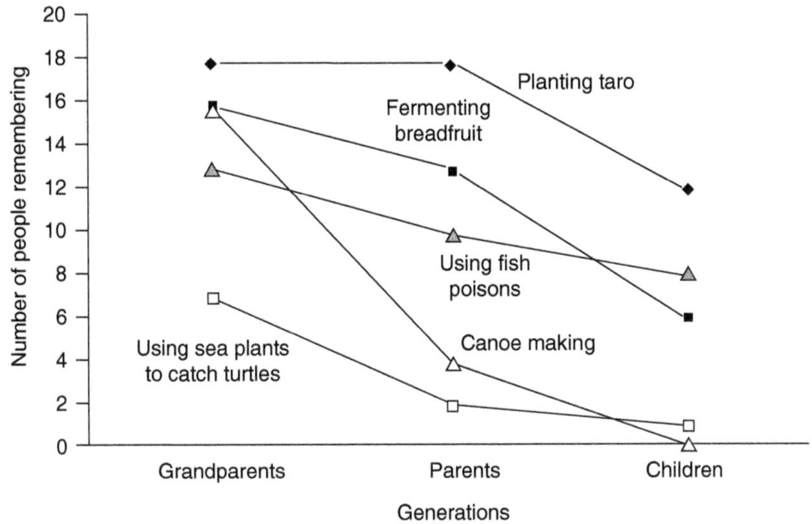

Figure 16.1 Erosion of traditional knowledge on Pohnpei, FSM

During an annual course on ethnobotanical techniques offered at the College of Micronesia (COM) in Pohnpei in 1999, we carried out an informal survey amongst the students. This involved a series of questions regarding how many students remembered seeing their grandparents and parents making canoes, and how many of the students had ever made a canoe. The results were extraordinary but not all that unexpected – not a single person on the course had experience in canoe making. One year later, during the next course we carried out a more formal survey about generational knowledge covering various components of Micronesian life: planting taro; using plants to stun and capture fish; fermenting breadfruit as a method to preserve it as a famine food; using marine plants as turtle bait; and constructing outrigger canoes. The results, presented in Lee *et al.* (2001) (Figure 16.1) showed the predicted loss of information between generations on this island. In addition, this paper developed a linear regression for the survey results from each set of traditional knowledge (Figure 16.2) and made a series of very tentative predictions about the time (expressed in generations) that each of these skill sets might become extinct. This regression showed that the traditional knowledge involving canoe making and turtle catching were at greatest risk of extinction, predicted to disappear in the generation represented by the college students.

Because the small and biased (e.g. limited to a college class) sample yielded such tentative results, the following year we carried out a survey of traditional knowledge in Pohnpei, involving an instrument that contained 72 questions, in Pohnpean, administered by Pohnpeans, with a sample size of 160 people, approximately 0.5 per cent of the entire island's population. This survey included a significant focus on canoe-making, patterns of sakau (*Piper methysticum* G. Forst.) consumption and quality of life questions. The results from this survey are currently being prepared for publication, but are consistent with the conclusions of the preliminary surveys – it is clear that there is a rapid rate of loss of traditional knowledge about canoe-making, along with other skills on Pohnpei. Through their quantitative approach, these studies have also demonstrated that some skills and knowledge are more vulnerable than others, thus offering the possibility that priorities could be developed and evaluated for possible remediation of this loss, based on the rate at which the skill is being lost as well as its importance.

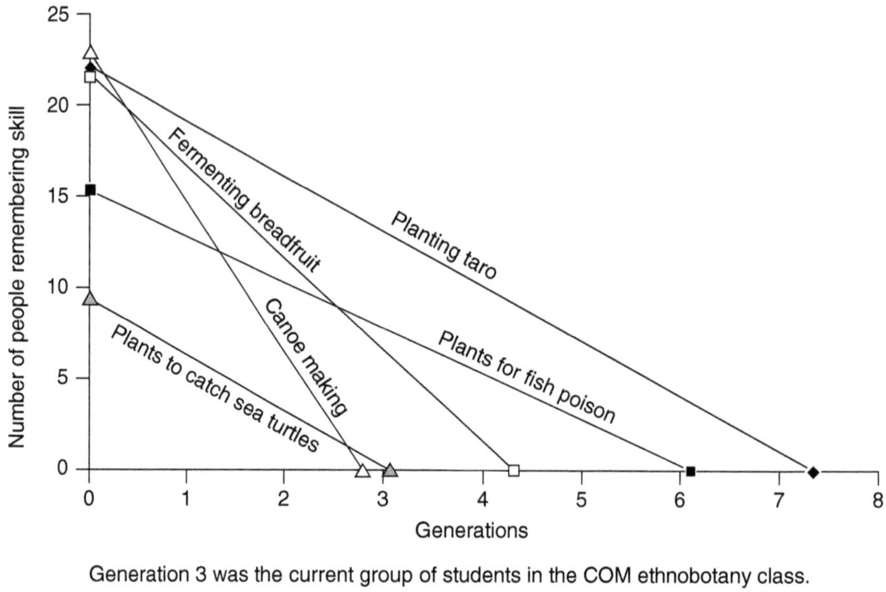

Figure 16.2 Predicted extinctions of traditional knowledge

The model developed to address this loss of cultural knowledge as well as the biodiversity, collectively called biocultural diversity, centered on three primary questions: (1) What is the plant and fungal diversity of Pohnpei, and what species are common, endangered or invasive? (2) What are the local uses of the native and introduced plants of Pohnpei, along with the common names and folk taxonomy of these plants? (3) How can the local flora, particularly those considered as having medicinal properties, be used to improve health care delivery on Pohnpei? Following more than a decade of field studies that involved dozens of local ethnobotanists on Pohnpei and a group of international researchers from many disciplines, we published three major volumes that addressed these questions. These were a specimen-based floristic checklist (Herrera, Lorence, Flynn and Balick, 2010), a manual of the ethnobotany of Pohnpei (Balick and collaborators, 2009) and a primary health care guide (Dahmer *et al.*, 2012). We have used this framework for gathering information on biocultural diversity in other areas of the Pacific and Central America, and these have helped to provide some degree of support for both cultural knowledge and biodiversity conservation.

Studying traditional healing in Belize

From 1988 to the early 2000s, a group of traditional healers and conservationists in Belize worked with the New York Botanical Garden on a project to inventory and catalogue the flora and ethnobotanical knowledge of that country. The objectives of the project include the preservation of cultural and traditional knowledge, natural products research through the National Cancer Institute, technology transfer, institutional development and training. The scope and flow of activities are illustrated in Figure 16.3, evolving from the establishment of an ethnobotanical inventory program. Collaborators have included eight governmental and non-governmental organizations in Belize, with over 120 individuals active in the project. Again, following the framework discussed above, the three questions have been asked

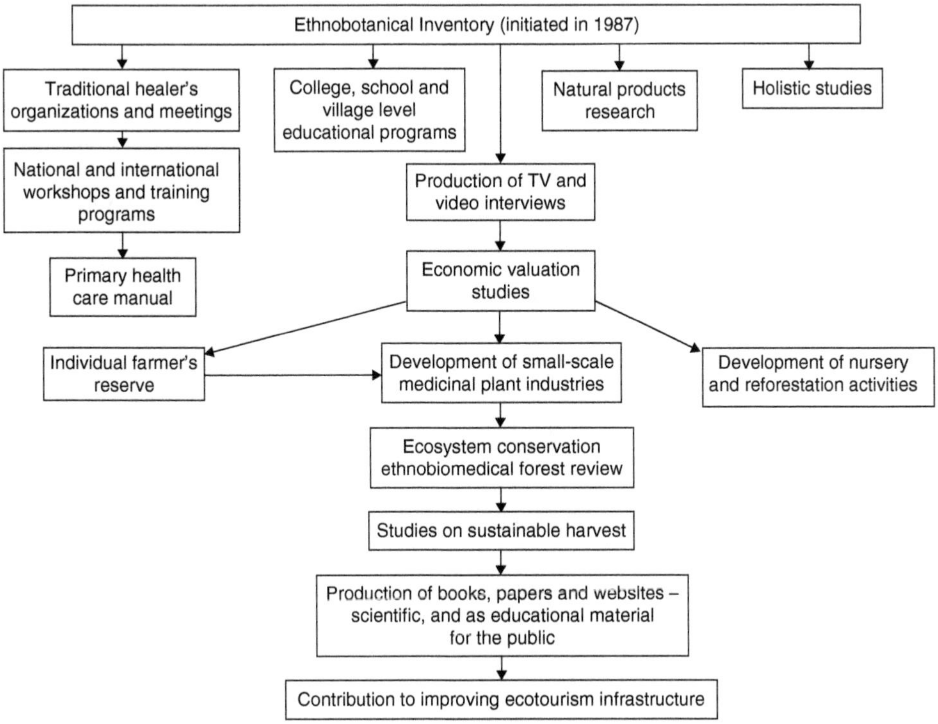

Figure 16.3 Chart of activities that developed as part of the Belize Ethnobotany Project

and answered with three publications concerning the flora (Balick, Nee and Atha, 2000), a primary health care manual (Arvigo and Balick, 1998) and a volume on the ethnobotany of Belize – contemporary and historic (Balick and Arvigo, 2015). As part of our work in that country, from the standpoint of traditional knowledge, one of the most important results has been the establishment of an association of traditional healers, allowing the development of a community of individuals dedicated to this practice. During fieldwork, over 8,000 plant specimens were collected, representing nearly 20 per cent of the holdings of the Forestry Department Herbarium in Belmopan, the capital of Belize. The project also promoted conservation of biodiversity, through various local initiatives including the establishment of an ethnobiomedical forest reserve, public displays, post-secondary classes, youth camps, school competitions, field trips and guest lectures. I will touch on some of the lessons learned during this project in this chapter.

Factors contributing to devolution

Based on experience derived from several projects in various regions over the past two decades, we can identify some of the reasons for the loss of traditional knowledge, and the constraints to addressing this devolution. Modernization is probably one of the foremost issues involved in changing the focus of people's educational endeavours. Emerging generations in many locations around the world now have new career trajectories, based on opportunities derived from modernization and globalization that are the result of the information age. This modernization has been accompanied by the inability of people, particularly the young, to recognize value in traditional ways, as related to their daily lives. In many cases there do not seem to be perceived economic returns from engaging in traditional activities. For example, in the early 1990s, I was in a taxi in Belize, and the driver, a young man in his twenties, asked the purpose of my visit. When I replied that I was a student of bush medicine, he enthusiastically launched into a monologue, laden with sentiment, about how his grandfather was a great bush doctor, who knew all of the uses of the plants in the forest, and would treat ill members of the family with great success. His father, the young man explained, knew much about the forest, and the uses of plants, but was not as skilled as his grandfather. The taxi driver himself had no interest in the forest, plants or traditional healing when he was growing up, and did not accompany either his grandfather or father in the forest when they went about their work. His goal, instead, was to have a vehicle, and a modern life filled with the most modern music, culture and food. Thus, as a result, at the point that his dream had been fulfilled, he felt trapped within it, as there were no longer options to learn family wisdom as both his grandfather and father had died without passing along their knowledge to family members. All too often, this is the case, and by the time young people begin to develop a passion for their roots, it can be too late.

In many places, the diffusion of the family as a unit has tended to reduce interest in traditional activities. As Hezel (2001) has pointed out in Micronesia, these changes have resulted in a complete reshaping of daily life in this region. On the island of Guam, the erosion of traditional culture has been linked to a rise in youth gang membership and criminal activity. Schmitz and Christopher (1997) noted that

> within Micronesian cultures, family kinship, community cohesion, folk knowledge, and religious pageantry have long shaped the cultural life of the community. Courtesy, respect, deference to elders, cooperation, and community hospitality are cultural virtues. Traditional society hinges on family cohesion and community consciousness.
>
> (Schmitz and Christopher, 1997)

In recent times, the ideology of modern Western society, individualism, leaves a vacuum in the lives of peoples accustomed to living as a community, and thus Schmitz and Christopher concluded:

> gang membership provides a perceived solution to disintegrating traditionalism and the unattractiveness, or unattainability, of modernism. Gang members ascribe to a moral vision based on traditional tribal warrior values.
>
> (Schmitz and Christopher, 1997)

In many locations, the introduction of television has become a substitute for family and village storytelling and conversations, during which traditional knowledge was formerly transmitted. In the early 1980s I worked with the Apinaje Indians of Northeastern Brazil, on a project involving the use of the babassu palm as an economic crop (Balick, 1988). We chose to go to these people because of their vast knowledge of the babassu, known as the tree of life in this region. The community would go into the forest surrounding their village at least three days a week, collecting babassu, Brazil nuts and jaborandi, the source of a leaf made into a pharmaceutical medication for glaucoma. As people sat in the forest and cracked the nuts, they would tell stories, transmit community information and gossip, and teach the younger members of the group about traditional life. This informal training would last for many hours during collection days, and continue around the fire at night. I was able to record stories, lore, songs and other information through my presence in the collection activity. Ten years later, in 1993, I returned for a brief visit to these people, to see how they had progressed now that a railway and highway were constructed near the reserve. Upon entering the village, the most striking new addition was the parabolic antenna, constructed next to the main communal palm thatched house. Inside, nearly all of the village's children were seated, facing a colour television, watching a blond-haired, blue-eyed entertainer based in Rio dance and sing along with her audience of children. There were many complaints from the village leaders about the lack of interest in traditional values and activities now that television had arrived. The major role model for the children had become a series of television shows, rather than the traditional leaders.

There is also the fear that traditional knowledge will be used to the advantage of groups outside the culture, perhaps as a new drug or food plant, and thus there is often hesitation to collaborate on projects that might yield benefit to the community as far as preservation of knowledge is concerned. Over the past decade or two, this fear has been exacerbated by outside forces whose stated objectives are to protect, guide and council indigenous groups in order that they might not be taken advantage of. While much of this guidance has been very useful in shaping local perspectives on the appropriate nature of the partnership and collaboration with outsiders, other efforts have resulted in the complete disenfranchisement of the cultures from potential opportunities that could rekindle interest in traditional knowledge and activities.

Finally, there is often the lack of a structure or support system for traditional knowledge and the activities related to its maintenance. For example, in many areas, people skilled in traditional activities cannot become part of a larger community of those with similar interests, nor are they supported by governmental, legal or educational institutions. Prior to the development of the Belize Association of Traditional Healers and the Traditional Healers Foundation, people in Belize who were skilled in this aspect of their culture acted alone as individuals. They were belittled by their families and friends, criticized by local educational and religious institutions, and in the most severe cases, persecuted by the law. Once a community had been established, with standards for membership, and seminars and workshops developed, more and more prestige was given to this group of elders in Belizean society. Through a series of television shows, educational videos distributed to the schools, and related activities, children in Belize now accept the fact that traditional knowledge about medicinal plants is an important subject to learn, and there is presently much greater interest shown in ethnomedicine than ever before.

One way of strengthening the position of the traditional healer employed in the Belize project has been to consider these people as colleagues and teachers, rather than as informants. The more traditional way of giving healers an unknown identity can be an insult to them, as in most cases it is their knowledge or intellectual property that guided the research (H. O'Brien, pers. comm.). By including traditional healers who provided information for research as co-authors, or providing acknowledgement using their names, all parties benefit. Indeed, copyright law recognizes the legal rights of co-authors to claim authorship and to exploit a co-authored work.[1]

An example of this is to be found in Glinski *et al.* (1995), a pharmacological research project. After discussing the interest in identifying bioactive compounds with Belizean healer Don Elijio Panti, he suggested a group of plants for testing in various screens by the Glinski lab. One of these, *Psychotria acuminata* Benth., was subsequently identified as a source of phenophorbide a, a green pigment that deactivates cell surface receptors. According to the resulting publication, 'our investigations suggest that the inactivation of cell surface receptors contributes not only to the antitumor effect of PDT [photodynamic therapy], but also to the systematic immunosuppression, a serious side effect of PDT'. It was found that an extract of this plant inhibited cytokinine and monoclonal antibody binding to cell surfaces, and this was attributed to the presence of phenophorbide a and pryophenophorbide a. This discovery was a contribution to the corpus of scientific information about natural products chemistry and bioactivity, but not relevant to the development of a new drug. Importantly, Don Elijio Panti was a co-author of this paper, published in *Photochemistry and Photobiology*, acknowledging, in the judgement of the research team and reviewers, that his discovery and utilization of the plant for many decades constituted a crucial and significant intellectual contribution to the paper. This is a standard that we, and increasingly more of our scientific colleagues, have attempted to adhere to in our ethnobotanical studies. Authorship of the paper was one of the achievements of which Don Elijio was quite proud, and the reprint was prominently displayed on the wall of his home for many years. It was also extremely useful in dispelling the gossip from the teachers and religious leaders of the village that because this man believed in the Maya spirits, and practised ancient medicine, he was not deserving of people's respect. After collaborating with us for over a decade, he passed away in 1996, at the age of 103. Today, Don Elijio's house is a small museum and shrine to this master of traditional knowledge, and younger people in the village now practise Maya healing.

Moving forward and letting go

In the realm of traditional knowledge, what are the parameters for deciding what skills and data survive and what goes extinct? Who makes this decision and what should it be based upon? Perhaps it would be useful to disengage this part of the discussion from the legal issues of intellectual property rights, and learn from traditional perspectives. In 1999, I recorded a conversation with Ashok Ripoche, a Tibetan monk who came to the US as an emissary of the Dalai Lama. In his role as Director of the main library at Dharamsala, India, he had recently been charged with a project to introduce Western science to Tibetan students, via the translation of significant textbooks and references. His group chose to concentrate on physics first, and biology second. We posed the question as to how a culture such as his can survive in the presence of another, more powerful culture that surrounds it. He replied:

> Tibetan culture will never be the same as it used to be before. It is always changing … It will never be the same culture after 10 years, after 15, or 20 years … sometimes of course I am disturbed, but sometimes we know that this is a phenomenon. It will never be the same, it keeps changing, look at history … and then whether we have any authority or the power to control the change or not – do we have it or not? Sometimes we think, yes we have some power or some control. And sometimes we find that there is no control. But if we could give some greater contribution – even though we know that it keeps changing – that includes the change of the culture from one point to the next … we can give a greater influence and the change will turn into favorable ways. And that way, maybe we can say we are preserving our culture … In many cases we have to say goodbye [to the past] but in many ways we have to cling on, hang on, and say, we give a good contribution so that the change will turn into a favorable way … the cultural aspect keeps changing in one form to another. Sometimes we see a loss. Another time we don't see a loss, we see an improvement, so we don't know exactly what's really improved and what's really lost – this is really difficult to see … before changing or losing whatever it is, we have to learn what it is – the heritage… At least we should have gotten the message from it, and then let it go. We cannot keep it.

Ashok Ripoche then recounted the story of coming to Dharamsala in 1959, and having a greater respect at that time for Western medicine as compared with traditional Tibetan medicine, which he and others considered primitive. Gradually, however, he learned via the interest

Table 16.1 Traditional skills on Pohnpei and their levels of importance

1. Very important skills
   Wiahda ihmw en Pohnpei (making local house)
   Wiepen sapwasapw (traditional farming system)
   Wiahda wahr (making canoe)
   Wiahda koal (making grass skirt)
   Wiahda likoutei sang kilin mahi-likoumeimei (making breadfruit bark skirt)
   Kaunopada wini sang tuhkehn Pohnpei (preparing local medicine from native plants)
   Pahda kahdeng sang ahlek (weaving curtain from ahlek plant)
   Pahda lirou ohng mehn didih ihmw (weaving lirau plant to be used in house construction)
   Wiahda pweten lihli (making local basket from coconut leaf for the preparation of *uhmw en lihli* [type of traditional breadfruit paste])
   Wiada kisin pwehl (making local rope from coconut husk)
   Wiahda kopwou sang idahnwel (weaving basket from the *idahnwel* plant)
   Wiahda litopw sang wahn ahis oh pwehl (making local paint from the *ahis* tree and soil)
   Wiahda kopwou sang tehn nih (making basket or local purse from coconut leaf)
   Wiadha uhk en laid sang dipenihd (making fishing net from coconut husk)
2. Important skills
   Wiahda lohs sang mwatal (weaving mat from the *mwatal* plant)
   Wiahda pwili ohng wie mar (making the seashell for traditional preparation of *mahr* [breadfruit])
   Wiahda padil sang kolou (making paddle from hibiscus)
   Wiahda kpennok sang dipenihd (making broom from coconut husk)
   Wiahda kilahs en du sang tuken Pohnpei (making diving/fishing goggles from native trees)
3. Not as important skills
   Wiahda spoon sang poundal (making spoon and fork from coconut shell)
   Charcoal sang pohndal (making charcoal from coconut shell)
   Wiahda lisoarop sang deipw (making local hat from pandanus)
   Waiahda mehn limalim sang kelou (making canoe bailer scoop from hibiscus)

Source: This list was prepared by Pelihter Raynor, Ally Raynor, Robert Gallen, Elpiana Amor and Mark Kostka following discussions with various people in Pohnpei.

that Western physicians showed in traditional Tibetan medicine that the latter had value, and 40 years later now feels 'more comfortable taking a Tibetan pill every day, rather than a Western chemical medicine'. His overall sense of the issue involved in the devolution of traditional knowledge was that people had to decide, on their own, or with outside help what subset of traditional knowledge to leave behind and what subset to move forward.

Back in Pohnpei, following the implementation of the various surveys, a group of Pohnpean elders and young people involved in the Micronesia Ethnobotany Project met to offer their perspectives on the importance of traditional information, including the development of a prioritized list of traditional skills (Table 16.1). It is interesting to note that many of the skills categorized as most important involve the construction of traditional structures such as houses and canoes, as well as the production of traditional dress, knowledge of traditional healing and fishing skills. Least important skills included items that were already made obsolete by the availability of inexpensive plastic and nylon substitutes on the island – spoons and forks, hats, and canoe bailers.

This series of exercises, including formal and informal surveys and grass-roots conversations and meetings, has helped Pohnpei to begin to set its priorities regarding the conservation of traditional knowledge. On other islands that are lacking in traditional leadership and interest, such knowledge is disappearing much more rapidly, a topic of concern that has been addressed by several recent conferences of Micronesian traditional leaders over the past few years.

Another factor often associated with traditional knowledge – particularly that concerning healing and medicinal plants – is the power that comes with its possession. In many cultures, including Western cultures, a person who can influence another person's health, whether by offering therapies or ameliorating the perceived cause of the condition, holds respect and thus power in the community. Amongst certain practitioners of Ayurvedic medicine that I observed in Sri Lanka, the concern was that the source of their influence and power to heal depended on a series of secret cures that they possessed. Furthermore, if these were to be taught to their students, then the teacher would lose his/her power, become ineffective at their profession, and thus they could no longer practise. In this type of setting, certain therapeutic regimens are expected, even obligated to become extinct following the death of the practitioner. In other systems and areas, this secret information is considered to be family or clan property, and is not taught outside of the group, but passed on to the younger generation as a valued inheritance. However, despite the expectation that family information will be preserved by the next generation, in many locations at present, there is often a lack of interest in carrying on the elder's work in healing, resulting in greater rates of disappearance of this type of information.

Rethinking strategies for protecting traditional knowledge in ethnomedicine

In thinking about how best to 'protect' traditional knowledge, it might be useful to examine the qualities of traditional knowledge that make it somewhat unique. Each of our disciplines looks at this topic through its own set of lenses, which can offer different vistas of the same image. In many cases, projects to protect knowledge have involved significant components that involve documentation. Many of these projects are based in part in academic settings, and an important requirement for funding natural and social science research projects is the use of the scientific method, where hypotheses are put forward and tested as a major component of the project. These hypotheses involve data gathering, then imply that the activity or knowledge can be reduced to a discrete collection of data points, gathered by the scientists or assistants. Once entered into the database, it can be analysed, evaluated and preserved, and publications and websites produced. This is the operating model for much ethnobotanical work, where hypotheses are proposed, use information is gathered based on interviews and observations, and results are evaluated. This reductionist viewpoint assumes that, using modern scientific tools, a collection of individual pieces of data can be reconfigured into a reconstruction, and therefore an understanding of the whole.

As an object of scientific study, perhaps ethnomedical systems are equally, or even more complicated than, say, the remarkably diverse puzzle that is the DNA of a fruit fly. If so, other models of analysis need to be developed that involve a more holistic understanding of the system, rather than one that seeks to reduce it to a collection of parts. For example, when a healer treats a patient complaining of lower back pain, the observing ethnobotanists' response is usually to collect the plant being used, identify it and assign it to a use category, and write a few words about the preparation of the medication. Then it is entered into a database, and, with increasing frequency, the process of collection or even treatment may be filmed.

However, from a medical viewpoint, lower back pain is a symptom of many different conditions. First, it must be categorized as either acute, mechanical lower back pain, as with a lumbar strain, degenerative disc disease or fracture; non-mechanical lower back pain, as with a neoplasia, infection or inflammatory arthritis; or, lower back pain with neurologic signs, such as a herniated disc. A patient history is called for and the patient is asked how motion, posture and rest affect the pain, whether there is fever, weight loss or rash, and whether the presence of visceral disease – vascular, gastro-intestinal or kidney – is evident. The physician has a wide range of possible diagnoses to contend with, including lumbar strain, spondylosis, fractures, congenital diseases, facet joint asymmetry, neoplasm, infection, renal diseases, infection, aortic aneurysm, pancreatitis, cholecystitis, penetrating ulcer and prostatitis, to name a few. A plant that might be used for 'cough' might actually be used to treat seasonal allergies, upper respiratory illness, gastroesophageal reflux, lung cancer, tuberculosis, asthma or chronic obstructive lung disease (R. Lee, pers. comm.). The physician has a lot more at stake than the ethnobotanist – after all, the outcome of a poor ethnobotanical interview is at most the eventual rejection of a manuscript, while in medicine, it may be the loss of the patient. What, then, is the best way to preserve this practice?

This is not to argue against the value of ethnobotanical inventories. In the same way that in many regions of the Earth there exist no inventories of the native and introduced biodiversity, the case is also the same for an ethnobotanical understanding of the area and its people. Just as a checklist of the plants and animals of the regions is a tool for conservation and preservation – not an actual conservation unit in and of itself – an ethnobotanical inventory is also a tool, not an endpoint for preservation of traditional knowledge. Additional actions are required. It would seem appropriately humble, in the case of the preservation of traditional knowledge, to admit that an effective, science-based methodology for ensuring its indefinite preservation does not yet exist. In essence, a study involving the documentation of traditional knowledge or skills is a snapshot in time, freezing our concept of its framework, technologies and use of raw materials. It could be argued that, due to the way in which the body of traditional knowledge is formed – constant experimentation and change, as well as its complexity, the snapshot approach can never be effective in achieving the goal of preservation. For example, under the paradigm utilized by many ethnomedical systems, patients seen by a traditional healer are treated individually and often with different modalities or plant species, even though their conditions might be the same. In many cases, our present efforts comprise little more than producing a list of ingredients that bears little resemblance to the actual product. Each of the modern collection techniques has a place in capturing bits of data, and some of that cache may be appropriate to direct other scientific research, such as in pharmacology and drug discovery, and thus give it immediate value to Western society. From our perspective, we are often interested in saving what we need, and there is certainly benefit to this. It is clear, however, that the most effective way of saving traditional knowledge as a dynamic, living and vital system is to keep it in practice – to encourage its practitioners, to give economic and other importance to its end products, to incorporate its teaching into formal and informal curricula, and to incorporate its ethical values into everyday lives.

How can a scientist contribute to this goal? Perhaps it is time to dissect ethnobotanical methodologies, and see where strengths and weaknesses are found. At a superficial level, greater emphasis needs to be put on data capture using the most modern available tools – including digital videos. We have made films in Belize that chronicle traditional knowledge and beliefs of bush doctors, and these films continue to inspire young people who view them, long after the elders have passed on. Books in local languages, geared to primary health concerns, are extremely important contributions and, again, help keep family lore alive. The creation of a cadre of local ethnobotanists is an extremely worthy goal. Outside scientists have the responsibility of being role models to people they interact with. This can include teaching people to gather data and appreciate the values found in their communities. Local institutions need to be supported as well, and initiated if they do not yet exist. Academic research projects should always leave something behind that has a perceived value to the community. Prior to the initiation of a project, thorough discussions with the community must be held, mutual expectations established and risks and benefits outlined.

In the arena of benefit-sharing, there is a great deal of room for developing innovative strategies that go beyond what is considered the gold standard – a royalty-sharing provision in the contract, along with up-front benefits. Others write about these mechanisms in this book, so I will not dwell on this topic. Cox (2001) outlines a novel benefit-sharing programme that has resulted from his work with an antiviral phorbal isolated from *Homalanthus nutans (G. Forst.) Guill.* in Samoa. Prior to the production of a commercial compound from the plant, there has been over $480,000 supplied to the village of Falealupo, the home of the two traditional healers who taught Cox the use of the plant, as part of what he refers to as the 'Falealupo Covenant'. If a drug is to be developed from the plant extract, the government of Samoa will receive 12.5 per cent of the net profits of the Aids Research Alliance, with 6.7 per cent going to Falealupo village, and 0.4 per cent going to each of the families of the two healers.

Another unusual benefit-sharing programme, resulting from the previously discussed ethnobotanical work in Belize, was derived from the publication and sale of a primary health care manual, *Rainforest Remedies: 100 Healing Herbs of Belize* (Arvigo and Balick, 1993, 1998). As outlined in Johnston (1998), a pension program was devised for the 11 traditional healers that contributed knowledge to the book. Proceeds from the sale of the book are distributed twice per year to the healers – in July and December – through the Traditional Healers Foundation. As of early 2000, the total distributed was over US$20,000. The publisher, Lotus Press, also contributed a portion of its profits from the sale of the book to the Traditional Healers Foundation. The book has been adopted as a primary health care reference by many people in Belize.

However, as the example of Don Elijio Panti given previously shows, there is much more to benefit-sharing than monetary value. The concept of cultural support, expressed by Wolff and Medin (2001) and discussed earlier in this chapter, becomes a very important part of benefit-sharing. It is not limited by the financial resources of the investigator, but rather only by their level of cultural sensitivity, understanding and desire to make a difference.

Finally, I would like to offer an unusual example of how traditional knowledge is being saved – through export to other regions – as people migrate to new islands, countries and continents. Most traditional cultures around the world have diseases or illnesses that are specific to their culture or region. Patients with these conditions seek treatments that are often traditional in their origins, but when symptoms become severe, also present at emergency rooms or to physicians in clinics. Such culture-bound diseases have received a great deal of attention, particularly in the field of psychiatry. The *DSM-IV-TR: Diagnostic and Statistical Manual of Mental Disorders* published by the American Psychiatric Association (2002) contains an appendix, 'Outline for Cultural Formulation and Glossary of Culture-Bound Syndromes', specifically discussing this topic. This document suggests that 'The term culture-bound syndrome denotes recurrent, locality-specific patterns of aberrant behavior and troubling experience that may or may not be linked to a particular DSM-IV diagnostic category'. It further suggests that:

> many of these patterns are indigenously considered to be 'illnesses' or at least afflictions and most have local names. Although presentations conforming to the major DSM-IV categories can be found throughout the world, the particular symptoms, course and social responses are very often influenced by local cultural factors. In contrast, culture-bound syndromes are generally limited to specific societies or culture areas and are localized, folk, diagnostic categories that frame coherent meanings for certain repetitive, patterned and troubling sets of experiences and observations. There

> is seldom a one-to-one equivalence of any culture-bound syndrome with a DSM diagnostic entity.
>
> (American Psychiatric Association, 2002, p. 898)

The manual lists 25 culture-bound syndromes, including *hwa-byung* from Korea, *koro* from South and East Asia, *locura* from Latin America, *mal de ojo* from Mediterranean cultures, *shenkui* from China and *taijin kyofusho* from Japan. *Hwa-byung* describes a Korean 'anger syndrome' and includes symptoms such as panic, dysphoric effect and indigestion. *Locura* refers to a severe form of chronic psychosis in which patients will exhibit incoherence, agitation and sometimes auditory and/or visual hallucinations. *Mal de ojo*, or 'evil eye' in English, is a Mediterranean condition in which children are most vulnerable. It presents with symptoms of crying without apparent cause, insomnia, diarrhoea and/or vomiting.

It is interesting that the American Psychiatric Association has recognized the traditional disease concepts that characterize different cultures as a significant problem in contemporary Western society. Our work on urban ethnobotany in New York City has included much experience with traditional healers in the Dominican community, centred in Washington Heights. Thousands of miles away from their island homes, Dominican traditional healers are practising their trade, and providing an effective, parallel system of health care in the heart of the allopathic medical community. Ethnomedical systems continue to be carried out, and elders are teaching the younger generation. Stores such as *botanicas* are well stocked with plants that are collected in the Dominican Republic and sent to New York City, or grown in local farms on the east coast of the US. Far from being destroyed by Dominican governmental programmes of past decades that minimized the value of traditional Dominican medicine, it is alive and well, both in the diaspora and at home. The same scenario is true for any number of ethnomedical systems of the multitude of cultures that flourish in the US and elsewhere in the world outside of their origins, a sort of 'reverse globalization' in which cultures can take great pride.

There are many actions that could contribute to the preservation of traditional knowledge, from a broad variety of disciplines. So far, top-down international mechanisms have been relatively ineffective. Grass-roots efforts seem to be working in some locations. Attempts aimed at preservation of traditional knowledge are constrained by the lack of significant funding, and the lack of agencies and institutions responsible for supporting this activity. The general nature of funding – for example short-term grants that need to be renewed every few years and contain an innovative 'twist' each time they are resubmitted for consideration – does not lend itself to addressing this problem. It is time for a sincere global commitment to the preservation of traditional knowledge, one that does not get caught up in layers of bio-political bureaucracy. Scientists must rethink how, if at all, their studies and other activities can contribute to keeping traditional practices alive. It is time to enlarge the group of disciplines typically involved in this topic, and identify new ways of approaching an age-old problem that is getting worse with time.

Traditional knowledge is rich in content and heritage, and an important legacy of those who have created it. We must also consider traditional knowledge the foundation on which to practise one's cultural belief system, and thus a basic human right – analogous to religious freedom – deserving of preservation and protection against the contemporary forces that seek to destroy it in so many parts of the world, as well as in our own backyards.

Acknowledgements

This chapter discusses the results from a number of multidisciplinary field projects studying the relationship between plants and people. In Micronesia, this effort focused on botany, ethnomedicine, traditional land management and resource systems, conservation and education. Collaborating groups included the College of Micronesia – FSM, the Continuum Center for Health and Healing at Beth Israel Medical Center, the National Tropical Botanical Garden, the Nature Conservancy, the New York Botanical Garden, Pohnpei Council of Traditional Leaders, Pohnpei State Government and the University of Arizona Program in Integrative Medicine. In Belize, the project focused on traditional medicine and culture, and has involved the collaboration of a number of organizations including the Ix Chel Tropical Research Foundation, Belize Center for Environmental Studies, Faculty of Agriculture and Natural Resources of the University of Belize, Agriculture Research and Development Station in Central Farm, the Belize Zoo and Tropical Education Center, Belize Forestry Department, Belize Association of Traditional Healers, Traditional Healer's Foundation of Belize and the Institute of Economic Botany of the New York Botanical Garden. Since 1977, in Brazil, we have collaborated with the Centro Nacional de Recursos Geneticos, the former Instituto Estadual do Babassu, the Conselho Nacional de Desenvolvimento Cientifico e Tecnologico, and the Fundaçao Nacional do Indio. Any set of long-term efforts, in this case measured in decades, requires the support and commitment of multiple sources in order to be successful. Gratitude is offered to the supporters of these projects over the years including the US National Institutes of Health/National Cancer Institute, the US Agency for International Development, the MetLife Foundation, the Overbrook Foundation, the Edward John Noble Foundation, the Prospect Hill Foundation, the Rex Foundation, the Rockefeller Foundation, the Healing Forest Conservancy, the John D. and Catherine T. MacArthur Foundation, the Gildea Foundation, the Nathan Cummings Foundation, CERC – the Consortium for Environmental Research and Conservation at Columbia University, as well as the Philecology Trust, through the establishment of the Philecology Curatorship of Economic Botany at the New York Botanical Garden. I am very grateful to the persons discussed in the chapter, who so freely provided me with their thoughts and opinions on the nature of ethnobotany and the preservation of traditional knowledge. My thanks go to Chuck Peters for his helpful comments on the original manuscript.

Note

1 See Committee for Creative Non-violence v. Reid, 490 U.S. 730, 753 (1989)(dictum).

References

American Psychiatric Association (2002) *DSM-IV-TR: Diagnostic and Statistical Manual of Mental Disorders*, 4th edn. American Psychiatric Association, Washington, DC.

Arvigo, R. and M.J. Balick (1993, 1998) *Rainforest Remedies: 100 Healing Herbs of Belize*. Lotus Press, Twin Lakes, WI.

Balick, M.J. (1988) 'The use of palms by the Apinaye and Guajajara Indians of Northeastern Brazil', in M.J. Balick (ed.), *The Palm – Tree of Life: Biology, Utilization and Conservation. Advances in Economic Botany*, Vol. 6. The New York Botanical Garden, New York, NY, pp. 1–282.

Balick, M.J. and R. Arvigo (2015) *Messages from the Gods: A Guide to the Useful Plants of Belize*. Oxford University Press/The New York Botanical Garden, New York, NY.

Balick, M.J., M.H. Nee, and D.E. Atha (2000) *Checklist of the Vascular Plants of Belize, With Common Names and Uses. Memoirs of The New York Botanical Garden*, Vol. 85. The New York Botanical Garden Press, Bronx, NY.

Balick, M.J. and Collaborators (2009) *Ethnobotany of Pohnpei: Plants, People and Island Culture*. University of Hawaii Press/The New York Botanical Garden, Honolulu.

Cox, P.A. (2001) 'Ensuring equitable benefits: The Falealupo Covenant and the isolation of the anti-viral drug Prostratin from a Samoan medicinal plant', *Pharmaceutical Biology*, Vol. 39, pp. 33–40.

Dahmer, S., M.J. Balick, A. Hillman Kitalong, C. Kitalong, K. Herrera, W. Law, R. Lee, V.R. Tadao, F. Rehuher, S. Hanser, K. Soaladaob, G. Ngirchobong, M. Besebes, F. Wasisang, D. Kulakowski and I. Adam (2012) *Palau Primary Health Care Manual: Health Care in Palau, Combining Conventional Treatments and Traditional Uses of Plants for Health and Healing*. The New York Botanical Garden, Ministry of Health, Republic of Palau, The Continuum Center for Health and Healing and Belau National Museum. Charleston SC.

Glinski, J.A. *et al.* (1995) 'Inactivation of cell surface receptors by Pheophorbide, a green pigment isolated from *Psychotria acuminata*', *Photochemistry and Photobiology*, Vol. 62, no. 1, pp.144–50.

Harrison, K.D. (2007) *When Languages Die: The Extinction of the World's Languages and the Erosion of Human Knowledge*. Oxford University Press.

Herrera, K., D.H. Lorence, T. Flynn and M.J. Balick (2010) 'Checklist of the Vascular Plants of Pohnpei, Federated States of Micronesia with Local Names and Uses', *Allertonia*, Vol. 10, pp. 1–204.

Hezel, F.X. (2001) *The New Shape of Old Island Cultures*. University of Hawaii Press, Honolulu.

Johnston, B. (1998) 'The new ethnobotany: Sharing with those who shared', *Herbalgram*, Vol. 42, pp. 60–63.

Lee, R., M.J. Balick, D. Lee Ling, F. Sohl, B.J. Brosi and W. Raynor (2001) 'Cultural dynamism and change – an example from the Federated States of Micronesia', *Economic Botany*, Vol. 55, no. 1, pp. 9–13.

McManis, C. (2004), 'Fitting Traditional Knowledge Protection and Biopiracy Claims in the Existing Intellectual property and Unfair Competition Framework', in Burton Ong (ed.), *Intellectual Property and Biological Resources*. Marshall Cavendish Academic, pp. 425–510.

Salick, J., K. Konchar and M. Nesbitt (2014) *Curating Biocultural Collections: A Handbook*. Kew Publishing, Royal Botanic Gardens, Kew.

Schmitz, S. and J.C. Christopher (1997) 'Troubles in Smurftown: Youth gangs and moral visions on Guam', *Child Welfare*, Vol. 76, pp. 411–28.

Vandebroek, I. and M.J. Balick (2012) 'Globalization and loss of plant knowledge: Challenging the paradigm', *PloS ONE*, Vol. 7, p. 5, www.plosone.org/article/info%3Adoi%2F10.1371%2Fjournal.pone.0037643.

Wolff, P. and D.L. Medin (2001) 'Measuring the evolution and devolution of folk-biological knowledge', in L. Maffi (ed.), *On Biocultural Diversity: Linking Language, Knowledge and the Environment*. Smithsonian Institution Press, Washington, DC, pp. 212–27.

17

BIOPROSPECTING AND TRADITIONAL KNOWLEDGE IN AUSTRALIA

Michael Blakeney

Introduction

Australia is one of seventeen countries described as being "megadiverse", supporting more than 70 per cent of the biological diversity on earth (Mittermeier *et al.*, 1989). Australia has some 44,000 species of plants of which 88 per cent occur only in Australia (Coordination Committee, 1994, 12). A report of the Coordination Committee on Science and Technology of the various Australian State and Federal Government Departments noted that 90 per cent of Australian flowering plants, more than 80 per cent of its mammals, 70 per cent of terrestrial birds, 89 per cent of its reptiles, and 93 per cent of its frogs, are endemic to Australia. Similarly, of the estimated 600 inshore fish species in the southern temperate zone, about 85 per cent are found only in Australia.

The isolation of the Australian continent has resulted in the survival of a number of simplified plant and animal types which provide a valuable reservoir of characteristics for the genetic modification of non-Australian plants and animals (e.g. see Beattie, 1995 and Garson, 1997). For example, the aridity and salinity of the continent has resulted in the development of organisms with particular resistance to these conditions (e.g. see Jaradat, 2010). The identification of useful plants and animals in Australia has also been informed by the traditional knowledge of Australia's Indigenous Aboriginal and Torres Strait Islander peoples (e.g. see Barr, 1988; Blakeney, 1997; O'Bryan, 2004; Evans *et al.* 2009).

At the date of the last national census, 30 June 2011, the estimated resident Aboriginal and Torres Strait Islander population of Australia was 670,000 people, out of a total population of 22.3 million people (approximately 3 per cent of the total Australian population) (ABS, 2011). Within the Aboriginal and Torres Strait Islander population, an estimated 606,000 people (90 per cent) were of Aboriginal origin, 38,100 people (6 per cent) were of Torres Strait Islander origin and 25,600 people (4 per cent) were of both origins. Approximately 32 per cent of Aboriginal and Torres Strait Islanders lived in New South Wales, 28 per cent in Queensland, 13 per cent in Western Australia and 10 per cent in the Northern Territory.

Nationally in 2013, Indigenous-owned or controlled land comprised 16 per cent of the area of Australia (SCRGSP, 2014, table 9A.2.1). Nearly all (97.9 per cent) Indigenous-owned or controlled land was in very remote areas of Australia (SCRGSP, 2014, table 9A.2.2), which are often areas of high biodiversity.

This chapter will examine the impact of national legislation and the legislation of the states and territories upon bioprospecting and the extent to which that legislation acknowledges the rights and entitlements of Indigenous Peoples. The chapter will examine some Australian case studies of bioprospecting utilizing the traditional knowledge of Aboriginal and Torres Strait Islander peoples and will conclude with a review of some bioprospecting protocols which have been developed by Australian Indigenous Peoples.

Australia's international bioprospecting obligations

Australia ratified the Convention on Biological Diversity (CBD) in 1993 and the International Treaty on Plant Genetic Resources for Food and Agriculture (International Treaty) in 2005. Both of these instruments required some action by signatories to acknowledge or protect traditional knowledge (TK). Article 8 of the CBD provides that each Contracting Party shall, as far as possible and as appropriate:

> (j) Subject to its national legislation, respect, preserve and maintain knowledge, innovations and practices of indigenous and local communities embodying traditional lifestyles relevant for the conservation and sustainable use of biological diversity and promote their wider application with the approval and involvement of the holders of such knowledge, innovations and practices and encourage the equitable sharing of the benefits arising from the utilization of such knowledge, innovations and practices.

Article 9.2(a) of the International Treaty requires that "each Contracting Party should, as appropriate, and subject to its national legislation, take measures to protect and promote Farmers' Rights", including the "protection of traditional knowledge relevant to plant genetic resources for food and agriculture".

For the most part, the provisions of the CBD and the International Treaty concerned with traditional knowledge have not been implemented in Australia.

In October 2010 the Conference of Parties to the CBD concluded the Nagoya Protocol on Access to Genetic Resources and the Fair and Equitable Sharing of Benefits Arising from their Utilization (ABS) to the Convention on Biological Diversity, which provided (in Art.5.5) that each Party "shall take legislative, administrative or policy measures, as appropriate", in order that the benefits "arising from the utilization of traditional knowledge associated with genetic resources" are shared in a fair and equitable way with Indigenous and local communities holding such knowledge. Australia signed the Protocol in January 2012, which came into force on 12 October 2014 following its ratification by 51 State parties to the CBD. Australia has yet to ratify it due to Commonwealth and State and Territory laws being currently inadequate to give effect to the Protocol domestically. The Federal Department of Sustainability, Environment, Water, Population, and Communities has issued an invitation to the public to provide its views on the best way to implement the Nagoya Protocol in Australia.[1] This invitation remains open.

Australia, together with New Zealand, Canada and the United States, voted against the Resolution adopting the United Nations Declaration on the Rights of Indigenous Peoples (UNDRIP) adopted by 143 member states in the General Assembly of the United Nations on September 13, 2007. Article 31 of UNDRIP recognized the rights of Indigenous Peoples to maintain, control, protect and develop their intellectual property over their cultural heritage, traditional knowledge and traditional cultural expressions. Australian Senator Marise Payne explained the various reasons for the Australian Government's opposition to the Declaration, pointing out that "as our laws here currently stand, we protect our Indigenous cultural heritage, traditional knowledge and traditional cultural expression to an extent that is consistent with both Australian and international intellectual property law, and we are not prepared to go as far as the provisions in the text of the draft declaration currently do on that matter" (Payne, 2007). She also indicated the Australian Government's opposition to "the inclusion in the text of an unqualified right of free, prior and informed consent for Indigenous Peoples on matters affecting them" because the text did "not acknowledge the rights of third parties – in particular, their rights to access Indigenous land and heritage and cultural objects where appropriate under national law". With the change of government in Australia, Prime Minister Kevin Rudd announced on 3 April 2009 Australian support for the Declaration (UNNC, 2009). Since that date, Australia has had a number of changes of government but with no action to implement UNDRIP.

Australia has been an active participant in the deliberations within the Intergovernmental Committee on Intellectual Property and Genetic Resources, Traditional Knowledge and Folklore ("IGC") of the World Intellectual Property Organization (WIPO) for treaties concerning TK and genetic resources (GRs). To inform its negotiating position, in November 2013 IP Australia, the federal intellectual property office, launched a public consultation on Indigenous Knowledge.[2] It asked "How should Australia protect Indigenous Knowledge?" However, the breadth of this question was qualified by the explanation that "We want your views about how Indigenous Knowledge can work with the intellectual property (IP) system". Further IP Australia explained that

> There's no point making change for change's sake. The IP system should be useful and fair for all Australians. The IP system needs to work for you. That's why we want to hear your views.[3]

In other words the consultation has been framed with an IP, although a soft law solution is suggested by the observation that

> Many communities and their representative bodies have protocols for managing Indigenous Knowledge. Some cultural, government and research organisations also have protocols. Do these protocols provide sufficient protection for Indigenous Knowledge? Can IP better protect Indigenous Knowledge?[4]

Australian bioprospecting legislation and policy initiatives

Australia is a federation of six states[5] and two territories.[6] Australia and each of its states and territories have their own parliaments. Bioprospecting in Australia is regulated by Commonwealth (Federal) legislation, together with statutes of some states and the Northern Territory, as well as by protocols which have been developed by some Aboriginal and Torres Strait Islander peoples.

The land area of Australia is made up of its states and territories, together with some federally controlled defence establishments and national parks. The seas around Australia from the low water mark are within the jurisdiction of the national Parliament. Superimposed upon land ownership in Australia are the "native title" rights of Aboriginal and Torres Strait Islander peoples, recognized for the first time by the High Court of Australia in *Mabo v Queensland (No. 2)*[7] to the occupation, use and enjoyment of their traditional lands. The recognition and protection of native title rights has been provided a legislative framework through the enactment of the Federal Native Title Act 1993 (Cth) and equivalent state and territory legislation.

An illustration of the bioprospecting implications of this native title legislation is the High Court of Australia's decision in *Northern Territory of Australia v Arnhem Land Aboriginal Land Trust*[8] which concerned the Aboriginal Land Rights (Northern Territory) Act 1976 (Cth). This Act conferred rights upon Aboriginal Peoples in the Northern Territory to exclude from tidal waters persons who wished to take fish or aquatic life in those waters, including persons holding a licence under the Fisheries Act 1988 (NT). The Yolngu people who were the traditional owners of parts of Blue Mud Bay in northeast Arnhem Land were confirmed in their rights to exclude fishermen and others from the waters of Blue Mud Bay.

An example on the native title rights of Aboriginal Peoples to participate in bioprospecting-like activities in the High Court decision in *Yanner v Eaton*[9] which concerned a member of the Gunnamulla clan of the Gangalidda tribe of Aboriginal Australians from the Gulf of Carpentaria area of Queensland. He used a traditional form of harpoon to catch two juvenile estuarine crocodiles in Cliffdale Creek in the Gulf of Carpentaria. He and other members of his clan ate some of the crocodile meat and the rest was stored at his home, in breach of the Fauna Conservation Act 1974 (Qld). A magistrate dismissed the prosecution on the basis that the hunter was exercising his traditional hunting rights. The High Court upheld the rights of the hunter as exercising or enjoying his native title rights and interests as preserved by the Native Title Act 1993 (Cth).

It should be noted that the right of Aboriginal Peoples to engage in bioprospecting is linked with their right of access to the relevant land. This principle is derived from the High Court's 2002 decision in *Mirriwung Gajerong* (*Western Australia v Ward*).[10] One of the issues which had been canvassed in that case was whether access to sacred sites, where artworks on rock were located, or ceremonies were performed by Aboriginal Peoples could be regarded as an incident of native title. This was rejected in the majority judgment of the High Court as the assertion of "something approaching an incorporeal right akin to a new species of intellectual property".[11] The judgment expressed the concern that the "'recognition' of this right would extend beyond denial or control of access to land held under native title".[12] The majority judgment noted that in relation to the cultural knowledge of Aboriginal peoples, "the law respecting confidential information, copyright, and fiduciary duties" provided some protection.[13] This judgment was taken to suggest that "only those native title holders with exclusive rights to land will ever be able to assert 'ownership' rights, including potential rights to protect cultural knowledge of the uses of bio-resources" (Meyers and Owoeye, 2013, p. 66).

Environment Protection and Biodiversity Conservation Act 1999 (Cth) (EPBC Act)

The piece of Commonwealth legislation most relevant to bio-prospecting is the Environment Protection and Biodiversity Conservation Act 1999 (the EPBC Act), which provides "a legal framework to protect and manage nationally and internationally important flora, fauna, ecological communities and heritage places" (DoE, 2015). Pursuant to an agreement with the Australian states and territories, the EPBC Act focuses Australian Government interests on the protection of matters of national environmental significance, with the states and territories having responsibility for matters of local significance (CAOG, 1997).

The only mentions of Indigenous Peoples in the EPBC Act are the continuation of the traditional use of Commonwealth reserves by Indigenous persons "for hunting or food-gathering (except for purposes of sale); or ceremonial and religious purposes".[14] Both of these categories of use are also excluded by the EPBC Act from Australia's obligations under the Convention on International Trade in Endangered Species (CITES).[15]

Regulations for the access provisions in section 301 of the EPBC Act were implemented on 1 December 2005 and form Part 8A of the Environment Protection and Biodiversity Conservation Regulations 2000 (Cth) (the EPBC Regulations). The purpose of these regulations is provided in regulation 8A.01 emphasizing the nationally consistent approach to accessing biological resources in Australia, "ensuring the equitable sharing of the benefits arising from the use of biological resources"[16] in Commonwealth areas, and "recognizing the special knowledge held by indigenous persons about biological resources".[17]

Permits are required to access biological resources,[18] but only access for commercial or potentially commercial purposes will require a benefit-sharing agreement. This must then be obtained with the prior informed consent of the owner of the land – where that land is Indigenous peoples' land, and where the access provider is the owner of that land.[19] Benefit-sharing agreements must provide for the recognition, protection and valuing of any Indigenous Peoples' knowledge that will be used as part of the access.[20] Among the other details, the benefit-sharing agreement must include:

(h) a statement regarding any use of indigenous people's knowledge, including details of the source of the knowledge;
(i) a statement regarding benefits to be provided or any agreed commitments given in return for the use of the indigenous people's knowledge;
(j) if any indigenous people's knowledge of the access provider, or other group of indigenous persons, is to be used, a copy of the agreement regarding use of the knowledge (if there is a written document), or the terms of any oral agreement, regarding the use of the knowledge.[21]

An applicant seeking an access permit must recognize and provide evidence of agreements to use the TK of the access provider and any other groups of Indigenous persons providing information in relation to the biological resource for which access is being sought. Thus far only a small number of permits have been granted for commercial purposes.[22] The majority are for non-commercial purposes which do not require a benefit-sharing agreement. It should also be noted that this regime only covers Commonwealth areas.

Plant Breeder's Rights Amendment Bill 2002 (Cth)

As part of the negotiations surrounding the passage through the Australian Senate of the Plant Breeder's Rights Amendment Bill 2002, the Australian Democrat representatives proposed amendments to the Plant Breeder's Rights Act 1994 (Cth) to "reduce the chances of bio piracy from indigenous land and increase the capacity of the indigenous community to object when bio piracy is occurring..."[23] A proposed section 42(4) was to provide that "If a plant variety is a traditional landrace variety or a variety which is essentially derived from such landrace plant variety, PBR must not be granted to that variety". "Traditional landrace variety" was defined as a: "variety developed over millennia by selecting favourable characteristics within a cultivated crop species, or a variety that is the outcome of Indigenous Peoples' traditional and customary innovations and practices". A further amendment was proposed to be made at section 42(5), which was intended to prevent private PBRs being created other than with the consent of Indigenous communities. That proposal was to apply in respect of plant varieties discovered on Crown lands, national parks, world heritage sites and native title lands. The proposed amendments were not able to attract sufficient support to be passed into law (see Woodall and Cox, 2014, p. 6).

Queensland

The state of Queensland has enacted the Biodiscovery Act 2004 (Qld), which commenced on 12 November 2004. The objectives of this Act were to provide for streamlined and sustainable access to the State's native biological resources for the purposes of biodiscovery while returning a fair and equitable benefit to the state, for the benefit of all Queenslanders.[24]

The objectives are achieved through the provision of: a regulatory framework for taking and using native biological resources; a contractual framework for benefit-sharing; a compliance code and collection protocols; and monitoring and enforcement provisions, but this legislation does not consider the use of traditional or Indigenous knowledge in its access or benefit-sharing provisions. This has been left to the Queensland Code of Ethical Practice for Biotechnology 2001 (Qld). The Code is mandatory for three types of organizations:

(a) Queensland Government agencies, research centres, laboratories and public hospitals that conduct biotechnological activities;
(b) private sector companies, academic institutions and research bodies that receive financial assistance from the Queensland Government to undertake biotechnological activities; and
(c) co-operative research centres (CRCs) that receive financial assistance from the Queensland Government to undertake biotechnological activities, and all CRCs that conduct biotechnological activities that have a Queensland Government body or officer as a participating member.[25]

These organizations are defined as "Queensland Biotechnology Organisations". All other organizations undertaking biotechnological activities without Queensland Government involvement can voluntarily subscribe to the Code. In relation to those organizations for which the Code applies, Article 11 deals with access to Queensland's biological resources. The prior informed consent of the landowner is to be obtained before samples are collected from privately owned land. Further, a reasonable benefit-sharing arrangement is to be negotiated with that landowner in return for access to those samples.

Section 121 of the Act stipulated that a review of its operation had to be conducted within five years of its commencement to decide whether its provisions remained appropriate. This initial review was conducted in 2009 and the government response was published in July 2010. In the 2009 review it was recommended, and the Government supported, subsequent five yearly reviews to ensure that the Act continued to meet the aims and accommodates emerging trends and international developments. The following terms of reference for the 2015 review were established with particular regard to:

(a) consideration of whether the current legislation is the most efficient means to achieve the policy objectives and if not, options for other mechanisms to achieve the objectives. In considering other options, gather evidence of the impacts of the other options on the regulated community to allow comparison to the current legislation and if there were no regulation;
(b) consideration of developments internationally and nationally in relation to the implementation of the Nagoya Protocol and Australia's Biodiversity Conservation Strategy (ABCS) since the commencement of the Act and alignment with and any impact on its application;
(c) examination of developments in native title law, traditional knowledge and changes to IP law that may affect ownership of genetic resources.[26]

Northern Territory

Section 122 of the Territory Parks and Wildlife Conservation Act 2002 (NT), permits Indigenous people in the Northern Territory to use their traditional country in accordance with Aboriginal tradition, for hunting, food gathering and for ceremonial and spiritual purposes. However, anyone taking native wildlife for commercial purposes requires authorization under the Act (s.55) and the failure to obtain a permit is a criminal offence under the Act (s.67). "Commercial purpose", in relation to an animal or plant, is defined in s.9(1) to mean "the keeping, breeding, displaying, moving or other dealing with or use of the animal or plant for the purposes of selling, trading or bartering with the animal or plant or of otherwise earning a livelihood or making a profit, and includes the use of the animal or plant for scientific purposes".

Bioprospecting in the Northern Territory is regulated by the Biological Resources Act, 2006 (NT). Bioprospecting is defined in section 5(1) of the Act as "the taking of samples of biological resources, existing *in situ* or maintained in an *ex situ* collection of such resources, for research in relation to any genetic resources, or biochemical compounds, comprising or contained in the biological resources".

Biological resources are defined in section 4(1) of the Act to include "genetic resources, organisms, parts of organisms, populations and any other biotic component of an ecosystem with actual or potential use or value for humanity" and "genetic resources" are defined in the same provision to mean "any material of plant, animal, microbial or other origin that contains functional units of heredity and has actual or potential value for humanity".

Excluded by section 5(2) from bioprospecting under the Act are the taking of biological resources from an area of land or water by Indigenous people who have traditionally used the area of land or water in accordance with aboriginal tradition for hunting, food-gathering (other than for sale) and for ceremonial and religious purposes; dealing with any biological material of human origin; taking samples of biological resources that have been cultivated or tended for a purpose other than biodiscovery and where the samples are not to be used for biodiscovery; taking samples of biological resources specified in a Ministerial declaration under section 10; taking samples of a biological resource that is: (i) a genetically modified organism for the purposes of section 10 of the Gene Technology Act 2000 (Cth); or (ii) a plant variety for which a Plant Breeder's Right has been granted under section 44 of the Plant Breeder's Rights Act 1994 (Cth); and taking harvested fish.

"Biodiscovery" is defined in section 4(1) to mean "research on samples of biological resources, or extracts from those samples, to discover and exploit genetic or biochemical resources of actual or potential value for humanity".

A person who wishes to engage in bioprospecting in the Territory is required by section 11 to apply for a permit. A criminal penalty is imposed under section 38(1) for bioprospecting without a permit and section 38(2) provides that a person is taken to engage in bioprospecting "if there is a reasonable prospect that biological resources taken by the person will be subject to research and development on any genetic resources, or biochemical compounds, comprising or contained in the biological resources".

Section 27(1) requires that a bioprospector must enter into a benefit-sharing agreement with each resource access provider in relation to the resources to be taken under a permit. Section 27(3) provides that a benefit-sharing agreement is not valid unless the resource access provider has given prior informed consent to the terms of the agreement.

In considering whether a resource access provider has given informed consent, section 28 requires the following to be considered: (a) whether the resource access provider had adequate knowledge of this Act and was able to engage in reasonable negotiations with the applicant for the permit about the benefit-sharing agreement and (b) whether the resource access provider was given adequate time to consult with relevant people; and with traditional owners if the biological resources are in an area that is Aboriginal land; and (c) whether the resource access provider has received independent legal advice about the application and requirements of the Act.

Western Australia

In December 2000, the Law Reform Commission of Western Australia (the Commission) received a reference to inquire into and report upon Aboriginal customary laws in Western Australia and consider whether, and if so how, those laws should be recognized within the Western Australian legal system. As part of the research gathering phase of the project 15 background papers on different areas of interaction between Australian law and Aboriginal law and culture were also commissioned. These included a background paper on Indigenous cultural and intellectual property and customary law (Janke and Quiggin, 2005 at 451–506) which was addressed in Part VI of the Final Report (LRCWA, 2006). In relation to bioprospecting the Commission recommended that, at the earliest opportunity, the Western Australian government should develop protocols aimed at "addressing issues that arise from the 'bioprospecting' of Aboriginal knowledge" and these protocols should aim to safeguard Indigenous cultural and intellectual property by ensuring that those who seek to benefit from traditional cultural knowledge:

(1) undertake direct consultation with Aboriginal people as to their customary law and other requirements;
(2) ensure compliance with Aboriginal peoples' customary law and other requirements;
(3) seek free, prior and informed consent for the use of any Aboriginal knowledge from the custodians of that traditional knowledge;
(4) seek free, prior and informed consent for access to Aboriginal land for any purposes, including collection;
(5) ensure ethical conduct in any consultation, collection or other processes;
(6) ensure the use of agreements on mutually agreed terms with Aboriginal people for all parts of the process;
(7) devise equitable benefit-sharing arrangements; and
(8) acknowledge the contribution of Aboriginal peoples. (LRCWA, 2006, Recommendation 81).

This recommendation arose out of the Commissions observation that it understood that this was important to Aboriginal people and that "unregulated bioprospecting could represent a lost opportunity for some Aboriginal communities to capitalize on their traditional knowledge and to develop the community's economic base".[27]

On 25 November 2015, the Biodiversity Conservation Bill 2015 was introduced to the Legislative Assembly of the Western Australia Parliament by the Minister for Environment, to repeal and replace the Wildlife Conservation Act 1950. The Bill was passed on 13 September 2016 and became the Biodiversity Conservation Act 2016 upon receiving Assent on 21 September 2016. Although biodiversity conservation was the primary focus of the Bill, it contained some provisions dealing with bioprospecting. Clause 256 envisages the promulgation of regulations granting bioprospecting licences subject to a condition "authorising bioprospecting activity that requires the licence holder to enter into an arrangement with the CEO or another person for the sharing of profits". "CEO" is defined in clause 5 of the Bill to mean the chief executive officer of the Department of Conservation and Land Management. "Bioprospecting activity" is defined in clause 5 to mean "an activity involving or related to the taking of fauna or flora for the purposes of identifying, extracting or recovering biological resources".

In a review of this Bill, published in February 2016, the Environmental Defender's Office (EDO) of Western Australia was critical of the fact that the Bill did not include as one of its objectives, "an object to ensure the fair and equitability sharing of benefits resulting from the use of genetic resources/bioprospecting" (EDO, 2016, p. 4). It instanced as best practice South Africa's National Environmental Management: Biodiversity Act 2004, which contains extensive provisions for benefit-sharing and which makes explicit that this was one of the objects of that Act (EDO, 2016, p. 30). Finally, the EDO pointed out that there were "no details provided on how benefit-sharing and necessary consultation will occur, including with Indigenous people, over bioprospecting" (EDO, 2016, p. 31).

National Strategy for the Conservation of Australia's Biodiversity

In 1996, the then Commonwealth Department of the Environment, Sport and Territories published a National Strategy for the Conservation of Australia's Biological Diversity (the National Strategy) (DEST, 1996). This Strategy was prepared by the Australian and New Zealand Environment and Conservation Council, a Ministerial Council that existed between 1991 and 2001. The National Strategy included a clear recognition of the contribution and rights of Indigenous Peoples over such genetic resources. The ninth objective of the National Strategy reflected the language of the CBD in declaring that:

> The close, traditional association of Australia's Indigenous peoples with components of biological diversity should be recognized, as should the desirability of sharing equitably benefits arising from the innovative use of traditional knowledge of biological diversity.

Objective 1.8 of the National Strategy required Australia to "[r]ecognise and ensure the continuity of the contribution of the ethno-biological knowledge of Australia's Indigenous Peoples to the conservation of Australia's biological diversity" (DEST, 1996, ch.1). The use of TK could proceed only "with the cooperation and control of the traditional owners of that knowledge". The National Strategy was reviewed in 2001 (DoE Review, 2001). It was observed that Objective 1.8 had not been achieved. The National Strategy was replaced with Australia's Biodiversity Conservation Strategy for 2010–30.[28] One of the priorities of the Biodiversity Conservation Strategy 2010–30 was the increased engagement of Indigenous people, pointing out that:

> Indigenous peoples have a special connection and relationship with Australia's natural environments. Accordingly, the important role of Indigenous traditional ecological knowledge in conserving Australia's biodiversity needs to be more actively promoted to other biodiversity managers. This transfer of knowledge needs to be two-way – it is also important that Indigenous peoples have access to scientific knowledge and best practice for natural resource management. In addition, traditional ecological knowledge is continually evolving and Indigenous peoples need support for the recording, ongoing development and intergenerational transfer of Indigenous knowledge.[29]

The three outcomes for increasing Indigenous engagement, envisaged in the Biodiversity Conservation Strategy 2010–30, were an increase in: (i) the employment and participation of Indigenous Peoples in biodiversity conservation activities; (ii) the use of Indigenous knowledge in biodiversity conservation decision-making; and in the extent of land managed by Indigenous Peoples for biodiversity conservation.[30] No reference was made to benefit-sharing from the utilization of the knowledge of Indigenous Peoples.

It was envisaged that the Biodiversity Conservation Strategy 2010–30 would be reviewed every five years. The 2015 review purports to align the National Biodiversity Strategy and Action Plan with the Aichi Biodiversity Targets[31] of the CBD's Strategic Plan for Biodiversity 2011–20.[32]

Research protocols

In May 1998 the then Australian Science, Technology and Engineering Council (ASTEC) produced its National Principles and Guidelines for the Ethical Conduct of Research in Protected and Environmentally Sensitive Areas. These Principles provide for research that is conducted in "protected and environmentally sensitive areas" to be accountable to the concerns and interests of Aboriginal and Torres Strait Islander peoples. This includes, amongst other things, ensuring proper negotiations with Traditional Owners, obtaining informed consent from Traditional Owners before commencing research, and acknowledging Aboriginal and Torres Strait Islander peoples' right to control use of their knowledge and intellectual property and to share in any profits derived from the use of their knowledge.[33]

The Australian Institute of Aboriginal and Torres Strait Islander Studies (AIATSIS)[34] Research Ethics Committee is responsible for reviewing all AIATSIS research projects involving Aboriginal and Torres Strait Islander participants to ensure the appropriate ethical standards have been met. In 2002 AIATSIS published its Guidelines for Ethical Research in Indigenous Studies.[35] These were subsequently revised in 2010 and 2012. The Guidelines import most of the objectives and principles of the traditional knowledge and traditional cultural expressions texts currently being negotiated by the World Intellectual Property Organization (see Blakeney, 2015).

The Guidelines require respect for Indigenous knowledge, practices and innovations (principle 5) and consultation, negotiation and free, prior and informed consent of Indigenous Peoples (principle 6) to ensure that Indigenous people are equal participants in the research process. Principle 7 provides that responsibility for consultation and negotiation is ongoing and that this should achieve mutual understanding about the proposed research (principle 8), resulting in a formal agreement for the conduct of a research project (principle 9).

The National Health and Medical Research Council (NHMRC) has developed Values and Ethics: Guidelines for Ethical Conduct in Aboriginal and Torres Strait Islander Health Research (2003).[36] The Guidelines point out that the construction of ethical relationships between Aboriginal and Torres Strait Islander peoples on the one hand and the research community on the other "must take into account the principles and values of Aboriginal and Torres Strait Islander cultures".[37] The Guidelines require "a fair distribution of the benefits and burdens of participation in research and, for any research participant, a balance of burdens and benefits".[38]

Bioprospecting case studies

Smokebush (Western Australia)

The *Background Paper on Indigenous cultural and intellectual property and customary law*, which was prepared for the Law Reform Commission of Western Australia, gave a number of instances of bioprospecting in Western Australia involving the TK of Indigenous people (Janke and Quiggin, 2005). The best-known example concerned the commercial exploitation of Smokebush

(*Conospermum sp.*), a flowering shrub commonly found in Western Australia which had been used by the Nyoongah people for generations as a traditional medicine (citing Fourmile, 1996). In the 1960s, the Western Australian government had granted the United States National Cancer Institute (NCI) a licence to collect plants for screening for the presence of cancer-fighting properties.

The specimens were found to be ineffective, but they were held in storage until the late 1980s when they were tested again in the quest to find a cure for AIDS. Smokebush was one of four plants found to contain the active property Conocurvone, which laboratory tests showed could destroy the HIV virus in low concentrations (Janke, 1999, p. 24).

The *Background Paper* reported that NCI had a licence to collect more samples in Western Australia, but that a contract could not be agreed with the Western Australian Department of Conservation and Land Management (CALM). CALM attempted unsuccessfully to negotiate a contract with the NCI (Janke and Quiggin, 2005, p. 487). It was reported that after the breakdown of these contract negotiations a collector was found at Melbourne Airport (in the east of Australia) attempting to leave the country with two cases filled with of smokebush and other plants (reported in Parke and Kendall, 1997 p. 13, citing Armstrong and Hooper, 1994 p. 14, quoted in Janke and Quiggin, 2005 p. 487). Apparently, the "claims of biopiracy and attendant negative publicity for the NCI that followed these events led ultimately to the NCI granting a Victorian pharmaceutical company, Amrad, an exclusive worldwide licence to develop Conocurvone" and it was reported that an agreement was made between CALM and Amrad for Western Australian scientists to be involved in the research on production and preparation of Conocurvone and for the receipt of royalties from any commercial drug development where the leading compound is Conocurvone (Parke and Kendall, 1997 pp. 13–14).

A search of the database of the US Patents and Trademark Office reveals a number of patents involving *Conospermum*. The abstracts of three patents, dating from 1993,[39] describe the respective inventions as providing "novel antiviral naphthoquinone compounds, which may be isolated from plants of the genus Conospermum" and which "may be used alone or in combination with other antiviral agents in compositions … to inhibit the growth or replication of a virus, such as a retrovirus, in particular a human immunodeficiency virus, specifically HIV-1 or HIV-2". The assignee of the inventions is "The United States of America as represented by the Secretary of the Department of Health and Human Services".

In the "detailed description of the preferred embodiments" each of the patent documents report that "all of the known species in the genus *Conospermum* are endemic to the Australian continent" and that the plants from which the extracts leading to the inventions were obtained "were collected in 1981 by R. Spjut (under an NCI/USDA interagency agreement) in the Gairdner mountain range of western Australia, midway between Geraldton and Perth".

These patents formed the basis of an international patent application under the Patent Cooperation Treaty.[40] This application designated Australia, Canada and Japan and the member countries of the European Patent Organization. The European patent was granted on 7 May 1997.[41] The application entered the national phase in Australia on 31 January 1994, but the application was not proceeded with and ceased 31 January 2009.[42]

Richard W. Spjut was a distinguished ethnobotanist employed by the NCI to collect plant specimens from around the world for its cancer screening programme, for ten years until 1982. He first visited Western Australia on 20 August until 31 October 1981, when he collected six species of smokebush (Spjut, 1981). In his report he made no reference to any Aboriginal assistance he might have received during his collecting expedition and only refers to the assistance of six students from Murdoch University and the University of Western Australia (Spjut, 1981 p. 7), as well as the taxonomic assistance he received from the WA Herbarium in Perth (Spjut,

1981 p. 8). In his report on this trip Spjut refers to the winding-down of the NCI anticancer screening program (Spjut, 1981 p. 7). In May 1983, following the termination in October 1982 of the cooperative agreement between the NCI and the US Department of Agriculture's Agricultural Research Service, Spjut founded World Botanical Associates (WBA) in Laurel, Maryland to collect plants for pharmaceutical research (Spjut, 2006).

Despite the US patent documents referring to the derivation of antiviral naphthoquinone compounds from plants of the genus *Conospermum* collected by Spjut in 1981, he wrote that it was the WBA procurement efforts during 1991–92 which facilitated the discovery of conocurvone (Spjut, 2006). He reported on a trip to Western Australia between Aug 25–Sep 25, 1992 on behalf of WBA when "ground and aerial reconnaissance has been conducted of the coastal areas of Western Australia from Perth north to Kalbarri to assess the available biomass of *Conospermum stoechadis, C. incurvum*, and closely related species" and to obtain samples of these species (Spjut, 1992). In his report of this trip Spjut wrote that "the first week was spent obtaining photocopies of locality data on herbarium records from Perth" and that specimens at the Perth Herbarium were photographed and their location marked on a road map. Field work was undertaken from Perth north to Kalbarri National Park by four-wheel drive land cruiser, supplemented by aerial surveys. No mention is made in this report of any discussion with Aboriginal persons on any matter.

The first mention of *Conospermum* as a potential source of HIV/AIDS medicine occurred in two publications in 1993, which was contemporaneous with the first US patent application (Decsoterd *et al.*, 1993; Cragg *et al.*, 1993). Almost all of the authors of these publications were named as inventors on this patent application.[43] No mention is made in them of Aboriginal medical knowledge being used as an aid to this research. The *New Scientist* in 1994 mentions the negotiating difficulties between the Western Australian government and NIC in relation to the acquisition of Smokebush samples (Miller and Dayton, 1993). The first mention of smokebush as an Aboriginal medicine is a 1994 article in the WA Department of Conservation and Land Management journal, *Landscope* (Armstrong and Hooper, 1994). This was followed by Henrietta Fourmile's 1996 paper on the subject, which was quoted by the WA Law reform Commission *Background Paper*. She is an Aboriginal Australian from the Yidinji tribe from the Cairns region in eastern Australia and was apparently working in Cairns and studying at Macquarie University in Sydney until 1997, when she was appointed to the United Nations Secretariat for the Convention on Biological Diversity (Bancroft, 1994).

Kerr (2010 p. 45) questions why it was not until 2000 that it was mentioned that smokebush was used as a traditional Aboriginal medicine. A 2006 scoping study on the Aboriginal Traditional Knowledge in remote central Australian areas refers to the use of smokebush by Aboriginal Peoples for "a variety of therapeutic purposes" (Smallacombe, Davis and Quiggin, 2006 p. 39). It refers to smokebush being "collected and screened for scientific purposes by the US National Cancer Institute, under license from the West Australian Government" in the 1960s (*ibid.*, pp. 39–40). However, there is no evidence of any collecting by the NCI in the 1960s, as the 1981 expedition by Spjut appears to have been the first such venture and there does not appear to be any documented evidence of this therapeutic use.

Although the claims about the relationship between Aboriginal traditional knowledge and the bioprospecting of smokebush appear to be unproven, the 2006 scoping study and the WA Law Reform Commission's *Final Report* indicate that these claims are playing an important role in supporting claims for the recognition of the role which Aboriginal TK might make in the identification of useful biological resources. In any event the government of Western Australia moved in 1993 (following the promulgation of the CBD) to amend the Conservation and Land Management Act 1984, by expanding the functions of the CEO of the public service department administering the Act to "promote and encourage the use of flora for therapeutic, scientific or horticultural purposes for the good of people in this State or elsewhere, and to undertake any project or operation relating to the use of flora for such a purpose".[44] Pursuant to this provision, subsection (6) was inserted into section 31 of the 1984 Act to permit the Minister and the CEO to

> enter into an agreement with another person under … this Act to grant, issue or refuse licences or permits to take or remove forest produce or other flora in a manner that has the effect of conferring on the other person an exclusive or preferential right to take or remove forest produce or other flora referred to in the agreement from land (other than private land) referred to in the agreement.

In his Second Reading Speech introducing the amending legislation to the Legislative Assembly, the Minister for the Environment, the Hon Kevin John Minson, who was also the Minister for Aboriginal Affairs, made no reference to any Aboriginal traditional knowledge being utilized in the identification of Western Australian biological resources.[45] He explained that

> There are numerous examples where the inability of Governments to control the utilization of native plants has resulted in their exploitation in a way that has resulted in no benefit to the country in which the plants exist, and in some cases has led to the destruction of the species because of over-exploitation. The amendments contained in this Bill are required to ensure that Western Australia retains control of the utilisation of native plants, particularly where they have the potential to yield pharmaceutical compounds.[46]

The Members of Parliament in the debate on the Bill in the Legislative Assembly and the Legislative Council referred to the necessity to preserve endangered species and to the bio-prospecting activities of NCI, but made no mention of Aboriginal traditional knowledge.[47]

The Kuuku I'yu Project, Central Cape York Peninsula, Queensland

The homelands of the Kuuku I'yu Northern Kaanju people are centred on the upper Wenlock and Pascoe Rivers in Central Cape York Peninsula, Queensland. Their homelands have been identified as containing "a huge diversity of flora and fauna, including several rare, threatened and endangered species" (Smith and Claudie, 2003, referred to in Claudie *et al.*, 2012 p. 30). In 2002 the Kuuku I'yu people formed the Chuulangun Aboriginal Corporation (CAC)[48] to represent their interests.

The CAC identified the natural resources of Kuuku I'yu homelands and the detailed traditional ecological knowledge held by its people as an opportunity "to address issues of social disadvantage, economic hardship and loss of connection amongst their people, and to sustain them into the future" (Claudie *et al.*, 2012, p. 36). A project was put together by the CAC between the Kuuku I'yu people in relation to the development of their traditional medicinal plants by researchers from the University of South Australia (UniSA). A research agreement was signed between the CAC and UniSA protecting the intellectual property rights of the Kuuku I'yu people and providing for benefit-sharing mechanisms in the event of successful commercialization (Claudie *et al.*, 2012, p. 39). The Kuuku I'yu people proceeded to work closely with the university-based researchers at all stages of the project from collecting plant materials, to preparing traditional medicines for testing, and making decisions to commercialize the medicines (Claudie *et al.*, 2012, pp. 50–51).

The first plant to be evaluated by UniSA researchers is Uncha (*Dodonaea polyandra* (Sapindaceae) used for medicinal purposes by particular Kuuku I'yu traditional owners to treat pain and discomfort in the mouth from toothache and infection (described in Simpson, 2010). The plant was first collected in 2006 and following laboratory tests for anti-inflammatory, anti-bacterial and cell-toxicity effects the research partners in 2009 decided to proceed with two provisional patent applications. The first of these covered an extract and extraction process for the Uncha plant.[49] The second covered the compounds themselves and likely derivatives for anti-inflammatory applications.[50] The CAC and UniSA were joint applicants. These applications have lapsed and have been replaced by an international application under the Patent Cooperation Treaty (PCT), made on 10 November 2011.[51] The patent claims pharmaceutical and cosmetic compounds "isolated from plant material from Dodonaea polyandra" which have anti-inflammatory activity, as well as "methods of treating inflammation using the compounds".[52]

Also being evaluated under the project are essential oils from some of the aromatic plants that grow on Kuuku I'yu homelands which are being tested for activity against micro-organisms (including bacteria, yeasts and viruses) that cause skin, oral, respiratory and gastrointestinal infections (Claudie *et al.*, 2012, p. 48).

This project is suggested as a model for interactions between Aboriginal peoples and commercial exploiters af their traditional knowledge (Smith, 2005).

Mudjala Project, Kimberley Region of Western Australia

In 1986, John Watson, a Senior Lawman of the Nyikina people of the Lower Fitzroy River, in the West Kimberley region of Western Australia, had half his finger bitten off by a crocodile when hunting. He used the bark of the "nyardoo mudjala" tree (*Barringtonia acutangula*)[53] to stem the pain, before travelling some hours to the nearest hospital (Mills, 2008, reported in Marshall *et al.*, 2013). Following news reports of this the Nyikina Mangala Elders appointed a former administrator from the Kimberley Land Council, the representative body of the Aboriginal peoples of the Kimberley region, to investigate the commercialization potential of mudjala. These negotiations led to a partnership between Griffith University (GU) and Jarlmadangah Burru Aboriginal Corporation (JBAC)[54] for the isolation and identification of the active analgesic compounds in mudjala.[55] Griffith University chemists reported that they had isolated nine triterpene saponins and a triterpene aglycone from the tree's bark (see Mills *et al.*, 2005).

The partnership arrangement provided for the joint holding of any patents between GU and JBAC and the joint sharing of commercialization benefits and IP Australia reported that patents have been registered in Australia, Japan, India, New Zealand and the USA (UTS Forum, 2014, p. 26). It was reported that the practical benefits gained by the Jarlmadangah Burru community from the project is that by participating in harvesting and monitoring trials and natural resource and sustainable wild harvest management, "the Nyikina Mangala Rangers have been trained in vegetation assessment and mapping of mudjula" (Marshall *et al.*, 2013, p. 19).

Kakadu plum, Northern Australia

The Kakadu plum (*Terminalia ferdinandiana*) is a traditional food and medicine source for Aboriginal Australian Peoples in Northern Australia.[56] It is rich in vitamin C and contains Gallic acid, which has anti-bacterial, anti-viral and anti-fungal, anti-inflammatory, anti-tumour, anti-mutagenic and anti-bronchodilatory applications (see Gorman *et al.*, 2006; Mohanty and Cock, 2012). On 10 March 2009 a question was asked of the Minister for Innovation, Industry, Science and Research in the Australian Senate concerning Australian patent applications by three companies concerning processes to extract vitamin C from the Kakadu plum and

> what steps are being taken to provide legal recognition of traditional knowledge for the purpose of giving Indigenous people the capacity to participate in the economic development of Australia, thereby generating income and wealth for them which will be independent of Government?[57]

The Minister provided details of Australian patent applications concerning skin care preparations, dietary supplements and for a food composition containing Kakadu Plum[58] and a granted US patent to a method for preparing dried powder from the Kakadu plum, and two granted US patents for anti-allergy compositions which may contain Kakadu plum and methods of using the composition.[59]

On 19 January 2007 the US-based cosmetics company Mary Kay Inc. applied for an international patent under the PCT for "compositions comprising kakadu plum extract or açaí berry extract".[60] The patent application entered the national phase in Australia on 22 July 2008. An Australian academic, Daniel Robinson, on 14 October 2009, filed a section 27 submission under the Patents Act, 1990 asserting the obviousness of many of the claims made in the patent application (Robinson, 2010a). The application attracted attention in the national press. For example an article in the *Sydney Morning Herald* on 4 December 2010, reported the opposition to the patent application of the Mirarr, the traditional Aboriginal owners of the north-east Kakadu National Park and parts of Western Arnhem Land. The Gundjeihmi Aboriginal Corporation, representing the Mirarr, said "the Kakadu plum has been an important source of food and medicine for the Mirarr", and that "[i]t also features in oral histories and 'dreaming' stories".

An Examination Report issued by IP Australia reported its finding that the "Aborigines have been using the Kakadu plum extract for around 40,000 years as a food source and a healing agent".[61] On 12 October 2011 the Australian application was withdrawn, although the granted patents, as well as patent applications remain on foot in a number of countries (Robinson, 2010b).

The bioprospecting issue which has been raised in relation to this case concerns the source of the Kakadu plum used by Mary Kay, Inc. It has been pointed out that if it was obtained by a commercial supplier, there would have been no obligation to share benefits with Indigenous communities under the Northern Territory Biological Resources Act 2006 or under any Nagoya Protocol-type legislation (Holcombe and Janke, 2012, pp. 309–11).

Titjikala Plant Database (Central Australia)

The Titjikala community is an Aboriginal community in the south of the Northern Territory of Australia with a population in 2011 of 201 persons.[62] In 2002, the Titjikala community decided "to capitalise on their TK about plants" as well as to "validate their medicinal knowledge through a comparative scientific laboratory study of selected plant species" (Evans *et al.*, 2009). A tribal elder met with a university on a research project named "Plants for People" (P4P). After initial discussions between the two an extension of the project was arranged through the Desert Knowledge Cooperative Research Centre (DKCRC). Among the objectives of this second phase were to: "validate TK about medicinal uses of local plants through laboratory studies on pharmacological, bioactive, antibiotic, anti-viral and anti-tumour properties of selected plants" and to develop a legally binding agreement that would ensure that TK about local plants was protected, and that the Titjikala community had a "seat at the table" in any discussions about the commercialization of the research findings (Evans *et al.*, 2009, p. 395).

The project was overseen by a Council of Elders comprising traditional owners of the land on which Titjikala is located and all major stakeholders groups participated in the decision-making process, including Curtin University of Technology (CUT) researchers, Titjikala community members and members of the Tapatjatjaka Community Government Council, which represents that community, as well as the Central Land Council, the regional Aboriginal representative body.

The DKCRC had an Indigenous Intellectual Property Protocol which dealt with access and benefit-sharing, among other things.[63] A draft final agreement was signed by the TCGC and CUT in February 2005 (Evans *et al.*, 2009, p. 397). In their account of these negotiations Evans *et al.* report that the draft agreement deferred the allocation of intellectual property rights, but that on the advice of the Central Land Council and researchers from the University of South Australia seeking to develop a similar agreement with Aboriginal communities in South Australia for other P4P projects, counselled against this deferral (Evans *et al.*, 2009, p. 398). In the final agreement equity from commercialization was to be divided equally between the TCGC and CUT and that any intellectual property arising from the project would also be divided equally.[64]

An issue which has been raised subsequent to the finalization of the agreement concerns the rights of Aboriginal groups outside the Titjikala community, whose plants would be identified for commercial exploitation. The proposed solution is for equal benefit-sharing for all Aboriginal peoples having cultural associations with particular medicinal plants proposed for commercial exploitation (Evans *et al.*, 2009, pp. 399–400).

Bawinanga Aboriginal Corporation, wildlife enterprise, Maningrida, Arnhem Land, Northern Territory

Wildlife enterprises have been identified as an important means of local economic development and Indigenous employment while providing Indigenous people with opportunities to continue their close connection with their country and to maintain customary wildlife harvesting practices (see Altman and Cochrane, 2005; Gorman *et al.*, 2008).

In 2006 the BAC, the commercial arm of the Indigenous population of the Maningrida township and surrounding region in Arnhem Land, Northern Territory, established a Wildlife Centre to promote the development of sustainable wildlife industries in the Maningrida region. In 2008 the BAC commissioned the Centre for Aboriginal Economic Policy Research (CAEPR) at the Australian National University (ANU) to report on the commercial feasibility of a wildlife industry focusing on the sale of eggs and hatchlings of the northern long-necked turtle (*Chelodina rugosa*), as well as the sale of tarantula spiders (*Selenotholus* sp.) to the domestic market and tarantula venom to pharmaceutical companies (reported in Fordham *et al.*, 2010).

The study commenced with gathering and documenting Indigenous knowledge relevant to the turtle and tarantula spider, through interviews with the Indigenous people of the region. Respondents were asked about their knowledge of the distribution and abundance of *C. rugosa* and *Selenotholus* sp., to obtain insights from Indigenous ecological knowledge for the Wildlife Centre in its enterprise development (Fordham *et al.*, 2010, pp. 4, 7).

The study reported that most Indigenous traditional owners agreed that access to the country required prior negotiation with the relevant landholders before collection or study of either species occurred and in this regard custodianship of the species was an important consideration when granting access (Fordham *et al.*, 2010, p. 14). One traditional owner explained that he was one of the keepers of the spider (*Mardayin*), which brought with it important responsibilities requiring consultation about the potential impact on the spider population of any use of spiders in commercial or educational development.

It was subsequently reported by CAEPR that the Queensland Parks and Wildlife Service had granted permits for the collection in 2009 of five female spiders with egg sacs and eggs to be hatched and grown for sale, but that permits were yet to be granted for the sale of venom to pharmaceutical companies "due to issues associated with Intellectual Property and royalty payments" (Fordham *et al.*, 2010, p. 4).

The integration of Indigenous ecological knowledge with Western science to develop the harvesting, animal husbandry and population monitoring techniques to underpin a sustainable wildlife industry raised the question of intellectual property ownership and the feasibility report suggested that further attention be paid to "critical issues surrounding intellectual property and publication rights in the development of wildlife industries" (Fordham *et al.*, 2010, p. 20). Before this commercial venture could be developed, in October 2012 the Bawinanga Aboriginal Corporation was placed under special administration due to financial problems.[65] This special administration was terminated in June 2014.[66]

Conclusion

It has been suggested that the concerns about biopiracy might be overblown (e.g. Heald, 2003), particularly since in the case of international examples such as the patenting of turmeric (Verma *et al.*, 2014) and Basmati rice (Lightbourne, 2003), as well as in the case of the Kakadu plum, the relevant patents were overturned or refused. However, the legal procedures involved in revoking patents are both costly and time-consuming and may well be beyond the resources of traditional communities.

It has been pointed out that instances of biopiracy are quite few. Oldham *et al.*, in a 2013 article reported on text mining 11 million patent documents published between 1976 and 2010 by the US Patents and Trademark Office, the EPO and by WIPO under the International Patent Cooperation Treaty. They found that "the bulk of patent activity is concentrated around a small number of well-known and cosmopolitan species" (Oldham *et al.*, 2013). In other words, the instances of biopiracy involve exotic exceptions to the bulk of patenting activity. However, in a political environment, such as in Australia where there is a history of bad treatment of Indigenous Peoples, even the few examples of biopiracy can have a disproportionate effect on the bioprospecting environment.

At the most recent session of WIPO's Intergovernmental Committee on Intellectual Property and Genetic Resources, Traditional Knowledge and Folklore (IGC), under the Chair of Ian Goss, the former head of IP Australia,[67] a Joint Recommendation on Genetic Resources and Associated Traditional Knowledge was submitted by the USA.[68] This proposed that WIPO Member States should provide legal, policy or administrative measures, as appropriate and in accordance with national law, to prevent patents from being granted erroneously with regard to claimed inventions that include genetic resources and traditional knowledge associated with genetic resources. This was adopted for transmission to the next session of the IGC.[69]

Notes

1 www.environment.gov.au/topics/science-and-research/australias-biological-resources/nagoya-protocol-convention-biological.

2 www.ipaustralia.gov.au/about-us/public-consultations/indigenous-knowledge-consultation/.

3 See www.ipaustralia.gov.au/about-us/public-consultations/indigenous-knowledge-consultation/have-your-say/.
4 *Ibid.*
5 New South Wales, South Australia, Tasmania, Queensland, Victoria and Western Australia.
6 Australian Capital Territory and Northern Territory.
7 (1992) 175 CLR 1.
8 [2008] HCA 29.
9 [1999] HCA 53.
10 *Western Australia v Ward* (2002) 213 CLR 1.
11 *Ibid.* at [59].
12 *Ibid.*
13 *Ibid.* at [61].
14 EPBC Act, s. 359A(1).
15 EPBC Act, s.303BAA.
16 EPBC Regulations, r. 8A.01(b).
17 *Ibid.*, r. 8A.01(c).
18 *Ibid.*, r. 8A.06.
19 *Ibid.*, r. 8A.07(1) and r. 17.03A(6)(a). This addresses one of the suggestions made by the Commonwealth Department of Agriculture, Fisheries and Forestry that benefit-sharing agreements be contingent on the commercialization of the materials accessed. See further Stoianoff (2009, p. 304).
20 EPBC Regulations, r. 8A.08.
21 *Ibid*, r. 8A.08.
22 Australian Government, National Parks, www.australia.gov.au/about-australia/australian-story/national-parks.
23 Cherry J, "Plant Breeder's Rights Amendment Bill 2002: Second Reading Speech", Senate Hansard, 21 October 2002, p. 5553.
24 See www.parliament.qld.gov.au/Documents/TableOffice/TabledPapers/2009/5309T1532.pdf.
25 *Code of Ethical Practice for Biotechnology in Queensland*, Pt 1, Preamble, Art. 2.
26 www.qld.gov.au/dsitia/assets/documents/biodiscovery-act-2015-review-tor.pdf.
27 *Ibid.*
28 Available at: www.environment.gov.au/biodiversity/publications/australias-biodiversity-conservation-strategy-summary.
29 *Ibid.* at 42.
30 *Ibid.* at paras 1.2.1–3.
31 See www.cbd.int/sp/targets/.
32 Available at www.environment.gov.au/biodiversity/conservation/strategy.
33 Available at www.dest.gov.au/archive/Science/astec/ethics/contents.html.
34 The Australian Institute of Aboriginal Studies (AIAS) was founded by a Federal statute in 1964. The AIAS Act was replaced by the Australian Institute of Aboriginal and Torres Strait Islander Studies Act 1989 which created a Research Advisory Committee. See http://aiatsis.gov.au/about-us/our-history.

AIATSIS relocated to the Acton Peninsula in 2001, heralding an exciting and challenging era. For the first time in the modern history of Australia an Aboriginal and Torres Strait Islander organization received national prominence on a site of international importance and created the ability to showcase Aboriginal and Torres Strait Islander cultures and the richness of AIATSIS collections in a manner never before possible.
35 Available at http://aiatsis.gov.au/research/ethical-research.
36 Available at www.nhmrc.gov.au/_files_nhmrc/publications/attachments/e52.pdf, replacing "Guidelines on Ethical Matters in Aboriginal and Torres Strait Islander Health Research" (issued in 1991).
37 *Ibid.* at para. 1.3.1.
38 *Ibid.* at para. 1.5.
39 US Patent 5,672,607, filed January 29, 1993 and granted September 30, 1997 and US Patent 5,783,598, filed March 8, 1996 and granted July 21, 1998 and US Patent number 5,869,522, filed May 31, 1995 and granted February 9, 1999.
40 WO1994017055 A1, filed January 31, 1994.
41 EP 0 681 578 B1.
42 Australian Patent Application 60991/94.

43 "Boyd; Michael R. (Ijamsville, MD), Cardellina, II; John H. (Walkersville, MD), Gustafson; Kirk R. (Mt. Airy, MD), Decosterd; Laurent A. (Nyon, CH), Parsons; Ian (Ithaca, NY), Pannell; Lewis (Silver Spring, MD), McMahon; James B. (Frederick, MD), Cragg; Gordon M. (Bethesda, MD)". US Patent 5,672,607.

44 Conservation and Land Management Amendment Act 1993 inserting para. (ca) into s.33(1).

45 Western Australia, 34th parliament, 1st Session, Western Australian Parliamentary Debates (Hansard), November 3, 1993, 6149–53.

46 *Ibid.*, p. 6151.

47 *Ibid.*, pp. 6149–53 (Legislative Assembly), 8160–8172 (Legislative Council).

48 The CAC was formed under the Commonwealth Aboriginal Councils and Associations Act 1976 (Cth) now replaced by the Corporations (Aboriginal and Torres Strait Islander) Act 2006 (Cth).

49 Australian patent application number 2009905499.

50 Australian patent application number 2009905498.

51 International Application No.: PCT/AU2010/001502.

52 *Ibid.*

53 Also known as the Marjalah tree.

54 Jarlmadangah Burru was established in 1987 by the traditional owners of the country and is within the Nyikina Mangala native title claim, registered by the Nyikina Mangala people with the National Native Title Tribunal and administered by the Kimberley Land Council. See www.planning.wa.gov.au/dop_pub_pdf/Jarlmadangah_Burru_LP2_Amendment_report.pdf.

55 See "Aboriginal medicine R&D project", www.triplebl.com.au/expertise/mudjala-project/index.html.

56 In the Northern Territory it is known as "Billy goat plum", in Western Australia as "Gubinge" or "Murungu" and in Queensland as "Kakadu plum". See T. Janke (2009) *Indigenous Ecological Knowledge and Natural resources in the Northern Territory: Report on the Current Status of Indigenous Intellectual Property*, Darwin, Natural Resource Management Board, 149.

57 Senator R. Siewert, Australian Senate Question 1172, March 10, 2009, www.aph.gov.au/hansard/senate/dailys/ds100309.pdf, pp. 247–48.

58 Australian Patent application 2007205838 by MARY KAY, INC. relates to a skin care product comprising Kakadu plum extract or acai berry extract (Claim 1); Australian Patent application 2004268233 by MANNATECH, INC. relates to a dietary supplement which may contain Australian bush plum (Claims 33–41); Australian Patent application 2005328670 by MANNATECH, INC. relates to a modified release dietary supplement comprises polysaccharides which is compressed at a pressure of greater than 100psi (Claim 1); Australian Patent application 2006237559 by MANNATECH INC., relates to a modified release dietary supplement which comprises polysaccharides which is compressed at a pressure of greater than 100psi (Claim 1); Australian Patent application 2004203276 by CORADJI Pty Ltd relates to a method of removing the seed from the fruit of the *Terminalia ferdinandiana* (i.e. bush plum) (Claim 1); Australian Patent application 2007231781 by EXIST MARKETING PTY LTD. to a method and compositions of treating bursitis which may contain Kakadu Plum (page 14); Australian Patent application 2007249801 by INTERLEUKIN GENETICS, INC. is to a food composition comprising rose hips and optionally Kakadu concentrate, from the Kakadu Plum (Claim 1); Australian Innovation patent application 2008100919 by GREENTASTE Pty Ltd. is to a herbal composition which optionally may contain Kakadu Plum (page 26).

59 US Patent 7175862 assigned to ACCESS BUSINESS GROUP INTERNATIONAL LLC is to a method of preparing dried powder from the Kakadu Plum. US Patent 7384654 assigned to ACCESS BUSINESS GROUP INTERNATIONAL LLC is to an antiallergy composition which may contain Kakadu concentrate. US Patent 7384656 assigned to ACCESS BUSINESS GROUP INTERNATIONAL LLC is to a method of inhibiting an allergic response by administering a composition which may which may contain Kakadu concentrate.

60 WO/2007/084998.

61 Examination Report available at http://pericles.ipaustralia.gov.au/ols/auspat/applicationDetails.do?applicationNo=2007205838.

62 Australian Bureau of Statistics, "Titjikala, NT (State Suburb)" (October 2012). Traditional languages are Luritja, Arrernte and Pijantjara.

63 This protocol was amended in 2007 and is mentioned on the DKCRC website, but cannot be accessed there.

64 *Ibid.*
65 See www.oric.gov.au/publications/media-release/bawinanga-seeks-assistance, accessed 8 November 2015.
66 www.oric.gov.au/publications/media-release/bawinanga-special-administration-ends.
67 IGC, 29th Session, Geneva, February 15 to 19, 2016.
68 On behalf of Canada, Japan, Norway, the Republic of Korea and the USA, document WIPO/GRTKF/IC/28/5.
69 Decisions of the twenty-ninth session of the committee, WIPO/GRTKF/IC/29/5.

References

ABS (Australian Bureau of Statistics) (2011) Estimates of Aboriginal and Torres Strait Islander Australians, June 2011, Canberra, AGPS, 2013, www.abs.gov.au/ausstats/abs@.nsf/mf/3238.0.55.001.

Altman, J.C. and M. Cochrane (2005) "Sustainable development in the indigenous-owned savannah: Innovative institutional design for cooperative wildlife management", 32 *Wildlife Research* 473.

Armstrong, J. and K. Hooper (1994) "Nature's Medicine", in Department of Conservation and Land Management *Landscope* 12.

Bancroft, R. (1994) "Fourmile, H.", in D. Horton, ed., *Aboriginal and Torres Strait Islander History, Society, and Culture.* Canberra, Aboriginal Studies Press.

Barr, A. (1988) *Traditional Bush Medicines: An Aboriginal Pharmacopoeia. Aboriginal Communities of the Northern Territory of Australia.* Richmond, Victoria, Greenhouse Publications.

Beattie, A.J. (1995) *Biodiversity: Australia's Living Wealth.* Sydney, Reed Books.

Blakeney, M. (1997) "Bioprospecting and the Protection of Traditional Medical Knowledge of Indigenous Peoples: An Australian Perspective", 6 *EIPR* 298.

Blakeney, M. (2015) "Protecting the Knowledge and Cultural Expressions of Aboriginal Peoples", 39 *University of Western Australia Law Review* 180.

Claudie, D.J., S.J. Semple, N.M. Smith and B.S. Simpson (2012) "Ancient but New: Developing Locally Driven Enterprises Based on Traditional Medicines in Kuuku I'yu Northern Kaanju Homelands, Cape York, Queensland, Australia", in P. Drahos and S. Frankel, eds, *Indigenous Peoples' Innovation: Intellectual Property Pathways to Development*, Canberra, ANU E Press, 30.

COAG (Council of Australian Governments) (1997) "Heads of agreement on Commonwealth and State roles and responsibilities for the Environment" 1997, www.environment.gov.au/resource/heads-agreement-commonwealth-and-state-roles-and-responsibilities-environment.

Coordination Committee on Science and Technology (1994) *Access to Australia's Biological Resources*, Canberra, AGPS.

Cragg, G.M., M.R. Boyd, J.H. Cardellina II, M.R. Grever, S.A. Schepartz, K.M. Snader and M. Suffness (1993) "Role of Plants in the National Cancer Institute Drug Discovery and Development Program", in A.D. Kinghorn and M.F. Balandrin (eds) *Human Medicinal Agents from Plants.* Washington, DC, American Chemical Society, 80–95.

Decosterd, L.A., I.C. Parsons, K.R. Gustafson, J.H. Cardellina II, J.B. McMahon, G.M. Cragg, Y. Murata, L.K. Pannell and J.R. Steiner (1993) "Structure, absolute stereochemistry, and synthesis of conocurvone, a potent, novel HIV-inhibitory naphthoquinone trimer from a Conospermum sp.", 115 *J. Amer. Chem. Soc.* 6673–79.

DEST (Commonwealth of Australia, Department of Environment, Sport and Territories) (1996) *National Strategy for the Conservation of Australia's Biological Diversity*, "Introduction", www.environment.gov.au/archive/biodiversity/publications/strategy/index.html.

DoE (Australian Government, Department of the Environment), www.environment.gov.au/epbc.

DoE Review (2001) Australian Government, Department of Environment, *Review of The National Strategy for the Conservation of Australia's Biological Diversity* (2001), www.environment.gov.au/resource/review-national-strategy-conservation-australias-biological-diversity.

EDO (2016) Environmental Defender's Office of Western Australia, White Paper: Western Australia's Biodiversity Conservation Bill 2015: Review and Recommendations, February 2016, p. 4, www.edowa.org.au/assets/Uploads/20160210-Biodiversity-Bill-Review-FINAL-1–2.pdf.

Evans, L., H. Scott, K. Muir and J. Briscoe (2009) "Effective intellectual property protection of traditional knowledge of plants and their uses: an example from Australia", 74 *GeoJournal* 391.

Fordham, A., W. Fogarty, B. Corey and D. Fordham (2010) "Knowledge Foundations for the Development of Sustainable Wildlife Enterprises in Remote Indigenous Communities of Australia" *CAEPR Working Paper No. 62/2010*, Canberra, Centre for Aboriginal Economic Policy Research, February.

Fourmile, H. (1996) "Protecting Indigenous Intellectual Property Rights in Biodiversity", Paper presented at Kaltja vs Business Conference, 28 August.

Garson, M. (1997) "Biodiversity and Bioprospecting", 4(2) *Australasian Journal of Natural Resources Law and Policy* 227.

Gorman, J.T., A.D. Griffiths and P.J. Whitehead (2006) "An analysis of the use of plant products for commerce in remote Aboriginal communities of Northern Australia", 60 *Economic Botany* 362.

Gorman, J.T., P.J. Whitehead, A.D. Griffiths and L. Petherman (2008) "Production from marginal lands: Indigenous commercial use of wild animals in northern Australia", 15 *International Journal of Sustainable Development and World Ecology*, 1.

Heald, P. (2003) "The Rhetoric of Biopiracy", 11 *Cardozo J. Int'l and Comp. L.* 519.

Holcombe, S. and T. Janke (2012) "Patenting the Kakadu Plum and the Marjarla tree: biodiscovery, intellectual property and Indigenous Knowledge", in M. Rimmer and A. McLennan, eds. *Intellectual Property and Emerging Technologies: The New Biology*, Edward Elgar, 293.

Janke, T. (1999) *Our Culture: Our Future, Report on Australian Indigenous Cultural and Intellectual Property Rights*, Sydney, ATSIC.

Janke, T. and R. Quiggin (2005) "Indigenous Cultural and Intellectual Property and Customary Law", in *Aboriginal Customary Laws Background Papers*, Background Paper 12, Perth: LRCWA, www.lrc.justice.wa.gov.au/2publications/reports/ACL/BP/BP-12.pdf.

Jaradat, A.A. (2010) "Genetic resources of energy crops: Biological systems to combat climate change", 4 *Australian Journal of Crop Science* 309.

Kerr, P.G. (2010) "Bioprospecting in Australia – Sound Biopractice or Biopiracy?", 29 *Social Alternatives* 44.

Lightbourne, M. (2003) "Of rice and men: An attempt to assess the basmati affair", 6 *The Journal of World Intellectual Property* 875.

LRCWA (Law Reform Commission of Western Australia) (2006) *Aboriginal Customary Laws. The interaction of Western Australian Law with Aboriginal Law and Culture.* Final Report, Project 94, www.lrc.justice.wa.gov.au/_files/P94_FR.pdf.

Marshall, V., T. Janke and A. Watson (2013) "Community Economic Development in Patenting Traditional Knowledge: A Case Study of the Mudjala TK Project in the Kimberley Region of Western Australia", 8 *Indigenous Law Bulletin* 17.

Meyers, G.D. and O.A. Owoeye (2013) "Intellectual Property Law and the Protection of Indigenous Australian Traditional Knowledge in Natural Resources", 22 *Journal of Law, Information and Science* 56.

Miller, S.K. and L. Dayton (1993) "Australia takes tough line on 'HIV plant'", 1880 *New Scientist* 4.

Mills, C., A.R. Carroll, and R.J. Quinn (2005) "Acutangulosides A-F, Monodesmosidic Saponins from the Bark of Barringtonia acutangula", 68 *Journal of Natural Products* 311.

Mills, J. (2008) "Aborigine, scientist find pain relief in marjarla tree", *Courier Mail* (online), www.couriermail.com.au/news/queensland/native-tree-bark-easespain/story-e6freoof-1111117814741.

Mittermeier, R.A.P., G. Robles and C.G. Mittermeier (eds.) (1989) *Megadiversity: Earth's Biologically Wealthiest Nations.* Mexico City, Cemex.

Mohanty, S. and Ian Cock (2012) "The chemotherapeutic potential of Terminalia ferdinandiana: Phytochemistry and bioactivity", 6 *Pharmacognosy Reviews* 29.

O'Bryan, Katie (2004) "The Appropriation of Indigenous Ecological Knowledge: – Recent Australian Developments", 1 *Mq Jnl ICEL* 29.

Oldham, P., S. Hall and O. Forero (2013) "Biological Diversity in the Patent System", 8(11) *PLoS ONE*: e78737. doi:10.1371/journal.pone.0078737.

Parke, M. and C. Kendall (1997) "Bioprospecting of Traditional Medical Knowledge", unpublished paper, Murdoch University School of Law, Perth.

Payne, Senator M. (2007) Commonwealth, Parliamentary Debates: United Nations Declaration on the Rights of Indigenous Peoples 53 http://parlinfo.aph.gov.au/parlInfo/genpdf/chamber/hansards/2007-09-10/0075/hansard_frag.pdf fileType=application%2Fpdf.

Robinson, D.F. (2010a) "The Biological Patent Predicament Traditional Knowledge and Biological Product Derivative Patents: Benefit-Sharing and Patent Issues Relating to Camu Camu, Kakadu Plum and Açaí Plant Extracts" Guest Article, United Nations University, Institute of Advanced studies, Traditional Knowledge Initiative, www.unutki.org/news.php?doc_id=174.

Robinson, D. (2010b) *Confronting Biopiracy: Challenges, Cases and International Debates*, London, Earthscan.

SCRGSP (Steering Committee for the Review of Government Service Provision) (2014) *Overcoming Indigenous Disadvantage: Key Indicators 2014*, Canberra Productivity Commission.

Simpson, B. (2010) "Evaluation of the anti-inflammatory properties of Dodonaea polyandra, a Kaanju traditional medicine", 132 *J. Ethnopharmacol* 340.

Smallacombe, S., M. Davis and R. Quiggin (2006) "Scoping Project on Aboriginal Traditional Knowledge", Report of a study for the Desert Knowledge Cooperative Research Centre, Desert Knowledge Cooperative Research Centre, Alice Springs.

Smith, B. (2005) "'We got our own management': Local Knowledge, Government and Development in Cape York Peninsula", 2 *Australian Aboriginal Studies* 4.

Smith, B. and D. Claudie (2003) "Developing a Land and Resource Management Framework for Kaanju Homelands, Central Cape York Peninsula", Discussion Paper 256, Centre for Aboriginal Economic Policy Research, Canberra, Australian National University.

Spjut, R.W. (1981) "Foreign Travel Report: Western Australia and Tasmania August 20–November 8, 1981", www.worldbotanical.com/images/WA-rpt-text.pdf.

Spjut, R.W. (2006) "World Botanical Associates (WBA) History", www.worldbotanical.com/who-we-are.htm.

Spjut, R.W. (1992) Spjut, "Preliminary Report on Conospermum (Proteaceae)", www.worldbotanical.com/images/Australia%20Conospermum%20report.pdf.

Stoianoff, N. (2009) "The Recognition of Traditional Knowledge under Australian Biodiscovery Regimes: Why Bother with Intellectual Property Rights?" in Antons, C. (ed.) *Traditional Knowledge, Traditional Cultural Expressions and Intellectual Property Law in the Asia-Pacific Region*, The Netherlands: Kluwer Law International.

UNNC (United Nations News Centre) (2009) "Experts Hail Australia's Backing of UN Declaration of Indigenous People's Rights", www.un.org/apps/news/story.asp?NewsID=30382.

UTS – Indigenous Knowledge Forum, North West Local Land Services (2014) *Recognising and Protecting Aboriginal Knowledge Associated with Natural Resource Management*, White Paper for the Office of Environment and Heritage, www.ipaustralia.gov.au/pdfs/UTS_-_Recognising_and_Protecting_Aboriginal_Knowledge.pdf, 26.

Verma, M., Chauhan, I., Kumari, R. and Sharma, M. (2014) "India –Victim of Bio-Piracy", 4 *Indo-American Journal of Pharmaceutical Research* 329.

Woodall, M. and D. Cox (2014) "Disclosure of Origin in Patent and PBR Applications: The Australian Experience" paper presented at the ASEAN Australia New Zealand Free Trade Area (AANZFTA) Economic Cooperation Work Programme (ECWP) Policy Workshop on IP and Genetic Resources (GR) and Traditional Knowledge (TK), June 2–5, 2014, Jakarta, Indonesia.

18

IF WE HAVE NEVER BEEN MODERN, THEY HAVE NEVER BEEN TRADITIONAL

'Traditional knowledge', biodiversity, and the flawed ABS paradigm

Graham Dutfield[1]

Introduction

Traditional knowledge and customary practices relate to all three objectives of the Convention on Biological Diversity: conservation, sustainable use and benefit-sharing arising from access.[2] The first two are largely taken for granted and there is a wealth of empirical evidence showing, without falling into the trap of romanticism, that indigenous peoples' knowledge, innovations and practices are enormously relevant to *in situ* conservation and sustainability, often in positive ways. However, most of the attention that traditional knowledge (TK) gets concerns the issue of how to return benefits from the use of traditional knowledge and associated genetic material in high-tech corporate discovery, research, development and manufacturing. In 2010, the Nagoya Protocol to the Convention on Biological Diversity (CBD) was adopted, and has since entered into force. The Protocol, whose full name is the Nagoya Protocol on Access to Genetic Resources and the Fair and Equitable Sharing of the Benefits Arising from their Utilization, seeks to further the third objective of the CBD: the fair and equitable benefit-sharing arising from the use of genetic resources and associated traditional knowledge (Greiber *et al.*, 2012; UNCTAD, 2014) – often contracted to ABS. However, the conventional wisdom behind the TK-related aspects of the 'ABS+plus IP'[3] model embodied by the Convention on Biological Diversity and the Nagoya Protocol is fundamentally flawed. This chapter identifies four main reasons why such approaches generally fail to provide solutions that are fair, equitable and efficient. We call these: (i) the hybrid nature of knowledge systems; (ii) the problem of origin and attribution; (iii) the overregulation and corporatization tendency; and (iv) the exchange value 'distraction'. In the meantime, traditional knowledge is being lost worldwide. International solutions are no substitute for legal and policy initiatives giving more power to local people to make their own decisions. However, whether they elect to retain, abandon, mix or transform their local biodiversity-friendly livelihoods and customary practices is ultimately for them to decide.

The (insufficiently appreciated) hybrid nature of knowledge systems

The political appeal of traditional knowledge and biopiracy discourses and rhetorics is manifold. National governments' regarding indigenous peoples as defenders of the national patrimony is one aspect of this (Conklin, 2002). As for the international environmental movement who see globalization as inherently harmful to the planet, traditional knowledge provides healthier alternatives to the supposedly arrogant and hegemonic brutalism of Western science and technology and therefore is something to be promoted. For those environmentalists of an economic bent TK might even help to finance biodiversity through its commercialization. It also fits neatly into rhetoric about information societies and knowledge-based economies (Strathern, 1999). Evidently, 'traditional knowledge' serves a number of different stakeholders whose motivations are hardly in close alignment. But what is it? Can we even say what it is? Can we say at least say what TK is *not*?

It is commonly supposed that tradition and modernity operate in separate spheres, except where the modern incorporates tradition in a decontextualized fashion and then claims it as its own. Such an opposition between tradition and modernity is a fundamental assumption of post Second World War modernization theory. Its appeal may have had much to do with its consistency with how Europeans over a considerable period of time have frequently distinguished their own societies from 'the other': barbarians, savages, primitive people and so forth – and with anthropologists being complicit in this apparently rather deep-seated tendency (Kuper, 2005).[4] A number of mostly US-based social scientists in the Post-War era identified fundamental social and cultural differences between traditional and modern societies and assigned to each a set of descriptive terms that were in opposition to each other. Accordingly, as they saw it, social and cultural evolution could be best understood in terms of progress that would entail the replacement of terms applicable to traditional societies such as 'community', 'patron–client relationship', 'routine' and 'solidarity', with their modern polar opposite counterparts 'individual', 'bureaucratic relationships', 'innovation' and 'competition', respectively. Since evidence of progress essentially entailed the latter terms applying rather than the former ones, there was little accommodation for hybridity including its positive aspects for both societies. Ironically, like the Marxism that modernization theory was in large part intended to be an alternative to, both approaches were susceptible to somewhat deterministic evolutionary stages notions (Olivier de Sardan, 2005; Rostow, 1960).[5]

Of course, the notion of inexorable progress was inherent to each, showing how equally rooted they were in the Enlightenment's spirit of idealism. However, whereas '[t]he emergence of sociology at the turn of the 20th century introduced ambivalence as a key theme in a discourse on modernity', and in doing so 'broke with the spirit of the Enlightenment … The celebration of "modernization" in the mid-20th century appears, with hindsight, unvarnished in its optimism' (Englund and Leach, 2000). Modernization theory has largely been discredited, at least in the academy, on a number of grounds. In hindsight we should have been more wary of such positive certainty and more attuned to the ambivalence and doubt that were also features of Enlightenment thinking. Two major criticisms, at least of the cruder versions of modernization theory, aside from its ahistoricism, are its determinism and its failure to provide much accommodation for the idea of mixing and hybridization and their potentially beneficial aspects also for the traditional sector. At its crudest, modernization theory saw social progress and economic development as the necessary transformation of traditional societies into modern ones. It is 'tradition' that was holding societies back.

If the academic version of modernization theory has fallen completely out of fashion some of its basic assumptions have proved to be highly resilient being expressed for example in the overhyping of genetically modified plants in developing countries as if they are the last word in agricultural modernization alongside the dismissal of landraces and local cultivars as old varieties that should be abandoned, and the blanket contempt for traditional medicine (Tallis, 2004). Perhaps modernization theory's potentially most dangerous supposition is that *all* of what is true for modernity becomes the opposite for tradition.[6] This should immediately be cause for reflection: that tradition has tended, and to a large extent still does tend, to be used negatively as something outmoded. The word 'tradition' does not help here insofar as it implies some kind of pure condition that is not adulterated with or diluted by elements from other societies or transformed in any way endogenously: TK just *is* – it does not *become*. Seen in binary opposition, it follows that the more tradition you have the more modernization you need; the less tradition the better. Over the years, some hard lessons were learned. For example, in the 1960s, Balinese farmers forced to plant Green Revolution modern high-yielding varieties and purchase industrial chemical inputs suffered diminished productivity and crop disease and pest outbreaks. However, according to Lansing's classic study, when they returned to their own varieties and their original management systems and practices of irrigation, fallowing and organic disease and pest management based on a network of water temples that had been in place for centuries, high productivity and sustainability both recovered (Lansing, 2007).

The contraction of 'knowledge, innovations and practices' in Article 8(j) of the Convention on Biological Diversity to just plain 'knowledge' in the Nagoya Protocol, albeit attached to genetic resources using the phrase 'associated with', hardly helps. It tends to downplay the creativity and adaptiveness of indigenous groups of each generation, as well as other societies with tradition-based bodies of knowledge that they wish to protect but that have changed culturally and in other ways from the groups and societies they were in the past (Sunder, 2006).

The use, borrowing, appropriation, misappropriation, or whatever name one chooses to call the inclusion of information, knowledge, methods and materials from one system of health or agriculture into another different and more dominant system tends to be seen as being unidirectional. As tradition wanes and the modern waxes, people assume the latter takes bits of tradition, used in the form of informational leads or raw materials, and gives nothing back in return. Modern *appears* no less modern for doing this because there is a translation and repackaging which generally strips tradition of its origins and cultural and spiritual entanglements or else denies it entirely. Accordingly, modernity is parasitic on tradition. An understanding of global power disparities reinforces such a perspective. It follows that the way to respond is to create a market for knowledge transactions so that access is exchanged for monetary or non-monetary forms to even things up. Enter the Nagoya Protocol.

Systems of knowledge tend to be hybrids because they are generally open, and they tend to have two-way 'valves': knowledge, techniques, practices and materials go both out and in. Chinese medicine, for example, was not 'traditional' until it was named as such a few decades ago largely for political reasons. Traditional Chinese Medicine co-evolved with Western scientific medicine (see below). As for experimenting and trying things out in a systematic way, it is not just white-coated laboratory scientists who do this; many traditional healers and farmers, who often breed modern varieties with their own, do so as well. Chinese 'traditional' medicine remains highly popular as do the classical traditional Indian systems such as Ayurveda, Siddha and Unani Tibb, not all of which originate – or are currently practised – entirely in the Subcontinent anyway. They may have very deep historical roots, but they are hardly devoid of novelty or innovation.

Treating tradition and modern (or scientific) as binary opposites blinds one to mutual interdependence and hybridization, which raises moral questions that few people seem willing to address, including the fundamental issue of who, if anybody, has a moral debt, if any, and to whom, as knowledge and associated biological material move around the world and get absorbed into different systems of knowledge and technology. We will discuss this matter further below by reference to historical experience. History shows us that free exchange of knowledge and material has been common in human history and has done much good for givers, recipients and humankind generally. It has also done harm to some, and that needs to be acknowledged too. Those who currently self-identify as being indigenous peoples have been among those most victimized. Apart from free exchange, there has been control, priority and attribution and of course secrecy too. These are just as much a part of human social interaction as free exchange, and probably have been for about as long, and can be justified on various grounds.

Adaptability and openness are the main drivers of innovation. So might there be advantages in no longer defining tradition as the polar opposite of modernity and instead seeing the two as actually related to each other and capable of interacting positively? In reality they do, and have done from the age of Enlightenment, if not before. That might sound counter-intuitive. But if it happens to be true then we should go with it and follow it to its logical conclusions.

Most biopiracy incidents have nothing to do with the pharmaceutical industry (Robinson, 2010). Nonetheless, of all industries this one is considered by many to have gained more than any other from open access to genetic material and associated knowledge from today's developing countries (RAFI, 1994). Indeed, one is tempted to agree with the critics that it has been parasitic on them pretty much continuously since its emergence. At the same time, its claims to being wholly science-based imply that it has never had such an intellectual or material dependency relationship. In fact, the history of the pharmaceutical industry exemplifies the subtleties of cross-cultural material and intellectual exchanges over time. It also raises questions as to the efficacy of adversarial approaches which assume that there will always be exploitation unless strong international rules are put in place. The transfer of genetic material and associated TK is not, and has never been, an inherently zero sum game (cf. Mooney, 2000).

The modern pharmaceutical industry really took off around the 1880s when scientists began to crack the problem of how to harness chemistry to other emerging scientific disciplines and practices, in order to solve hitherto intractable health problems on a regular and systematic basis. However, things did not change overnight for the majority of doctors and patients: well into the twentieth century, the number of non-traditional drugs available for doctors to prescribe was still pretty limited (Le Fanu, 1999).

Pharmaceutical products over the 150 or so years of the industry's existence have typically been single molecules, usually small ones. They work by binding to certain proteins and causing a change in their behaviour (Stockwell, 2011). Historically speaking, which type of protein was bound to and why a therapeutic effect ensued was generally unknown, as were the reasons why some people suffered from side-effects while others were unaffected. There is still often much that is unknown concerning mode of action and the way drugs work differently on different people, hence the immense attention being given nowadays to personalized medicine.

Scientifically speaking, where did all this start? From about 1805 to the early 1830s, numerous therapeutically significant alkaloids were isolated from plants. Among the most important were morphine from opium (by Sertürner), emetine from ipecacuanha (by Pierre-Joseph Pelletier and François Magendie), quinine from *Cinchona cordifolia* (by Pelletier and Caventou) and codeine, also from opium (by Robiquet). François Magendie's highly influential work displays a clear understanding of the importance of obtaining a consistent formulation and producing the right dosage, including of the fever-reducing cinchona alkaloids which of course include quinine (Greene, 2014). To Pelletier has been attributed the notion that purity has therapeutic value. Note that all of these natural products were used for medicinal and other purposes by indigenous peoples and other non-European populations.

Sneader explains the wider significance of the 1820 paper that Pelletier published with Caventou on their isolation of quinine (Pelletier and Caventou, 1820). As he put it, the article 'marked a new departure, for previously the active principles of plants had been isolated for analytical rather than therapeutic purposes. By urging their medical colleagues to study pure plant principles, Pelletier and Caventou set the course of drug discovery in a new direction' (Sneader, 2005).

However extreme the theoretical and epistemological differences might appear, *there is no perfect traditional and modern divide marking out two quite separate worlds*. Nor was there from the moment that Europeans began to ply the seven seas in search of new lands, products and trading opportunities. One might assume that European colonials and traders had nothing but disdain for local *materia medica*, but this was not necessarily the case. From the early modern era, Europeans collected, documented, traded across the world and imported into Europe medicinal plants from the East and the New World.

The great international trading companies founded at the start of this age, such as the Dutch East India Company, were instrumental in these botanic transfers (Cook, 2007). Late eighteenth-century introductions to the European pharmacopoeia coming from the New World included Jesuit's bark (quinine) and ipecacuanha (Bynum, 1994; Gänger, 2015). American plants also reached Muslim regions and Portuguese possessions and trading posts in Asia, and in the other direction westwards to the Philippines. Apart from trading companies, merchants and Jesuits were also involved in moving these plants around the world. Other plants during the age of European exploration and colonization came into Europe from Asia and Africa. These included foods as well as drugs. Spices from the East not only flavoured otherwise tasteless or disgusting food, but possessing humoral effects were included in the *materia medica*: '*The history of spices is in part the prehistory of the pharmaceutical industry*' (Arikha, 2007).[7] Global bioprospecting efforts provided a large proportion of the drugs available to Europe in the later colonial era. Of the 175 plants in the 1885 British pharmacopoeia, 40 per cent were of European origin, 25 per cent each were from Asia and the Americas, 9 per cent were African and 1 per cent was from Australia (Osseo-Asare, 2008).

What was the attraction of these plants coming from distant lands? It is frequently claimed that their exotic origins and names, as well as the colourful stories concerning their sources, the local uses of them, and the means by which they were 'discovered' by Europeans gave them much of their appeal. According to one recent historical work, their uses by native populations was likely to have been a factor in their popularity during the late seventeenth century, although we cannot be sure how important this was (Gänger, 2015, *op. cit.*).[8]

On the other hand, some fitted neatly into existing *material medica* and treatment practices because in certain senses they were *not* exotic. Their use may have been compatible with humoral approaches to sickness and health, or else they were related biologically to already known plants. In some cases, as trade expanded and populations moved on a greater scale, so did disease. A treatment used for a disease in one part of the world was perhaps presumed often to work for the same affliction, or similar symptoms, in very distant places.

Osseo-Asare makes an intriguing connection between high colonialism and modern pharmaceutical development, one that is worth closer inspection: 'The rise of pharmaceutical chemistry in Europe at the end of the nineteenth century dovetailed with the wars of imperial expansion in Africa' (Osseo-Asare, 2008). The 1880s are the decade when British and French colonial expansion had reached its highest point and new imperial nations like Germany and Belgium had just joined the global land grab, the United States following a decade later taking Cuba, Puerto Rico and the Philippines from the Spanish. Nor were the Dutch and the Portuguese entirely inactive. The infamous Berlin Conference which carved up Africa for division among the European powers was concluded in 1885, the same decade as the dyestuff industry's pharmaceutical turn (Dutfield, 2009) and the appearance of Antifebrin, Antipyrin, Pyramidon and Sulfonal, the first pharmaceutical industry products, all of which were synthetic and had nothing to do with traditional knowledge however defined. Admittedly none of these was for a tropical disease. Nonetheless, the fact that the industry emerged simultaneously with the notorious European scramble for Africa and domination of the world is certainly intriguing. It might have done this by turning colonies into giant medical laboratories to take advantage of, but also to further advance, the progress being reached in chemistry, physiology, disease causation, etc. (Tilley 2011). Or did colonialism have some other significance, such as that it stimulated mass outbreaks of biopiracy (as we now call it) of benefit to this emerging industry?

Running empires required plenty of manpower and, in a reverse direction to today's population movements, substantial numbers of European peoples moved to the tropics, becoming exposed to the same diseases as the native people. Economic and political interests are of course very important in determining where government support and private investment are directed in terms of pharmaceutical research and development. Colonialism certainly did affect which diseases should be studied, hence the interest in finding cures for tropical diseases and other ailments especially common in the colonies, such as malaria, trypanosomiasis (sleeping sickness), yellow fever and plague. Numerous schools of tropical medicine were opened in Britain, Germany, other European colonial nations and the United States (Bynum, 2006). That imperialism stimulated the growth of the industry, if not its initial emergence, seems plausible. The colonies were sources of plants and ethnobotanical information, and markets for products. In addition, the colonies served effectively as scientific laboratories including for medical doctors (Amster, 2013; Tilley, 2011). Medical research facilities were also established in the colonies (primarily, of course, for the benefit of the colonizers, not those being colonized; Chakrabarti, 2012).

What is true of the past is partly true also of the present. As mentioned above, plants together with micro-organisms remain the primary source of at least a quarter of new medicines being approved. In some cases as extracts or mixtures these were known about and used before industrial chemists and drug companies ever got their hands on them. To name a few drugs in the modern pharmacopoeia sourced from traditional medicine, reserpine, the vinca alkaloids and the opiates spring to mind. Recent additions include artemisinin, arsenic trioxide and nicosan.

Apparently the industry's initial existence does indeed owe something to traditional knowledge. But was the relationship between industrial biomedicine and traditional medicine purely parasitic? Or was it also symbiotic? A recent historical work on plant-based medicine in colonial and post-colonial Africa convincingly asserts that 'herbal medicine and pharmaceutical chemistry have mutually supportive, simultaneous histories up to the present' (Osseo-Asare, 2014). Indeed, the author even goes so far as to claim that biomedicine and African traditional healing 'were, in fact, actually adapted from one another' (*ibid.*). This may go a little too far. But it is certainly more in step with the view that the former imperial nations of Western Europe have been shaped far more by their encounters with the people, societies and the biodiversity of their former colonies than traditional Eurocentric (or Anglocentric) and positivistic histories tended to admit to, as if there was nothing much to be learned (Drayton, 2000). As Drayton explains, the development of European science is intimately related not just to imperialism and commerce in natural products, but also to what many nowadays call traditional knowledge: 'what we may call the sciences of collection and comparison – among which we may include botany, zoology, and geology – depended on Europeans becoming exposed to the planet's physical and organic diversity, and often to the scientific traditions of non-European people' (*ibid.*) In turn, as he argues, 'the sciences shaped the pattern of imperial expansion'. A major consequence of this is that new economies came to arise 'on the basis of the discovery of the raw materials for food, medicines, dyes, and perfumes' (*ibid.*). This sounds mostly quite bad. Europeans exploited the rest and benefited from it. As is well known too, the diseases Europeans spread were catastrophic in many places, especially in the Americas.

But as we move towards the present one more clearly sees gains as well for the exploited. Life-expectancy rates in most developing countries have rapidly improved since historical times, and Western biomedicine has played a part in this, including acceptance of the germ theory of disease. Traditional medicine practitioners who are open to certain biomedical perspectives and practices to supplement their own ones have also contributed. Most likely, competition from biomedicine has also discredited some of the less plausible traditional treatments and led to them being abandoned, as happened in Europe with the practice of bleeding.

Traditional medicines are of course very much in use today. Typically, they consist of processed or unprocessed single or mixed natural products of plant, animal or mineral origin, administered orally in solid or liquid form. Whole plants may be used, or else plant or animal parts or their products. Unlike pharmaceuticals, they are not single chemicals obtained through industrial processes. The notion of the *active principle*, that is, the specific compound having the therapeutic effect, was, and remains, alien to traditional healers, whose treatments are inherently *impure*, allowing for the possibility of synergisms between the various ingredients. Further, their usage was, and still is, justified on the basis of theories of health, sickness, well-being and efficacy, as well as cultural and spiritual values, which most modern medical practitioners and pharmaceutical scientists understandably find impossible to accept. They are subject to a very different regulatory system and tend to be sold over the counter by retailers.

Encounters between European chemistry and non-European scientific traditions have had long-term repercussions in various different ways right up to the present. Thus, in both India and China, there is a great deal of hybridization going on in terms of describing, formulating, making, testing, evaluating, commercializing, in the ways that therapeutic claims are justified, and also of the growing centrality of 'the drug' in healthcare (Lei, 2014; Pordié and Gaudillière, 2013). Western biomedical ways are impacting on *traditional* medicine in other ways, as the latter's patient base expands globally. As the former aims to become more personalized, traditional medicines as they enter mainstream markets including over-the-counter outlets increasingly target more generalized use with standardized formulations and dosage instructions.

One must, however, distinguish between traditional remedies and traditional knowledge-*derived* treatments, the latter being traditional-modern hybrids. Indeed, some modern pharmacologists are re-investigating old herbal medicines (Adams *et al.*, 2009; Everett and Gabra, 2014). It remains to be seen whether they will come up with some treatments to benefit today's patients. The very existence of the discipline of ethnopharmacology underlines the argument being made here, that biomedicine and ethnobiology can and do interact – as they should. Nowadays, there is a consensus that such cross-cultural exchanges should be subject to fair procedures of consent and benefit-sharing, at least where ethnobiological knowledge and the plants used are current rather than merely historical; hence the CBD, Nagoya and the recent activities at the World Intellectual Property Organization concerning genetic resources, traditional knowledge and traditional cultural expressions.

Much has been lost, but has tradition really gained nothing from its exposure to other modes of understanding sickness and health? Indeed, traditional knowledge in health has not gone away, nor has it remained unchanged (Hsu, 2001; Pordié and Gaudillière, *op. cit.*). A much-cited figure from the World Health Organization is 80 per cent for the proportion of the developing country population that relies on traditional medicine to meet its primary healthcare needs. Chinese 'traditional' medicine remains highly popular, as are the classical traditional South Asian systems such as Ayurveda. These are well documented and the systems themselves are officially sanctioned with their own recognized training facilities and registered practitioners. They may have historical roots going back a very long time, but they are hardly devoid of novelty or innovation. As mentioned, Chinese medicine, for example, was not 'traditional' until it was named as such. Traditional Chinese Medicine co-evolved with Western scientific medicine and has accommodated elements of modern science, for example, the germ theory of disease (Lei, *op. cit.*). This largely sums up why the word 'tradition' is misleading and problematic, especially when applied broadly.

Of course, in other parts of the world, traditional systems of health have in no way been mainstreamed. Isolated indigenous peoples in places like the Amazon possess localized knowledge of flora and ecosystems, enabling them to meet many of their healthcare concerns. But it is unlikely that all of the biota they exploit or the knowledge they apply are entirely local, or have ever been. Although uncontacted groups still exist in the Amazon (Lawler, 2012; Wallace, 2011), most human societies do not stay rooted to one spot over centuries *and* over a substantial period turn their backs on the world outside their own little part of it.

This might all seem esoteric or at least irrelevant. It is not. One of the difficulties we have is that once we identify disparities in wealth and power, we understandably see the presence of an injustice and then clamour that something be done about it. Responses may be realistic and effective, but they may also take the form of poorly designed laws and regulations. This happened with many of the national and regional ABS regimes that have mostly failed to entice commercial users of genetic resources and traditional knowledge to engage in equitable partnerships with TK holding groups. I do not wish to be misunderstood about this point. There *is* injustice. It is done to indigenous peoples especially, but it is done to the rest of us too, albeit in a less obvious or tangible way. The causes are not the above-mentioned disparities in wealth and power alone, but also these legal and regulatory measures intended to alleviate them. Pharmaceutical scientists *can* and *do* learn from shamans and healers, even if not usually directly or even consciously. Notwithstanding the view that most of the low-hanging fruit has most likely been gathered already, one can still wonder how much more could be learned if healers and biomedical researchers got together more often than they do – which is almost never. At the same time, indigenous peoples need much better access to the fruits of biomedicine. Legal monopolies and excessive pricing get in the way. Perhaps we need to deal with both problems at the same time. Indigenous peoples, like the rest of us, get cancer and all the other diseases afflicting humans around the world. If so many health products have arisen over centuries from exchanges of knowledge and material between different societies, even under the worst circumstances of colonial domination, then we should be encouraging interaction, not discouraging it. To the extent that intellectual property rights and the assertion of bureaucratic access regulations lock up and separate knowledge and materials, we are all the poorer for it. If for once we were to look beyond the Manichean zero-sum view that views dominance and subjugation as inherent conditions, we might see a rich potential for positive interaction between traditional medicinal knowledge and biomedicine, just as there has famously been with artemisinin.

The problem of origin and attribution

Inherent in much of the international debate and the national and regional schemes adopted to implement access and benefit-sharing including prevention of TK misappropriation, is a basic conceptual mistake that would frequently make a correct assignment of rights in a clearly defined 'piece' of intangible property in the form of TK somewhat quixotic. The belief that TK is (i) clearly identifiable, pure, unadulterated and all the better for that; and (ii) mostly solely attributable to a specific group or country, is persistent but largely false. One consequence is that groups or governments claiming sole benefit allocations or extended sovereignty rights that others may be equally, or even more, entitled to, or perhaps even that nobody is entitled to.

Construing 'traditional knowledge' in its broadest conceivable sense renders origin and attribution especially problematic. Where does traditional knowledge end if anything done to or with, say, powdered turmeric (or some other product deemed to be a national heritage) by people outside India is treated as misappropriation? And for that matter, where also does national sovereignty end? An excessively broad interpretation will unreasonably lock up vast amounts of publicly available knowledge which no identifiable group of people or nation could make any credible claim to, and whose circulation can no longer realistically be controlled anyway. The way that farmers in South Asia use neem tree seeds to protect their crops, to give one example, really is public domain information, and compensation is due to nobody. The author is yet to hear of a compelling argument for saying it belongs to the farmers of this generation or to the government of India, none of whom actually came up with the idea of using neem in this way. The custodianship argument, that generations 'invest' in the responsibility of caring for resources and associated knowledge for future generations and should have rights on that basis, cannot take us very far in this particular context (though it might in others). It really is too widespread for that. Accordingly, constructing either an origin or custodianship-based moral case for compensation from others' commercial use can be a difficult if not impossible challenge. A general presumption behind many of the attacks on neem-related patents in Europe and the United States was that India was a victim on the basis that (a) neem is an Indian tree, (b) the knowledge being 'stolen' is Indian and (c) neem-related patents are essentially theft of India's biocultural heritage (Shiva, 1997). There is a problem with this. First, research suggests the species is native to a broad area, probably large enough to span Afghanistan and Myanmar. Second, the relevant 'traditional knowledge' is mostly very commonly known and is most unlikely to be bounded by the frontiers of modern India. Third, the tacit assertion that all neem-related patents are biopiracy, with India as victim, is tantamount to the assertion of reach-through claims over all global neem-related innovations. This is hard to justify legally, morally or on policy grounds. India, as with all countries, is not biologically or intellectually self-sufficient. It is of some relief, at least to this author, that in recent years, complaints about neem patenting have largely died down.

Even with ostensibly culturally quite distinct indigenous peoples, attributing knowledge to one group and one group alone can be controversial. To name one example, the use and knowledge of hoodia as a thirst and appetite suppressant is almost universally attributed to the San people of Southern Africa. However, recent research suggests the situation is far from clear. While the San may well be the original discoverers, many of them did not consume it, while various non-San and mixed populations have used it in recent centuries, and some of them cultivated it too (Kamau, 2013; Osseo-Asare, 2014).

The overregulation and corporatization tendencies

There are many who accept this point, but assume that top-down regulatory regimes on the lines of the CBD and Nagoya remain the least-worst answer. Unfortunately, national and regional ABS regimes have generally done little to generate officially sanctioned bioprospecting arrangements, especially when compared to the enormous number of transactions under the global genetic resource facilitated access regime of the FAO International Treaty on Plant Genetic Resources for Food and Agriculture. If ABS regimes are supposed to be promoting fair deal-making, they are not succeeding. Bilateralism itself seems to be the problem.[9] To the extent they encourage anything much, they tend to assume impersonal transactions based on conventional business practices albeit with the aim that these be 'fair and equitable' to be the normal mode of exchange. Experience strongly suggests that after twenty years of trying these have done far less to encourage fair exchange – or any exchange for that matter – than we expected. There are certainly strong arguments favouring the need to have governments play the role of correcting the power disparities when these are so extreme. However, a consequence of government paternalism, expansive understandings of TK and the North-South slant on the TK debate is typically a downplaying of the independent claims of specific national subpopulations. As a result, some governments seem inclined to treat traditional knowledge as national property, akin perhaps to mineral resources, with little consideration as to whether certain minority or indigenous groups may have perfectly legitimate claims to ownership or at least to claim independent benefits from inalienable usage rights which in turn may have several plausible legal bases. But whether property claims are made by corporation or by state it is still biopiracy. This is a somewhat controversial position, and one must admit that it is likely to be uncommon for national ownership of TK *explicitly* to be provided in a country's statutory law. However, implicit in much of the rhetoric and even some of the practices of national governments is the assumption that traditional knowledge existing in a country belongs to that country and that the government on that basis has the moral authority to pursue claims as if it had a property interest. When a government agency initiates a legal challenge against a wrongly granted patent claiming common knowledge of that country, as the Indian government did when it successfully opposed a US patent on the use of turmeric in wound healing, what lies behind an apparently straightforward legal request that the patent be revoked for lack of novelty is effectively an effort to enforce a right over the 'biopirated' knowledge. If this were not so, why would it take the trouble of challenging the patent (which itself had no economic value but was heavily laden with symbolism)? Another example: if a government sets up a competent authority to collect benefits from use of TK that cannot be traced to a specific indigenous group *and* which is common knowledge, is this not tantamount to *de facto* nationalization of common knowledge whose distribution may not even be confined to that country anyway?[10]

The desire to protect TK from aggressive, inappropriate and unwanted modernity is entirely understandable, and is a sentiment that the Nagoya Protocol seeks to give legal expression to. Accordingly, countering the otherwise inevitable TK erosion is to be done by valorizing it, creating a market for knowledge transactions so that access is exchanged for benefits. What else could be done? Eliminating all exchange of knowledge and materials is not the only alternative. It would almost certainly not work anyway. Knowledge is inherently 'leaky', plants get around pretty easily, and duplicates of both are likely to exist elsewhere. In comparison ABS regulation allowing conditional access seems at first blush to be perfectly reasonable and sensible.

To the extent that Nagoya provides for enhanced negotiating powers for indigenous peoples and local communities, this is of course welcome. However, indigenous peoples, inhabiting their own territorial spaces or elsewhere, as well as rural communities, are not companies. They lack suitable governance structures and the necessary business nous to negotiate business deals. Corporations engage in impersonal knowledge transactions because intellectual property rights make doing this possible, and such exchanges work for them because intellectual property rights are the common currency that they all use and understand and which fits into their business models and the governing commercial law legal frameworks.

Embedding TK protection in ABS regulations reduces a broad range of possible exchange norms to a single one, that of top down bureaucratic frameworks which vest strong regulatory powers in governments but that also permit conditional private contracting. CBD-inspired access and benefit-sharing frameworks tend to imply that mediating between the different interests of providers and users can only be done through the oversight of state bureaucracy and the more the better. None of this is to deny that misappropriation of genetic resources and TK happens, that state action may be necessary in some form or another, nor that there is anything wrong with the principle that commercial users should share benefits with original providers of biological material and local and indigenous knowledge and that exchanges should be based on mutually agreed terms and prior informed consent. Indeed the present author is wholly supportive of the latter principle. But regulation needs to be grounded on a much higher degree of realism and be based on the cautious encouragement of fully consensual partnerships under norms that are appropriate both for providers and users. Between free exchange at one end and secrecy enforced by threats of violence at the other is a broad spectrum of possibilities which fall mostly outside what is envisaged in the CBD and Nagoya.

The exchange value 'distraction'

A fourth problem is that the current approaches to TK protection are concerned primarily with the exchange values of genetic resources and TK as if the importance is purely for their industrial potential. This is entirely understandable and nothing in this chapter should be taken as an argument that misappropriation, unequal exchange and blatant exploitation of TK holding individuals, groups and people should be deemed to be acceptable. However, there is a consequent tendency to downplay their intrinsic and local (commercial and non-economic) values which are in fact much more significant especially for indigenous peoples for whom knowledge is an essential aspect of the cultural identify that binds them together and ensures their well-being. It also largely disregards the context for the inclusion in the CBD of 'knowledge, innovations and practices of indigenous and local communities embodying traditional lifestyles': their contribution to *in situ* conservation of biodiversity and to sustainable use, rather than their supposed commercial value. These approaches tend also to push them towards acceptance of legal instruments and property rules that businesses have much more experience of using (see above). Again, terminology and rhetoric do not help. By treating traditional knowledge as a unified, discrete counter-modern stock of useful knowledge, it gets reduced to an array of 'raw' inputs for life science corporations, which is then regulated accordingly. In doing so, we devalue it, essentially reducing it to a random compilation of leads, hints, hopes, errors, deceptions and *cul de sacs* from which the useful needs to be separated from the supposedly useless. The rhetoric might suggest it is something more worthy and significant than that but close inspection of how TK gets inputted into commercially oriented scientific research reveals that TK has those diverse and generally rather limited qualities in that particular context. Anyway, the persisting hopes that TK has genuine value in that setting leads regulators and policymakers to focus their attention on the instrumental value of TK to others, and away from the holders themselves within their own communities and among others with which they socialize and otherwise interact. This has negative practical implications, and the approaches being considered internationally in their current form will not help, especially as these aforementioned leads, hints and hopes will not in most cases be reducible to traceable and enforceable single legal claims justified by having made a tangible contribution to a commercial product.[11] Meanwhile, the land and other rights of indigenous groups within the borders of countries whose representatives in Geneva clamour for international protection of TK continue often to be denied.

Conclusion

Where tradition ends and modern formal systems of knowledge start is far less clear-cut a matter than many suppose. Both have and continue to borrow from each other often in beneficial ways. If a global access and benefit-sharing regime and/or a treaty on traditional knowledge protection can assist such healthy cross-fertilization in fair and equitable ways all well and good. This chapter has identified basic conceptual problems that currently make such a noble goal hard to achieve.

Policymakers debating the Nagoya Protocol and seeking ways to implement it must at least face up to the futility of confining the norms of exchange to intellectual property rights, contracts, top-down government regulations and nothing more than those. Those are the laws of the powerful. Instead, the rules and principles of the weaker party should apply in the first instance. The weaker party is not the corporation, nor is it the government but the indigenous peoples. That is a matter of fairness but it is also the only practical basis for mutually advantageous relationships. We need to strengthen the role of customary law as a third source of regulatory norms that facilitates rather than stops two-way exchange but in ways that are culturally compatible with indigenous peoples' values and that further their interests (Coombe, 2001; Tobin, 2014, 2015). Either that, or to develop 'hybrid approaches that interweave elements of Western law and local, traditional rules for the circulation of knowledge' (Brown, 2005). These laws are local or national but they are not universal in their scope. What *is* universal though is that customary norms are far more ubiquitous than people assume.

Ultimately, it is highly unlikely that anything other than piecemeal locally driven and controlled solutions can provide much satisfaction for those keen for justice to be seen to be done. Indigenous peoples should be allowed to enjoy the full value of their knowledge *to themselves* first. It might be that a necessary first step is to improve their ability to profit from local and regional sales of their own goods. Branding is relevant here. In Peru, communities inhabiting the Potato Park use an attractive logo on their products which include herbal teas and textiles (Argumedo, 2013). Given that the cultural heritage of many indigenous groups attracts many consumers seeking authentic and exotic products including souvenirs and can generate immediate benefits, it might be that marks certifying qualities of authenticity, sustainability and heritage could be of real benefit (Swiderska *et al.*, 2016), at least in areas frequented by tourists who are more likely than most locally residing customers to be willing to pay good prices. Once indigenous peoples are successful in retaining most or all of the income generated from local and regional trade in their *own* goods, one would expect them to be in a better position to exploit the exchange value of their knowledge with international scientific institutions and commercial partners through formal agreements than if they were unable even to trade locally in goods without being exploited. For less visited or more isolated areas the fact remains that they are unlikely to be able without such experiences and considerable outside support to negotiate positive outcomes with life science and biotechnology firms even with the most apparently friendly regulatory framework. Local control with secure land rights and application of enforceable customary norms would appear to be necessary conditions. The operative international ABS instrument have little that is concrete to say about any of these.

Notes

1 This chapter benefited from comments on an earlier draft by Peter Harrison (University of York). The title's allusion to Bruno Latour's *We Have Never Been Modern* (Cambridge: Harvard University Press, 1993) is not incidental.
2 See Arts 8(j) and 10(c), Convention on Biological Diversity.
3 The 'ABS+plus IP' model is one that envisages bilateral negotiating as a mutually beneficial approach that empowers the weaker access-providing party, and intellectual property in current or modified form as the primary supplemental legal means to govern the distributional elements of such deals. The conventional discourse of CBD debates and proposals including Nagoya are framed in this way.
4 Rather provocatively, Kuper also suggests that this tendency to falsely dichotomize is being perpetuated by use of the modern term indigenous peoples.
5 In the nineteenth century, the famous Oxford comparativist jurist Sir Henry Sumner Maine offered a rather similar perspective. Maine envisaged a shift 'from status to contract' as societies evolve into 'progressive societies' (Maine, 1861).
6 For an excellent early critique highlighting the fallacies and harms caused by the use of tradition and modern as if they are binary opposites, see Gusfield (1967).
7 Emphasis added.
8 References deleted from quote.
9 The economist Joseph Vogel is a long-time critic of bilateralism, and has proposed that genetic resource provider countries form cartels (Vogel, 1997). However, his main focus is genetic resources as what he and others call 'natural information' rather than TK *per se*. See also Ruiz Muller (2015); Stoll (2013).
10 Iraq's Draft Law on the Protection and Exchange of Genetic Resources for Food and Agriculture states that information related to plant genetic resources is owned by the state, and the clear implication is that this includes TK (Khalaf, 2016).
11 Albeit expressed rather differently, a similar argument is made by Angerer (2011). Related to this difficulty is the issue of potentially extensive distance in material and cognitive terms between biological material and associated TK, the invention claimed in a patent, and a final product. Should benefit-sharing obligations be calibrated so as to be in proportion to distance according to some kind of measurement? Accordingly the shorter the distance, the greater would be the benefits. See Harrison (2015).

References

Adams, M., Berset, C., Kessler, M. and Hamburger, M. (2009) 'Medicinal herbs for the treatment of rheumatic disorders – A survey of European herbals from the 16th and 17th century', *Journal of Ethnopharmacology* Vol. 121, pp. 343–59.

Amster, E.J. (2013) *Medicine and the Saints: Science, Islam, and the Colonial Encounter in Morocco, 1877–1956*, University of Texas Press, Austin.

Angerer, K. (2011) 'Frog tales – on poison dart frogs, epibatidine, and the sharing of biodiversity', *Innovation: The European Journal of Social Science Research* Vol. 24 no. 3, pp. 353–69.

Argumedo, A. (2013) *Collective Trademarks and Biocultural Heritage: Towards New Indications of Distinction for Indigenous Peoples in the Potato Park, Peru*, International Institute for Environment and Development, London.

Arikha, N. (2007) *Passions and Tempers: A History of the Humours*, HarperCollins, New York.

Brown, M.F. (2005) 'Heritage trouble: Recent work on the protection of intangible cultural property', *International Journal of Cultural Property*, Vol. 12, pp. 40–61.

Bynum, W. (1994) *Science and the Practice of Medicine in the Nineteenth Century*, Cambridge University Press, Cambridge.

Bynum, W.F. (2006) The rise of science in medicine, 1850–1913, in W.F. Bynum, A. Hardy, S. Jacyna, C. Lawrence and E.M. Tansey (eds) *The Western Medical Tradition: 1800 to 2000*, Cambridge University Press, Cambridge, pp. 111–239.

Chakrabarti, P. (2012) *Bacteriology in British India: Laboratory Medicine and the Tropics*, University of Rochester Press, Rochester.

Conklin, B. (2002) 'Shamans versus pirates in the Amazonian treasure chest', *American Anthropologist* Vol. 104, no. 4, pp. 1050–61.
Cook, H.J. (2007) *Matters of Exchange: Commerce, Medicine, and Science in the Dutch Golden Age*, Yale University Press, New Haven.
Coombe, R. (2001) 'The recognition of indigenous peoples' and community traditional knowledge in international law', *St. Thomas Law Review* Vol. 14, pp. 275–85.
Drayton, R. (2000) *Nature's Government: Science, Imperial Britain, and the 'Improvement' of the World*, Yale University Press, New Haven.
Dutfield, G. (2009) *Intellectual Property Rights and the Life Science Industries: Past, Present and Future*, World Scientific, Singapore.
Englund, H. and Leach, J. (2000) 'Ethnography and the meta-narratives of modernity', *Current Anthropology* Vol. 41, no. 2, pp. 225–48.
Everett, N. and Gabra, M. (2014) The pharmacology of medieval sedatives: The 'Great Rest' of the Antidotarium Nicolai, *Journal of Ethnopharmacology*, Vol. 155 no. 1, pp. 443–49.
Gänger, S. (2015) 'World trade in medicinal plants from Spanish America, 1717–1815', *Medical History*, Vol. 59 no 1, pp. 44–62.
Greene, J.A. (2014) *Generic: The Unbranding of Modern Medicine*, Johns Hopkins University Press, Baltimore.
Greiber, T., Peña Moreno, S., Åhrén, M., Nieto Carrasco, J., Chege Kamau, E., Cabrera Medaglia, J., Oliva, M.J., Perron-Welch, F. in cooperation with Ali, N. and Williams, C. (2012) *An Explanatory Guide to the Nagoya Protocol on Access and Benefit*, IUCN, Gland.
Gusfield, J.R. (1967) 'Tradition and modernity: Misplaced polarities in the study of social change', *American Journal of Sociology* Vol. 72 no. 4, pp. 351–62.
Harrison, P.S. (2015) 'Distal horizons: an investigation of the justifiable downstream limits to the positive protection of traditional knowledge associated with genetic resources within drug discovery', PhD thesis, University of York, UK.
Hsu, E. (ed.) (2001) *Innovation in Chinese Medicine*, Cambridge University Press, Cambridge.
Kamau, E.C. (2013) 'Common pools of traditional knowledge and related genetic resources: A case study of San-Hoodia', in E.C. Kamau and G. Winter (eds), *Common Pools of Genetic Resources: Equity and Innovation in International Biodiversity Law*, Routledge, Abingdon, pp. 40–54.
Khalaf, N.M. (2016) 'Green economy in the Arab region: Regulating agricultural biodiversity and plant related innovation in Egypt and Iraq', PhD thesis, Bangor University, UK.
Kuper, A. (2005) *The Reinvention of Primitive Society: Transformations of a Myth*, Routledge, Abingdon.
Lansing, J.S. (2007) *Priests and Programmers: Technologies of Power in the Engineered Landscape of Bali*, Princeton University Press, Princeton.
Lawler, A. (2012) 'Mashco Piro tribe emerges from isolation in Peru', *Science* Vol. 349, p. 679.
Le Fanu, J. (1999) *The Rise and Fall of Modern Medicine*, Little, Brown and Co., London.
Lei, S.H.-L. (2014) *Neither Donkey nor Horse: Medicine in the Struggle over China's Modernity*, University of Chicago Press, Chicago.
Maine, H.S. (1861) *Ancient Law: Its Connection with the Early History of Society, and Its Relation to Modern Ideas*, John Murray, London.
Mooney, P.R. (2000) 'Why we call it biopiracy', in H. Svarstad and S.S. Dhillion, *Responding to Bioprospecting: From Biodiversity in the South to Medicines in the North*, Spartacus Forlag, Oslo, pp. 37–44.
Olivier de Sardan, J.P. (2005) *Anthropology of Development: Understanding Contemporary Social Change*, Zed Books, London and New York.
Osseo-Asare, A.D. (2008) 'Bioprospecting and resistance: Transforming poisoned arrows into strophantin pills in colonial Gold Coast, 1885–1922', *Social History of Medicine* Vol. 21 no. 2, pp. 269–90.
Osseo-Asare, A.D. (2014) *Bitter Roots: The Search for Healing Plants in Africa*, University of Chicago Press, Chicago and London.
Pelletier, J. and Caventou, J. (1820) 'Recherches chimiques sur les quinquinas', *Annales de Chimie et Physique* Vol. 15, pp. 289–318.
Pordié, L. and Gaudillière, J.-P. (2013) 'The reformulation regime in drug discovery: Revisiting polyherbals and property rights in the Ayurvedic industry', *East Asian Science, Technology and Society: An International Journal*, Vol. 8, pp. 1–23.
Robinson, D.F. (2010) *Confronting Biopiracy: Challenges, Cases and International Debates*, Earthscan, London.
Rostow, W.W. (1960) *The Stages of Economic Growth: A Non-Communist Manifesto*, Cambridge University Press, Cambridge.

Ruiz Muller, M. (2015) *Genetic Resources as Natural Information: Implications for the Convention on Biological Diversity and Nagoya Protocol*, Routledge, Abingdon.
Rural Advancement Foundation International (RAFI) (1994) *Conserving Indigenous Knowledge: Integrating Two Systems of Innovation. An Independent Study by the Rural Advancement Foundation International*, United Nations Development Programme, New York.
Shiva, V. (1997) *Biopiracy: The Plunder of Nature and Knowledge*, South End Press, New York.
Sneader, W. (2005) *Drug Discovery: A History*, John Wiley and Sons, Chichester.
Stockwell, B.R. (2011) *The Quest for the Cure: The Science and Stories Behind the Next Generation of Medicines*, Columbia University Press, New York.
Stoll, P.-T. (2013) 'ABS, justice, pools and the Nagoya Protocol', in E.C. Kamau and G. Winter (eds), *Common Pools of Genetic Resources: Equity and Innovation in International Biodiversity Law*, Routledge, Abingdon, pp. 305–14.
Strathern, M. (1999) 'What is intellectual property after? *The Sociological Review* Vol. 47, pp. 156–80.
Sunder, M. (2006) 'The invention of traditional knowledge', *UC Davis Legal Research Paper No. 75.*
Swiderska, K., Mead, A., Dutfield, G. and Argumedo, A. (2016) 'Designing a labelling system for biocultural heritage-based products', *Policy Matters* no. 21, pp. 140–48.
Tallis, R. (2004) *Hippocratic Oaths: Medicine and its Discontents*, Atlantic Books, London.
Tilley, H. (2011) *Africa as a Living Laboratory: Empire, Development, and the Problem of Scientific Knowledge, 1870–1950*, Chicago University Press, Chicago.
Tobin, B. (2014) *Indigenous Peoples, Customary Law and Human Rights: Why Living Law Matters*, Routledge, Abingdon.
Tobin, B. (2015) 'Traditional Knowledge sovereignty: The fundamental role of customary law in the protection of traditional knowledge', in M. Rimmer (ed.) *Intellectual Property: A Handbook of Contemporary Research*, Edward Elgar, Cheltenham.
United Nations Conference on Trade and Development (2014) *The Convention on Biological Diversity and the Nagoya Protocol: Intellectual Property Implications. A Handbook on the Interface between Global Access and Benefit-sharing Rules and Intellectual Property*, UNCTAD, Geneva.
Vogel, J.H. (1997) 'Bioprospecting', *Biopolicy Journal*, Vol. 2 no. 5.
Wallace, S. (2011) *The Unconquered: In Search of the Amazon's Last Uncontacted Tribes*, Crown Publishers, New York.

19

WHERE CUSTOM IS THE LAW

State and user obligations to 'take into consideration' customary law governing traditional knowledge and genetic resources

Brendan Tobin

Introduction

The protection of traditional knowledge is widely seen as being crucial for achieving fair and equitable regulation of access to and sharing of benefits arising from use of both genetic resources and traditional knowledge itself (Nijar, 1996; Posey and Dutfield, 1996a; Laird and Noejovich, 2002). Significant advances in national, regional and international law beginning with the adoption of the Convention on Biological Diversity (CBD) in 1992 have confirmed the collective and individual rights of Indigenous Peoples and local communities over their traditional knowledge and genetic resources.[1] The United Nations Declaration on the Rights of Indigenous Peoples (UNDRIP) recognizes their rights over their traditional knowledge, traditional medicines, seeds, plants and intellectual property.[2] With the entry into force of the Nagoya Protocol on Access to Genetic Resources and the Fair and Equitable Sharing of Benefits Arising from their Utilization to the CBD (Nagoya Protocol), in 2014, these rights were given binding legal status, obliging States Parties to adopt and enforce measures to protect rights over traditional knowledge.

A central part of the new legal order is composed of obligations under UNDRIP and the Nagoya Protocol requiring States to give due respect, recognition and/or consideration for the laws, primarily customary (i.e., uncodified, oral) laws, of Indigenous Peoples and local communities[3] in their implementation of national law and policy on traditional knowledge. Customary law has also figured prominently in draft international instruments on the protection of traditional knowledge under negotiation at the World Intellectual Property Organization (WIPO) Intergovernmental Committee on Intellectual Property, Genetic Resources, Traditional Knowledge and Folklore (IGC). At the national level, a majority of countries give constitutional recognition to the rights of Indigenous Peoples and/or local communities to govern all or part of their affairs in accordance with their own customary laws (Cuskelly, 2011). Responsibility for ensuring compliance with these customary laws is being passed on to users of traditional knowledge in national and regional measures implementing the Nagoya Protocol.

This chapter focuses on the significance of customary law for the international regulation of traditional knowledge. It begins with consideration of the importance of traditional knowledge for Indigenous Peoples and local communities and the challenges posed by scientific and commercial use. It then examines the nature and status of Indigenous Peoples and local communities' legal regimes under international law. It goes on to discuss national and international regulation of traditional knowledge and the role of customary law in its protection, with particular attention to the experience of Andean countries. Finally, it explores the potential of biocultural protocols to bridge the gaps between customary legal regimes, national laws and international law.

The chapter concludes that action by both source and user countries, as well as by users, to ensure due respect, recognition and consideration of customary law, is vital if legal certainty is to be secured for both traditional custodians and the users of traditional knowledge and associated genetic resources. It further concludes that Indigenous Peoples and local communities have an important opportunity to influence the nature of national and international regulation of traditional knowledge through the exercise of their rights as lawmakers in the development of community, national, regional and international protocols on access to and use of traditional knowledge. It suggests the international community provide support for Indigenous Peoples and local communities in this endeavour with a view to securing the effective and culturally appropriate protection of their knowledge and resource rights.

Traditional knowledge rights and wrongs

Traditional knowledge includes, but is not limited to, knowledge of astrology, history, art, song, dance, stories, mathematics, language, land and resource management, agriculture, fisheries, human and animal health, education, religion, law, selection of traditional authorities, and the governance of intragenerational and intergenerational caring and sharing of knowledge (COICA, 1999). It is central to the social, cultural, economic and spiritual lives of Indigenous Peoples and local communities and their realization of a wide range of human rights including rights to food, health, life, development, human dignity, culture and self-determination. At the same time, the realization of rights to human dignity, culture, self-determination, as well as to traditional lands, resources, and legal regimes is dependent upon the maintenance, development and continuing use of traditional knowledge (Tobin and Taylor, 2009). In 1995, Special Rapporteur Erica Irene Daes presented a proposal for Principles and Guidelines for the Protection of the Heritage of Indigenous Peoples, which state that

> International recognition and respect for the Indigenous Peoples' own customs, rules and practices for the transmission of their heritage to future generations is essential to [their] enjoyment of human rights and human dignity.[4]

Not surprisingly, Indigenous Peoples view the protection of traditional knowledge as being 'just as important as the struggle for self-determination'.[5]

Traditional knowledge of plants and animals has for centuries served as an important source of medicinal and agricultural knowledge to the wider world (Juma, 1989; Githae, 2009). Advances in biotechnology have led to increased interest in its potential for the identification of plants and animals with active compounds for pharmaceutical and natural product research (Sanchez and Juma, 1994; Ten Kate and Laird, 1999). While this has, undoubtedly, led to some innovative agreements promoting fair and equitable sharing of benefits with Indigenous Peoples and local communities (Laird and Noejovich, 2002), numerous cases exist of what has come to be known as biopiracy, i.e., the unapproved and or uncompensated use of traditional knowledge or genetic resources. Iconic cases of biopiracy include those involving patents relating to turmeric, neem, ayahuasca, maca and hoodia (Krieger, 2008; Robinson, 2010). More recently, French researchers have sought to patent the anti-malarial properties of *Quassia amara* from French Guiana,[6] while in Australia, concern has been expressed over patents related to Gumbi Gumbi or *P. phylliraeoides*, which has recorded Australian Aboriginal usage to treat bacterial and fungal diseases.[7]

Successful actions have been brought against turmeric and neem patents in Europe, and a benefit-sharing deal has been negotiated for the hoodia plant (Wynberg, 2004). However, Indigenous Peoples and local communities have received few of the benefits generated from the multi-billion dollar trade in products developed utilizing traditional knowledge and associated genetic resources (Reid, 1993; Nijar, 1996; Robinson, 2010). In fact, they are often worse off as a result of such trade. In the case of maca, a medicinal plant used by Andean communities, for example, over-exploitation for international trade has impeded traditional free local access, while all the rewards go to Peruvian and foreign commercial enterprises.

International regulation of traditional knowledge

Recognition of Indigenous Peoples' rights to their traditional knowledge is found in more than a dozen international instruments (Dodson, 2007), including the CBD and UNDRIP.[8] UNDRIP recognizes Indigenous Peoples' rights over their genetic resources, traditional knowledge and intellectual property, as well as their rights to their own legal regimes and traditional authorities.[9] Article 8(j) of the CBD obliges States to promote wider use of traditional knowledge with the consent of indigenous and local communities. The Nagoya Protocol has expanded on Article 8(j), establishing binding legal obligations for Parties to take measures to ensure that access to and use of the genetic resources (Article 6.2) and traditional knowledge (Article 7) of 'indigenous and local communities' is subject to their prior informed consent and on mutually agreed terms. The Protocol requires Parties to adopt measures to secure compliance with the relevant national laws of the country where genetic resources (Article 15) and traditional knowledge (Article 16) were legally obtained in accordance with the Protocol, and where the indigenous and local community custodians of traditional knowledge are located.

To date, the implementation of the Nagoya Protocol by developed nations has tended to focus on the establishment of 'due diligence' requirements for users of genetic resources and traditional knowledge.[10] Switzerland, for example, requires users to exercise 'due diligence' in ensuring that traditional knowledge and genetic resources are legally obtained and are the subject of fair and equitable benefit-sharing.[11] The European Union requires users to exercise due diligence to ensure compliance with national laws regulating access to genetic resources and traditional knowledge in providing countries.[12] Where Indigenous Peoples and local communities have the jurisdiction to govern, wholly or in part, access to and/or use of their traditional knowledge and associated resources, then 'due diligence' studies will need to ensure that there has been compliance with their legal regimes.

A weakness with the European legislation is that it only extends protection to traditional knowledge that is legally protected in the country from which it is sourced. As there are less than a dozen specific national regimes in place, the extent of traditional knowledge covered by the European regime would be very limited were it not for the important role of customary law at the national level. At least 115 countries already grant some form of constitutional recognition to customary law in areas such as rights to culture, land, resources and traditional knowledge as well as to autonomy, traditional authorities and/or legal systems (Cuskelly, 2011).

Therefore, users will need to exercise 'due diligence' to ascertain the status of customary law over traditional knowledge wherever it is sourced.

Under the provisions of Article 12 of the Nagoya Protocol, States are obliged to 'take into consideration' customary law in implementing its provisions relating to traditional knowledge. This has clearly not been done in the EU Regulation, which makes no reference to customary law. Furthermore, the EU process leading to the adoption of its implementing regulations shows no effort to examine the challenges and opportunities for consideration of customary law and its role in protection of traditional knowledge (Tobin, 2014a). This is clearly a case of non-compliance with both the word and the spirit of the Protocol, which, if not rectified, should be brought to the attention of the Protocol's Compliance Committee.

For countries desirous of protecting the rights of the Indigenous Peoples and local communities over their traditional knowledge, strengthening and making known the extent of national recognition of the role of customary law in regulating traditional knowledge and associated genetic resources (e.g., providing information to the CBD clearing house mechanism) offers an important means to prevent the unapproved and uncompensated transfer of rights and value of traditional knowledge to third parties. In this way, States can help ensure the application of due diligence requirements under implementing legislation in foreign jurisdictions, such as that of the European Union and Switzerland, to the traditional knowledge of their citizens.

The nature and status of customary law

All dominant legal regimes have been influenced in their development by customary law and all legal regimes display aspects of customary practice (Glenn, 2014). Unlike formal instruments of positive (stipulated) law, customary law is generally unwritten, informal, spontaneous, conservative, status based and reliant upon its own enforcement procedures (Tomtavala, 2005). Amongst its principal attributes are legitimacy, flexibility and adaptability. It draws its legitimacy and strength from the willingness of indigenous and local community members to be bound by its unwritten laws and their continuing reliance on traditional authorities to resolve internal disputes. In contrast to dominant legal regimes, customary law tends to seek restoration of community harmony rather than retribution (Tobin, 2014b).

One of the principal reasons for the survival of customary law among Indigenous Peoples and local communities has been its link to their subsistence strategies and the fact that national legal systems are not considered efficient, accessible or appropriate for their cultural realities (Yrigoyen, 1999; Van Cott, 2003). In addition, customary legal regimes tend to be less costly, and faster in providing remedies (Van Cott, 2003). Many Indigenous Peoples and local communities continue to rely on customary law and traditional knowledge to regulate land and resource use, define sacred sites and areas of taboo, and place permanent or temporary restrictions on hunting, fishing and resource collection over all or a specified group of species of plant or animal or over specific areas, such as spawning grounds for fish (Caillaud *et al.*, 2005). Although there is a tendency to consider both customary law and traditional knowledge as static, they are, in fact, dynamic systems of law and knowledge generation, which inform decision-making with attention to external environmental, political and legal factors over time. It is important, therefore, to avoid the tendency to either demonize or idealize such systems of law and knowledge and to appreciate their true nature, strengths and weaknesses (Johnson, 2012).

Historically, customary law has tended to be stronger where the State is weak and in many places it continues to be the dominant body of law applicable to the day-to-day governance of civil and criminal law issues (Chanock, 2005). In post-colonial countries, decolonization of their legal regimes has frequently resulted in greater recognition and enforcement of custom. Specific examples of the foregoing include: the recognition of custom as standing alongside the common law under the Constitution of South Africa (Cornell and Muvangua, 2012),[13] the development of the underlying law in Papua New Guinea which seeks to place customary law above inherited British common law (Corrin, 2014), and the elevation of customary justice to stand alongside national law in Bolivia.[14]

International recognition of Indigenous Peoples' rights to regulate their affairs in accordance with their own laws is grounded in Article 1 of the 1966 United Nations International Covenant on Civil and Political Rights and Article 1 of the 1966 United Nations International Covenant on Economic, Social and Cultural Rights, which recognize the rights of all peoples to self-determination. ILO Convention 169, the Convention on the Prevention of all Forms of Racial Discrimination, the Convention on the Rights of the Child, regional human rights instruments and UNDRIP have all advanced awareness and acceptance of the role of customary law as both a source of law and a primary basis of self-standing legal regimes governing the affairs of large sectors of the global populace.[15]

The central premise of UNDRIP is that Indigenous Peoples have a right of self-determination[16] and are entitled to autonomy or self-government over their internal affairs.[17] This includes the right to maintain and strengthen their distinct political, legal, economic, social and cultural institutions[18] and to promote, develop and maintain their juridical systems or customs, in accordance with international human rights.[19] UNDRIP gives specific recognition to Indigenous Peoples' rights over human and genetic resources, seeds, medicines, and knowledge of the properties of fauna and flora, as well as over their cultural heritage and intellectual property.[20] It requires States to provide redress, including compensation, for breaches of Indigenous Peoples' resource rights[21] and requires that they have access to prompt, just and fair procedures to resolve disputes with States or other parties.[22] These procedures and any decision taken under them must give due regard to the Indigenous Peoples' customary laws and traditions and to international human rights.[23]

Although not legally binding of itself, UNDRIP is widely seen as reflecting the status of Indigenous Peoples' human rights under treaty law and customary international law (Stamatopoulou, 2010). Unlike treaties, which are only binding on states that have ratified them, customary international law is, subject to few exceptions, binding on all states (Smelcer, 2006). According to Anaya, customary international law has crystallized around Indigenous Peoples' rights to self-determination, traditional lands and cultural integrity (Anaya, 2004). It has also been claimed to exist with regard to their resources, language, sacred sites and cultural artefacts (Weissner, 1999), as well as to their own legal regimes (Perry, 2011; Tobin, 2011) and the obligation of states to respect and recognize Indigenous Peoples' legal regimes in order to secure their human rights (Tobin, 2011).

The most important instrument for the protection of Indigenous Peoples' human rights is the International Labour Organization Convention 169 on Indigenous and Tribal Peoples in Independent Countries (Convention 169), which is legally binding on State Parties. The Convention, which has been ratified by 22 countries,[24] requires States to secure Indigenous Peoples' participation in decision-making processes that affect them.[25] It also requires States 'in applying national laws and regulations to the peoples concerned' to give due regard to their customs or customary laws.[26] It provides that Indigenous and tribal peoples 'shall have the right to retain their own customs and institutions, where these are not incompatible with fundamental rights defined by the national legal system and with internationally recognized human rights'.[27] The binding nature of Convention 169 means that States Parties must ensure that other international and regional laws, to which they are party, are developed and implemented with due regard to the customs and customary laws of relevant Indigenous Peoples.

Customary law and the protection of traditional knowledge

The relationship between the protection of traditional knowledge and customary law was first recognized in Agenda 21, adopted in 1992, which recommended that governments '[a]dopt or strengthen appropriate policies and/or legal instruments that will protect indigenous intellectual and cultural property and the right to preserve customary and administrative systems and practices'.[28] The Conference of the Parties (COP), the governing body of the CBD, has signalled the importance of ensuring that States Parties respect the customary laws and practices of Indigenous Peoples and local communities in the adoption and implementation of strategies for the protection of traditional knowledge.[29] The Bonn Guidelines on Access to Genetic Resources and Fair and Equitable Sharing of the Benefits Arising out of their Utilization, states that the prior informed consent of indigenous and local communities 'should be obtained, in accordance with their "traditional practices"'.[30] As noted earlier, Article 12 of the Nagoya Protocol obliges all Parties in implementing the Protocol to 'take into consideration the customary laws and community protocols and practices' of Indigenous Peoples and local communities. This applies to the customary laws of both Indigenous Peoples and local communities within a country and those whose traditional knowledge may be imported into the country.

Although, the CBD has a specific mandate for addressing issues of traditional knowledge related to biodiversity[31] it has tended to defer to the WIPO IGC on the issue of creation and/or recognition of property rights over traditional knowledge in favour of Indigenous Peoples and local communities. Established in 2001 the WIPO IGC was tasked from the outset, by the WIPO General Assembly, with investigation of the role of customary law in protection of traditional knowledge (WIPO, 2001).

At the IGC, Indigenous Peoples and local communities have continuously stressed that any regime should be based upon customary law. In the words of a representative of the Kaska Dena Council, an objective of any regime should be to

> [m]eet the actual needs of communities ... be guided by the aspirations and expectations expressed directly by Indigenous Peoples and local communities, [and] respect their rights under customary law, including indigenous national, regional and international law.[32]

At the IGC, recognition of the importance of customary law for protection of traditional knowledge has come from a wide range of countries, including China,[33] India,[34] New Zealand,[35] Nigeria,[36] South Africa[37] and the United States,[38] as well as the African Group[39] and the Like Minded Group of countries.[40] From the start, the IGC draft instruments recognized a central role for customary law in, among other things, identifying traditional knowledge,[41] ensuring equitable benefit-sharing,[42] defining cases of misappropriation[43] and unfair trade[44] and securing customary usage.[45] This recognition of customary law by the IGC demonstrated a collective commitment by a majority of states to the development of a *sui generis* (i.e. of its own kind) regime for the protection of traditional knowledge. Support has not, however, been unanimous and a small number of States, including Australia and Japan, have sought to prevent recognition of any significant role for customary law. Canada has argued against what it called the 'extraterritorial application of indigenous customary laws and protocols',[46] which it claimed could have legal implications for Canada beyond the role of intellectual property law.[47] These differing views led to calls by the IGC for research into the role of customary law in protecting traditional knowledge.

In 2006 WIPO prepared and circulated a draft issues paper outlining a wide variety of ways in which customary law has been recognized in intellectual property related cases and national legislation (WIPO, 2006; Taubman, 2005). It also published a report from Latin America on the role of customary law in regulating traditional knowledge (Cruz, 2007). WIPO was, however, criticized by New Zealand for 'focusing on existing legal frameworks rather than on the holistic needs of indigenous and local communities. As a result [of which] indigenous and local communities' priorities, customary laws, protocols and values associated with their traditional knowledge and traditional cultural expressions were not the focus or main premises of the analyses.'[48] Between 2004 and 2006 the United Nations University, Institute of Advanced Studies (UNU- IAS), in collaboration with WIPO, the International Union for the Conservation of Nature (IUCN), Townsville University, and a range of local and international partners, coordinated a series of regional and sub-regional workshops on the role of customary law in regulating access to genetic resources and traditional knowledge in the Andean and Pacific Island countries. This involved participatory workshops with representatives of Indigenous Peoples, local communities, government, non-governmental organizations and research institutions. One of the central conclusions of a 2008 report on the findings of this programme was a call from participants for a more bottom-up approach to research of the actual role of customary law in protecting traditional knowledge (Tobin, 2013).

Despite the importance given by IGC parties to research into the role of customary law in protection of traditional knowledge, the WIPO secretariat held off publishing both the final issues paper (WIPO, 2013) and the Report of the Andean and Pacific Islands investigation (Tobin, 2013) until 2013, by which stage almost all references to customary law in the draft negotiating texts had been deleted or replaced with the term 'cultural norms' – a term unknown to international law.

The roll back in recognition of customary law at the IGC, when taken together with the failure by the European Union to 'take into consideration' customary law in any meaningful sense in the implementation the Nagoya Protocol (Tobin, 2014a), would seem to reflect a growing resistance by developed nations, in particular ex-colonial powers and settler states, to the resurgence of the status of customary law under constitutional and international human rights law (Borrows, 2002; Cornell and Muvangua, 2012; Tobin, 2014b). This runs counter to the approach being taken by developing countries at the regional level where the Andean Community of Nations, the African Regional Intellectual Property Organization, the Organization of African States and the Pacific Islands Forum, have all adopted laws and/or policies that recognize a central role for customary law in the governance of traditional knowledge.[49]

Customary law and protection of traditional knowledge in Andean countries

In 1996, the Andean Community of Nations adopted Decision 391, establishing a regional access and benefit-sharing regime which requires prior informed consent of indigenous, local and Afro-American communities, as a pre-condition for obtaining approval for bioprospecting agreements involving the collection of resources on their land or use of their traditional knowledge.[50] Decision 391 recognizes the authority of 'native, Afro-American and local communities to decide about their know-how, innovations and traditional practices associated with genetic resources and their by-products'.[51] Andean Community Decision 486 obliges Andean countries to ensure that applications for patents involving the use of traditional knowledge of Indigenous Peoples from any Andean Community country have been secured with their prior informed consent.[52] These provisions provide Indigenous Peoples and local communities with opportunities to require compliance with their customary laws as a condition for access to, or use of, their traditional knowledge.

In 1997, Bolivia became the first Andean country to adopt legislation implementing Andean Community Decision 391. *Decreto Supremo No. 24676* guarantees the recognition of the rights of Indigenous Peoples and *campesino* (peasant) communities as the providers of the intangible component (traditional knowledge) associated with genetic resources.[53] In 2007, Bolivia made UNDRIP part of its national law[54] and in 2009, a new Constitution was adopted giving extensive recognition to Indigenous Peoples' rights to self-determination.[55] The Constitution recognizes Indigenous Peoples' rights to self-governance and requires that 'negotiation, subscription and ratification of international relations' is carried out with due respect for indigenous rights.[56] Under the Constitution, customary law is now held to be superior to executive decrees, regulations and other resolutions[57] and community justice systems have been placed on a par with national law for matters affecting indigenous communities'.[58] Municipalities and provinces with majority indigenous populations are entitled to be recognized as autonomous indigenous regions,[59] whose governance is a matter for traditional authorities according to statutes developed by them based on their own norms and practices.[60] Consultations regarding decisions that may affect Indigenous Peoples' environments are to be carried out with respect for their norms and procedures.[61]

The Constitution recognizes Indigenous Peoples' rights to have their traditional knowledge, traditional medicine, languages, rituals, symbols and vestments valued, respected and promoted.[62] They are also collectively entitled to intellectual property in their knowledge and sciences.[63] The State is obliged to promote and guarantee respect, use and investigation of traditional medicine and to protect associated knowledge as the intellectual, historic and cultural property and patrimony of Indigenous Peoples.[64] Traditional knowledge and technologies are recognized to be the patrimony of Indigenous Peoples and the State is obliged to protect their rights through the Bolivian intellectual property system.[65] In support of this end, the *Servicio Nacional de Propiedad Intelectual* (National Intellectual Property Service) is developing a register of traditional knowledge, expressions of culture, rites, rituals and crafts, which may be used to ensure recognition of rights and oppose the granting of property rights over traditional knowledge to third parties.[66]

In 2012, Bolivia adopted the *Ley de Vivir Bien* (Law of Living Well) establishing a new paradigm for national development grounded on the Aymara and Quechua Indigenous Peoples' concept of *sumac kawsay*, which promotes harmony with *pachamama* (Mother Earth).[67] Taken together, the Constitution of 2009, *Ley de Vivir Bien* and the incorporation of the UNDRIP as part of national law place Bolivia at the forefront of legal developments in the recognition of Indigenous Peoples' rights at the highest level (De Sousa Santos and Rodriguez, 2012).

In Colombia, the Constitution of 1991 grants Indigenous Peoples wide-ranging powers to exercise autonomy in their territories, subject to the constitution and national law. This includes rights to exercise jurisdictional functions in accordance with their own norms and procedures,[68] administer and govern their territories[69] and to be governed by their own authorities and administer their interests in accordance with their own customs.[70]

Colombia's constitutional court has been highly active in the protection of Indigenous Peoples' rights, adopting several key decisions on issues of self-determination, land and resource rights, and obligations relating to consultation and prior informed consent. In 1993, the Court held that the exploitation of natural resources on the territories of Indigenous Peoples must be carried out without impairment of their cultural, social and economic integrity.[71] It also held that in making decisions regarding resource exploitation, the government must facilitate participation of the relevant indigenous communities (Olsen, 2008). In one of its most far-reaching decisions, the court held that the right of Indigenous Peoples to prior consultation regarding exploration or exploitation of resources on their lands was a fundamental right, due to its importance to the protection of their cultural, social and economic integrity.[72]

Legislation adopted in Colombia in 1997 recognizes the right of ethnic communities to conserve, enrich and diffuse their cultural patrimony and identity and generate knowledge over these in accordance with their own traditions.[73] This legislation also seeks to protect the languages, traditions, customs and knowledge of ethnic groups.[74] The cumulative effect of these and related legislative provisions is to provide a solid basis for the recognition and application of Indigenous Peoples and local communities' customary laws relating to traditional knowledge (Cruz, 2007).

From 2009 to 2013, the *Ministerio del Ambiente y Desarrollo Sostenible* (Ministry of the Environment and Sustainable Development) coordinated a highly participative process involving a National Inter-Ethnic Committee and regional workshops to develop a national policy on traditional knowledge. By mid-2017 the policy had still not been adopted leaving traditional knowledge vulnerable to unapproved access and use in Colombia.[75]

Influenced by Indigenous Peoples' spiritual relationship with the earth, Ecuador adopted a new constitution in 2008, which gives recognition to the rights of *Pachamama* (Mother Earth) to 'her existence and the maintenance and regeneration of [her] vital cycles, structure, functions and evolutionary processes'.[76] The Constitution embraces the Quechua concept of *sumac kawsay,* or *buen vivir*, promoting alternative holistic development models sensitive to the importance of 'human relationships with their own history and their natural surroundings' (Davalos, 2008). Article 57 of the Constitution is ground-breaking, prohibiting all extractive activities in the traditional territories of Indigenous Peoples living in voluntary isolation.[77] It obliges the State to take measures to guarantee their lives, self-determination and right to maintain their isolation. Violation of these rights constitutes the crime of ethnocide. The granting of rights to, or carrying out of, bioprospecting activities on the traditional lands of isolated Indigenous Peoples could therefore lead to a constitutional action for breach of Article 57, including ethnocide if contact arising from such activities proves fatal.[78]

The Ecuadorian Constitution provides recognition and guarantees for the protection of Indigenous Peoples collective rights 'to uphold, protect and develop collective knowledge; ... and knowledge about the resources and properties of fauna and flora'.[79] It prohibits any form of appropriation of collective knowledge, in the fields of science, technology and ancestral wisdom.[80] It also prohibits granting of rights, including intellectual property rights, to by-products or synthetics obtained from collective knowledge associated with national biodiversity.[81] This prohibition is repeated in national regulations on access to genetic resources adopted in 2011,[82] which require the prior informed consent of Indigenous Peoples for access to their traditional knowledge associated with genetic resources.[83] Under these regulations, the *Autoridad Nacional Ambiental* (National Environmental Authority), in coordination with Indigenous Peoples' organizations, is responsible for ensuring recognition of the rights of local communities as the providers of traditional knowledge associated with genetic resources.[84]

On November 29, 2016, Ecuador adopted the *Código Orgánico de Economía Social del Conocimiento, la Creatividad y la Innovación* (Code of Social Economy of Knowledge, Creativity and Innovation) which, according to the Ecuadorian Intellectual Property Office, will shift national development away 'from a primary export economy to a social economy of infinite resources'.[85] The Code seeks to move from a traditional monopolistic intellectual property dominated framework for protection of innovation to a system more attuned to recognizing and defending the interests of innovators of all sorts, including Indigenous Peoples and local communities. The new Code builds upon the principles of *buen vivir* and is an important example of the way indigenous legal and social principles are informing the development of a corpus of intercultural justice and equity (Tsosie, 2007; Tobin, 2014b), and promoting new development paradigms which place a premium on the protection of environmental, social and cultural integrity.

In Peru, the Constitution of 1993 provides that Indigenous Peoples and local communities – referred to as *campesino* communities (Coastal and Andean regions) and native communities (Amazonian region) – are entitled, subject to law, to exercise autonomy with regard to their organization, economy, administration, communal work and in the free disposition of their territories.[86] They are also entitled to exercise judicial functions within their territories in accordance with their customary laws, as long as they do not violate the fundamental rights of the person.[87]

In 2002, Peru became the first country to adopt a comprehensive legal regime regulating Indigenous Peoples' rights over their traditional knowledge relating to biological diversity.[88] Law 27811 is declaratory in nature, recognizing that rights over traditional knowledge spring not from any act of government but from the existence of the knowledge itself. The law recognizes traditional knowledge to be the cultural patrimony of Indigenous Peoples,[89] requiring their prior informed consent for access and a licence for commercial use.[90] Benefits arising from the use of traditional knowledge are to be shared not only with contracting indigenous communities but also with the wider indigenous community through an Indigenous Development Fund, managed by Indigenous Peoples.[91] Under the law, third parties are entitled to access and use traditional knowledge in the public domain without the need to secure prior informed consent.[92] However, Indigenous Peoples are entitled to fair and equitable sharing in the benefits derived from use of their traditional knowledge for a period of twenty years from its entry into the public domain.[93] Peru has also established a National Commission Against Biopiracy with a mandate to investigate and challenge cases of biopiracy, many of which are related to resources with long historical use by Indigenous Peoples and local communities.

Since 2004, the National Institute for consumer affairs and intellectual property (INDECOPI) has actively promoted registration of traditional knowledge, carrying out visits to local communities. They have, however, only visited a miniscule fraction of the 7,599 communities of 55 Indigenous Peoples found in Peru. At this rate, hampered by time and funding constraints, it is likely to take hundreds of years to complete this work.

In 2014, Peru became the first Andean country to ratify the Nagoya Protocol, followed by Bolivia in 2016 and Ecuador in September 2017. To ensure compliance with the Protocol, both as countries home to Indigenous Peoples and as importers of traditional knowledge from other countries, Bolivia, Ecuador and Peru will need to review the adequacy of their national legislation with respect to the recognition of customary law and requirements for prior informed consent for access to, and use of, traditional knowledge. They will also need to review the relationship between the Protocol and Andean Community Decisions 391 and 486 and ensure that any Free Trade Agreement to which they are party does not undermine their effective implementation.

Consideration of customary law and community protocols

Article 12 of the Nagoya Protocol requires States Parties, in implementing their obligations relating to protection of rights over traditional knowledge associated with genetic resources, to 'take into consideration' the 'customary laws, community protocols and procedures of indigenous and local communities'. This creates obligations to ensure adequate measures are in place to regulate prior informed consent in countries where indigenous and local communities are located and in countries into which their traditional knowledge or genetic resources associated with their traditional knowledge are imported or in which they are utilized. It also implies obligations to ensure consideration of their laws, protocols and procedures in relevant judicial or alternative dispute resolution procedures for the adjudication of conflicts arising from access and use of traditional knowledge. A court may, for example, be required to take into consideration the relevant customary law of an Indigenous People or local community, whose traditional knowledge has been utilized for commercial or scientific purposes, to determine the validity of prior informed consent for access to and use of that traditional knowledge. Similarly, a national patent authority may need to refer to any relevant customary law to determine whether use of traditional knowledge in the development of a product, the subject of a patent application, complies with restrictions on its use under the customary laws of its traditional custodians.[94]

Across Africa, Latin America, Asia, Oceania and North America, customary law is already playing a significant role in addressing issues of land rights, resources and cultural heritage. In the process, issues of evidence of customary law, its status and interrelation with national and international law are reforming the notion of legal pluralism and the development of intercultural justice (Tobin, 2014b). Consideration of customary law will be a challenge for courts and administrative authorities unaccustomed to doing so. The challenge will be even greater in cases where the customary law in question is that of Indigenous Peoples or local communities from foreign jurisdictions. This is, however, the commitment that parties to the Nagoya Protocol have made. Although, neither the European Union nor Switzerland make reference to considerations of customary law or community protocols in their implementing legislation, their obligations to do so persist, and will need to be met sooner or later. The sooner the parties to the Nagoya Protocol begin to address these challenges, the sooner fair and equitable benefit-sharing and protection of traditional knowledge rights can be achieved.

The elaboration and amendment of national and regional laws to implement the Nagoya Protocol will need to address issues relating to the attribution of rights and the sharing of benefits in accordance with the customs, traditions and principles (e.g., reciprocity and equilibrium) of its custodians (Zamudio, 2013). Concepts such as prior informed consent, mutually agreed terms and benefit-sharing are common to many customary law systems and the challenge for states is to work with Indigenous Peoples to build functional interfaces between positive and customary law regimes so that national laws and judicial bodies are supportive of customary governance practices.

Indigenous Peoples and local communities around the world are increasingly drafting their own community protocols as a means for defining the scope of their rights over their lands, resources, cultural expressions and traditional knowledge. A wide variety of what are now commonly referred to as biocultural protocols have been adopted by a range of groups, including traditional healers, farming communities, livestock keepers and Indigenous Peoples, to help govern issues of access to genetic resources and traditional knowledge (Swiderska *et al.*, 2012). Article 12 of the Nagoya Protocol requires States to support Indigenous Peoples and local communities, including women within these communities, in the development of their own community protocols.[95] Support for this process may also come from the CBD, Global Environmental Facility, international development agencies, non-governmental organizations and international organizations. In doing so, however, they should be careful to avoid adopting a one-size-fits-all approach to the development of protocols.

The customary legal regimes of Indigenous Peoples and local communities are characterized by their dynamic character, locality-distinctiveness and how they are attuned to their environmental, economic, social, cultural and spiritual realities. If protection of traditional knowledge is to be firmly based on recognition and respect for the legal systems under which it was developed and maintained over centuries, it will be important to avoid any impulse to promote biocultural protocols as a magic wand solution driven by external 'experts' responding to the agendas, concerns and interests of the state, research, non-governmental and other sectors. Research of customary law and traditional knowledge protection in the Pacific Island countries warns that, even those state-based legal structures that reference custom 'may not reflect the complex, tiered, socially organised, and semi-exclusionary rights customarily applied in these countries' (Robinson and Forsyth, 2016). For this reason, Indigenous Peoples and local communities have called for a bottom-up process for the development of laws and protocols based on customary law (Tobin, 2013).

Parties to the Nagoya Protocol are required to cooperate, with the involvement of concerned Indigenous Peoples and local communities, to secure the implementation of the Protocol in regard to traditional knowledge shared by Indigenous Peoples and local communities across national borders.[96] Examples of situations involving shared knowledge include the Inuit in the USA, Canada, Russia and Greenland, the San in South Africa, Namibia and Botswana, and the Jivaro peoples on the borders of Peru and Ecuador. Awajun communities in Peru, who form part of the Jivaro peoples, have suggested using overlapping community protocols to protect rights over traditional knowledge and genetic resources, including shared resources and knowledge (Tobin and Taylor, 2009). Protocols may also be developed by all members of a specific Indigenous People wherever they are based. Such 'people-wide' protocols, developed irrespective of national borders, offer opportunities for Indigenous Peoples to define common criteria for dealing with their shared resources and knowledge. A step further may see the development by Indigenous Peoples, as lawmakers, of international biocultural protocols as a form of 'soft' law, serving as a logical step in the progressive evolution of legal pluralism and intercultural law.

Conclusion

The entry into force of the Nagoya Protocol has established binding legal obligations for all States Parties to adopt measures for the protection of traditional knowledge. A key aspect of these obligations is a requirement to 'take into consideration' the customary laws and community protocols of Indigenous Peoples and local communities. To date, developed nations have shown very little attention to this obligation and continuing failure to do so will entitle concerned parties to bring the matter before the Protocol's Compliance Committee.

National and regional implementation of the Nagoya Protocol in developed countries has tended to focus on establishing 'due diligence' requirements for users to show compliance with the laws of source countries and the countries in which the traditional custodians are located. Although very few countries have adopted specific laws on traditional knowledge, many countries have recognized the rights of Indigenous Peoples and local communities to govern their own affairs in accordance with their own largely customary legal regimes. Where it forms part of national law – whether recognized as such by custom, constitution, national law or international law – then users' due diligence requirements will extend to ensuring compliance with relevant customary laws.

The slow pace of negotiations at the WIPO IGC, and weak implementation of the Nagoya Protocol, means that traditional knowledge remains largely unprotected at the global level. The burden for adopting measures for protection of traditional knowledge continues, therefore, to fall primarily on Indigenous Peoples, local communities and the countries in which they are located. States concerned to protect traditional knowledge are able to take administrative and legislative measures that need not be overly burdensome to protect these valuable resources. European legislation, for example, requires users to demonstrate due diligence in complying with national laws relating to access and use of traditional knowledge. States seeking to protect the traditional knowledge rights of their Indigenous Peoples and local communities have the opportunity to take advantage of these provisions by informing the CBD clearing house mechanism where their national laws directly or indirectly recognize a role for customary law in regulating, at the local and or national level, the management of traditional knowledge and associated genetic resources. States may also consider adopting new legal or policy measures to strengthen, or give recognition to, customary law relating to the access and use of genetic resources and traditional knowledge as a means to ensure its protection both inside and outside the country.

The Andean community has already taken preliminary steps towards the development of regional legislation on traditional knowledge through the commissioning of a study by indigenous experts (Cruz *et al.*, 2005). All countries of the region have developed, or are in the process of developing, legal and policy measures relating to the protection of traditional knowledge. They are all parties to ILO Convention 169 and, therefore, obliged to consult with Indigenous Peoples and local communities in the development of legislation that affects them. They have also recognized the rights of Indigenous Peoples and local communities' to their own customary laws and practices. By resuscitating its efforts to develop regional legislation on traditional knowledge, the Andean Community can continue to show leadership at the international level by implementing its international legal obligations to secure the protection of Indigenous Peoples and local communities' rights over their traditional knowledge. This must be done with due respect, recognition and consideration for their own customary and other laws, community protocols and practices.

To date, the development of international instruments for the protection of traditional knowledge has been a largely state-driven affair, in which Indigenous Peoples and local communities have been treated as observers and, at best, tolerated discussants. The result has been weak legislation, poor implementation and suspension of negotiations at the whim of governments. It has also meant a progressive drift away from *sui generis* forms of protection towards individualistic intellectual property style systems which are at odds with the collective stewardship systems of Indigenous Peoples and local communities (Tobin, 2017).

Indigenous Peoples are recognized as lawmakers under international human rights law. As such, they are entitled to develop their own protocols, defining the parameters, elements and modalities for the *sui generis* protection of traditional knowledge. Protocols may be developed by local communities, an indigenous community within a specific territory, Indigenous People across state boundaries, or by specific community sectors such as farmers at the sub-regional, regional or global levels. The Nagoya Protocol requires States to support the development of community protocols. Scaling up to the international level, the United Nations Permanent Forum on Indigenous Issues, UNESCO, the Convention on Biological Diversity and its Nagoya Protocol, FAO and/or the WIPO IGC, should seriously consider supporting the development by Indigenous Peoples and local communities of a series of sub-regional, regional and/or global biocultural protocols on traditional knowledge.

In the development of biocultural protocols, Indigenous Peoples and local communities can draw upon their own laws, including statutory tribal laws, customary laws, traditions and existing protocols, as well as any relevant declarations (Posey, 1999), statements and drafting submissions to international bodies and negotiating forums, made by Indigenous Peoples and local communities from around the world. In the process of developing these biocultural protocols, it may also be helpful if reference was made to aspects of national and international law, as well as relevant human rights jurisprudence, that address Indigenous Peoples and local communities' human rights to self-determination, to their lands, resources and knowledge, as well as to restitution, free prior informed consent and social, economic, cultural and spiritual integrity.

By taking the initiative to exercise their rights as lawmakers in the development of protocols of this nature, Indigenous Peoples and local communities have the opportunity to indicate the nature, scope and form of protection of traditional knowledge they consider appropriate, thereby setting clear boundaries for the development of appropriate national, regional and international law. This will also help ensure compliance with international human rights laws on participation and consultation, secure more effective consideration and enforcement of customary laws, ensure legal certainty, and increase the legitimacy and likelihood of successful law-making and its effective implementation at all levels. It will also help bring about global recognition that, in many cases, custom is indeed the law.

Notes

1 In this chapter I will refer frequently to genetic resources associated with traditional knowledge; this is the reverse of the formulation in the Nagoya Protocol which refers to traditional knowledge associated with genetic resources. The intention is to focus the readers' attention on the call by Indigenous Peoples and local communities for protection of their rights over those the resources they have conserved, cultivated and managed for centuries as reflected in their traditional knowledge of the same.
2 United Nations Declaration on the Rights of Indigenous Peoples (UNDRIP), Arts 4, 5, 11, 24 and 34.
3 The Nagoya Protocol and the Convention on Biological Diversity use the term 'indigenous and local communities'. International law, however, creates legal distinctions between Indigenous Peoples as 'Peoples' entitled to rights of self-determination and autonomy, and local communities of farmers, fisher folk and ethnic groups who do not fall under the international legal definition of Indigenous Peoples who have not been recognized with the same level of collective rights (Posey and Dutfield, 1996b). In this chapter, the term 'Indigenous Peoples' is used throughout, except when citing specific articles of international legal instruments. This is in recognition of their status as peoples under international law as recognized by the United Nations Declaration of Human Rights and by human rights Treaty bodies. It is not intended to downplay the arguments in favour of the recognition of similar rights for local communities. See the draft declaration of the United Nations working group on the rights of peasants and other people working in rural areas: www.ohchr.org/EN/HRBodies/HRC/RuralAreas/Pages/3rdSession.aspx.
4 E/CN.4/Sub.2/1995/26, 21 June 1995.
5 UNDP, 'Consultation on the Protection and Conservation of Indigenous Knowledge, Sabah, East Malaysia, 1995, Basic points of agreement on the issues faced by the indigenous peoples of Asia', reprinted in Darrell Posey (ed.) (1999), p. 575.
6 'When Indigenous Knowledge is Patented for Profit', *The Conversation*, March 8, 2016, http://theconversation.com/biopiracy-when-indigenous-knowledge-is-patented-for-profit-55589.
7 Correspondence with John Hunter and Daniel Robinson, Sydney, Australia, June 2016.
8 United Declaration on the Rights of Indigenous Peoples (UNDRIP) Art. 31.
9 UNDRIP, Arts 4, 5, 11, 24 and 34, www.un.org/esa/socdev/unpfii/documents/DRIPS_en.pdf.
10 See Regulation of the European Parliament and of the Council on compliance measures for users from the Nagoya Protocol on Access to Genetic Resources and the Fair and Equitable Sharing of Benefits Arising from their Utilisation in the Union, Adopted by the European Parliament 11 March 2014 (EU Regulation), http://register.consilium.europa.eu/doc/srv?l=EN&f=PE%20131%202013%20INIT.

See also Amendments to Federal Act on the Protection of Nature and Cultural Heritage adopted by the Swiss Parliament on 21 March 2014 in order to implement the Nagoya Protocol (Swiss ABS Law), www.sib.admin.ch/en/nagoya-protocol/implementation-in-switzerland.
11 Swiss ABS Law.
12 EU Regulation.
13 *Richtersveld Community and Others v. Alexkor Ltd and Another* (488/2001) [2003] ZASCA 14; [2003] 2 All SA 27 (SCA).
14 www.hunter.cuny.edu/sociology/faculty/john-hammond/repository/files/HRQFINAL.PDF.
15 The United Nations Permanent Forum on Indigenous Issues (UNPFII, 2007) has estimated there are at least 5,000 Indigenous Peoples worldwide, each of which may have their own distinct legal regime. Many local communities and ethnic minorities also rely on their own laws for internal governance.
16 UNDRIP, Art. 3.
17 UNDRIP, Art. 4.
18 UNDRIP, Art. 5.
19 UNDRIP, Art. 34.
20 UNDRIP, Art. 31.

21 UNDRIP, Art. 28.
22 UNDRIP, Art. 40.
23 UNDRIP, Art. 40.
24 For a list of Treaty parties see www.ilo.org/ilolex/cgi-lex/ratifce.pl?C169.
25 ILO Convention 169, Art. 6.
26 ILO Convention 169, Art. 8.1.
27 ILO Convention 169, Art. 8.2.
28 Agenda 21, Chapter 26.4.
29 Decision VI/10F. 33.
30 Bonn Guidelines on Access to Genetic Resources and Fair and Equitable Sharing of the Benefits Arising out of their Utilization, Secretariat of the Convention on Biological Diversity, Montreal, 2002, para. 31.
31 Convention on Biological Diversity, Decision IV/10.
32 WIPO/GRTKF/IC/9/14 Prov. 2, para. 119.
33 WIPO/GRTKF/IC/9/14 Prov.2, para. 124.
34 WIPO/GRTKF/IC/11/15 Prov. para. 47.
35 WIPO/GRTKF/IC/14/12, para. 47.
36 WIPO/GRTKF/IC/9/14 Prov.2, para. 103.
37 WIPO/GRTKF/IC/9/14 Prov.2, para. 111.
38 WIPO/GRTKF/IC/9/14 Prov.2, para. 230.
39 WIPO/GRTKF/IC/13/9 African Group Proposal on the Protection of Traditional Knowledge, Traditional Cultural Expressions and Genetic Resources, Geneva 2008.
40 WIPO/GRTKF/IC/11/5(a) para. 58.
41 WIPO/GRTKF/IC/10/5 Annex, Revised provisions for the Protection of Traditional Knowledge, Art. 6.
42 *Ibid.*, Art. 5.
43 *Ibid.*, Art. 1(5). The application, interpretation and enforcement of protection against misappropriation of traditional knowledge, including determination of equitable benefit-sharing and distribution of benefits, should be guided, as far as possible and appropriate, by respect for the customary practices, norms, laws and understandings of the holder of the knowledge, including the spiritual, sacred or ceremonial characteristics of the traditional origin of the knowledge.
44 *Ibid.*
45 *Ibid.*
46 WIPO/GRTKF/IC/8/15, para. 93.
47 WIPO/GRTKF/IC/8/15, para. 93.
48 WIPO/GRTKF/IC/13/11, para. 81.
49 See, Andean Community Decision 391: Establishing the Common Regime on Access to Genetic Resources, (1996) and Andean Community Decision 486, Establishing the Common Industrial Property Regime (2000), African Model Legislation for the Protection of the Rights of Local Communities, Farmers and Breeders and for the Regulation of Access to Biological Resources (OAU Algeria 2000), Swakopmund Protocol on the Protection of Traditional Knowledge and Expressions of Folklore Within the Framework of the African Regional Intellectual Property Organization (ARIPO) Adopted by the Diplomatic Conference of ARIPO at Swakopmund (Namibia) on August 9, 2010, Guidelines for Developing Legislation for the Protection of Traditional Biological Knowledge, Innovations and Practices Based on the Traditional Biological Knowledge, Innovations and Practices Model Law, Pacific Islands Forum Secretariat 2010.
50 See Andean Community of Nations Decision 391: Common Regime on Access to Genetic Resources.
51 *Ibid.*, Art. 7.
52 Andean Community, Decision 486, Common Intellectual Property Regime, Art. 26(i).
53 *Decreto Supremo No. 24676*, Art. 5(c).
54 Law No 3760.
55 Political Constitution of the Pluricultural State of Bolivia, 2009 (Art. 2).
56 *Ibid.*, Art. 225 II(4).
57 *Ibid.*, Art. 410.
58 www.hunter.cuny.edu/sociology/faculty/john-hammond/repository/files/HRQFINAL.PDF.
59 Law 300, *De La Madre Tierra y Desarrollo integral Para Vivir Bien,* Art. 267.
60 *Ibid.*, Art. 190 II.

61 *Ibid.*, Arts 341, 349 and 350.
62 *Ibid.*, Art. 30(9).
63 *Ibid.*, Art. 30(11).
64 *Ibid.*, Art. 42.
65 *Ibid.*, Art. 100.
66 www.servindi.org/actualidad/1701/1701.
67 Law 300, *De La Madre Tierra y Desarrollo Integral Para Vivir Bien* promotes equality and interdependence among four areas of rights: Rights of Mother Earth as a subject of collective public interest; collective and individual rights of Indigenous Peoples, AfroBolivian and intercultural communities; fundamental civil, political, economic, social and cultural rights of the Bolivian people to *Vivir Bien* through integral development; and the right of the urban and rural population to live in a society of justice, equity and solidarity. The law sets out orientation for the establishment of policies, plans, programs and projects for the maintenance of genetic patrimony and traditional knowledge.
68 Political Constitution of Colombia, 1991, Art. 246.
69 *Ibid.*, Art. 286.
70 *Ibid.*, Art. 287.
71 Sentence T-188, Constitutional Court, Colombia (12 May 1993).
72 Sentence SU-039/97, Constitutional Court, Colombia (3 February 1997), section II.3.2. See also Olsen (2008), pp. 17–18.
73 Ley 397, 1997, implementing Arts 70,71 and 72 and other related articles of the Constitution over cultural patrimony. Art. 1(6), www.boyaca.gov.co/SecGeneral/images/ArchivoDepartamental/Normatividad/Nacional/LEY%20397%20DE%201997.pdf.
74 *Ibid.*, Art. 13.
75 Avanza discusión sobre política de protección de los conocimientos tradicionales asociados a la biodiversidad. Ministerio del Ambiente, 30 de agosto de 2013, www.minambiente.gov.co/contenido/contenido.aspx?catID=1358andconID=8795.
76 Political Constitution of Ecuador, 2008, Art. 71.
77 Political Constitution of Ecuador, 2008, Chapter 4, Art. 57: 'The territories of the peoples living in voluntary isolation are an irreducible and intangible ancestral possession and all forms of extractive activities shall be forbidden there. The State shall adopt measures to guarantee their lives, enforce respect for self-determination and the will to remain in isolation and to ensure observance of their rights. The violation of these rights shall constitute a crime of ethnocide, which shall be classified as such by law.' Available at http://pdba.georgetown.edu/Constitutions/Ecuador/english08.html.
78 In Peru the entry by illegal loggers using paths opened by Shell during exploration in the Camisea region led to the death of up to 60% of the Nahua who were living in voluntary isolation (Tobin, 2014a, p. 190).
79 *Ibid.*, Art. 57(12).
80 *Ibid.*, Arts 57(12) and 322.
81 *Ibid.*, Art. 402.
82 Reglamento Nacional de Acceso a los Recursos Genéticos (Executive Decree N° 905, 3 October 2011), Art. 6.
83 Decreto Ejecutivo N° 905, Art. 20.
84 *Ibid.*, Art. 8.
85 Código Orgánico de Economía Social del Conocimiento, la Creatividad y la Innovación – widely referred to as the 'Codigo Ingenios' (Code of the Ingenious), Published in Official Register 9 December 2016, www.asambleanacional.gob.ec/sites/default/files/private/asambleanacional/filesasambleanacionalnameuid-29/Leyes%202013–2017/133-conocimiento/ro-cod-econ-conoc-899-sup-09-12-2016.pdf.
86 Constitution of Peru, 1993, Art. 89.
87 Constitution of Peru, 1993, Art. 149.
88 Law 27811, Regime for the protection of the collective knowledge of Indigenous Peoples associated with biological diversity.
89 Law 27811, Art. 11.
90 Law 27811, Art. 6.
91 Law 27811, Art. 8.
92 Law 27811, Art. 13.

93 For more detailed discussion of Ley 27811 and the role of customary law in regulation of traditional knowledge in Peru see Tobin and Taylor (2009).
94 An agreement between Washington University and three federations representing Awajun communities from Peru for the use of biological resources and traditional knowledge for the development of new pharmaceutical products excluded rights for development and patenting of life forms where such activity ran counter to the mores and ethical beliefs of the community. Tobin, Bannister and Render unpublished manuscript 2003 on file with author.
95 Nagoya Protocol, Art. 12(3)(a).
96 Nagoya Protocol, Art. 11.2.

Bibliography

Anaya, J. (2004) *Indigenous Peoples in International Law* (2nd edn), Oxford University Press, Oxford.

Borrows, J. (2002) *Recovering Canada: The Resurgence of Indigenous Law*, University of Toronto Press, Toronto.

Caillaud, A. Boengkih, S., Evans-Illidge, E.A., Genolagani, J., Havemann, P., Henao, D., Kwa, E., Llewell, D., Ridep-Morris, A. (2004) 'Tabus or not Taboos? How to use traditional environmental knowledge to support sustainable development of marine resources in Melanesia', *SPC* Traditional *Marine Resource Management and Knowledge Information Bulletin* No.17, December 2004.

Chanock, M. (2005) 'Customary Law, Sustainable Development and the Failing State', in P. Orebech F. Bosselman, J. Bjarup, D. Callies, M. Chanock, H. Petersen (eds) *The Role of Customary Law in Sustainable Development*, Cambridge University Press, Cambridge.

COICA (1999) *Biodiversidad, derechos colectivos y regimen* sui generis *de propiedad intelectual*. COICA, OMAERE, OPIP. Quito.

Cornell, D. and Muvangua, N. (2012) *Ubuntu and the Law: African Ideals and Postapartheid Jurisprudence*, Fordham University Press, New York.

Corrin, J. (2014) 'Getting down to business: developing the underlying law in Papua New Guinea', *Journal of Legal Pluralism and Unofficial Law* Vol. 46, pp. 155–71.

Cruz, R. de la (2007) 'Customary law in the Protection of Traditional Knowledge', WIPO/GRTKF/LIM/07/5, WIPO, Geneva.

Cruz, R. de la, Muyuy, G., Viteri, A., Flores, G. (2005) 'Elementos para la Protección sui generis de los conocimientos tradicionales colectivos e integrales desde la perspectiva indígena', Comunidad Andina y Corporación Andina de Fomento, Caracas.

Cuskelly, K. (2011) *Customs and Constitutions: State Recognition of Customary Law Around the World*, IUCN, Bangkok.

Dávalos, P. (2008) 'Reflexiones sobre el Sumak Kawsay (el buen vivir) y las teorías de desarollo', ALAI, America Latina en Movimiento, http://alainet.org/active/25617andlang=esBo.

De Sousa Santos, B. and J.L., Exeni Rodríguez (eds) (2012) *Justicia indígena, plurinacionalidad e interculturalidad en Bolivia*. Fundación Rosa Luxemburg/Abya Yala, Quito.

Dodson, M. (2007) 'UNPFII Study on Customary Laws Pertaining to Indigenous Traditional Knowledge and to What Extent Such Customary Laws Should be Reflected in International and National Standards Addressing Traditional knowledge'. E/C.19/2007/10. UNPFII, New York.

European Union (2012) Proposal for a Regulation of the Parliament and of the Council on Access to Genetic Resources and the Fair and Equitable Sharing of Benefits Arising from the Utilization in the Union, COM(2012) 576 final.

Forsyth, M. (2012) 'Do you want it giftwrapped? Protecting traditional knowledge in the Pacific Island countries', in Drahos, P. and Frankel, S. (eds) *Indigenous Peoples' Innovation: IP Pathways to Development*, ANU Press, Canberra.

Githae, J.K. (2009) 'Potential of TK for Conventional Therapy – Prospects and Limits', in Kamau, E.C. and G. Winter (2009) *Genetic Resources and Traditional Knowledge and the Law: Solutions for Access and Benefit-sharing*, Earthscan, London and Sterling VA.

Glenn, H.P. (2014) *Legal Traditions of the World* (5th edn) Oxford University Press, Oxford.

Johnson, S. (ed.) (2012) *Indigenous Knowledge*, White Horse Press, Cambridge.

Juma, C. (1989) *The Gene Hunters: Biotechnology and the Scramble for Seeds*, Zed Books. London.

Krieger, M.J. (2008) *Intellectual Property Rights and Traditional Knowledge: Biopiracy or Bioprospecting?* The Berkeley Electronic Press, www.bepress.com/ndsip/reports/art15.

Laird, S. and Noejovich, F. (2002) 'Building Equitable Research Relationships with Indigenous Peoples and local communities', in Laird, S. (ed.) *Biodiversity and Traditional Knowledge: Equitable Partnerships in Practice*, Earthscan. London and Sterling, VA.

Nijar, G.S. (1996) 'In Defense of Indigenous Knowledge and Biodiversity: A Conceptual Framework and Essential Elements of a Rights Regime', Third World Network, Biodiversity Convention Briefings, TWN, Penang.

Olsen, V. (2008) 'Marco legal para los derechos de los pueblos indígenas en Colombia', *Human Rights Everywhere*, Santander.

Perry, R. (2011) 'Balancing Rights or Building Rights? Reconciling the Right to Use Customary Systems of Law with Competing Human Rights in Pursuit of Indigenous Sovereignty', *Harvard Human Rights Journal*, Vol. 24, pp. 71–114.

Posey, D.A. (assisted by G. Dutfied, K. Plederleith, E. da Costa e Silva, and A. Argumedo) (1996) *Traditional Resource rights: International Instruments for Protection and Compensation of Indigenous Peoples and Local Communities,* IUCN.

Posey, D.A. (ed.) (1999) *Cultural and Spiritual Values of Biodiversity: A complementary contribution to the Global Biodiversity Assessment*, United Nations Environment Program and Intermediate Technology Publications, Nairobi and London.

Posey, D.A. and Dutfield, G. (1996a) *Beyond Intellectual Property: Toward Traditional Resource Rights for Indigenous Peoples and Local Communities*, International Development Research Centre, Ottawa.

Posey, D.A. and Dutfield, G. (1996b) 'Mind the Gaps: Identifying Commonalities and Divergences Between Indigenous Peoples and Farmers Groups'. Paper presented at the Global Biodiversity Forum, Buenos Aires, Argentina, November 1996.

Reid, W., Laird, S., Gamez, R., Sittenfeld, A., Janzen, D.H., Gollin, M.A. and Juma, C. (1993) 'A New Lease on Life', in Reid, W., Laird, S., Gamez, R., Sittenfeld, A., Janzen, D.H., Gollin, M. A. and Juma, C., *Biodiversity Prospecting: Using Genetic Resources for Sustainable Development*, World Resources Institute, Instituto Nacional de Biodiversidad (INBio), Rainforest Alliance, African Centre for Technological Studies, Washington DC.

Robinson, D. (2010) *Confronting Biopiracy: Challenges, Cases and International Debates*, Earthscan. London and Washington.

Robinson, D. and M. Forsyth (2016) 'People, plants, place, and rules: the Nagoya Protocol in pacific island countries', *Geographical Research* Vol. 54(3), pp. 324–35.

Sanchez, V. and Juma, C. (eds) (1994) *Biodiplomacy: Genetic Resources and International Relations*, ACTS Press, Nairobi.

Smelcer, J.D. (2006) 'Using International Law More Effectively to Secure and Advance Indigenous Peoples' Rights: Towards Enforcement in US and Australian Domestic Courts', *Pacific Rim Law and Policy Journal* Vol. 15(1) pp. 301–30.

Stamatopoulou, E. (2010) 'Taking Cultural Rights Seriously', in S. Allen and A. Xanthaki, *Reflections on the UN Declaration on the Rights of Indigenous Peoples*, Hart Publishing Ltd, Oxford.

Swiderska, K., Shrumm, H., Hiemstra, W., Oliva, M.J., Kohli, K. and Jonas, H. (eds) (2012) 'Biodiversity and Culture: Exploring Community Protocols, Rights and Consent', *Participatory Learning and Action*, Vol. 65, IIED, London.

Taubman, A. (2005) 'Saving the Village: Conserving Jurisprudential Diversity in the International Protection of Traditional Knowledge', in K.E. Maskus and J.H. Reichmann (eds), *International Public Goods and Transfer of Technology under a Globalized Intellectual Property Regime*, Cambridge University Press, Cambridge.

ten Kate, K. and Laird, S.A. (1999) *The Commercial Use of Biodiversity: Access to Genetic Resources and Benefit Sharing*, Earthscan, London.

Tobin, B. (2011) 'Why Customary Law Matters: The Role of Customary Law in the Protection of Indigenous Peoples' Human Rights', PhD Thesis, National University of Ireland, Galway.

Tobin, B. (2013) *The Role Of Customary Law in Access and Benefit-Sharing and Traditional Knowledge Governance: Perspectives from Andean and Pacific Island Countries*, UNU-IAS/WIPO, Geneva.

Tobin, B. (2014a) 'Biopiracy by Law: European Union Draft Law Threatens Indigenous Peoples' Rights over their Traditional Knowledge and Genetic Resources', *European Intellectual Property Review*, Vol. 36(2), pp. 124–36.

Tobin, B. (2014b) *Indigenous Peoples, Customary Law and Human Rights – Why Living Law Matters*, Routledge, London.

Tobin, B. (2017) 'Now You See It Now You Don't: The Rise and Fall of Customary Law in the WIPO IGC', in Robinson, D., A. Abel-Latif and P. Roffe (eds), *Protecting Traditional Knowledge: The WIPO Intergovernmental Committee on Intellectual Property and Genetic Resources, Traditional Knowledge and Folklore*, Routledge, London.

Tobin, B. and Taylor, E. (2009) *Across the Great Divide: Complementarity and Conflict Between Customary Law and National sui generis TK Law in Peru*. SPDA/IDRC, Lima.

Tomtavala, Y. (2005) 'Customary Laws in Pacific Island Countries and Their Implications for the Access and Benefit-sharing Regime', Powerpoint Presentation made at the Pacific Regional Workshop on Access and Benefit-sharing, Traditional Knowledge and Customary law, Organized by UNU-IAS, 21–24 November 2005, Cairns, Copy with UNU-IAS.

Tsosie, R. (2007) 'Cultural Challenges to Biotechnology: Native American Genetic Resources and the Concept of Harm', *Journal of Law, Medicine and Ethics* (Fall), pp. 396–411.

UNPFII (2007) 'Report of the Secretariat on Indigenous Traditional Knowledge' (New York, 14–15 May) E/C.19/2007/10.

Van Cott, D.L. (2003) 'From Exclusion to Inclusion: Bolivia's 2002 Elections', *Journal of Latin American Studies*, Vol. 35, no. 4, pp. 751–75.

Weissner, S. (1999) 'Rights and Status of Indigenous Peoples: A Global Comparative and International Legal Analysis', *Harvard Human Rights Journal*, Vol. 12, pp. 57–128.

WIPO (2001) 'Matters Concerning Intellectual Property and Genetic Resources, Traditional Knowledge and Folklore: Document prepared by the Secretariat,' WIPO/GRTKF/IC/1/3, WIPO, Geneva.

WIPO (2006) 'Draft Customary Law Issues Paper 1: Customary Law and the Intellectual Property System in the Protection of Traditional Cultural Expressions and Traditional Knowledge: Issues Paper' – version 3.0, WIPO, Geneva.

WIPO (2013) *Customary Law, Traditional Knowledge and Intellectual Property: An Outline of the Issues*, WIPO, Geneva.

Wynberg, R. (2004) 'Rhetoric, Realism and Benefit-Sharing: Use of Traditional Knowledge of Hoodia Species in the Development of an Appetite Suppressant', *J. World Intell. Prop.*, Vol. 7, p. 851.

Yrigoyen Fajardo, R. (1999) *Pautas de Coordinación entre el Derecho Indígena y el Derecho Estatal*. Fundación Myrna Mack. Guatemala.

Zamudio (2013) 'Los conocimientos tradicionales y el régimen legal de acceso y distribución de beneficios', *Revista de la Facultad de Derecho de la PUCP*, Vol. 69, pp. 259–79.

PART V

Biodiversity and intellectual property protection

20

BIODIVERSITY, INTANGIBLE CULTURAL HERITAGE AND INTELLECTUAL PROPERTY

Christoph Antons

Introduction: biodiversity, the "traditional knowledge" of "indigenous and local communities" and cultural heritage

In 1992, the Convention on Biological Diversity (CBD) signified a paradigm shift in thinking about the "stewards" and "custodians" of biodiversity conservation. In its Article 8 j., the CBD famously included the "traditional knowledge and practices" of local and indigenous communities in its ambit. To include communities and give them a significant role in biodiversity conservation constituted a major change from the state-centred models that had dominated international treaties and development policies for many decades. Especially in the developing world, where much of the world's biodiversity was to be found, post-World War II modernization and development programmes were shaped by the strategic planning of government agencies, often with financial support and advice from UN agencies and the international financial institutions of the World Bank and the IMF. The "local and indigenous communities" mentioned in Article 8 j. CBD were often regarded as obstacles to modernization and development in such earlier programmes. Continuing the discriminative policies of colonial powers towards people variously described as "remote living", "backward" or "in need of development" (Persoon, 2009, p. 196; Cramb, 2007, p. 8), development planners of the 1950s or 1960s showed little interest in their attitudes towards their environment, which, if anything, were regarded as destructive (Dove *et al.*, 2005, p. 2; Duncan, 2004, pp. 13–15).

It was the work of ecologists and environmental activists such as Darrell Posey and Fikret Berkes that brought about a re-thinking of the contribution of such peoples over the following decades (Posey, 1985; Berkes, 1999).[1] These fresh approaches combined with an emerging transnational movement of indigenous peoples, who were becoming aware of the very similar issues they were facing in their respective relationships with national governments, which typically were designing conservation programmes that included resettling them in areas away from their original habitat and thereby denying them any rights to their native lands and resources or to the continuing practice of their cultures (Dove *et al.*, 2005, p. 3; Duncan, 2004; Sissons, 2005, pp. 23–24). As with any political movement and rights discourse, however, the progress in establishing such rights has been slow and uneven. Many governments have been slow in recognizing the relevance of the concept of "indigenous peoples" for their respective territories or they have argued that in their particular settings, the term has to be interpreted differently (Kingsbury, 1999; Benjamin, 2017). The UN Declaration on the Rights of Indigenous Peoples of 2007 is widely seen as an important step forward for the international indigenous movement, yet it is a non-binding declaration and the absence of a treaty shows the continuing differences in this field. And while the CBD has recognized the important contribution that local and indigenous communities are able to make to the environment (Article 8 j., CBD), it has also reinforced the national sovereignty of governments in controlling their genetic resources in Article 3 of the treaty. This compromise between the international embrace of local stewardship of environmental principles and national sovereignty in the interest of development has been regarded as a "somewhat confusing hybrid" (Ariffin, 2007, p. 216).

Now, more than two decades after the CBD, the general public's perception about terminology and concepts has clearly begun to change in a positive way, but significant issues remain. Environmental anthropologists and indigenous activists have cautioned against a romanticizing of indigenous life that disregards the economic and social aspirations of local and indigenous people and keeps them in a position as the exotic "other" and as an object of curiosity for national mainstream cultures (Lowenhaupt Tsing, 2008; Sissons, 2005, pp. 39–43; Lowenhaupt Tsing *et al.*, 2005, p. 7). Research by legal academics in developing countries shows that colonialism has distorted many customary law regimes to an extent that they often have to be resurrected and revived if they are to play a significant role in the allocation of rights and responsibilities (Chanock, 2005, 2009). Under the circumstances, the scope to be given to such customary laws in legally pluralist settings continues to be debated (Forsyth, 2013) and remains mostly limited and subordinated to national laws related to environmental protection and resource allocation. It comes as no surprise then that critics have argued that the prior informed consent of and benefit-sharing with local and indigenous communities, required by the CBD, has brought few tangible results (De Schutter, 2014). Equally, a harmonization of the principles of the CBD with the strong intellectual property rights impacting on biodiversity granted by the WTO TRIPS Agreement has been urged by the governments of developing countries, but not been supported by the industrialized world. Discussions at the World Intellectual Property Organization about international instruments to protect what intellectual property lawyers call "traditional knowledge" and "traditional cultural expressions" as well as associated genetic resources have made only modest progress since they started in 2001 (Antons, 2012a).

In this situation of a temporary impasse in the discussions about the "protection" of emerging forms of intellectual property and resource rights, wider notions of the protection of cultural landscapes under the 1972 World Heritage Convention (WHC) or of the "safeguarding" of intangible cultural heritage as promoted in the UNESCO Convention of 2003 have come to the fore. They are often seen as more promising in gaining recognition for forms of knowledge and epistemologies that continue to be at odds with the still ruling modernization paradigms in general, and the ethos and aims of national intellectual property systems in particular. The following section will show the paradigm shift in UNESCO treaties towards recognition of natural and "biocultural" heritage that has taken place since the 1970s. In a further section, the interplay of national heritage laws and intellectual property laws will be analysed, using examples from Asian developing countries. The section shows that many national governments seek to overcome the limitations of the soft law character of international law related to intangible cultural heritage by supplementing international law with intellectual property legislation for the "protection" of forms of intangible cultural heritage. This results in particularly strong intellectual property claims exerted by national governments to forms of culture that can lead to international disputes in particular where such cultures are shared and practised across borders (Antons, 2009; 2013a; 2015). The section also shows similar claims from regional governments in countries like Indonesia, which ultimately, however, remain subordinated to national intellectual property laws. The aim is to show the many confusing overlaps between heritage and intellectual property laws, which ultimately regulate the same subject matter, variously described as "intangible cultural heritage" or "traditional knowledge". While some hold hopes that this overlap will work for the benefit of communities and their cultural rights (Geismar, 2013), this chapter will argue that government involvement remains strong in both heritage and intellectual property schemes and that the combination of the two frequently facilitates an appropriation process that turns local forms of heritage into national heritage under the administration of the national government.

UNESCO and cultural heritage: from a "discourse on monumentality" to "intangible cultural heritage" and "biocultural heritage"

As Nafziger *et al.* (2010, p. 206) have noted, there is an often confusing interchangeable use of the terms "cultural property" and "cultural heritage". Of greater concern is the frequent confusion of "cultural property" and "intellectual property" in nationalistic discourses and rhetoric, which is observable in particular in younger nation states making monopolistic claims to culture as basis for their national identity (Antons, 2009, 2012b, 2013a, 2013b, 2015; Chong, 2012; Aragon and Leach, 2008; Aragon, 2012). Vrdoljak (2008, pp. 209–2011) has traced the "cultural property" concept to the Vienna Congress of 1815, at which the restitution of art works plundered by the Napoleonic army across Europe was discussed. After World War II, the Convention for the Protection of Cultural Property in the Event of Armed Conflict of 1954 was concluded for the same purpose. The cultural property terminology reached its high point in 1970 in the UNESCO Convention on the Means of Prohibiting and Preventing the Illicit Import, Export and Transfer of Ownership of Cultural Property. Centred on national forms of heritage (Vrdoljak, 2008, pp. 209–2011; Merryman, 2005, p. 22; Francioni, 2008a, p. 3), the 1970 Convention defined such cultural property as "property which, on religious or secular grounds, is specifically designated by each state as being of importance for archaeology, prehistory, history, literature, art or science".

In the following years, the fixation on "property" was criticized (Prott and O'Keefe, 1992). Although intensively using the term "property" situated in a state's territory, the 1972 World Heritage Convention nevertheless shifted from cultural property to cultural heritage (Yusuf, 2008, p. 27). The 2003 UNESCO Convention for the Safeguarding of Intangible Cultural Heritage no longer uses the cultural property terminology. Still, it continues to be used even today, for example in the updated operational guidelines for the 1972 World Heritage Convention, and it remains important because the older conventions are referenced in the more recent ones. The importance of the "cultural property" terminology for indigenous peoples seeking recognition of their forms of cultural heritage has also been emphasized (Tsosie, 2012, pp. 237–39; Carpenter *et al.*, 2009).

As for natural heritage, the 1972 World Heritage Convention became a watershed moment in bringing closer the separate spheres of cultural and natural heritage at a time when contemporaneous approaches towards environmental protection, such as the 1972 Stockholm Declaration on the Human Environment and work by the International Union for the Conservation of Nature (IUCN) on an instrument for the conservation of the world's natural heritage, supported such developments (Francioni, 2008a, pp. 4–5; 2008b, p. 14; Redgwell, 2008, p. 64; Forrest 2010, p. 227).

The 1972 Convention provides separate definitions of "cultural" and "natural heritage", but recognized the relationship between the two by including "sites" within the concept of "cultural heritage", defined in Article 1 as "works of man or the combined works of nature and man, and areas including archaeological sites which are of outstanding value from the historical, aesthetic, ethnological or anthropological point of view". In addition, in 1992, the concept of "cultural landscapes" was introduced into the Operational Guidelines. They are defined as "cultural properties" representing the "combined works of nature and man". The guidelines require them to be "illustrative of the evolution of human society and settlement over time, under the influence of the physical constraints and/or opportunities presented by their natural environment and of successive social, economic and cultural forces, both external and internal" (para. 47 of the Operational Guidelines). In Guidelines on the Inscription of Specific Types of Properties on the World Heritage List in an Annex to the Operational Guidelines, it is further explained that "Cultural landscapes often reflect specific techniques of sustainable land-use … and a specific spiritual relation to nature. Protection of cultural landscapes can contribute to modern techniques of sustainable land-use … The protection of traditional cultural landscapes is therefore helpful in maintaining biological diversity".

The Guidelines on the Inscription of Specific Types of Properties divide cultural landscapes into three categories: (a) landscapes designed and created intentionally by man; (b) organically evolved landscapes; (c) associative cultural landscapes. The fundamental changes in the thinking about the relationships between nature and man is particularly visible in the last two categories. An "organically evolved landscape" can be a "relict (or fossil) landscape", but also a "continuing landscape" with "an active social role in contemporary society closely associated with the traditional way of life". An "associative cultural landscape" finally is one inscribed "by virtue of the powerful religious, artistic or cultural associations of the natural element rather than material cultural evidence, which may be insignificant or even absent" (para. 10 of Annex 3). The main advisory body for the evaluation of cultural landscapes is the International Union for Conservation of Nature (IUCN). According to the evaluation procedures of the advisory bodies for nominations in Annex 6 of the WHC operational guidelines, the IUCN concerns itself with a number of factors in the process, including with the conservation of natural and semi-natural systems, the conservation of biodiversity within farming systems, sustainable land use, *ex-situ* collections and "outstanding examples of humanity's inter-relationship with nature" (para. B. 8. of Annex 6).

Although the linking of culture and nature could be seen as addressing some of the concerns of indigenous communities, forest dwellers and other minorities living at the margins of nation states, the World Heritage Convention has been criticized as Eurocentric, favouring elite notions of heritage values and as overly concerned with the notion of "property" (Smith, 2006, pp. 95–102). As early as 1980, an expert group charged with the development of a global strategy found Europe, historic towns and religious buildings, Christianity, historical periods and "elitist" architecture over-represented in the World Heritage List (Yusuf, 2008, pp. 33). A further expert group convened in 1992 finally drafted the Global Strategy for a balanced, Representative and Credible World Heritage List, which was formally adopted by UNESCO in 1994. UNESCO also adopted a revision of the cultural heritage criteria. Yusuf (2008, pp. 36–37) points to the important consequences of this policy shift: the setting aside of artistic achievement as an important criterion; the stronger recognition of the link between cultural and natural heritage, the opening up to forms of heritage previously not recognized, such as cultural landscapes, seascapes, traditional settlements and their environments, sites of spiritual significance and properties associated with events and living traditions. He concludes that the revision "constituted a clear move away from a purely monumental view of cultural heritage towards a more anthropological, comprehensive and diversified conception of the wealth and diversity of human cultures" (Yusuf, 2008, p. 37).

A certain bias remains nevertheless visible from the requirement that "sites" or "natural sites" must have "outstanding universal value" from either a "historical, aesthetic, ethnological or anthropological" point of view (Article 1, WHC) or from "the point of view of science, conservation and natural beauty" (Article 2, WHC). Further, anthropologists working on biodiversity issues in indigenous and local communities have urged that programs of governments and international organizations seeking to provide incentives for biodiversity conservation activities should better ask the question of what local communities are already doing and how to strengthen, or at least not weaken, such activities. In other words, the maintenance of already existing activities must be made possible rather than trying to bring in new activities and concepts from outside (Dove *et al.*, 2005). The WHC seems to miss this point when it requires state parties "to adopt a general policy which aims to give the cultural and natural heritage a function in the life of the community and to integrate the protection of that heritage into comprehensive planning programmes" (Article 5(a), WHC). As the anthropological literature points out, cultural and natural heritage already has this function in the lives of communities and government planning programmes should foster existing practices rather than construct theoretical incentives for different ones. If designed without community involvement, as is often the case (Tuck-Po, 2005), such government planning programmes may be counterproductive. Studies of UNESCO World Heritage Sites in Southeast Asia have found "indifference or even antipathy" of local villagers to sites, in which "tourism-related businesses were owned by incomers" rather than locals (Cochrane, 2016, p. 333). Writing on the Puerto Princesa Underground River National Park on Palawan Island in the Philippines, Dressler *et al.* (2006) and Fross (2016, pp. 359–60) show how environmental restrictions, introduced after the declaration of the National Park, conflicted with indigenous tenure and resource rights granted to the indigenous peoples of the area in 1997 under the Philippines' progressive Indigenous Peoples Rights Act.

The revised criteria for the World Heritage Convention also paved the way for greater recognition of intangible cultural heritage by requiring "properties" to be "directly or tangibly associated with events, living traditions, beliefs, ideas or artistic and literary works of outstanding universal significance" (Yusuf, 2008, p. 37). Over the following years, the importance of intangible cultural heritage increased further, especially after the recognition of traditional knowledge in the 1992 Convention on Biological Diversity and of traditional farming knowledge in the FAO sponsored 2001 International Agreement on Plant Genetic Resources for Food and Agriculture (Forrest, 2010, pp. 365–66; Aikawa-Faure, 2008, p. 15). These developments culminated in the adoption of the 2003 UNESCO Convention for the Safeguarding of the Intangible Cultural Heritage. This Convention brought a considerable progress in recognizing, in the preamble, the "important role" of "communities, in particular indigenous communities, groups and, in some cases, individuals" in the "production, safeguarding, maintenance and recreation of intangible cultural heritage". Because the "domains" it focused on included, among others, "oral traditions and expressions, ... performing arts, ... knowledge and practices concerning nature and the universe" and "traditional craftsmanship", the Convention also brought a significant overlap with intellectual property and the activities of WIPO in developing an international instruments covering traditional cultural expressions, traditional knowledge and affiliated genetic resources (Antons 2013a, 2013b, 2015; Forrest, 2010). International lawyers specialized in heritage protection have distinguished intangible cultural heritage from IP by arguing that the latter "focuses on the end product of a specific artistic or cultural tradition, rather than on the societal structures and processes from which the cultural product is derived" (Francioni, as quoted in Aikawa-Faure, 2008, p. 29). As the case studies in the next section demonstrate, however, a number of developing country governments have ignored such fine distinctions and instead rather merged the two concepts to embed the "soft law" concept of intangible cultural heritage in the "hard law" commercial-type rights of intellectual property. This has triggered frenetic activities to claim heritage at national and international levels, before others can claim it. These activities in turn have led to diplomatic tensions between neighbouring countries with similar traditions and they are also increasingly leading to rival claims to cultural heritage between regional governments within the same country or between a regional government and the national government.

Cross-border and cross-regional claims to biocultural heritage

It has often been pointed out in UNESCO treaties concerned with heritage, and in the deliberations at WIPO concerning traditional knowledge, that such knowledge or cultural heritage is not static, but constantly evolving. Article 2(1) of the 2003 UNESCO Convention, for example, explains that the intangible cultural heritage "is constantly recreated by communities and groups in response to their environment, their interaction with nature and their history". In the latest Glossary of Key Terms related to Intellectual Property and Genetic Resources, Traditional Knowledge and Traditional Cultural Expressions of WIPO (WIPO, 2016a), the "Like-Minded Countries' Contribution to the Objectives and Principles on the Protection of Genetic Resources and Preliminary Draft Articles on the Protection of Genetic Resources" defines "Associated Traditional Knowledge" as "knowledge which is dynamic and evolving". The Glossary refers to a definition of "traditional environmental knowledge" of the Dene Cultural Institute as "both cumulative and dynamic". Elsewhere, however, the Glossary admits that there is as yet no accepted definition of traditional knowledge at the international level. The latest draft articles on the protection of traditional knowledge in its "use of terms" section states that the knowledge "may be dynamic and evolving" (WIPO, 2016b). In the Preamble, it is recognized that traditional knowledge systems are "frameworks of ongoing innovation" and, importantly, "have equal scientific value as other knowledge systems". In this sense, it is problematic to term forms of knowledge "traditional" (Sillitoe, 2006) that are essentially an "admixture of local folk knowledge and extra-local scientific knowledge" that could be termed "peasant science" (Dove, 2000, p. 215; Frossard, 2005; Winarto, 2004) and to juxtapose "traditional" and "scientific" forms of knowledge (Agrawal, 1995).

Apart from continuing discussions about the definition of the subject matter of "local", "indigenous" or "traditional knowledge", the debate over who are the "beneficiaries" or "stakeholders" also continues, as is visible from the very different alternatives currently offered in Article 2 of the WIPO Draft Articles on Traditional Knowledge (WIPO, 2016b). Most importantly, if the knowledge is current and dynamic, it is obvious that it moves with the relevant knowledge holders. Such knowledge holders are as diverse as the knowledge itself. As far as agricultural biodiversity knowledge is concerned, for example, the knowledge holders can be settled farming communities or swidden agriculturalists, whether or not they are indigenous. Even so-called "remote living" communities are far from homogenous. Their members are on the move across provincial, state and national boundaries, especially in regions that have only relatively recently come under greater control by nation states (Eilenberg, 2012; Eghenter, n.d.). They are also frequently moving back and forth between their "traditional" communities and urban life in regional capital cities. From a historical perspective, many of the current borders of developing nations were drawn by colonial powers for economic and political reasons. Such borders often divided ethnic communities, so that their languages and knowledge systems continued to be practiced on both sides of the border (Antons, 2013a).

With the recent attempts at national and international levels to list, classify and regulate heritage and forms of knowledge, this widespread dissemination of heritage has proven difficult to deal with. National governments attempting to legislate for heritage and forms of traditional knowledge could find themselves in conflict with neighbouring countries aiming to do the same for similar forms of heritage. Disputes developed initially about artistic expressions between neighbouring countries such as Thailand and Cambodia or Indonesia and Malaysia (Aragon and Leach, 2008; Chong, 2012). Soon, however, such controversies extended to local knowledge about plant material shared between neighbouring countries, as in the case of the patenting by the Massachusetts Institute of Technology in collaboration with the Malaysian government of the bioactive fraction of *Eurycoma longifolia*, a plant used in traditional medicine (*jamu*) in both Indonesia and Malaysia and known in Malaysia as *tongkat Ali* and in Indonesia as *pasak bumi* (Budiningsih, 2007; Antons and Antons-Sutanto, 2009, p. 383). More recently, the Indonesian press celebrated a registration of a geographical indication for organic rice, known in Indonesia as *Adan Krayan* rice, which is grown in the highlands of Borneo on both sides of the Indonesia-Malaysia border by related ethnic communities using similar wet rice cultivation techniques (Ardhana *et al.*, 2004; Langub, n.d.). One press article misunderstood and overstated the effect of the Indonesian GI registration in stating that "what was temporarily claimed by Malaysia was now the property of Indonesia" (Saputra, 2012). The claim by Malaysia referred to the Malaysian GI registration for what is known in Malaysia as *Bario* rice, which is grown by the same ethnic group, leading some Indonesian government officials to believe that all rice sold on the Malaysian side of the border in fact originated in the Krayan region of the Indonesian province of North Kalimantan (Kompas, 2012).

In countries that have introduced decentralization policies, claims to biocultural and other heritage are now also exerted at the national level between regions and provinces and by regions/provinces vis-a-vis national governments. Decentralization and community-based natural resource management in developing countries have been strongly supported by international donors for many years (Dressler *et al.*, 2006, p. 790; Vandergeest and Wittayapak, 2010; Brosius *et al.*, 2005). However, the often complex and complicated local and national politics that accompany such reforms have been neglected. The archipelagic nation of Indonesia, with its many languages and ethnic groups, is a particularly good example. Already centred on the main island of Java in colonial times, centralization reached its high point during many years of rule of the autocratic and military-backed government of former General Suharto. During the "reformation period" that followed Suharto's rule, his successor Habibie granted widespread autonomy in administrative and financial matters to the districts (Abdullah, 2009, p. 536). These laws were revised in 2004 and 2014, and decentralization principles were also incorporated into the revised Indonesian Constitution, although without changing the basic premise that Indonesia was a unitary and not a federated state. Also recognized in the revised Indonesian Constitution are "customary law communities" (*masyarakat hukum adat*) at the local level, "as long as they are still in existence and in accordance with the development of society and the principles of the unitary Republic of Indonesia as regulated by law" (Article 18B of the Indonesian Constitution of 1945).

Decentralization in Indonesia has led to difficult bargaining processes between the central and regional governments and the customary law communities as to who should be in control of, and allowed to profit from, natural resources, cultural heritage, tourism and local products. The customary law or *adat* communities, a term which includes communities elsewhere referred to as indigenous peoples, are still in a comparatively weak position and need recognition from the authorities under the qualifications listed in Article 18B. Nevertheless, such communities have long started to organize under the banner of the nation-wide operating Alliance of the Adat Communities of the Archipelago (*Aliansi Masyarakat Adat Nusantara* (AMAN).[2] In 2013, they successfully submitted the Indonesian Forestry Act for review to the Constitutional Court. The Court ruled in favour of the petition by AMAN and declared that "*adat* forests" were no longer generally forests under the control of the state, but forests subject to rights, by deleting the word "state" in Article 1.6 of the Forestry Law No. 41 of 1999 (Rachman and Siscawati, 2017).

While AMAN is thus testing the ground for stronger customary rights of communities in the courts, Law No. 23 of 2014 on Regional Government Administration puts the central government in charge of community intellectual property rights in the field of culture and all government levels, depending on their location, for the empowerment of *adat* communities. The recognition of such *adat* communities, their traditional knowledge and related rights is the responsibility of the local government authorities, unless the communities live in several provinces, in which case the central government is responsible. For the conservation of biodiversity, the law equally makes different government levels responsible depending on where a biodiversity conserving community is based.

An interesting development is the enactment of local intellectual property regulations with specific rules for intellectual property rights related to local heritage and resources in the provinces of Papua[3] and West Java,[4] with further regulations reportedly in preparation for Bangka Belitung (Belitong Ekspres, 2014) and East Java (University of Surabaya, 2015). Although intellectual property laws are normally national laws and should actually fall under the responsibility of the central government for "justice", the division of responsibilities in the case of Papua fall under a special autonomy law negotiated in 2001. This autonomy law makes a distinction between indigenous Papuans and other residents of the province, and defines "indigenous Papuans" as "descending from Melanesian racial stock, who are members of indigenous tribes or people recognized as indigenous by Papuan *adat* communities". Accordingly, Papua enacted the Special Regional Regulation of the Province of Papua No. 19 of 2008 on the Protection of Intellectual Property Rights of Indigenous Papuans. Biodiversity-specific regulations include "indigenous Papuan geographical indications" focusing on the "specific indigenous Papuan circumstances" and the specific Papuan character and quality of the GIs as well as specific rules for the protection of original and local Papuan plant varieties. An "original Papuan plant" is defined as a plant, which grows – presumably exclusively or originally – in the province of Papua. Local government officials are responsible for the inventory and collection of such plants. Removing original Papuan plants from the Papuan territory is prohibited without permission of the Governor and such permission is equally required for breeding and other activities using such plants. Where plant variety protection for such plants is obtained, the provincial government is entitled to a share in the royalties and all licensing agreements must be reported to the Governor.

In view of the vague definition of an "original Papuan plant" the effectiveness of these provisions must be questioned, although the definition section of the Regulation defines a "local variety" further as an "original Papuan variety which has been in existence and cultivated for generations by Papuan communities and become the property of Papuan communities". The wording is the same as used in the national Indonesian plant variety protection law for "local varieties" more generally. According to Article 7 of this national law, local varieties are "owned by the community", but "controlled by the state". The register for local varieties in the Plant Variety Protection Centre of the Ministry of Agriculture shows that, in accordance with an implementing government regulation, they have been registered exclusively by government officers such as mayors and the governors of regencies and provinces (Antons, 2015, p. 470).

The Papuan Special Provincial Regulation foresees the same process and thus adds little from the viewpoint of community control over such resources.

There are similar concerns about the Regional Regulation of the Province of West Java No. 5 of 2012 on the Protection of Intellectual Property. It regulates intellectual property rights and "neighbouring rights", in which it includes, deviating from the standard copyright terminology, genetic resources, geographical indications and indications of origin and biodiversity (*keanekaragaman hayati*). Also unusual for intellectual property legislation, it promises to protect, among other things, cultural heritage, genetic resources for food and agriculture, traditional cultural expressions and products of West Javanese communities from claims by other parties and it promises to do this for both tangible and intangible subject matter (Article 2 d.). Interestingly, in an implementing regulation on the facilitation of registration issued by the Governor of West Java in 2015,[5] the term "neighbouring rights" is used differently and now comprises "communal intellectual property", in which the regulation includes geographical indications and indications of origin, trade secrets and plant variety rights. As for the biodiversity protection, the Regional Regulation of 2012 extends the term "neighbouring rights" to plants for traditional medicine, rare plants, plants "with a West Java identity" and "certain plants". For plants with a "West Java identity", the Regulation gives the example of Gandaria (*Bouea Macrophylla*), for "certain plants" it is concerned with specific plants that are in need of conservation.

Although the 2012 Regional Regulation promises implementing provisions for the biodiversity and genetic resources "neighbouring rights", these are not provided in the Governor Regulation of 2015, because the terminology has changed and those matters are no longer included in the term. What the promised "protection" amounts to remains, therefore, unclear. Much of what is described in the Governor Regulation of 2015 is local government assistance to facilitate registrations at the Directorate General for Intellectual Property Rights in Jakarta. This assistance is offered in particular to local farmers.

What can further be observed in the 2012 Regional Regulation is a conflation of classical intellectual property material and material normally covered in heritage laws, such as *adat* rituals and clothing, heirlooms, traditional food and even "cultural landscapes" (Article 21 of Regional Regulation No. 5 of 2012). As I have argued elsewhere (Antons, 2009, 2012b, 2013a, 2013b, 2015), this conflation is rather frequent. It turns local knowledge or traditional knowledge, as used in intellectual property debates, into heritage and heritage material that is normally the subject of conservation laws only, into intellectual property.

Conclusion: the transformation of "local knowledge" into "biocultural heritage"

A paradigm shift following the conclusion of the CBD in 1992 promised greater recognition of "local" or "traditional" knowledge related to biodiversity and greater involvement of communities practising such knowledge. Almost twenty-five years later, the results are rather disappointing. The prior informed consent and access to benefit-sharing schemes have brought few tangible results for traditional/local knowledge holders. The harmonization of the CBD principles with the rules of the WTO TRIPS Agreement has not received sufficient support and the WIPO deliberations about an international instrument to protect traditional knowledge, traditional cultural expressions and associate genetic resources have not progressed very much. The "soft law" provisions of UNESCO treaties have equally been criticized as elitist and studies show that the transformation of their natural surroundings into UNESCO World Heritage sites has been a mixed blessing for the locals.

Of particular concern is that the overlap between heritage and intellectual property terminologies, as visible in the UNESCO Convention for the Safeguarding of the Intangible Cultural Heritage, is leading to a conflation of the two in the national and regional laws of developing countries. The result is exaggerated claims, in which heritage is claimed as exclusive and often given an ethnic dimension supporting such exclusiveness. As with UNESCO schemes, however, the communities actually practising the heritage/traditional knowledge remain on the margins of such approaches in national and regional laws. The approaches envisage registration, listing and classification of the material in the interest of national development or the defence of heritage against foreign interests. The Indonesian Plant Variety Protection Law, according to which local varieties are owned by the communities, but controlled by the state, is a good example. The extent of the control by the state means, in the end, that local communities leave more or less empty-handed. Although the discussion of the last few decades has changed the terminology and perception about local knowledge for the better, it remains, therefore, important to listen to communities and respect the contributions they are already making to biodiversity conservation rather than to present them with newly constructed theoretical concepts that may or may not work on the ground.

Notes

1 For the rethinking among anthropologists of the role of humans in shaping landscapes in different parts of the world and relevant literature see Dove *et al.* (2005, p. 3).

2 See also Rachman and Siscawati (2017), who use the term "Alliance of Indigenous Peoples of the Archipelago" rather than a literal translation. They also point to the colonial origins of the official term "customary law communities".

3 Peraturan Daerah Khusus Provinsi Papua Nomor 19 Tahun 2008 Tentang Perlindungan Hak Kekayaan Intelektual Orang Asli Papua.

4 Peraturan Daerah Provinsi Jawa Barat Nomor 5 Tahun 2012 Tentang Perlindungan Kekayaan Intelektual.

5 Peraturan Gubernur Jawa Barat Nomor 73 Tahun 2015 Tentang Petunjuk Pelaksanaan Fasilitasi Pendaftaran Hak Kekayaan Intelektual.

References

Abdullah, T. (2009) *Indonesia: Towards Democracy*, Institute of Southeast Asian Studies, Singapore.

Agrawal, A. (1995) "Dismantling the divide between indigenous and scientific knowledge", *Development and Change*, Vol. 26, pp. 413–39.

Aikawa-Faure, N. (2008), "From the Proclamation of Masterpieces to the Convention for the Safeguarding of Intangible Cultural Heritage", in L. Smith and N. Akagawa (eds.), *Intangible Heritage*, Routledge, London and New York, pp. 13–44.

Antons, C. (2009) "What is 'traditional cultural expression'? – International definitions and their application in developing Asia", *The WIPO Journal*, No. 1, pp. 103–16.

Antons, C. (2012a) "Intellectual Property Rights in Indigenous Cultural Heritage: Basic Concepts and Continuing Controversies", in C.B. Graber, K. Kuprecht and J.C. Lai (eds), *International Trade in Indigenous Cultural Heritage: Legal and Policy Issues*, Edward Elgar, Cheltenham, UK and Northampton, MA, pp. 144–74.

Antons, C. (2012b) "Geographies of Knowledge: Cultural Diffusion and the Regulation of Heritage and Traditional Knowledge/Cultural Expressions in Southeast Asia", *The WIPO Journal*, Vol. 4, Issue 1, pp. 83–91.

Antons, C. (2013a) "Asian Borderlands and the Legal Protection of Traditional Knowledge and Traditional Cultural Expressions", *Modern Asian Studies*, Vol. 47, Issue 4, pp. 1403–33.

Antons, C. (2013b) "At the Crossroads: The Relationship Between Heritage and Intellectual Property in Traditional Knowledge Protection in Southeast Asia", *Law in Context*, Vol. 29, No. 1, 74–94.

Antons, C. (2015) "Epistemic communities and the 'People without History': The Contribution of Intellectual Property Law to the 'Safeguarding' of Intangible Cultural Heritage", in I. Calboli and

S. Ragavan (eds), *Diversity in Intellectual Property: Identities, Interests and Intersections*, Cambridge University Press, New York, pp. 453–71.
Antons, C. and Antons-Sutanto, R. (2009) "Traditional medicine and intellectual property rights: A case study of the Indonesian *jamu* industry", in C. Antons (ed.), *Traditional Knowledge, Traditional Cultural Expressions and Intellectual Property Law in the Asia-Pacific Region*, Kluwer Law International, Alphen a.d. Rijn, pp. 363–84.
Aragon, L.V. (2012) "Copyright Culture for the Nation? Intangible Property Nationalism and the Regional Arts of Indonesia", *International Journal of Cultural Property*, Vol. 19, pp. 269–312.
Aragon, L.V. and Leach, J. (2008) "Arts and Owners: Intellectual Property Law and the Politics of Scale in Indonesian Arts", *American Ethnologist*, Vol. 35, pp. 607–31.
Ardhana, I.K., Langub, J. and Chew, D. (2004) "Borders of Kinship and Ethnicity: Cross-border Relations Between the Kelalan Valley, Sarawak, and the Bawan Valley, East Kalimantan", *Borneo Research Bulletin*, Vol. 35, pp. 144–79.
Ariffin, Y. (2007) "Developmental and environmental policies: Past trends, present issues, future prospects", in P. de Senarclens and A. Kazancigil (eds), *Regulating Globalization: Critical Approaches to Global Governance*, United Nations University Press, Tokyo and New York, pp. 205–48.
Belitong Ekspres (2014) "Kanwilkumham Babel Ajukan Perda HAKI", 13 December 2014.
Benjamin, G. (2017) "Indigenous peoples: Indigeneity, indigeny or indigenism", in C. Antons (ed.), *The Routledge Handbook of Asian Law*, Routledge, London-New York, pp. 362–77.
Berkes, F. (1999) *Sacred Ecology: Traditional Ecological Knowledge and Resource Management*, Taylor and Francis, Philadelphia-London.
Brosius, J.P., Lowenhaupt Tsing, A., Zerner, C. (2005) *Communities and Conservation: Histories and Politics of Community-Based Natural Resource Management*, Altamira Press, Lanham, MD and Plymouth, UK.
Budiningsih, C.R. (2007) "Menyoal Paten Pengetahuan Tradisional", *Pikiran Rakyat*, 30 July 2007.
Carpenter, K.A., Katyal, S.K. and Riley, A.R. (2009) "In Defense of Property", *Yale Law Journal*, Vol. 118, pp. 1022–1125.
Chanock, M. (2005) "Customary law, sustainable development and the failing state", in P. Ørebech, F. Bosselman, J. Bjarup, D. Callies, M. Chanock and H. Petersen (eds), *The Role of Customary Law in Sustainable Development*, Cambridge University Press, Cambridge, pp. 338–83.
Chanock, M. (2009), "Branding Identity and Copyrighting Culture: Orientations towards the Customary in Traditional Knowledge Discourse", in C. Antons (ed.), *Traditional Knowledge, Traditional Cultural Expressions and Intellectual Property Law in the Asia-Pacific Region*, Kluwer Law International, Alphen aan den Rijn, pp. 177–93.
Chong, J.W. (2012), "Mine, Yours or Ours? The Indonesia-Malaysia Disputes over Shared Cultural Heritage", *SOJOURN: Journal of Social Issues in Southeast Asia*, Vol. 27, No. 1, pp. 1–53.
Cochrane, J. (2016) "It's a Jungle out There: Contestation and Conflict at Indonesia's Natural World Heritage Sites", in V.T. King (ed.), *UNESCO in Southeast Asia: World Heritage Sites in Comparative Perspectives*, NIAS Press, Copenhagen, pp. 313–46.
Cramb, R.A. (2007) *Land and Longhouse: Agrarian Transformation in the Uplands of Sarawak*, NIAS Press, Copenhagen.
De Schutter, O. (2014) "The transformative potential of the right to food", Final report of the Special Rapporteur on the right to food, UN Doc. A/HRC/25/57, 24 January 2014.
Dove, M.R. (2000) "The Life-Cycle of Indigenous Knowledge, and the Case of Natural Rubber Production", in R. Ellen, P. Parkes and A. Bicker (eds), *Indigenous Environmental Knowledge and its Transformations: Critical Anthropological Perspectives*, Routledge, London-New York, pp. 213–51.
Dove, M.R. and Carpenter, C. (2008) "Introduction: Major Historical Currents in Environmental Anthropology", in M.R. Dove and C. Carpenter (eds) *Environmental Anthropology: A Historical Reader*, Blackwell Publishing, Malden, MA-Oxford, UK-Carlton, Victoria, 1–85.
Dove, M.R., Sajise, P.E. and Doolittle, A.A. (2005) "Introduction: The Problem of Conserving Nature in Cultural Landscapes", in M.R. Dove, P.E. Sajise and A.A. Doolittle (eds), *Conserving Nature in Culture: Case Studies from Southeast Asia*, Yale University Southeast Asia Studies, New Haven, pp. 1–21.
Dressler, W.H., Kull, C.A. and Meredith, T.C. (2006) "The politics of decentralizing national parks management in the Philippines", *Political Geography*, Vol. 25, pp. 789–816.
Duncan, C.R. (2004) "Legislating Modernity among the Marginalized", in C.R. Duncan (ed.), *Civilizing the Margins: Southeast Asian Government Policies for the Development of Minorities*, Cornell University Press, Ithaca and London, pp. 1–23.

Eghenter, C. (n.d.) "Borneo: One Island One Sustainable Future", in *The Human Heart of Borneo*. The Heart of Borneo Initiative, World Wildlife Fund, Forum of the Indigenous Peoples of the Highlands in the Heart of Borneo, p. 45.

Eilenberg, M. (2012) *At the Edges of States: Dynamics of State Formation in the Indonesian Borderlands*, KITLV Press, Leiden.

Forrest, C. (2010) *International Law and the Protection of Cultural Heritage*, Routledge, London and New York.

Forsyth, M. 2013, "How can traditional knowledge best be regulated? Comparing a proprietary rights approach with a regulatory toolbox approach", *The Contemporary Pacific*, Vol. 25, no. 1, pp. 1–31.

Francioni, F. (2008a) "The 1972 World Heritage Convention: An Introduction", in F. Francioni (ed.), *The 1972 World Heritage Convention: A Commentary*, Oxford University Press, Oxford, pp. 3–7.

Francioni, F. (2008b) "The Preamble", in F. Francioni (ed.), *The 1972 World Heritage Convention: A Commentary*, Oxford University Press, Oxford, pp. 11–21.

Fross, J.K. (2016) "Natural World Heritage Sites and Local Communities: A Conflict of Interests? Two Case Studies from Palawan, the Philippines", in V.T. King (ed.), *UNESCO in Southeast Asia: World Heritage Sites in Comparative Perspective*, NIAS Press, Copenhagen, pp. 347–66.

Frossard, D. (2005) "In Field or Freezer? Some Thoughts on Genetic Diversity Maintenance in Rice", in M.R. Dove, P.E. Sajise and A.A. Doolittle (eds), *Conserving Nature in Culture: Case Studies from Southeast Asia*, Yale University Southeast Asia Studies, New Haven, CT, pp. 144–66.

Geismar, H. (2013) *Treasured Possessions: Indigenous Interventions into Cultural and Intellectual Property*, Duke University Press, Durham and London.

Hoffman, B.T. (2006) "Introduction: Exploring and Establishing Links for a Balanced Art and Cultural Heritage Policy", in B.T. Hoffman (ed.), *Art and Cultural Heritage: Law, Policy and Practice*, Cambridge University Press, New York, pp. 1–18.

Kingsbury, B. (1999) "The Applicability of the International Legal Concept of 'Indigenous Peoples' in Asia", in J.R. Bauer and D.A. Bell (eds), *The East Asian Challenge for Human Rights*, Cambridge University Press, Cambridge, pp. 336–77.

Kompas (2012) "Beras Adan Resmi Dipatenkan", 15 January 2012.

Langub, J. (n.d.) "Lati'ba: the process of wet-rice cultivation", in *The Human Heart of Borneo*, The Heart of Borneo Initiative, World Wildlife Fund, Forum of the Indigenous Peoples of the Highlands in the Heart of Borneo, p. 44.

Lowenhaupt Tsing, A. (2008) "Becoming a Tribal Elder and Other Green Development Fantasies", in M.R. Dove and C. Carpenter (eds), *Environmental Anthropology: A Historical Reader*, Blackwell Publishing, Malden, MA-Oxford, UK-Carlton, Victoria, pp. 393–422.

Lowenhaupt Tsing, A., Brosius, J.P. and Zerner, C. (2005) "Introduction: Raising Questions about Communities and Conservation", in J.P. Brosius, A. Lowenhaupt Tsing and C. Zerner (eds), *Communities and Conservation: Histories and Politics of Community-Based Natural Resource Management*, Altamira Press, Lanham, MD-Plymouth, UK, pp. 1–34.

Merryman, J.H. (2005) "Cultural Property Internationalism", *International Journal of Cultural Property*, Vol. 12, pp. 11–39.

Nafziger, J.A.R., Kirkwood Paterson, R. and Dundes Renteln, A. (2010) *Cultural Law: International, Comparative and Indigenous*, Cambridge University Press, New York.

Niezen, R. (2003) *The Origins of Indigenism: Human Rights and the Politics of Identity*, University of California Press, Berkeley.

Persoon, G. (2009) "'Being Indigenous' in Indonesia and the Philippines", in C. Antons (ed.), *Traditional Knowledge, Traditional Cultural Expressions and Intellectual Property Law in the Asia-Pacific Region*, Kluwer Law International, Alphen aan den Rijn, pp. 195–216.

Posey, D.A. (1985) "Indigenous Management of Tropical Forest Ecosystems: The Case of the Kayapó Indians of the Brazilian Amazon", in M.R. Dove and C. Carpenter (eds), *Environmental Anthropology: A Historical Reader*, Blackwell Publishing, Malden-Oxford-Melbourne, pp. 89–101.

Prott, L.V. and O'Keefe, P.J. (1992) "'Cultural Heritage' or 'Cultural Property'?" *International Journal of Cultural Property*, Vol. 1, pp. 307–20.

Rachman, N.F. and Siscawati, M. (2017) "Forestry Law, *Masyarakat Adat*, and Struggles for Inclusive Citizenship in Indonesia", in C. Antons (ed.), *The Routledge Handbook of Asian Law*, Routledge, London and New York, pp. 224–49.

Redgwell, C. (2008) "Article 2 – Definition of Natural Heritage", in F. Francioni (ed.), *The 1972 World Heritage Convention: A Commentary*, Oxford University Press, Oxford, pp. 63–84.

Saputra, A. (2012) "Sempat Diklaim Malaysia, Beras Adan Krayan Kini Milik Indonesia", *detikNews*, 13 January 2012, http://news.detik.com/read/2012/01/13/195550/1815286/10.

Sillitoe, P. (2006) "Ethnobiology and applied anthropology: *rapprochement* of the academic with the practical", in R. Ellen (ed), *Ethnobiology and the Science of Humankind*. Blackwell Publishing, Malden, MA-Oxford-Melbourne, pp. 147–75.

Sissons, J. (2005) *First Peoples: Indigenous Cultures and Their Futures*, Reaktion Books, London.

Smith, L. (2006) *The Uses of Heritage*, Routledge, London and New York.

Tsosie, R. (2012) "International trade in indigenous cultural heritage: an argument for indigenous governance of cultural property", in C.B. Graber, K. Kuprecht and J.C. Lai (eds), *International Trade in Indigenous Cultural Heritage: Legal and Policy Issues*, Edward Elgar, Cheltenham, UK – Northampton, MA, pp. 221–45.

Tuck-Po, L. (2005) "Uneasy Bedfellows? Contrasting Models of Conservation in Peninsular Malaysia", in M.R. Dove, P.E. Sajise and A.A. Doolittle (eds), *Conserving Nature in Culture: Case Studies from Southeast Asia*, Yale University Southeast Asian Studies, New Haven, CT.

University of Surabaya (2015) "Pemprov Diminta Terbitkan Perda HKI", 11 June 2015, www.ubaya.ac.id/2014/content/news_detail/1548/Pemprov-Diminta-Terbitkan-Perda-HKI.html.

Vandergeest, P. and Wittayapak, C. (2010), "Decentralization and Politics", in C. Wittayapak and P. Vandergeest (eds), *The Politics of Decentralization: Natural Resource Management in Asia*, Mekong Press, Chiang Mai, 1–20.

Vrdoljak, A. (2008) *International Law, Museums and the Return of Cultural Objects*, Cambridge University Press, Cambridge.

Winarto, Y. (2004) *Seeds of Knowledge: The Beginning of Integrated Pest Management in Java*, Yale University Southeast Asia Studies, New Haven, CT.

WIPO (2016a) *Glossary of Key Terms Related to Intellectual Property and Genetic Resources, Traditional Knowledge and Traditional Cultural Expressions*, WIPO/GRTKF/IC/32/INF/7 of 4 October 2016.

WIPO (2016b) *The Protection of Traditional Knowledge: Draft Articles*, WIPO/GRTKF/IC/32/4 of 3 October 2016.

Wittayapak, C. and Vandergeest, P. (2010) *The Politics of Decentralization: Natural Resource Management in Asia*, Mekong Press, Chiang Mai.

Yusuf, A.A. (2008) "Article 1: Definition of Cultural Heritage", in F. Francioni (ed.), *The 1972 World Heritage Convention: A Commentary*, Oxford University Press, Oxford, pp. 23–50.

21

INTELLECTUAL PROPERTY, BIODIVERSITY AND FOOD SECURITY

Brad Sherman

Introduction

According to recent data, one in eight people in the world (one billion people) go hungry every day, while more than two in eight (over two billion people) live in fear of hunger or starvation (FAO, 2014). A number of factors have contributed to this problem, including population growth,[1] under-investment in agricultural research and innovation, increasing energy costs, changing climatic conditions, the degradation of natural resources, and political and economic instability. While food security is not a major issue in developed countries (although it is an issue), domestic agriculture is under threat from similar factors. For example, as the Australian Government's 2014 *Agricultural Competitiveness* Green Paper recognizes, there is an urgent and ongoing domestic need to improve agricultural yields, increase sustainability, enhance and facilitate the breeding and development of new plant varieties, and assist local agriculture in adapting to climatic and environmental changes. Similar problems arise in other developing counties around the world. This is important for global food security because these issues undermine the ability of these countries to help meet the global demand for food.

Over the last decade or so there has been a growing interest in the potential role that intellectual property law might be able to play in helping to ensure a secure and consistent supply of food (Chiarolla, 2011; Haugen, Muller and Narasimhan, 2011). For some, intellectual property has the potential to enhance agricultural productivity and build food security. In particular, it is thought that intellectual property should, if used strategically, be able to improve conservation and access to genetic resources, help to protect indigenous knowledge and traditional farming practices, stimulate investment in agricultural research and innovation, foster research and development, improve the uptake and adoption of the resulting research results, and protect the reputational value of farmers' products. In so doing, it is suggested that intellectual property has the potential to lead to new food security solutions such as the development of novel food sources, the improvement in crop yields, nutritive value, climatic adaptability and livestock health. It is also suggested that intellectual property laws can also be used to protect and sustain traditional food production, build markets for food products, and improve the way food is delivered, transported, and stored. It seems that this trend is set to continue given that CGIAR (previously known as the Consultative Group on International Agricultural Research) – which is the leading international agricultural research agency in the world and which has traditionally been opposed to intellectual property protection of public good research – recently embraced the use of intellectual property in a limited number of situations (Sherman, 2012).

A number of factors have contributed to the growing interest in the role that intellectual property might play in building food security. These include changes in intellectual property laws that expanded the scope of what can be protected (primarily as a consequence of countries changing their laws to become TRIPS-compliant), changes in the international regimes that regulate genetic resources, greater private sector involvement in agricultural research, a growing interest by publically funded organizations (particularly universities and State-funded research bodies) in using intellectual property to recoup investment in research, and a corresponding growth in public-private partnerships. The growing interest in the role that intellectual property potentially plays in stimulating agricultural innovation and promoting food security has also been a product of changes in the way that intellectual property is conceptualized (something to which I return below).

At the same time, there has also been a corresponding chorus of complaints that intellectual property not only undermines attempts to improve agricultural productivity, but that it also threatens local and global food security. For example, it is suggested that intellectual property has the potential to hinder the breeding and development of new varieties, that it restricts access to genetic resources, and that it slows down the uptake and adoption of research results.[2] The use of plant breeder's rights for new plant varieties, for example, is said to limit access to seed and undermine age-old and effective methods of saving, storing and using seed. Intellectual property also carries 'the risk of restricting farmers' rights to reuse, exchange and sell seed … practices which form the basis of their traditional role in conservation and development' (Commission on Intellectual Property Rights, 2002, p. 13; Braunschweig *et al.*, 2014). The same is increasingly true for livestock breeds where the use of trade secrets and licensing agreements, rather than patents, has led to a consolidation of the pig and poultry sectors.[3]

The aim of this chapter is to look at one specific area where, for better or worse, intellectual property law potentially plays a role in shaping food security; namely, in relation to the conservation, preservation and use of genetic resources (The Crucible Group, 1994). This is important because ready access to genetic resources is essential to the development of new plants and animals and thus to the ongoing sustainability and viability of agriculture: a need which is exacerbated by changing climatic conditions, the advent of new pests and diseases, and declining yields.

International intellectual property

Most of the discussions about intellectual property and the way it impacts on biodiversity and food security that have occurred to date have focused almost exclusively on the international legal regimes that operate in this area: namely, the Convention on Biological Diversity, the International Treaty on Plant Genetic Resources for Food and Agriculture, and the Agreement on Trade-related Aspects of Intellectual Property Rights.

The Convention on Biological Diversity (CBD) is a landmark international treaty that attempts to change the way that biological resources are used. While providing access to a secure and consistent food source is not a primary goal of the Convention, it does have a number of important ramifications for food security (Lawson, 2012; Crookshanks and Phillips, 2012). Of these, two stand out. The first is the effort to protect traditional knowledge associated with genetic resources. The second is the goal of encouraging States to recognize property-style rights in relation to genetic resources. This is done, in part, as a way of ensuring that benefits are shared with the parties that provide access to genetic resources. While the goal of preserving and protecting biodiversity, along with associated traditional knowledge, has long-term ramifications for food security, other aspects of the Convention on Biological Diversity are potentially at odds with the global flow of genetic resources that sustain and underpin agricultural productivity.

The importance of providing access to genetic resources cannot be overstated. Most, if not all, countries not only import food, but much of the food that is grown locally is based on germplasm derived from other countries. For example, 98% of Australia's wheat crop is sown to varieties with genetic material derived from the International Maize and Wheat Improvement Centre (CIMMYT) based in Mexico (Braidotti, 2013). The potential tension that exists between the Convention on Biological Diversity and the desire to ensure that genetic resources are readily available for agricultural purposes was acknowledged in Article 8(c) of the Nagoya Protocol which provides that when developing and implementing access and benefit-sharing legislation or regulatory requirements, each Party shall 'Consider the importance of genetic resources for food and agriculture and their special role for food security'. While we are still awaiting a sustained empirical study into the effect of biodiscovery (or bioprospecting) laws (cf. Coolsaet *et al.*, 2015), the potential that exists within the Convention on Biological Diversity to restrict access has been given some credence by the growing number of anecdotal stories that tell about the way in which the bureaucratic frameworks that have been developed to regulate biodiscovery research have prevented or slowed down the productive use of biological resources (Bosse, 2016; Ruiz, 2015).

While the Convention on Biological Diversity does not directly deal with intellectual property, nonetheless intellectual property is still an important part of the Convention. This is because the Convention on Biological Diversity was prompted, in part, by concerns raised about bioprospecting and biopiracy, and the role that intellectual property played in this process. Although many of the examples of biopiracy that have been raised involve medical inventions, they also often extend to include agricultural produce, such as Basmati rice, the Gnali Nut, or the Enola bean.

The International Treaty on Plant Genetic Resources for Food and Agriculture, which came into effect in 2004, provides for the conservation and sustainable use of plant genetic resources for food and agriculture (Frison and Esquinas-Alcazar, 2011a, 2011b). It also provides for the fair and equitable sharing of benefits that derive from using these resources. The International Treaty covers the plant materials listed in Annex 1 of the Treaty that are under the management and control of governments and in the public domain, as well as the *ex situ* collections of CGIAR (Moore and Frison, 2011).[4]

Intellectual property plays a much more prominent role in the International Treaty than it does in the Convention on Biological Diversity. As with the CBD, the International Treaty attempts to respond to a potential problem created by intellectual property. Specifically, it responds to the concern that intellectual property protection, primarily patent and plant breeders rights, might be used to restrict access to plant genetic materials. Two techniques are relied upon to prevent this from happening. The first are the doctrinal rules of intellectual property law that specify that for inventions derived or based on plant genetic material to be protected, there must be an discernible gap between the underlying biological material and the resulting invention. Here, the space between the underlying material and the derivative invention ensures that the invention is new and, as such, does not take anything away from the public. A similar logic applies in plant breeders rights.

The second way plant materials are protected under the Treaty is by way of a standard material transfer agreement (SMTA) that governs the transfer of plant materials (Correa 2011). The standard material transfer agreement deals with plant materials 'in the form received' from the multilateral system, materials derived from the original material that is being modified (materials 'under development'), and 'products' that incorporate the plant material received, or any of its genetic parts or components ready for commercialization (excluding commodities and other products used for food, feed and processing). The standard material transfer agreement is essentially a contract between the provider and the recipient of the plant materials that sets out the terms and conditions of access, use and benefit-sharing, and a number of intellectual property matters. The standard material transfer agreement only covers access to materials for use or conservation for the purpose of research, breeding and training for food and agriculture. Any other uses, including chemical, pharmaceutical and other industrial uses, are outside the bounds of the standard material transfer agreement and require a separate agreement.

The standard material transfer agreement prohibits recipients from claiming intellectual property where the claim would restrict access to the plant materials 'in the form received' (and their genetic parts or components) for use or conservation for research, breeding and training for food and agriculture.[5] This leaves open the possibility of claiming intellectual property protection over any materials received so long as access to the materials for research, breeding and training for food and agriculture is allowed. This includes developments to the plant materials or genetic parts or components of the plant materials received. In the event that a recipient improves and commercializes material *and* the improved material is not available to others for research or breeding, the material transfer agreements provide that the recipient must make the material available to other parties in accordance with the principles of fair and equitable benefit-sharing.

The second way in which the International Treaty on Plant Genetic Resources for Food and Agriculture interacts with intellectual property is in its recognition of farmers' rights (Cubero, 2011; Pelegrina and Salazar, 2011). The Treaty recognizes the contribution that the local communities, Indigenous groups and farmers, particularly those in centres of origin and of crop diversity, have made and will continue to make to the conservation and development of plant genetic resources which constitute the basis of global food and agricultural production (Andersen and Winge, 2008). In reflection of this, Article 9 of the Treaty implores Member States to take measures to protect and promote farmers' rights, including: protection of traditional knowledge relevant to plant genetic resources for food and agriculture; the right to participate equitably in sharing benefits arising from the utilization of plant genetic resources for food and agriculture; and the right to participate in making decisions, at the national level, on matters related to the conservation and sustainable use of plant genetic resources for food and agriculture. The notion of farmers' rights was developed primarily to secure recognition of the farmers' role in the conservation and continuing development of local plant varieties and to protect their rights to these varieties and to save, sell and exchange seeds from their harvest. As the Preamble to the Treaty notes, the 'rights to save, use, exchange and sell farm-saved seed, and other propagating material' is fundamental for the realization of farmers' rights. Originally conceived in the 1980s as a means to minimize the impact of plant breeders' rights on local farmers, farmers' rights are enshrined in a variety of legal instruments regulating access to and use of genetic resources and traditional knowledge. Farmers' rights are also now the subject of specific legislation in a growing number of countries, especially those with large farming communities, such as Bangladesh and India (Tobin, 2013).

The third international treaty that figures in discussions of intellectual property, biodiversity and food security is the Agreement on Trade-related Aspects of Intellectual Property Rights (TRIPS) (Haugen, 2005). Typically, the main consideration here is the way in which TRIPS has forced developing countries to change their laws to provide more and stronger protection for intellectual property. One of the notable things about this area is that unlike other parts of the

TRIPS Agreement, which are fairly prescriptive in terms of the nature of the legal rights that need to be introduced, Article 27(3)(b) – which deals with botanical innovations – provides Member States with much more flexibility. In particular, Member States are required to provide for the protection of plant varieties either by patents, by an effective *sui generis* system, or by any combination thereof. As we will see below, this flexibility has important consequences for the way that we think about intellectual property and how it impacts on biodiversity and food security.

There are a number of noteworthy things about the international legal regimes that impact upon intellectual property, biodiversity and food security. It is clear, for example, that the Convention on Biological Diversity, the International Treaty on Plant Genetic Resources for Food and Agriculture, and (to a lesser extent) the TRIPS Agreement, have played an important role in heightening awareness about the importance of biodiversity for food security. The same is also true in relation to the role that intellectual property plays in this context. This is particularly the case in respect of the ways in which the treaties have heightened awareness of the ways in which intellectual property potentially restricts access to biodiversity and, as a result, the way it negatively impacts on food security. While this much is clear, what is less clear, however, is the effectiveness of the various national and international legal regimes that have been established to respond to the problems attributed to intellectual property (Kloppenburg, 2014).

A primary goal of the Convention on Biological Diversity and the International Treaty was to change the way that genetic resources are conserved, used and exploited. They also attempted to change the way that the parties involved in these processes relate to each other. One of the notable things about the regulatory schemes that were established to achieve these goals was that they presuppose that genetic resources circulate in a particular manner. They also presuppose that this network is populated by a number of key actors and that these actors relate to each other in a particular way.

There are two features of these regulatory networks that stand out. The first relates to the presumptions made about the motives of the parties accessing genetic resources. While it was accepted that parties seeking to access and use genetic resources can be motivated by either commercial or non-commercial goals, nonetheless the regulatory schemes introduced under the Convention on Biological Diversity and the International Treaty were shaped by the belief that the parties had to be treated as if they were motivated by commercial goals. This is because while the Convention on Biological Diversity and the International Treaty both aspire to noble and admirable goals, they both build upon a regulatory framework that works on the basis of worst-case scenarios. That is, they operate on the basis that even if a party is motivated by non-commercial goals, they need to be policed on the basis that they could, at any time, change their mind and commercialize the genetic resources. While biodiscovery and access and benefit-sharing laws may, for example, distinguish between parties who collect biological materials for commercial purposes from parties who collect for non-commercial purposes, nonetheless the regulations dealing with non-commercial uses are inherently shaped by the *potential* for commercial use.

The decision to frame the laws on the basis that the parties seeking access are potentially motivated by commercial ends has a number of consequences. One of the most important was that it immediately put access providers at odds with the interests of end users. In so doing it instituted a model of creation, production and (to a lesser extent) consumption that was premised on the idea that the relevant actors were motivated by conflicting interests and concerns. In turn, this had an important impact on the role that was assigned to the legal regimes. In particular, it meant that the role of intellectual property law was limited to the all-too familiar task of mediating and balancing these competing interests. Alternatively, intellectual property law was seen as an inherently negative tool that was used to exploit access providers and, as such, something that they needed to be protected from.

A second notable characteristic of the regulatory schemes that are used in the CBD and the International Treaty concerns the way that the parties associate with each other. Previously, the parties involved in the exchange and circulation of genetic resources were connected to each other in a myriad of ways (including through social, professional, or familial ties). After the CBD and the International Treaty came into force, new legal arrangements were layered on top of, and in some cases dominated, the pre-existing relationships. As a result, and irrespective of the relationships that existed previously, the intention of the treaties is to ensure that the parties are connected juridically.[6]

Beyond international law

Most of the discussions that have looked at the relationship of intellectual property, biodiversity and food security have done so through the lens of the international legal frameworks. While the situation is slowly changing, there is relatively little research that looks at how these international regimes operate at the local level (Coolsaet *et al.*, 2015). There has been even less attention given to local initiatives or to how other regimes operate in this area. One of the notable things about the research that looks at intellectual property, biodiversity and food security is how it has embraced the oppositional logic that underpins the international treaties. While it is clear why civil society groups, corporations and lobby groups would adopt this partisanal way of thinking, it is less clear why commentators feel compelled to adopt a similar approach when thinking and writing about the law in this area.

While a lot of important work has taken place in this area, there are many areas that have been neglected and ignored. To some extent, this is a product of the way that intellectual property has been conceptualized. While it is tempting to suggest that this is attributable to the oppositional and legal logic that is enshrined in and encouraged by the CBD and the International Treaty, it seems that its roots are more widespread.

Most accounts of intellectual property as it pertains to biodiversity and food security adopt what might be called a mechanical view of creation and innovation. There are a number of characteristics of this way of thinking about intellectual property: one of the more important being the way innovation is understood. With the human actor as the source of innovation, creation and innovation is seen as a process of unfolding, as one in which the inventive ideas are applied or reduced to practice. This way of thinking is reflected in many of the doctrinal rules and procedures, most notably in the requirement that the invention must not only be able to be reduced to a written format, but that it must also be able to be recreated from that written format. This way of thinking about intellectual property not only presupposes that the human agent is the source of creation, it also presupposes that these originating ideas can be distinguished and separated from the physical format in which they manifest themselves: that is, they presume that the intangible can be decoupled from the tangible. One of the lessons that we can take from the history of intellectual property law is that biological subject matter has never neatly fitted into this model. The law has never been able to separate the intangible from the tangible. As a result, the law has always had to incorporate and deal with the physical manifestation of the invention rather than, for example, a written description. In this situation, the roles of inventor and nature are reversed. Unlike the case with the archetypal invention – where the human agent exerts his or her intellectual skill to mould and shape the resulting invention – with biological inventions, it is nature that does the innovating, while the role of the inventor is reduced to recognizing and preserving that innovation.

Recognizing the distinctive nature of biological innovation has ramifications for how we think about intellectual property as it applies to biodiversity and food security. It leads us, for example, to reconsider the role and place of the standard material transfer agreements and the doctrinal notion of derivation; which, in order to regulate the movement of plant materials under the International Treaty, build upon a particular view of the nature of the relationship between the intangible and the tangible. It also leads us to question what we mean by intellectual property law. Typically, when questions about intellectual property and biodiversity arise, the focus of attention is on patent law, plant breeders rights and (to a lesser extent) trademarks. Recognizing that intellectual property protection for biological innovations necessarily includes the physical form in which the innovation manifests itself suggests that we should broaden our focus of attention to include legal regimes that focus on material objects (such as seeds), rather than those that (ostensibly) focus on the immaterial (such as the intellectual property associated with seeds). While this gives rise to many questions – perhaps the most important being what is intellectual property – it does suggest that we should broaden our scope of interest to include, for example, regimes such as seed certification laws, the one-variety laws adopted in the United States, the seed storage laws established under French and British colonial rule in West Africa, as well as the practice by the US Patent and Trademark Office in the early twentieth century of distributing seeds.[7] Seeing intellectual property in this way provides us with a better understanding of the way that the law affects biodiversity and food security. Broadening the scope of what we take intellectual property to mean is also important for countries implementing TRIPS in so far as it opens up a range of alternative ways of complying with Article 27(3)(b).

Another factor that has limited discussions about intellectual property as it applies to biodiversity and food security relates to what is expected or demanded of the law. Typically, intellectual property is seen as a resource that is used by private sector (commercial) organizations to recoup investment in innovation, as a mechanism to exploit creative outputs, or as a means to protect business reputation and goodwill. From this perspective, there is little scope for intellectual property outside of the commercial sphere; at best it is able to protect the reputation of authors and the integrity of their works; at worst, it positively hinders public good goals. While this logic has been applied in many contexts, it has been particularly prevalent in relation to publicly funded agricultural research that aims to improve food security and reduce poverty. Under this stereotypical view, there is little that intellectual property can do to assist these goals: at best it is irrelevant, at worst it is a positive hinderance. In this context, intellectual property is perceived as being largely alien to public good research. Under this traditional way of thinking, intellectual property and public good research are diametrically opposed to each other. This is because, on the one hand, the goal of publicly funded research is the generation of research results which are freely available; they are, in legal terms, part of the public domain. On the other hand, while intellectual property law recognizes the public domain, it is fundamentally opposed to free and open access. This dichotomy, so familiar in law, sets up what at first sight is essentially an unresolvable tension. The tension inherent in this apparent conflict manifests itself in a myriad of ways. One of which is the potential clash between the ideal of open access to the genetic resources and the situation where a third party acquires intellectual property protection over inventions developed from public domain materials and so undermines and limits access.

One of the notable things that has occurred over the last decade or so is the innovative way in which the international agricultural research organization, CGIAR, has attempted to resolve this tension (Sherman, 2012). What has happened within CGIAR, in effect, is that this problematic dichotomy has been unpacked. This has been done in two ways. The first change occurred when the focus of attention shifted away from research results as ends in their own right towards maximizing the impact of that research. In effect, the cult of the public domain was replaced by a desire to maximize the positive impact that research has on end-users. At the same time, with the introduction of the CGIAR Principles on the Management of Intellectual Assets in 2012, intellectual property was reconceptualized so that instead of being seen as something that primarily functioned to promote and serve commercial interests, as something that could only protect public interests negatively (for example by limiting the scope or length of protection), it also came to be seen as something that did have something valuable to offer in achieving goals such as reducing poverty and improving food security.

The CGIAR Principles on the Management of Intellectual Assets apply to the CGIAR Consortium, its 15 member Centers, and to all their activities and programs. The focus of the Principles are the 'intellectual assets' that are produced or acquired by Centers; that is, the results and/or products of research and development activities that are produced or acquired by Centers. These include, but are not limited to, knowledge, publications and other information products, databases, improved germplasm, technologies, inventions, know-how, processes, software and distinctive signs, whether or not they are protected by intellectual property rights.

The Principles confirm that the ultimate goal of CGIAR is the optimal impact on food security and global poverty. To this end, the Principles provide that Centers are under an obligation to manage the intellectual assets that they produce or acquire so as to maximize their accessibility and to ensure that they have the broadest possible impact on target beneficiaries. Centers are also under a general obligation to manage their research results and products with integrity, fairness, equity, responsibility and accountability. They are also under an obligation to comply with the International Treaty, the Convention on Biological Diversity and relevant human rights conventions.

While the Principles confirm that the ultimate goal of *all* CGIAR activities is to reduce global poverty and improve food security, they also recognize that in some cases the method that is used to achieve this goal has changed. That is, they recognize that in some situations, the goals of food security and poverty reduction can only be achieved by *limiting* access to CGIAR research (particularly by allowing the research results to be protected by intellectual property) (Sherman, 2012). To this end, the Principles set out a number of exceptional situations where Centers are able to limit or restrict access to research results.

The Principles aim to provide guidance about how Centers should manage research results and products in this new situation. Specifically, they provide guidance about when access to research results can be restricted (either by intellectual property, contract or other means). The Principles also aim to ensure that the management of intellectual assets is clear, consistent and concise. To allay fears that CGIAR's engagement with intellectual property specifically, and the private sector more generally, threatened to undermine the goals of food security and poverty reduction, the Principles aim to ensure that the management of intellectual assets is as transparent and open as possible and, as a result, that Centers are accountable for the way that they spend public funds. The Principles also aim to accommodate differences between Centers, respect Center autonomy and, at the same time, allow CGIAR to develop and exploit the benefits of system-wide approaches on important issues.

Underpinning these changes was the realization that exclusivity in general, and intellectual property in particular, were not necessarily antithetical to the goals of poverty reduction and food security (CGIAR, 2000; cf. RAFI, 2000). This realization has seen CGIAR Centers grant private sector organizations the exclusive right to use research products in order either to improve the products or to ensure that they are put into use. In some cases, Centers have also made strategic use of intellectual property themselves in order to enhance the impact that their research has on target beneficiaries (Sherman, 2012). Thus, for example, a situation might arise in which a Center develops a promising crop variety but does not have the resources to disseminate it to farmers in developing Country A. Country A's national public research and extension agencies inform the Center that they too lack the means to get the variety out to farmers. A few small seed companies might be interested in marketing the variety in Country A if they are granted an exclusive license for a limited period of time. In this situation, the Principles would allow the Center to grant exclusive access to the crop variety to the seed companies for a limited period of time. While opening up a space in which intellectual property is able to operate to improve food security created a number of problems (not least that it seemed to undermine the goals of promoting food security and poverty reduction), CGIAR has developed a number of mechanisms that attempt to negate or minimize the negative consequences that often accompany intellectual property. Although it may be some time before we can judge how successful this experiment has been, it is clear that a reconceptualized intellectual property law has an important role to play in helping to improve food security and to reduce poverty and hunger. For my purposes here, one of the most interesting things about the Principles is that they go beyond the increasingly tired and repetitive debates that characterize at least aspects of contemporary scholarship. While there are many unanswered questions, the Principles are important in so far as they offer an alternative way of thinking about intellectual property as it relates to biodiversity and food security and, as such, a possible model for how the law might develop in the future.

A third factor that has limited discussions about intellectual property as it applies to biodiversity and food security relates to the way that food production is seen. More specifically it relates to the fact that much of the existing research fails to take account of the fact that intellectual property potentially affects all aspects of the food chain, from the preservation of genetic resources used for breeding through to the marketing of foodstuffs. This is reflected in the fact that the research tends to focus on a limited number of specific aspects of the food chain, while neglecting others. Thus, while there have been a number of important and valuable studies in a few specific areas, notably in relation to traditional knowledge, access to genetic resources and the role that intellectual property plays in stimulating private sector investment in agricultural research, there are many important areas that have been neglected (particularly in relation to post-farm activities such as processing, transportation, packaging, and storage).

Another related limitation with the existing research is that it does not consider the way that different aspects of the food chain impact on one another. That is, it fails to take account of the way that different parts of the agricultural network interact with each other and then how this affects both the operation of intellectual property and the way that biodiversity is treated. This oversight limits the relevance of the research. The one exception to this is the focus on farmers' rights and the role that farmers play in protecting and building biodiversity. While this is important, there are many other areas that deserve attention including, for example, the way that patented production technologies or the purchasing power of supermarkets not only shape breeding practices, but also the movement of genetic materials.[8] To overcome these limitations, it is important to take a more holistic approach towards understanding the impact of intellectual property on all aspects of the food supply chain – from the genetic resources used in breeding, the breeding of new plants and animals, and on-farm practices (growing, harvesting and storage), through to the processing, transportation, and storage of food, the techniques used in the marketing of new foodstuffs and products, and the consumption and disposal of food.

Intellectual property potentially impacts on biodiversity and food security in a myriad of different ways. It has the potential both to promote and to hinder the conservation, preservation and use of biodiversity. Given that access to biodiversity is crucial to the production of new plants and animals and thus to providing a stable and secure food source, it is important that we critically review the various experiments that have taken place under the auspices of the

International Treaty and the CBD over the last twenty or so years. It is also important that we think more creatively about how we might maximize the benefits while minimizing the negative impact that intellectual property laws might have on biodiversity and food security.

Notes

1 Worldwide demand for food is projected to rise by 75% in the first half of the twenty-first century, with three-quarters of this growth in Asia. (Australian Government, 2014, p. 13).

2 This was prompted by events such as the granting of a patent, later overturned, over the Enola beans in the US. This led to the exclusion of Mexican farmers' products from US markets for almost ten years, with severe implications for food security. See Reichman (2016, p. 104).

3 Contracts preventing the buyer from selling breeding material from the purchased animals or requiring the payment of a royalty when breeding stock are sold are commonly used by pig and poultry breeding companies. For poultry in the United States, see Bugos (1992).

4 While the Treaty only applies to a proportion of the genetic resources held by the CGIAR Centers, nonetheless it has been adopted as the model by which *all* of the genetic resources held in Center collections are managed. A preliminary question concerned the legal status of the CGIAR germplasm collections. On this, see Siebeck and Barton (1991).

5 Standard Material Transfer Agreement, Art. 6.2. Available at www.fao.org/3/a-bc083e.pdf.

6 This was also reflected in the 'in-trust' agreements that the CGIA Centers signed with the FAO in 1994. The agreements brought the collections into line with the 1992 Convention on Biological Diversity. To this end, the agreements provided that Centers held 'designated germplasm in trust for the benefit of the international community' which constituted formal legal recognition of the ideal of open access. See CGIAR (1994).

7 On this see Sherman and Chapman (forthcoming).

8 Where parties are able to exert control over the food chain, whether through the use of patented technologies or market concentration (as with supermarkets), they are able to influence not only what is grown, but also where and how the end-products are stored and transported.

Bibliography

Australian Government (2014) *Agricultural Competitiveness Green Paper*, Canberra.

Andersen, R. and T. Winge (2008) 'Success Stories from the Realization of Farmers' Rights Related to Plant Genetic Resources for Food and Agriculture', The Farmers' Rights Project Background Study 7.

Antons, C. (2009) (ed.), *Traditional Knowledge, Traditional Cultural Expressions and Intellectual Property Law in The Asia- Pacific Region*, Kluwer Law International.

Braunschweig, T., F. Meienberg, C. Pionetti, S. Shashikant, and C. Dommen (2014) *Owning seeds, accessing food: A human rights impact assessment of UPOV 1991 based on case studies in Kenya, Peru and the Philippines*, The Berne Declaration, Zurich.

Bosse, J. (2016) 'Biopiracy in Queensland: A broken record that needs repair', *Trade Insight*, Vol. 12, No. 2.

Bugos, G. (1992) 'Intellectual Property in the American Chicken-Breeding Industry', *Business History Review*, Vol. 66, p. 127.

Braidotti, G. (2013) 'The international nature of germplasm enhancement', *Partners*, November, p. 27.

CGIAR (1994) *Joint Statement of FAO and the CGIAR Centers on the Agreement Placing CGIAR Germplasm Collections under the Auspices of FAO.*

CGIAR (2000) *Consultative Group on International Agricultural Research: Charting the CGIAR's Future – Reshaping the CGIAR's Organization*, Report of the 11th Meeting of the Genetic Resources Policy Committee, ICW/00/09. Rome: CGIAR Secretariat, pp. 7–8.

Chiarolla, Claudio (2011) *Intellectual Property, Agriculture and Global Food Security: The Privatization of Crop Diversity*, Edward Elgar Publishing, Cheltenham.

Commission on Intellectual Property Rights (2002) *Integrating Intellectual Property Rights and Development Policy: Report of the Commission on Intellectual Property Rights.*

Coolsaet, B., F. Batur, A. Broggiato, J. Pitseys, and T. Dedeurwaerdere (2015) *Implementing the Nagoya Protocol: Comparing Access and Benefit-Sharing Regimes in Europe*, Brill, Netherlands.

Correa, C. (2011) 'An Innovative and Transparent Option for Royalty Payment under the Treaty – Implementing Article 6.11 Crop Related Modality of the Standard Material Transfer Agreement', in Frison, Lopez, and Esquinas-Alcazar (eds) *Plant Genetic Resources and Food Security: Stakeholder Perspectives on the International Treaty on Plant Genetic Resources for Food and Agriculture*, FAO, Bioversity International and Earthscan, New York.

Crookshanks, R. and P.W.B. Phillips (2012) 'A Comparative Analysis of Access and Benefits-Sharing Systems', in Bubela and Gold (eds) *Genetic Resources and Traditional Knowledge: Case Studies and Conflicting Interests*, Edward Elgar Publishing, Cheltenham.

Crucible Group, The (1994) *The Impact of Intellectual Property on Biodiversity, Conservation, Trade and Rural Society*, International Development Research Centre, Ottawa, Canada.

Cubero, J. (2011) 'Plant Breeders: The Point of View of a Plant Breeder on the International Treaty on Plant Genetic Resources for Food and Agriculture', in Frison, Lopez, and Esquinas-Alcazar (eds) *Plant Genetic Resources and Food Security: Stakeholder Perspectives on the International Treaty on Plant Genetic Resources for Food and Agriculture*, FAO, Bioversity international and Earthscan, New York.

De Schutter, O. (2009) *Report of the Special Rapporteur on the Right to Food: The Right to Food: Seed policies and the right to food: enhancing agrobiodiversity and encouraging innovation* A/64/170.

FAO (2014) *The State of Food Insecurity in the World*, FAO, Rome.

Frison, C., F. Lopez and J. Esquinas-Alcazar (2011a) 'General Conclusions: Summary of Stakeholders' Views and Suggestions to Cope with the Challenges in the implementation of the International Treaty on Plant Genetic Resources for Food and Agriculture', in Frison, Lopez, and Esquinas-Alcazar (eds) *Plant Genetic Resources and Food Security: Stakeholder Perspectives on the International Treaty on Plant Genetic Resources for Food and Agriculture*, FAO, Bioversity International and Earthscan, New York.

Frison C., F. Lopez, and J. Esquinas-Alcazar (2011b) *Plant Genetic Resources and Food Security: Stakeholder Perspectives on the International Treaty on Plant Genetic Resources for Food and Agriculture*, FAO, Bioversity International and Earthscan, New York.

Halewood, M., I.L. Noriega, and S. Louafi (2013) (eds) *Crop Genetic Resources as a Global Commons: Challenges in International Law and Governance*, Bioversity International and Earthscan, New York.

Haugen, H.M. (2005) 'The Right to Food, the Right to Benefit from Science and the Trips Agreement', in Eide and Kracht (eds), *Food and Human Rights in Development: Legal and Institutional Dimensions and Selected Topics*, Vol. 1, Intersentia, Oxford.

Haugen, H., M.R. Muller and S. Narasimhan (2011) 'Food Security and Intellectual property Rights: Finding the Linkages', in Wong and Dutfield (eds), *Intellectual Property and Human Development: Current Trends and Future Scenarios*, Cambridge University Press, New York.

Kloppenburg, J. (2014) 'Re-purposing the master's tools: the open source seed initiative and the struggle for seed sovereignty', *The Journal of Peasant Studies*, p. 1.

Lawson, C. (2012) *Regulating Genetic Resources: Access and Benefit-sharing in International Law*, Edward Elgar Publishing, Cheltenham.

Moore, G. and E. Frison (2011) 'International Research Centres: The Consultative Group on International Agricultural Research and the International Treaty', in Frison, Lopez, and Esquinas-Alcazar (eds) *Plant Genetic Resources and Food Security: Stakeholder Perspectives on the International Treaty on Plant Genetic Resources for Food and Agriculture*, FAO, Bioversity International and Earthscan, New York.

Pelegrina, W.R. and S. Salazar (2011) 'Farmers' Communities: A reflection on the Treaty from Small Farmers' Perspectives', in Frison, Lopez, and Esquinas-Alcazar (eds) *Plant Genetic Resources and Food Security: Stakeholder Perspectives on the International Treaty on Plant Genetic Resources for Food and Agriculture*, FAO, Bioversity International and Earthscan, New York.

RAFI (2000) *Rural Advancement Foundation International: The Intellectual Property Challenges to Public Agricultural Research and Human Rights and 28 Alternative Initiatives: In Search of Higher Ground*, The Occasional Paper Series Vol. 6(1).

Reichman, J., P. Uhlir, and T. Dedeurwaerdere (2016) *Governing Digitally Integrated Genetic Resources, Data, and Literature*, Cambridge University Press, Cambridge.

Ruiz, M. (2015) *Genetic Resources as Natural Information: Implications for the Convention on Biological Diversity and the Nagoya Protocol*, London and New York, Routledge.

Sherman, B. (2012) 'The CGIAR Principles on the Management of Intellectual Assets', CGIAR.

Sherman, B. and S. Chapman (eds) (forthcoming) *Intellectual Property and Agriculture*, Edward Elgar, Cheltenham.

Siebeck, W. and J. Barton (1991) *The Legal Status of CGIAR Germplasm Collections and Related Issues*, CGIAR.

Tobin, B. (2013) 'Open Access Seeds and Breeds: Protection of farmers and livestock keepers' rights and commons', in Lawson and Sanderson (eds) *Intellectual Property and Food*, Ashgate Publishing, Farnham.

22

SISYPHUS REDIVIVUS? THE WORK OF WIPO ON GENETIC RESOURCES AND TRADITIONAL KNOWLEDGE

Nuno Pires de Carvalho

There, too, the hard-tasked Sisyphus I saw, / Thrusting before him, strenuous, a vast rock. / With hands and feet struggling, he shoved the stone / Up to a hill-top; but the steep well-nigh / Vanquished, by some great force repulsed, the mass / Rushed again, obstinate, down to the plain, / Again, stretched prone, severe he toiled, the sweat / Bathed all his weary limbs, and his head reeked.

Homer, *The Odyssey*, Book XI, transl. William Cowper (1791), www.gutenberg.org, e-book ≠24269, 2008

Introduction

Like Sisyphus, the WIPO Intergovernmental Committee on Intellectual Property and Genetic Resources, Traditional Knowledge and Folklore (IGC) keeps pushing its agenda on the protection of traditional knowledge (TK), traditional cultural expressions (TCEs) and their intersection with genetic resources (GRs), and it seems that there is no end to its endeavor. Initially, the IGC's mandate was to seek a multilateral consensus on how to create an internationally acceptable mechanism for protecting those intangible outputs of traditional cultures with an economic value. No predeterminations as to the specific type of protection were made initially, but several WIPO Member States were clearly searching for a *sui generis* regime of intellectual property protection, while others would be satisfied with a simple recognition that the mechanisms currently available would suffice.

This chapter explains the failure of the IGC in finding at least a beginning of a consensus in obtaining a final product of its debates. Put in mythological terms, the insistent attempts of the IGC to achieve something – anything – can be illustrated with the story of Sisyphus and the curse bestowed upon him to keep pushing a heavy boulder uphill, only to see it roll down as soon as it reaches the top. Like Sisyphus, who was cursed because he dared to offend Zeus with his tricks, IGC Members are paying for offending the fundamental rationale of intellectual property (IP). The first section of the chapter will explain that the IGC was set up for the wrong reasons. Its creation was not motivated by a common understanding of the international IP community that there was a need to move toward protecting TK and TCEs, but rather as a practical solution to save the negotiations of another treaty, the Patent Law Treaty (PLT), which had nothing to do with the IGC's purpose and subject matter. It will also describe how Sisyphus' rock was shaped (i.e., how the scope of the IGC's mandate was defined) and how his curse was put in motion. The following section will describe the efforts of the IGC to achieve a positive output, when various efforts looking like real negotiations took place. In this phase, Sisyphus was pushing, pushing, pushing. The chapter goes on to show where the IGC is now – after thousands of working hours dedicated to the Committee by WIPO civil servants, delegates, non-governmental organization employees, experts and facilitators, in the course of the last 16 years, after dozens of documents being elaborated and circulated, after several millions of Swiss francs spent, i.e., after so many efforts to push the rock uphill, IGC Members see it rolling down again, and again, and again. The chapter will explain the second major offense that WIPO Member States made to IP and which makes it impossible for the IGC (i) to achieve an output, or (ii) if it achieves it, to reach a significant output, such as a binding instrument or, at least, a recommendation by WIPO General Assembly, or (iii) in the improbable eventuality of obtaining this output, to achieve one solution that really works. This second – and, perhaps, the greater – offense to Zeus is that the IGC is offending the very core rationale of IP, which is primarily an instrument of entrepreneurship as far as it primarily ensures the promotion and preservation of differentiation of products and services of undertakings in their commercial rivalry with other undertakings. At this point in time, but for a few remarkable exceptions, there is no entrepreneurship involved in the making and trading of TK. Debates on the protection of TK involve many other aspects, such as the respect for human rights, the protection of sustainability and the environment, bioprospecting, the respect for traditional cultures, the preservation of languages and ways of living – all these aspects begging for great attention and respect, but not justifying the interference of IP mechanisms, because IP is about differentiation and rivalry between enterprises with the aim of luring or maintaining clients. And enterprises and clients have been conspicuously absent from the debates in the IGC, for the simple reason that those debates, at least for now, are not their concern.

In order to avoid an excessively long text, this chapter will keep the discussion focused on TK. But, in general terms, what will be said about the IGC discussion on TK will apply, in a large measure, to TCEs.

Shaping the rock

The adoption of international norms on IP requires a consensus among States in three different and sequential moments. First, States must agree that something has to be done as regards the absence of rules concerning the protection of certain intangible assets, having in view particular circumstances and goals, such as the need to rid international trade of certain obstacles and distortions caused by differences in the levels of national protection, or the wish to use certain IP assets as tools for the promotion of technological and/or industrial development. In the WIPO context, this consensus generally emerges in the course of discussions set in its various Standing Committees. After this initial consensus emerges, debates, discussions and negotiations will ensue.

Second, States, gathered in some negotiating forum – such as the WIPO Standing Committees – must find a consensus on the norms that will be approved. This may take years. Delegations will meet formally and informally to set the principles and iron out their differences on the rules. Once a consensus or a quasi-consensus is reached, States are ready to move to a Diplomatic Conference.

In the Diplomatic Conference, the last divergences are overcome, and States reach a third consensus on the terms of an Agreement or Treaty that will be approved by all.

The story could end here. But it does not. Once a consensus is achieved in a Diplomatic Conference, States must still agree – now, not in the form of a collective decision, but individually – to ratify the agreement. The lack of adherence to the agreement by a representative number of States may dictate its premature death. This has happened with the WIPO Treaty on the International Recording of Scientific Discoveries, of 1978, which was ratified by very few Contracting Parties, and never entered into force (WIPO, 1981). This has happened also with the Washington Treaty on Intellectual Property in Respect of Integrated Circuits, which, given the fierce opposition of Japan and the United States to its clauses on compulsory licences, never entered into force. Several of its provisions were, nevertheless, incorporated by reference by the WTO Agreement on Trade-Related Aspects of Intellectual Property Rights (the TRIPS Agreement).

The inception of the IGC did not follow this logical scheme. The matter of the protection of TK was introduced in the work of WIPO after the election of Kamil Idris as its Director General, in 1997. Eager to introduce new topics for discussion by WIPO Member States, Dr Idris flagged the possibility of requesting that the Secretariat explore new dimensions of IP. One possibility – linked to the approval of the Convention on Biological Diversity (CBD), in 1992 – was to bring to WIPO the then emerging debates on the protection of "indigenous knowledge" (WIPO, 2007, pp. 13–14). In 1998 Dr Idris created a division in charge of doing exploratory work on two topics. The Division – named the Global Issues Division, in which I worked for several years from 1999 on – had two sections, one composed by copyright experts, handling TCEs (then still designated as "expressions of folklore"), and the other formed by industrial property experts, namely patent experts, handling TK (then still designated "indigenous knowledge") and genetic resources.

The Global Issues Division undertook three important activities during its existence, until it was transformed and expanded in 2001 so as to serve as the Secretariat of the IGC: (i) it carried out nine fact-finding missions to 28 countries in all regions of the world, and produced a remarkable report on findings and expectations of indigenous peoples as regards the use of IP to protect their intangible assets (WIPO, 1999b); (ii) in 1998 and 1999 it organized two international seminars on IP and traditional knowledge, both held in Geneva; and (iii) in 1999 it carried out a survey on the protection of biotechnological inventions, in which a number of issues involving the consideration of traditional knowledge as prior art for the examination of patent applications as well as the need to disclose the origin of genetic resources as a patentability requirement were taken into consideration (WIPO, 1999a; WIPO, 2007, pp. 12–13).

These initiatives raised the awareness among WIPO Member States of the growing importance of the protection of TK and were the conceptual seed for the establishment of the IGC.

In parallel with this work, WIPO Member States were negotiating the adoption of the Patent Law Treaty (PLT) in the WIPO Standing Committee on the Law of Patents (SCP). In 2000, they gathered in Geneva, for the Diplomatic Conference that would approve it. That was when Colombia introduced, at the last minute, a proposal to add a provision to the PLT requiring the identification of the origin of genetic resources (and, in some versions of the requirement, evidence of prior informed consent by the holders of associated TK) (hereinafter, the disclosure requirement), when applicable, as a condition of patentability. This move led to the stalling of the PLT negotiations, in view of the fact that some delegations viewed the disclosure requirement as a matter of substantive law – in sharp conflict, therefore, with the exclusively formal tone of the PLT. Colombia, later, softened its position, admitting that the disclosure requirement could be introduced in the PLT as an option for Contracting Parties, depending on national statutes. Nevertheless, this was not sufficient to overcome the impasse.

The Director General conducted the informal negotiations that led to the establishment of the IGC. That was a solution of compromise: the IGC would not be a Standing Committee, as many Members wished, in the sense that every two years the General Assembly of WIPO would have to decide whether it should continue, but it would not be an *ad hoc* body either, as a few Members wanted, this meaning that, within each biennium the IGC would have autonomy to set its own agenda and schedule of work.

The discussions that followed the submission of the Colombian proposal focused on access to genetic resources, but certain Members wished to extend the debate to associated traditional knowledge and, while they were at that, to folklore as well.

At the WIPO General Assembly, Members agreed on the establishment of the IGC (WIPO, 2000b, p. 23).

However, it is important to note that, from the outset, there was a radical difference in expectations of WIPO Member States in a possible future outcome of the IGC work. The statements made during the 2000 WIPO General Assembly reflect that difference in a very clear way. At one end of the spectrum, various developing biodiversity-rich countries were ready for discussions aiming at developing legal regimes of IP *protection* for TK and TCEs (WIPO, 2000b, pp. 7–22). A few of those Members informed the Assembly that they had already started work on the adoption of such regimes (Algeria, Peru and Panama) (WIPO, 2000b, pp. 7–8, 10–11 and 12–13).

In general, it can be said that GRULAC (the Group of Latin American and Caribbean Countries) and African Countries were in favor of using the IGC to develop IP statutory regimes to protect TK and TCEs. Asian countries, while supportive of work aiming at preventing/repressing unauthorized exploitation of TK and TCEs, were not as clear on how to achieve such outcome. Nevertheless, they formed part of the consensus toward a positive, bold agenda for the IGC.

China seemed to have a different idea: it was still supportive of a normative IGC output but wished to deepen the conceptual discussions on how to protect TK and folklore (WIPO, 2000b, p. 10).

Later, the group of Asian countries and China would clarify (and harmonize) their common position as regards the protection of TK by proposing a double solution: either establishing "specific standards for the legal protection of traditional knowledge subject to [countries'] respective legal system" or improving "the application of existing standards" (WIPO, 2001b, p. 9).

At the other end of the spectrum, Japan, on behalf of Group B (i.e. the group that gathers developed countries), advocated a more cautious approach, putting emphasis on discussing the issues thoroughly, and, in particular, establishing a liaison between the IGC with the WIPO Standing Committee on the Law of Patents as well as with other intergovernmental bodies that also dealt with this issue (WIPO, 2000b, p. 10). Group B was thus looking forward to having discussions only, without any prejudgment as to possible outcomes – one of which could be, as a matter of course, the absence of output.

This was the first curse: WIPO Member States established the IGC on the basis of an ill-conceived consensus, limited to the agreement to participate, as a way to overcome the stalemate at the PLT Diplomatic Conference. In this regard, the success of the IGC was in its very creation. Unlike most developing countries, Group B countries had not committed themselves to look for an IGC output. In other words, there was no agreement as to the shape of the rock.

Pushing the rock uphill

Reporting to the General Assembly of 2003, the IGC disclosed significant exploratory work done in that three-year period, namely in the form of a number of outputs, such as "strengthened understanding of the policy basis and legal mechanisms for protection of TK, TCEs and IP aspects of genetic resources" and "specific steps to strengthen the defensive protection of TK and associated genetic resources", among others. Moreover, the IGC had started addressing the issue of the participation of representatives of indigenous and local communities in debates and negotiations in the IGC (WIPO, 2003a).[1]

The brevity of the 2003 report may be misleading as to the quantity of information produced, particularly the work undertaken by the Secretariat, which should not be underestimated. In those three years, as well as in the years that followed, dozens of papers, studies, surveys, reports, and technical visits were carried out by a small number of WIPO servants, with the cooperation of national authorities and experts, with a view to support WIPO Member States decisions as well as any national efforts towards the protection of TK and TCEs.

But, in spite of this (apparent) progress, the report also reveals that it "was unable to reach agreement on a recommendation to the General Assembly in this regard [of its future mandate]" (WIPO, 2003a).

This means that after a biennium of exploratory discussions, the IGC did not know exactly what to do next – the intervention of the General Assembly was therefore necessary, so that political forces could be put in motion to overcome the stalemate.

After intense negotiations, the 2003 General Assembly took a bold step: for the first time, the IGC admitted the possibility of moving to a negotiating phase as far as it accepted that "no outcome of its work is excluded, including the possible development of an international instrument or instruments" (WIPO, 2003b, pp. 18–19).

For the first time, it was unequivocally stated that the IGC could lead anywhere, from just a recognition of total failure to the adoption of one or eventually two treaties (one on TK, the other on TCEs). As an intermediate result, Members could settle on some sort of soft law, namely by adopting a recommendation that could be sent for approval by the General Assembly in the form of guidelines or model provisions. Those Members that so wished could then incorporate that recommendation into national law.

So, from this point on, the work of the IGC seemed to change in tone, becoming more similar to the normative agendas of WIPO Standing Committees.

Did Zeus, for the first time, show mercy to the IGC?

In 2006, at its ninth session, the IGC took note of a long series of topics that it had been pursuing, in particular the collection of comments by Member States on two important documents elaborated by the Secretariat that aimed at designing the future legal regimes of TK and TCEs protection (documents WIPO/GRTKF/IC/9/4 and WIPO/GRTKF/IC/9/5), and, equally importantly, extended the duration of the next meeting from five to seven working days (WIPO 2006). The relevance of these facts is that they show that Sisyphus, five years after starting to push the rock uphill, was not yet tired. On the contrary, he looked forward to continuing pushing and pushing – without, let us not forget, knowing the outcome.

Five sessions later, in 2009, the IGC had to admit (again) that no progress had been made. At its fourteenth session, it took note of those documents reflecting the status of substantive discussions on possible normative aspects concerning TK, TCEs and GRs (in this case, aspects exclusively concerned with unlawful access and the disclosure requirement) and that, as far as future work was concerned, no agreement had been reached (WIPO, 2009a). The IGC was again at a stalemate. The rock was rolling back down again.

So, in 2009 – nine years after its establishment – the IGC did not know what result it intended to reach, and yet several Members continued pushing for nothing less than a (binding) multilateral instrument.

One year later, at its seventeenth session, the IGC was still pushing. There was progress in the clarification of concepts and principles both as regards TK and genetic resources (GRs) (WIPO, 2010a; WIPO, 2010b). These two documents contain a sort of a menu of possible solutions for addressing issues concerning normative protection. The idea was not to pick up all the options suggested, because many of them were reciprocally incompatible, but rather for IGC Members to identify those solutions that might call for consensus. A few Group B members (Australia, Canada, New Zealand, Norway and the United States) submitted to the same meeting a document highlighting those points that, regarding GRs, they would accept as objectives and principles. Basically, they would accept the idea of using information on GRs as elements of prior art (to avoid the granting of patents in error) as well as components of the enabling disclosure (so that corresponding patent applications could teach how to reproduce them), when applicable and relevant. Access to genetic resources and obtaining prior informed consent from related TK holders were deemed matters of national legislation, and corresponding measures should not undermine the role of patents in promoting invention (WIPO, 2010c). The position of those five countries (Australia having been replaced by Korea), presented as a "Joint Recommendation", has been resubmitted to each IGC session since the twenty-third session (for the thirty-first session, see WIPO, 2016d).

After nine years of exhaustive work, dozens of papers elaborated and circulated, and, importantly, several million Swiss francs spent in organizing regular and intersessional meetings, seminars and workshops, and paying the travel expenses to Geneva of delegates from 26 Member States (five from each region plus one from China) for each IGC meeting, the Committee was as much lost in its work as it was in 2000. Members still stuck to the inaugural agreement to meet and discuss – which permitted the founding of the IGC – but nothing else could be predicted.

Sixteen years later, still pushing

The essence of Sisyphus' curse is that he should continue pushing the boulder for the sake of pushing. His fate was in achieving nothing, by incessantly making endless and futile efforts.

Between the seventeenth and the eighteenth sessions, the IGC worked in the form of intersessional meetings, in an attempt to push further and faster the practical work of delegations and observers, namely by elaborating normative texts on the three areas covered by the Committee: TCEs, TK and GRs. At the second intersessional session, Members agreed to establish six open-ended informal drafting groups for that purpose (WIPO, 2011a). The work was based on the texts elaborated by the Secretariat concerning the objectives and principles of the legal treatment of each specific field. Those texts on TCEs and TK were to be presented as working documents at the next session, but, as regards IP and GRs, Members were not clear as to the nature of future work (WIPO, 2011 b, p. 3). In other words, the IGC seemed to get closer (again?) to becoming a negotiating WIPO body, identifying common points and sorting out divergences, towards a draft text that, eventually, could be debated in a Diplomatic Conference.

To cut short this very long story, let us see where the work of the IGC on GR-related TK stands today, 16 years after its establishment by the WIPO General Assembly of 2000.

From the IGC's twenty-sixth session onwards, the Secretariat developed consolidated documents reflecting the status of discussions on normative language and changes introduced after each session (generally, those discussions take place in informal meetings parallel to the formal sessions, conducted by facilitators hired by the Secretariat, rather than in the Plenary). The consolidated documents have been submitted to the General Assembly, in the hope (by some Members) that a Diplomatic Conference could be convened to adopt a binding text, or, at least, that the Assembly could make a decision on how to move on with the IGC's work.

Let us examine just two disclosure provisions out of the consolidated documents concerning GRs (and associated TK), the first having specific reference to access and benefit-sharing (ABS) requirements and prior informed consent (PIC).

On GRs, Article 2 of the consolidated document requires that a disclosure requirement be adopted. But Article 6 of the same document has the following heading: "No new disclosure required", this meaning that no disclosure would be required besides the enabling one, which was already a standard requirement under the PCT and the TRIPS Agreement (WIPO, 2016a, Annex p. 8).

Two elements of this text should catch the reader's immediate attention: the enormous amount of square brackets – including those that embrace the texts in their entirety – indicating the components in respect of which delegations have not yet reached an agreement; and the absolutely contradictory tone of Articles 2 (supporting a mechanism of disclosure requirement) and 6 (rejecting such a mechanism, unless it is set as an element of the determination of the prior art or of the enabling disclosure). The fact that the rejection of Article 2 is expressed in a specific provision (Article 6) is very significant. Delegations that support Article 6 could simply keep Article 2 between square brackets, as it is now. Instead they preferred a provision that prohibits the adoption of the disclosure requirement as a condition of patentability in itself. Divergence between the two groups of delegations – one a vast majority of biodiversity-rich developing countries, the other a minority of Group B countries – could not be wider and deeper.

As regards a proposed regime for the protection of TK, divergences may be less radical, but they are nevertheless considerable. Just to refer to one element of protection (eventually, the most important) – the scope of protection, i.e., the type of rights to be accorded to TK holders – the last version of the draft Articles on TK protection submitted to the thirty-first session of the IGC, in September 2016, includes in its Article 3 (on "[Criteria for and] Scope of Protection") a menu of the rights to be granted, subject to a number of requirements (WIPO, 2016e, Annex p. 9).

This draft provision is not fully within square brackets, which means that all delegations believe that TK holders should be accorded some sort of right in their knowledge. But what right? This is where consensus ends and divergences (deep and wide) begin.

The approach taken in this draft provision has been designated as a "tiered approach", in the sense that it provides for various categories of rights corresponding to various types of TK, namely an economic sort of right for TK with a commercial nature/value, and a moral type of right for TK that is sacred. Moreover, rights are different for TK that is under control of its creators and for TK that has been widely disclosed (i.e. fallen into the public domain, even if this concept is problematic for many TK holders). This approach has been defended by the Chair of the IGC, in its thirty-first session (September 19 to 23, 2016).[2]

It should be noted that Article 3 refers to criteria of protection (between square brackets) implying some basic conditions for protection – namely, that the TK in question should be "[sacred], [secret] or [otherwise known] [closely held]".[3] There is no consistency between these four options. Secrecy and closely holding imply some sort of control of the subject matter by its holder(s). But the sacred nature of TK links it to moral aspects of protection, rather than to material or proprietary elements. The term "otherwise known" is confusing and eventually means every piece of TK that is somehow in possession of a community, regardless of its sacred or secret nature.

This text follows two lines of reasoning. Initially (subparagraph (a)), the text leans toward a more traditional IP regime in the sense that it supports a right to exclude, whether through private or governmental enforcement. But, by contrast with this approach, the text of subparagraph (b)(ii) opts for a mechanism of entitlement to remuneration, which implies that third parties may not be excluded from using the TK in question provided they pay for it.

The inconsistency between the two regimes is inevitable. Of course, if these were just draft texts to be followed as recommendations or guidelines for national laws, there would be no problem, because each country would be free to adopt the regime it would deem most appropriate. But if Members are contemplating the possibility of a multilateral instrument – be it a binding treaty or soft law (e.g., in the form of a recommendation to be approved by the WIPO General Assembly) – the incompatibility has to be eliminated.

Moreover, the draft text above introduces aspects of collective moral rights that are linked to the cultural dignity of TK holders. In this respect, it is requested that TK not be used in a way that disrespects the culture and traditions of the holders.

Besides the profusion of square brackets across the text, the aspect that is most striking to an observer linked to the world of business is the absolutely legal insecurity that stems from these draft articles. Of course, it would be unfair to expect that, as a draft text, the proposal could already contain the answer to the crucial questions that companies engaged in bioprospecting may have as regards investing significant amounts of money and time in products and processes embodying or bearing TK. But, after 16 years of IGC's intensive work, one could expect at least that it could provide some guidance on how GRs and associated TK are to be translated into the dimension of entrepreneurship and free competition – i.e. the dimension that IP has both in WIPO and the WTO.

In spite of the IGC's incapacity to narrow down the points of contention and simplify the texts to be discussed/negotiated, WIPO's General Assembly continues not only to renew its mandate but also to extend the frequency of its meetings. In October 2015, the Assembly decided that the IGC should continue working in the biennium 2016/2017, and attributed it a very intensive work programme. In 2017 the Assembly "will take stock of progress made, and decide on whether to convene a diplomatic conference or continue negotiations" (WIPO, 2016d).

Sisyphus will continue pushing, pushing, pushing.

IGC's real curse: rather than a difficult task, an impossible one

From the previous section it can be seen that Zeus, instead of giving our latter-day Sisyphus a regularly round shaped boulder, gave him one with sharp protuberances instead, which make the task of pushing it uphill much tougher.

But, unfortunately, the curse of the IGC does not stop at that. It if did, WIPO Member States, in the course of the last sixteen years, would eventually have been able to work on the boulder's shape and, with the help of the Secretariat, to find ways to bring it uphill – and make it stay there in a sustainable position.

But there is another curse that not only makes it difficult and painful for Sisyphus to push the rock uphill – it makes his task virtually impossible. The other curse – an unsolvable one, for that matter – is that the core features of the IP systems in force throughout the world were not designed to protect TCEs or TK (associated with GRs). The reason is simple, and yet it seems that IGC Members have not acknowledged it: the IP system protects those intangible assets that differentiate products and services, as well as the enterprises that sell them. IP, therefore, is inherently and inextricably linked to the notion of entrepreneurship, whereas TK and TCEs are not.

The core function of differentiation cuts across the whole spectrum of intellectual property and gives it consistency. The protection of all of its components has always been based on a condition of differentiation, in one way or another. This is evidenced by the fact that the terms that in general make certain intangible assets eligible for intellectual property protection, such as novelty, inventive activity (or non-obviousness), originality, distinctiveness, confidentiality (or secrecy or non-disclosure), are all synonymous with differentiation. Conversely, the prohibition of practices that constitute or lead to confusion, imitation, plagiarism, deceit, reputation tarnishing, is another expression of differentiation.

In a nutshell: IP performs its primary function by differentiating enterprises, whenever they exert their rivalry in luring clients from competitors and keeping their clients from being lured by rivals.

The holders of TK are not generally seen in this sort of scenario. TK holders (usually, traditional communities) do not seek IP protection so as to enforce it against competitors. Even if a traditional community eventually could see itself engaged in rivalry for the market of a certain piece of TK with another community, that would be an extremely rare situation. The romantic idea of the *bon sauvage* is just that – a romantic idea. Rivalry between communities is a reality, but when it exists, is not expressed in a commercial way. TK holders typically do not compete for the market. Therefore, any exclusivity in TK that a system may accord them is not to be enforced against competitors, but rather against their potential clients (i.e., those companies that engage in bioprospecting or acquire genetic resources embodying TK).

Undoubtedly, TK has economic value. Such value may be difficult to evaluate but it is nonetheless expressed in practical terms in initiatives like the Brazilian company Natura, a large corporation specialized in manufacturing cosmetics derived from genetic resources available in the Brazilian rainforest,[4] and the various International Cooperative Biodiversity Group (ICBG) arrangements (Rosenthal, 1997; Sampath, 2005; on the ICBG-Peru, McManis, 2003).[5] These initiatives (some more successful than others) show the possibility of a flow of useful knowledge, susceptible of being measured in monetary and other material gains, being transferred from traditional communities to enterprises and research institutions involved in basic science or in the manufacture of pharmaceuticals, cosmetics, or foodstuffs.

So, the problem of Sisyphus is not the absence of TK's economic value.

Moreover, the problem is not with a sometimes perceived incompatibility between Western legal regimes of private property and community rights, as some commentators allege (Kuruk, 1999; Obaldia, 2005; Bratspies, 2006/2007; Welkowitz, 2013). The incompatibility could exist if the flow of TK took place within a traditional environment. But this is not the problem – what the IGC is discussing is how to protect TK that leaves its traditional environment and reaches the business world. In this case, some adjustments would need to be made as to the subject matter protected, taking into consideration the special nature of TK (collectively owned and of a holistic nature), but the rights granted would be the same as those that exist already.

The real problem of Sisyphus resides in the inconsistency between the core function of IP and the expectations that TK holders have from an IGC solution: IP does not exist to be enforced against clients (potential or real) who steal or misappropriate intangible assets controlled by suppliers. IP can have that use, but incidentally only. As we have seen, IP essentially exists to prevent enterprises from reaping the intangible advantages sown by their competitors. In other words, the exclusive rights accorded by IP operate mainly in a horizontal, not a vertical, way.

This does not mean that traditional communities are unaware of the use of legal mechanisms to ensure the exclusivity of differentiating intangible assets in an environment of commercial rivalry. The creation of magic rituals, including taboos, has, in its origin, an intent to appropriate technical knowledge (Suchman, 1989). Under the social rules that preside over those rituals, besides their cultural and religious implications, shamans acquire and enforce exclusivity in medicinal knowledge. Such exclusivity ensures that they will keep a privileged status in their communities, and may transfer it to those members that they will designate (the initiated). The above-mentioned WIPO Report on the fact-finding missions confirms that traditional communities are not entirely strangers to a sense of ownership in their knowledge vis-à-vis competitors (WIPO, 1999b, p. 186).

But IP exists in a world of entrepreneurship, where it is used to ensure the differentiation of firms vis-à-vis their rivals, and of products and services vis-à-vis the products and services of other undertakings.

However, this is not the sort of mechanism that is being sought for TK. The IGC has been seeking a system that ensures protection against misappropriation (by clients) rather than against infringement (by competitors), with a moral and cultural dimension, rather than with a commercial one. In this token, the draft text that the IGC is currently discussing (WIPO, 2016e, p. 6) proposes to condition the enforceability of TK rights upon its use (in its various modalities) being made "beyond the traditional context". This means that those who seek to obtain TK protection do not intend to enforce its against their direct rivals – other communities or, in the event the TK holder is one individual, other individuals of the same community or of other communities – but against third parties that acquire (without authorization) TK to introduce it in the manufacture of products. The infringing acts are thus carried out by undertakings that have incorporated TK into their products and services, in general because of their association to genetic resources.

In a legitimate relationship, those undertakings (like NATURA and the ICBG examples above) appear as buyers of information – mainly of pieces of TK that serve as leads to the existence of useful active ingredients in genetic resources. So, in a strict sense, those buyers do not *use* TK in the manufacture of products. They use it as a shortcut, a source of information, based on traditional experience, that certain ingredients have determined properties, which can be useful in the production of certain goods. Second, those undertakings are the clients and the TK holders, the vendors. So, the relationship between TK holders and users is a vertical one, not a horizontal one. But, as said above, this does not correspond to the primary aim of IP, which is, at its core, enforceable in a horizontal relationship.

It follows that the sort of rights sought by the IGC – exclusivity against use, or, alternatively, prohibition of misappropriation – does not fit the majority of circumstances in which companies that use GRs obtained through TK-led bioprospecting take advantage of TK. Those companies do not "use" TK in the manufacture of products. They simply – and yet with great economic gains – benefit from the leads provided by TK.

Therefore, the appropriate regime should be one of a *sui generis* nature but not based on usual IP concepts, which rely on exclusivity enforceable against competitors. TK holders should be given protection in their knowledge against practices that lead to unjust enrichment, i.e., to gains derived from their knowledge without an appropriate compensation. The only common element with IP that one can find here is that the rights arise from an intangible asset – knowledge. But the similarity stops there, because the spirit of such a regime would not be the protection of competitive advantages.

In other words, the IGC is trying to achieve a wrong goal.

So, the rock confronting the IGC is not only difficult and pointless to push: it is virtually *impossible* to move. The IGC has spent these 16 years discussing the application of a legal mechanism to TK without noticing that TK does not fit in IP.

As we have seen, the reason is not because IP is about private individual rights, as there is no incompatibility between IP and collective ownership. Actually, most IP is owned by legal entities, many of them having more shareholders than indigenous communities have individual members. The reason is that IP is about commercial rivalry, and it was in that spirit that it emerged around the Mediterranean, along with the emergence of the first civilizations that grew economically based on free trade, carried by private merchants, in Mesopotamia (Carvalho, 2009).

Will the IGC ever reach the stage of calling for a Diplomatic Conference? Yes, it may, unlikely though it is. But Diplomatic Conferences do not necessarily reach an end. The Diplomatic Conference gathered at The Hague, in 1991 (after five years of debates in a WIPO Group of Experts), to negotiate and approve the Draft Treaty Supplementing the Paris Convention as far as patents are concerned was opened and never closed (WIPO, 1990).

Could such a Diplomatic Conference agree on a treaty on a *sui generis* IP regime for TK? Yes, at least in theory. But as we have seen in respect of the Treaty on the International Recording of Scientific Discoveries (WIPO, 1981), the acceptance of a Treaty is not a guarantee that it will ever enter into force or become applicable.

At least the Sisyphus of myth was able to move the rock uphill. The IGC, though constantly pushing, has scarcely moved the rock at all. And neither Sisyphus nor the IGC will be able to make the rock stand, in stability at the top of the hill. Both are cursed with endlessly and futilely repeating the same impossible task.

Conclusion

In his well-known essay, *The Myth of Sisyphus*, the French philosopher Albert Camus says that the whole point of Sisyphus' ordeal is his struggle. The fact that he will never be able to reach his goal of pushing the boulder uphill does not matter. Our role, as observers of Sisyphus' drama, is to imagine that he is happy.

Likewise we are also led to believe that the IGC is satisfied in looking for an international legal solution for the protection of TK, regardless of the lack of results this far – 16 years later. The reason for such a conspicuous failure is that the IGC's intense efforts and sometimes remarkable conceptual achievements are useless because its search for an IP regime of TK protection is not based on the need to respond to the pressure of entrepreneurship and commercial rivalry. Traditional communities do not compete on the market, except for a few economically insignificant exceptions of a predominantly local dimension, such as certain handicrafts that are sold to tourists. If indigenous peoples do not compete for clients, they do not need to differentiate their products and services. Hence, they do not need an IP mechanism. A legal system that focuses on unjust enrichment would be more adequate.

Eventually, when some piece of TK emerges that has commercial value to the extent it can be used as an enterprise differentiating asset and, eventually, it is tradeable, then existing IP mechanisms can protect it. After all, this is the role that IP has played in the last 4,000 years, and will continue playing.

Notes

1 Later, the WIPO General Assembly would establish a Voluntary Fund for Accredited Local and Indigenous Communities. Funds provided by donors, including a few Member States, are used in supporting the participation of representatives of indigenous and local communities in the work of the IGC. The objectives and operation of the Fund are detailed in the Annex to document WIPO/GRTKF/IC/9/3.

2 Information Note for IGC 31 – prepared by Mr Ian Goss, the IGC Chair. Informal note circulated among delegations, available at www.ip-watch.org.

3 Another draft provision (draft Art. 3.2) aims at taking care ofTK that is already publicly available (WIPO 2014, Annex, at 9–10).

4 Natura's motto is "Our mission is to promote the Well Being Well" ("*Nossa missão é promover o Bem Estar Bem*"). Its website explains this motto as follows: "Our products are the greatest expression of our essence. To create them we mobilize a network of people able to integrate scientific knowledge and the use of the rich Brazilian botanical biodiversity". www.natura.com.br/a-natura. One of Natura's departments is named Ekos, and pursues objectives relating to the protection and sustainable use of biodiversity. *Ibid.*

5 Rosenthal describes the International Cooperative Biodiversity Groups (ICBG) Program as a US Government funded effort to promote equitable sharing of biodiversity benefits in the context of integrated research and development toward drug discovery, biodiversity conservation and economic development. ICBG Programs involve the payment of royalties, advance monetary payments, capacity building (Rosenthal, 1997; Sampath, 2005; on the ICBG-Peru, McManis, 2003).

References

Bratspies, R. (2006/2007) "Theoretical Approaches to International Indigenous Rights: The New Discovery Doctrine: Some Thoughts on Property Rights and Traditional Knowledge", 31 *Am. Indian L. Rev.* 315.

Carvalho, N. (2009) *A Estrutura dos Sistemas de Patentes e de Marcas: Passado, Presente e Futuro [The Structure of the Patent and Trademark Systems: Past, Present, and Future]*, Lumen Juris, Rio de Janeiro.

Kuruk, P. (1999) "Protecting Folklore under Modern Intellectual Property Regimes: A Reappraisal of the Tensions between Individual and Communal Rights in Africa and the United States", 48 *Am. U.L. Rev.* 769.

McManis, C. (2003) "Intellectual Property, Genetic Resources and Traditional Knowledge Protection: Thinking Globally, Acting Locally", 1 *Cardozo J. Int'l and Comp. L.* 547.

Obaldia, I. (2005) Note, "Western Intellectual Property and Indigenous Cultures: The Case of the Panamanian Indigenous Intellectual Property Law", 23 *B.U. Int'l L.J.* 337.

Rosenthal, J. (1997) "A Benefit-sharing case study for the Conference of Parties to Convention on Biological Diversity", www.researchgate.net/profile/Joshua_Rosenthal3/publications/2.

Sampath, P. (2005) *Regulating Bioprospecting: Institutions for Drug Research, Access, and Benefit-Sharing*, United Nations University Press.

Suchman, M. (1989) "Invention and Ritual: Notes on the Interrelation of Magic and Intellectual Property in Preliterate Societies", 89 *Colum. L. Rev.* 1264.

Welkowitz, D. (2013) "Privatizing Human Rights? Creating Intellectual Property Rights from Human Rights Principles", 46 *Akron L. Rev.* 675.

WIPO (1981) *Records of the Diplomatic Conference for the Conclusion of a Treaty on the International Recording of Scientific Discoveries, 1978*, WIPO, Geneva.

WIPO (1990) "The 'Basic Proposal' for the Treaty and the Regulations", *Diplomatic Conference for the Conclusion of a Treaty Supplementing the Paris Convention as Far as Patents are Concerned*, PLT/DC/3, of December 21, 1990.

WIPO (1999a) Working Group on Biotechnology, November 8 and 9, 1999, WIPO/BIOT/WG/99/1, of October 28, 1999.

WIPO (1999b) Intellectual Property Needs and Expectations of Traditional Knowledge Holders – WIPO Report on Fact-Finding Missions on Intellectual Property and Traditional Knowledge (1998–1999), WIPO Publication No. 768(E).

WIPO (2000a) *Matters Concerning Intellectual Property and Genetic Resources, Traditional Knowledge and Folklore*, WO/GA/26/6, of August 25, 2000.

WIPO (2000b) *Report adopted by the Assembly*, WO/GA/26/10, of October 3, 2000.

WIPO (2001a) *Information Provided by WIPO Member States Concerning Practices Related to the Protection of Biotechnological Inventions*, WIPO/GRTKF/IC/1/6, of April 6, 2001.

WIPO (2001b) *Position Paper of the Asian Group and China*, WIPO/GRTKF/IC/2/10, of December 3, 2001.

WIPO (2002a) *Elements of a Sui Generis System for the Protection of Traditional Knowledge*, WIPO/GRTKF/IC/3/8, of March 29, 2002.

WIPO (2002b) *Report adopted by the Committee*, WIPO/GRTKF/IC/3/17, of June 21, 2002.

WIPO (2003a) *Matters Concerning the Intergovernmental Committee on Intellectual Property and Genetic Resources, Traditional Knowledge and Folklore*, WO/GA/30/5, of August 15, 2003.
WIPO (2003b) *Report Adopted by the Assembly*, WO/GA/30/8, October 1, 2003.
WIPO (2006) IGC, Ninth Session, Geneva, April 24 to 28, 2006, *Decisions of the Ninth Session of the Committee* (unnumbered document, of April 28, 2006).
WIPO (2007) *World Intellectual Property Organization 1992–2007* (Geneva, 2007), WIPO Publication No. 1011E.
WIPO (2009a) IGC, Fourteenth Session, Geneva, June 29 to July 3, 2009, *Decisions of the Fourteenth Session of the Committee* (unnumbered document).
WIPO (2009b) *The Protection of Traditional Knowledge: Overview*, WIPO/GRTKF/IC/14/5, of April 29, 2009.
WIPO (2009c) *Summary of Options for the International Dimension of the Committee's Work*, WIPO/GRTKF/IC/14/6, of April 29, 2009.
WIPO (2009d) *African Group Submission on Document WIPO/GRTKF/IC/13/9*, WIPO/GRTKF/IC/14/9, of June 26, 2009.
WIPO (2010a) *The Protection of Traditional Knowledge: Revised Objectives and Principles*, WIPO/GRTKF/IC/17/5, of September 15, 2010.
WIPO (2010b) *Genetic Resources: Revised List of Options and Factual Update*, WIPO/GRTKF/IC/17/6, of September 15, 2010.
WIPO (2010c) *Submission by Australia, Canada, New Zealand, Norway and the United States of America*, WIPO/GRTKF/IC/17/7, of September 15, 2010.
WIPO (2011a) *Draft Articles on the Protection of Traditional Knowledge Prepared at IGW 2*, WIPO/GRTKF/IC/18/7, of March 17, 2007.
WIPO (2011b) *Decisions of the Eighteenth Session of the Committee*, of May 13, 2007.
WIPO (2012) *Joint Recommendation on Genetic Resources and Associated Traditional Knowledge – Document submitted by the Delegations of Canada, Japan, Norway, the Republic of Korea and the United States of America*, WIPO/GRTKF/IC/20/9 REV., of February 21, 2012.
WIPO (2013) *Joint Recommendation on the Use of Databases for the Defensive Protection of Genetic Resources and Traditional Knowledge Associated with Genetic Resources – Document submitted by the Delegations of Canada, Japan, the Republic of Korea and the United States of America*, WIPO/GRTKF/IC/23/7, of February 5, 2013.
WIPO (2016a) *Consolidated Document Relating to Intellectual Property and Genetic Resources*, WIPO/GRTKF/IC/30/4, of March 19, 2016.
WIPO (2016b) *Report adopted by the General Assembly*, WO/GA/47/19, of February 5, 2016.
WIPO (2016c) *The Protection of Traditional Knowledge: Draft Articles*, WIPO/GRTKF/IC/31/4, of May 13, 2016.
WIPO (2016d) *Joint Recommendation on Genetic Resources and Associated Traditional Knowledge – Document submitted by the Delegations of Canada, Japan, Norway, the Republic of Korea and the United States of America*, WIPO/GRTKF/IC/31/5, of August 23, 2016.
WIPO (2016e) *Joint Recommendation on the Use of Databases for the Defensive Protection of Genetic Resources and Traditional Knowledge Associated with Genetic Resources – Document submitted by the Delegations of Canada, Japan, the Republic of Korea and the United States of America*, WIPO/GRTKF/IC/31/6, of August 23, 2016.

23

IS THE WHOLE GREATER THAN THE SUM OF ITS PARTS?

A critical reflection on the WIPO IGC

Daniel F. Robinson

Introduction

The World Intellectual Property Organization (WIPO) established the Intergovernmental Committee on Intellectual Property and Genetic Resources, Traditional Knowledge and Folklore (IGC) in 2000. The IGC emerged out of a number of developments both outside WIPO and within it. As Abdel-Latif (2017) explains, a proposal on genetic resources put forward by developing countries threatened to derail the Diplomatic Conference convened to adopt the Patent Law Treaty (PLT). The developing countries involved ultimately withdrew their proposal, enabling a successful adoption of the PLT, in exchange for a commitment from developed countries that discussions regarding the issues they had raised would continue at WIPO in the framework of an especially dedicated body. The work of WIPO and UNESCO on folklore since the 1960s resulted in multiple Conventions and model provisions that also influenced the IGC. The relationship between intellectual property (IP), genetic resources (GRs) and traditional knowledge (TK) also came into focus following the adoption of the Convention on Biological Diversity (1992) and of the Agreement on Trade Related Aspects of Intellectual Property Rights (TRIPS, 1995) (see Lawson, 2017 and Abdel-Latif, 2017).

The work of the WIPO IGC has evolved since its inception; this chapter will highlight some of its achievements. However, throughout its existence, the WIPO IGC has felt numerous crises, and its work has often faltered due to the different aspirations and expectations for the forum. There has been a palpable divergence in political will between "demandeur" and "non-demandeur" countries (Wendland, 2017). In addition to this, indigenous peoples have often questioned the legitimacy of the forum, which is frequently negotiating elements of their cultural heritage and what they see as their resources, without their having an equal seat at the table. Others have been critical of the fact that the forum focuses on GRs, TK and Traditional Cultural Expressions (TCEs) within a wealthy UN IP forum which has historically concentrated predominantly on the protection of private industrial and commercial property rights (e.g. see Drahos and Braithwaite, 2002). State-centric law-making thus may be dividing up aspects of Indigenous cultural heritage into legal categories as "GRs" and "TK" which align against patent laws, and "TCEs" and folklore against copyright protections, because these divisions reproduce the pre-existing legal categories from which IP rights are protectable (Coombe,

2009; Groth, 2012). Last, there have been concerns that forum-shifting negotiations on GRs and TK in particular out of the TRIPS Council was a deliberate tactic by developed countries (Helfer, 2009). Many developing countries had seen negotiations in TRIPS as preferable for dealing with issues like erroneous patents and misappropriations because the WTO carries the force of a dispute settlement mechanism and sanctions in cases of non-compliance. As Okediji has noted: "WIPO is neither formally charged nor structurally designed to accomplish the kind of linkage bargains now associated with the WTO" (Okediji, 2008, p. 73). However, others have suggested that developing countries welcomed the shift because it was seen as a less hostile forum than the WTO TRIPS Council. The appointment of a Director General in 1997 – Kamil Idris – the first from a from a developing country (Sudan), followed by the establishment of the WIPO Development Agenda in 2004, highlights the shift in WIPO towards global policy and development issues in the early 2000s (Abdel-Latif, 2017).

Despite these issues and challenges, this chapter in the next section highlights some of the many outcomes and achievements of the IGC and from the WIPO Secretariat, including the useful materials generated, and the still evolving draft instruments on IP and GRs, TK and TCEs. The following section considers the current stalemate and possible ways forward, including some discussion of the scale at which TK can best be protected, and national legal developments and local projects that are emerging (e.g. community protocols). The final section considers some rapid global political changes (e.g. Brexit and the election of President Trump) against the backdrop of a proliferation of bilateral and regional free trade agreements. Given the prospect of further withdrawal from multilateral institutions by major powers like the US and the UK, the paper considers implications and risks for IP and GRs, TK and TCEs and for the WIPO IGC.

Outcomes of the evolving WIPO IGC

Despite much cynicism about the IGC and its lengthy process, there are at least several useful and important outcomes that have emerged from the WIPO IGC. Probably the most important outcome from the WIPO IGC has been the development of three texts of Draft Articles which have the potential to become international legal instruments on IP and GRs, TK and TCEs.[1] Since 2009 the IGC has had a mandate to negotiate these texts. However, these texts have been heavily bracketed and heatedly debated, and remain contested even now at the 33rd Session of the IGC. The debate on the texts is informative of different perspectives, national and local contexts and agendas. For example, Group B has staunchly opposed a disclosure of origin patent requirement, instead putting forward proposals for the use of databases to prevent the grant of erroneous patents. Many "demandeurs" seeking a disclosure requirement have disputed the effectiveness of databases on several grounds: they are likely to always be incomplete; they require the collection of information that might be secret or sacred to Indigenous Peoples and local communities; the databases generally deal with TK associated with GRs but they may not identify or prevent the misappropriation of GRs; and relating to this, databases would not necessarily support access and benefit-sharing (ABS) compliance. Despite having no clear outcome, this has been an important debate that has generated significant policy-knowledge of the merits and limitations of the different approaches. The WIPO IGC discussions about disclosure of origin (DOO) have no doubt influenced the policy considerations of many countries, and there are now more than twenty CBD member countries that already have genetic resource DOO requirements for patent applications (Bagley, 2017).[2]

The WIPO Secretariat also undertook extensive consultations across the globe in national contexts and with Indigenous Peoples and local communities in the late 1990s. These have resulted in extensive fact finding mission (FFM) documents which provided a wealth of in-depth and comparable information about GRs, TK and folklore from different regions. These FFMs informed and contributed to the establishment of the IGC in 2000. As Abdel-Latif (2017) explains, the FFMs raised WIPO's profile, and gave it some credibility and leadership showing that it could make an important substantive contribution to international discussions on GRs, TK and folklore, despite the forum's traditional focus on IP.

The Intersessional Working Groups and Expert Groups of the WIPO IGC, as well as the many side-events run by industry, IGOs, Indigenous organizations and NGOs, have also brought a wealth of information to the delegates and representatives that attend the IGC meetings. This has then contributed to the evolution of the bracketed Draft Articles, including informing the different options for protection of different types of TK, for example, with recent proposals from some experts for a "tiered approach" with different levels of protection (narrow or broad) possible for different types of TK (see Dutfield, 2017 and Ruiz, 2017). Such a tiered approach pays recognition to the significant evidence that TK may be widespread and widely held in a country or within a region, but it may sometimes be secret or locally held, and in other cases the knowledge may be sacred or have sacred connections to a particular genetic resource. For example, a patent landscape by Robinson and Raven (2017) presented at the WIPO IGC Seminars in 2015, showed that there are 314 patents or patent applications relating to the species *Morinda citrifolia*. The species, often known as "noni", is found in Australia, the Pacific and throughout parts of Asia, and correspondingly there is TK associated with the plant in all of these regions. Assigning TK rights to uses of *Morinda citrifolia* would be extremely challenging, given its wide distribution and the extensive knowledge of its multiple uses. On the other hand, Robinson and Raven (2017) identified patents relating to Kakadu plum (*Terminalia ferdinandiana*) which is endemic to tropical parts of Northern Australia. The allocation of rights in TK could be more easily assigned for a plant like this, but even in this case there are numerous Indigenous Peoples who traditionally and currently use the plant, and some Indigenous Australian informants were concerned that the plant may be involved in peoples' Dreaming stories or may be a totemic species, making it sacred for certain families.[3]

The WIPO IGC has also considered different forms of TK protection including "positive" and "defensive" protection measures. Defensive protection refers to a set of strategies to ensure that third parties do not gain illegitimate or unfounded IP rights over TK (WIPO Secretariat, 2017).[4] There are some specific outcomes adopted by the WIPO Secretariat including defensive measures such as the amendment of WIPO-administered patent systems: the International Patent Classification system and the Patent Cooperation Treaty (PCT) minimum documentation. In effect, this means that PCT minimum documentation has included TK materials (mainly ethno-biological journals) in searchable prior art in international search reports.

The WIPO Secretariat has researched and produced a huge volume of reports and background material to inform and support the negotiations. For example, there are ten TK background briefs which are regularly updated and help inform new delegates and observers. They also provide an inventory of databases on TK and GRs, as well as an inventory of laws relating to TK, GRs and TCEs. Importantly, the Secretariat has also delved into research and analysis on customary law, given that it has been and continues to be an important basis for Indigenous and local regulation of GRs, TK and TCEs.[5] This research emerged out of the FFMs, and the need to include recognition of customary laws was emphasized in Indigenous Panels at the IGC (discussed below), and included reports on customary law in the Pacific and Andean regions by Tobin (2008). While customary law has been included within brackets in various versions of the Draft Articles texts, Tobin (2017) notes that it has had to compete with the term "cultural norms", which has been put forward by some countries, but which is essentially legally meaningless for Indigenous Peoples.

The WIPO Voluntary Fund, approved by the thirty-second WIPO General Assembly in 2005, facilitates the participation of Indigenous Peoples and local communities in the work of the IGC. Accredited observers which are representatives of local or Indigenous Peoples may apply for financial support for participation in IGC sessions and meetings of the Intersessional Working Groups (IWGs).[6] Since approximately 2005 there have been "Indigenous Panels" comprised of presentations delivered by a panel of representatives of indigenous and local communities at the beginning of sessions of the WIPO IGC.[7] These provide indigenous experiences and perspectives from specific peoples and nations. These are in addition to the interventions of indigenous people as "observers" in the sessions of the IGC. However, there are longstanding criticisms by these Indigenous representatives, that they are not given equivalent standing as "delegates" to make changes to the negotiating text. In addition, monetary contributions to the Voluntary Fund have been dwindling, with some recent sessions only able to support a handful of Indigenous representatives.

WIPO has also established an Indigenous Fellowship Program since 2009, allowing members of Indigenous Peoples and local communities to work in WIPO's Traditional Knowledge Division on issues relevant to indigenous peoples.[8] WIPO Indigenous Fellows have participated in the IGC, provided outreach to indigenous peoples and local communities, engaged in WIPO's cooperation with the United Nations Permanent Forum on Indigenous Issues, the Expert Mechanism on the Rights of Indigenous Peoples, and the Special Rapporteur on the Rights of Indigenous Peoples.

The work of the WIPO IGC and WIPO Secretariat therefore has provided a number of benefits in relation to the negotiation of issues relating to GRs, TK and TCEs, and the facilitation of Indigenous involvement in the negotiations. However, the main substantive outcomes – the Draft Articles on GRs, TK and TCEs – remain heavily bracketed and hotly contested to this day. Despite nearly seventeen years of negotiations, there still does not seem to be political will on the part of some countries to narrow their perspectives on the protection of GRs, TK and TCEs.

A question of scale: how and where can GRs and TK best be protected?

There are persistent questions in the WIPO IGC and its Inter-sessional Working Groups about how and where to best protect TK associated with GRs. From one perspective, given that the issues in question – misappropriations, erroneous patents, culturally offensive uses – are often transnational or transboundary in nature, working towards a global system for protection of GRs and TK makes sense. But because of the local and Indigenous sources of the GRs and knowledge, there are also voices that think they should be protected locally – indeed there are often already customary laws and rules made by Indigenous Peoples and local communities (Tobin, 2014). However, the recognition of Indigenous customary laws and Indigenous rights in general has always been problematic within the colonial legal constructs and sovereignty of states in which Indigenous Peoples now struggle. Balancing the needs and rights of Indigenous and local communities with the sovereign claims of states, and the transboundary issues that are arising relating to GRs and TK is the huge challenge that the IGC faces. As the former head of the WIPO Traditional Knowledge Division has described it, the global protection and recognition of the heterogeneous qualities of TK leads to a paradox: "to globalise diversity holistically" (Taubman, 2005, p. 525).

So if we consider the different types of TK (widely shared, locally held, secret and sacred), there might be different ways to protect it at different scales. As Ruiz has pointed out in his detailed papers about widely shared or dispersed TK, the main limitations affecting bilateral, contractual negotiations when TK is shared are:

(a) How can a contract be negotiated, or prior informed consent (PIC) obtained, when there is no single nor even defined right holder?
(b) Even if this was possible to determine, assuming TK is in the public domain or publicly available, is it feasible, and economically viable, to negotiate an advantageous contract?
(c) What are the effects of economic pressures (impacting benefit potential) when TK is in practice found and accessible from various sources? (Ruiz, 2017, p. 128.)

For widely shared TK, Ruiz proposes a number of different potential protections: defensive protection measures like databases that use public domain or registered TK to help prevent the granting of erroneous patents and "biopiracy"; the use of *domaine public payant* compensation (based on a copyright principle) to the state of their origin; TK and GR funds such as that administered by the International Treaty on Plant Genetic Resources for Food and Agriculture (ITPGRFA) (Ruiz, 2017). Each of these mechanisms requires the establishment of internationally administered or accessible databases, international compensation mechanisms and international funds. So for widely shared or dispersed TK, it seems highly likely that an international mechanism is required, reinforcing the role of the WIPO IGC and other relevant international forums or mechanisms like the CBD and Nagoya Protocol. In addition, as many "demandeur" countries have argued in the WIPO IGC, biopiracy and the granting of erroneous patents could be addressed with a patent disclosure of origin requirement. Even though many states have unilaterally decided to implement a disclosure of origin patent requirement,[9] such a mechanism would arguably be more effective at preventing these problems if it was coordinated internationally through WIPO under the PCT.

Where TK is locally held and associated with local or endemic GRs, there may be several relevant approaches to protect them. Certainly ABS regulations may require PIC and mutually agreed terms (MAT) for access to GRs and associated TK, helping prevent misappropriations and controlling R&D and subsequent commercial transactions relating to them. But Indigenous Peoples and local communities may also need to protect local TK and particularly secret and sacred TK from even entering the public domain, where it may end up being utilized without permission. The use of customary laws as well as community protocols have been discussed as one option for the localized control over TK and its disclosure, as well as for the local regulation of GRs. However, as has been pointed out by Tobin (2014), Dutfield (2017) and many others, Indigenous customary laws and rights have been poorly recognized within state legal systems in many countries and damaged by colonial impacts and dispossessions, meaning that the effectiveness of these forms of protection has been eroded. So local-scale TK protection is to a large degree contingent upon state-scale recognitions of rights.

Much of this discussion so far has focused on defensive mechanisms for protecting TK. Also important is the positive protection of TK and its promotion more broadly. There are some examples whereby national governments have actively sought to provide positive protections for TK, including the Cook Islands. The Cook Islands Traditional Knowledge Act 2014 has been developed with the support of the WIPO Secretariat and also the Pacific Islands Forum Secretariat (the latter of which has developed similar Model Laws on TK). As the relevant policy indicates, the purpose of the policy and law is to:

> Ensure mechanisms are in place to protect, preserve and promote the Cook Islands Traditional Knowledge nationally, in the Pacific region, and internationally. This

> policy recognizes that traditional knowledge belongs to the indigenous communities of the Cook Islands whose ties to the 15 islands can be traced back traditionally and over many generations. This policy guides further work in relation to Traditional Knowledge by signalling key principles and objectives. This policy is a first step towards Government acknowledging its social responsibility to work in partnership with its indigenous communities to protect, preserve and promote this knowledge for the sake of current and future generations.
>
> (*Cook Islands Herald*, 2011)

The Act would apply retrospectively to traditional knowledge existing before it comes into force, but does not affect contracts or IP rights relating to traditional knowledge existing before the act. The law encourages the registration of traditional knowledge (and includes an expansive definition of TK such as forms of TCEs and folklore) and would assert rights of the traditional knowledge to the creators of the knowledge or their successors. This includes rights of transmission, use, commercialization, acknowledgement (essentially as moral rights) and others. The law then restricts non-rights holders against similar actions in relation to registered knowledge (e.g. performing or exhibiting the work in public without permission). Registered traditional knowledge is protected in perpetuity, the rights are inalienable, and purport not to limit or affect other intellectual property rights (Robinson, 2015).

Interestingly, the law prohibits certain actions relating to registered secret-sacred knowledge, including use, transmission, receiving commercial benefit for use or development, without written authorization of the rights-holder. The law also anticipates multiple and overlapping registrations. If two or more registrations for ostensibly the same knowledge are made, then each applicant is able to view the others' application and must come to an agreement before registration is accepted for protection under the law (Robinson, 2015). In this case the law seems likely to be useful for locally held TK where there are only a handful of rightful claimants. In a small country like the Cook Islands, this might be practicable, but in a larger country and where there is widely shared TK it is not clear how such provisions would operate or be effective. In addition, as this is a very recently developed law, it is yet to be seen how it might be effective in relation to external and foreign actors. Enforcing such a law extraterritorially is likely to be extremely difficult, if not impossible, again raising the question of whether adequate protections for TK (as well as GRs and TCEs) can solely be achieved at local and national scales. Even though the substantive progress of the WIPO IGC has been limited – arguably the forum does fill an important gap in international policy-making, particularly as other relevant forums have been struggling and multilateralism in general may be under threat.

The withdrawal from multilateralism: what implications and risks for the WIPO IGC?

Multilateralism appears to be in crisis. Trade negotiations have been forum shifting as new economic priorities and opportunities for leverage emerge or are created. For several years now, the WTO has struggled to achieve any major new outcomes, unable to reach consensus, with negotiations of the 2001 Doha Round stalling for more than a decade. The 2013 Bali and 2015 Nairobi packages restored some faith in the Ministerial process, with some new agreements and rules of benefit for least developed countries (LDCs) and on agricultural subsidies. Progress in the TRIPS Council has largely reached a stalemate on several core issues,[10] and having achieved the standards that advanced economies desired in TRIPS, they have been seeking TRIPS-plus minimum standards and enforcement through bilateral and regional negotiations (some are even talking about TRIPS++). As tradeable "resources", knowledge and expressions, GRs, TK and TCEs are embroiled in the regulatory implications of this political economic trend. After 16 years and 33 sessions the WIPO IGC also still has heavily bracketed texts on its draft instruments for IP and GRs, TK and TCEs. If we look to recent bilateral and regional trade agreements there are generally very weak protections for GRs and TK, with recent agreements such as the Trans Pacific Partnership (TPP) noting side agreements which acknowledge the desires of countries to ratify and implement the Nagoya Protocol (Roffe, 2017). For example, Australia has side agreements with Malaysia and Peru which state:

> The Governments of Australia and [Malaysia/the Republic of Peru] recognize that access to traditional knowledge associated with genetic resources from providers, as well as the equitable sharing of benefits that may result from the utilization of that traditional knowledge, may be addressed through contracts that reflect mutually agreed terms between users and providers.
>
> (Australian Department of Foreign Affairs and Trade, 2015)

Considering that this goes beyond the CBD text's focus on ABS for GRs, it suggests that Australia and Malaysia are considering ratification of the Nagoya Protocol (Peru has ratified and Australia has signed and was recently undergoing consultations with the intent to ratify). Nevertheless, the language is relatively weak ("may be addressed") and makes no real commitment to protect GRs and associated TK. In virtually all bilateral and regional trade agreements to date, there have been minimal or no commitments on GRs and TK (Roffe, 2017).

We are also noticing a political rhetoric which signals a broader dissatisfaction with multilateralism, particularly in some parts of the developed world. This is most obvious in the Brexit vote and withdrawal of the UK from the EU, as well as the anti-elitist/anti-establishment rhetoric of President Trump prior to his election and in some of his actions since. There is the possibility that the US will withdraw funding from the UN for example (Borger, 2017). If this were to occur, it would pose a clear risk to the ongoing activities of UN agencies, particularly those seen as "non-core" activities which Group B countries have been known to characterize the IGC negotiations on GRs, TK and TCEs, against the backdrop of the core IP treaties under WIPO.

Given that the demandeurs of international legal instruments for GRs, TK and TCEs are primarily developing countries, these countries generally have greater hope of achieving their aims in forums where they can form coalitions and alliances. When isolated in trade negotiations, these countries have less leverage (in terms of balance of trade in goods, services and IP assets) from which to negotiate their priorities. This means that multilateral forums like the WIPO IGC are likely to play an important ongoing role for promoting *development imperatives and agendas* (as noted by Oguamanam, 2017). Even if the WIPO IGC is unable to achieve binding legal instruments, the forum and Secretariat have produced an extensive corpus of materials and have already helped develop international *policy concepts* for GRs, TK and TCEs that have hopefully been of benefit to countries, researchers and Indigenous Peoples. In the context of the Nagoya Protocol, for example, there are certainly challenges relating to the protection of GRs and TK that might not adequately be resolved by national regimes on ABS (e.g. transboundary or widely shared TK). While ABS might be able to regulate bioprospecting and other forms of R&D on GRs, is it really likely to be able to positively protect or promote TK? Furthermore, the ABS regime is incapable of dealing with erroneous patents and "biopiracy" concerns, which have been discussed in the WIPO IGC. While it is still polarized between demandeurs seeking a patent disclosure of origin requirement for GRs (and associated TK) and non-demandeurs preferring the use of databases to improve patent screening, the IGC at least provides a forum for incremental improvements such as the screening of TK databases noted earlier.

The risk for the WIPO IGC and for forums like the WTO is that they will become increasingly irrelevant. There are already discussions in Geneva about south-south plurilateral trade agreements to counter other mega-regionals like the Transatlantic Trade and Investment Partnership (TTIP) and TPP, and because of the static situation in the WTO. If developing countries end up pursuing plurilateral agreements because they can no longer rely on multilateral forums to promote development agendas, this could signal a worrying trend for cosmopolitan global negotiations and law-making. Some issues – particularly transboundary and trade-related issues – need to be negotiated internationally. While aspects of GR regulation and TK protection might best be dealt with at the local or national level, there are still aspects (e.g. patent disclosure of origin) that are best resolved at the international level. Otherwise there will just be a patchwork of national systems with a domestic focus but unable to properly deal with extra-territorial issues like non-compliant foreign actors (e.g. bioprospectors without valid permits; GR and TK related patent-holders without PIC and MAT).

Notes

1 www.wipo.int/tk/en/igc/draft_provisions.html 10/3/2017.
2 E.g. Belgium, Brazil, Chile, China, Egypt, Switzerland, Denmark, Norway, Bolivia, Colombia, Ecuador, Peru, Costa Rica, India, Cuba, Djibouti, Namibia, Kyrgyzstan, the Philippines and South Africa.
3 Authors' consultations in 2015, also noted in Robinson (2015).
4 www.wipo.int/tk/en/tk/ accessed 10/3/2017.
5 www.wipo.int/tk/en/indigenous/customary_law/ accessed 10/3/2017.
6 www.wipo.int/tk/en/igc/participation.html accessed 10/3/2017.
7 See: www.wipo.int/tk/en/igc/panels.html.
8 www.wipo.int/tk/en/indigenous/fellowship/.
9 A disclosure requirement is likely to support the implementation of the Nagoya Protocol, providing a checkpoint (Art. 17) to ensure GRs and associated TK have been accessed with PIC and MAT.
10 An amendment to the TRIPS agreement entered into force in January 2017 improving for developing countries a legal pathway to access affordable medicines under WTO rules, which has a long negotiation history since the 2001 Doha Ministerial Declaration on the TRIPS agreement and public health was lauded but then struggled to have effect because of import restrictions (see Roffe *et al.*, 2006).

References

Abdel-Latif, A. (2017) "Revisiting the creation of the IGC: the limits of constructive ambiguity?" in Robinson, D., Roffe, P., Abdel-Latif, A. (eds) *Protecting Traditional Knowledge: The Future of the WIPO Intergovernmental Committee on IP and Genetic Resources, Traditional Knowledge and Folklore*. Routledge, Oxon, pp. 10–30.

Australian Department of Foreign Affairs and Trade (6 October 2015) "TPP text and associated documents", http://dfat.gov.au/trade/agreements/tpp/official-documents/Pages/official-documents.aspx.

Bagley, M. (2017) "Of disclosure 'straws' and IP system 'camels' Patents, innovation and the disclosure of origin requirement" in Robinson, D., Roffe, P., Abdel-Latif, A. (eds) *Protecting Traditional Knowledge: The Future of the WIPO Intergovernmental Committee on IP and Genetic Resources, Traditional Knowledge and Folklore*. Routledge, Oxon, pp. 85–107.

Bavikatte, K. and D.F. Robinson (2011) "Towards a People's History of the Law: Biocultural Jurisprudence and the Nagoya Protocol on Access and Benefit-sharing" *Law, Environment and Development Journal*, Vol. 7, No. 1 p. 35, www.lead-journal.org/content/11035.pdf.

Borger, J. (2017) "UN funding: alarm at reports Trump will order sweeping cuts", 26 January 2017, *The Guardian*, www.theguardian.com/us-news/2017/jan/26/un-funding-alarm-at-reports-trump-will-order-sweeping-cuts.

Cook Islands Herald (2011) "Traditional Knowledge to be Protected", 20 April 2011, www.ciherald.co.ck/Articles/h560d.htm.

Coombe, R.J. (2009) "The expanding purview of cultural properties and their politics", *Annual Review of Law and Social Science*, Vol. 5, No. 3, pp. 393–412.

Drahos, P. and Braithwaite, J. (2002) *Information Feudalism: Who Owns the Knowledge Economy?* Earthscan, London.

Dutfield, G. (2017) "Should we protect turmeric lattes?" in Robinson, D., Roffe, P., Abdel-Latif, A. (eds) *Protecting Traditional Knowledge: The Future of the WIPO Intergovernmental Committee on IP and Genetic Resources, Traditional Knowledge and Folklore*. Routledge, London, pp. 332–38.

Groth, S. (2012) *Negotiating Tradition-The Pragmatics of International Deliberations on Cultural Property*. Göttingen University Press, Göttingen.

Helfer, L.R. (2009) "Regime shifting in the international intellectual property system", *Perspectives on Politics*, Vol. 7, No. 1, pp. 39–44.

Lawson, C. (2017) "Relationships between the work of the WIPO IGC and other forums" in Robinson, D., Roffe, P., Abdel-Latif, A. (eds) *Protecting Traditional Knowledge: The Future of the WIPO Intergovernmental Committee on IP and Genetic Resources, Traditional Knowledge and Folklore*. Routledge, London, pp. 56–82.

Oguamanam, C. (2017) "Ramifications of the WIPO IGC for IP and development" in Robinson, D., Roffe, P., Abdel-Latif, A. (eds) *Protecting Traditional Knowledge: The Future of the WIPO Intergovernmental Committee on IP and Genetic Resources, Traditional Knowledge and Folklore*. Routledge, London, pp. 339–46.

Okediji, R.L. (2008) "WIPO-WTO relations and the future of global intellectual property norms", *Netherlands Yearbook of International Law*, Vol. 39, No. 1, pp. 69–125.

Robinson, D.F. and Raven, M. (2017) "Identifying and Preventing Biopiracy in Australia: Patent Landscapes and Legal Geographies for Plants with Indigenous Australian uses", *Australian Geographer*, Vol. 48, No. 3, pp. 311–31.

Robinson, D.F. (2015) *Biodiversity, Access and Benefit-Sharing: Global Case Studies*. Routledge/Earthscan, London.

Roffe, P. (2017) "Preferential trade agreements and the protection of genetic resources and associated traditional knowledge" in Robinson, D., Roffe, P. Abdel-Latif, A. (eds) *Protecting Traditional Knowledge: The Future of the WIPO Intergovernmental Committee on IP and Genetic Resources, Traditional Knowledge and Folklore*. Routledge, London, pp. 303–16.

Roffe, P., Tansey, G., and Vivas-Eugui, D. (eds) (2006) *Negotiating Health: Intellectual Property and Access to Medicines*, Earthscan, London.

Ruiz, M. (2017) "The legal protection of widely shared and dispersed traditional knowledge" in Robinson, D., Roffe, P., and Abdel-Latif, A. (eds) *Protecting Traditional Knowledge: The Future of the WIPO Intergovernmental Committee on IP and Genetic Resources, Traditional Knowledge and Folklore*. Routledge, London, pp. 122–40.

Taubman, A. (2005) "Saving the Village: Conserving jurisprudential diversity in the international protection of traditional knowledge", in Maskus, K.E. and Reichman, J.H. (eds) *International Public Goods and Transfer of Technology: Under a Globalized Intellectual Property Regime*. Cambridge University Press, Cambridge, pp. 521–64.

Tobin, B. (2008) *The Role of Customary Law in Access and Benefit-Sharing and Traditional Knowledge Governance: Perspectives from Andean and Pacific Island Countries*, UNU, Yokohama and WIPO, Geneva, www.wipo.int/export/sites/www/tk/en/resources/pdf/customary_law_abs_tk.pdf.

Tobin, B. (2014) *Indigenous Peoples, Customary Law and Human Rights: Why Living Law Matters*. Routledge/Earthscan, London.

Tobin, B. (2017) "Now you see it now you don't. The rise and fall of customary law in the IGC" in Robinson, D., Roffe, P., and Abdel-Latif, A. (eds) *Protecting Traditional Knowledge: The Future of the WIPO Intergovernmental Committee on IP and Genetic Resources, Traditional Knowledge and Folklore*. Routledge, London, pp. 192–215.

Wendland, W. (2017) "The evolution of the IGC from 2001 to 2016: an insider's perspective" in Robinson, D., Roffe, P., and Abdel-Latif, A. (eds) *Protecting Traditional Knowledge: The Future of the WIPO Intergovernmental Committee on IP and Genetic Resources, Traditional Knowledge and Folklore*. Routledge, London, pp. 31–55.

WIPO Secretariat (2017) "Traditional Knowledge" www.wipo.int/tk/en/tk/.

PART VI

The ethics, economics and science-policy interface of biodiversity protection

24
NATURALIZING MORALITY

Ursula Goodenough

Most arguments for the preservation of biodiversity are consequentialist arguments – specifically cost/benefit analyses – seeking to demonstrate that the benefits of preserving biodiversity outweigh the cost of doing so, or alternatively that the cost of forgoing preservation of biodiversity outweighs any benefit derived from its destruction. An alternate ground – specifically an argument from antecedent moral precepts – for preserving biodiversity, however, can be derived from what may be thought of as 'religious naturalism'.

Moral precepts are commonly assumed to derive from the teachings of traditional religions. Currently, a comprehensive world view, albeit by definition a work in progress, is on offer as the outcome of scientific inquiry. This chapter explores the project of articulating an Ecomorality based on a science-based world view.

Religions can be said to have three strands: a theological strand, concerned with such matters as Meaning and Purpose and often including god(s); a spiritual strand, entailing subjective experiences of the sacred; and a moral strand, dealing with how best to be good. A mature religious tradition interweaves these in the context of a unifying story or Myth, but each can nonetheless be teased out and analyzed separately.

There is growing interest in an orientation that I will call here religious naturalism, wherein our scientific understandings of who we are and how we got here – the Epic of Evolution – serves as the unifying story or Myth. In my book *The Sacred Depths of Nature*, I suggest ways that this story can elicit such spiritual sensibilities as belonging, communion, gratitude, humility, assent, and awe. Recently, the Religious Naturalist Association was been constituted.[1]

Some of my current work considers how morality might be considered in the context of religious naturalism. A recent book by Larry Arnhart, *Darwinian Natural Right* (Arnhart, 1998), gives a thorough and thoughtful account of the intellectual history of ethical naturalism, and Terrence Deacon and I are developing a perspective on this question in the context of emergentism (Goodenough and Deacon, 2003, 2007). Here I offer an overview of the project and its trajectory, adapting material in part from previous writings (Goodenough and Woodruff, 2001; Goodenough, 2003, 2009).

Morality in religious naturalism

Any religious orientation worth talking about is also concerned with morality. As theologian John Haught has remarked (Haught, 2001): 'I would say that in this recent flurry of news about brain and religion, what is often left out is that religion means much more than a state of mind or an ecstatic or mystical mood. It's a commitment over a lifetime to what a person considers to be good'.

So how do we talk about moral thought and moral action as religious naturalists? What do we say to our children about how best to be good, and on what basis do we ground what we say?

My starting premise, working with understandings developed by Foot (2001), Hursthouse (1999) Woodruff (2001), and their school of contemporary ethicists, is that morality describes that which allows humans to flourish in community. And given the relentlessly social context of our lineage, it is vital that we generate flourishing communities.

Most organisms have no mandate to flourish in community. For most organisms, their purpose can be said to be to survive to produce offspring. To say that the purpose of life is to survive to produce offspring is, for some, an uninspiring and perhaps even bleak and depressing notion. For others of us, however, it is freighted with wonder and meaning. That there is life at all, that it is so poignantly purposive, is foundational to the matrix of my own religious life.

That being said, we in fact need not use such a minimalist word as 'survive'. For the mandate is not so much to survive as to flourish. An organism that manages to eke out survival and reproduction in a given ecosystem is far less likely to be the ancestor of a large lineage than an organism that flourishes and produces flourishing progeny in that ecosystem. 'Flourishing' is not a synonym for that old misunderstanding of 'fittest'. To flourish is to be well adapted to the particular environmental circumstance in which one finds oneself, to be healthy and resilient and resourceful. We can also introduce here the word 'good'. A flourishing bacterium or tree or mouse can be said to be a good bacterium or tree or mouse. A good willow maximizes the potential for willowness in all its manifestations: bark quality, disease resistance, pollen production, and so on.

So to return to morality. Most organisms, like bacteria and willows and mice, carry out their purpose – to flourish – with adaptive traits and behaviors, but their biological mandate is carried out in the context of self-interest. The project is an individual project or, in the case of sexual organisms, individuals and their genetic offspring who require some sort of nurture (seed coats, egg shells, nests, milk).

Social animals like ourselves (and unlike the social insects[2]) remain self-interested, but we also cooperate in various vital activities such as food acquisition or protection from predators. Therefore, the mandate is both to flourish as an individual and to flourish in community. A good wolf is a flourishing animal and a member of a flourishing pack; he is genetically scripted both to take care of his own needs and to cooperate with others in the hunt. A good schooling fish participates in schooling; a good bird joins others in chasing off the circling hawk. In flourishing social lineages, adaptive genetic scripts navigate the tensions between self-interest and group cooperation.

Genetic scripts can specify 'instinctive' behaviors, such as schooling, but they can also specify the capacity to *learn* adaptive behaviors. That is, the evolutionary process does not 'care' whether behavior is hardwired or learned; it only 'cares' about an adaptive outcome. For primates, whose brains undergo profound transitions from immaturity to maturity, much of what is inherited is in the form of capacities. Of interest to us here are capacities for morality, capacities that, when cultivated, allow the individual to flourish in community. These capacities are cultivated in the context of learning, that is, in the context of culture, and religious traditions have served as important cultural venues for moral education throughout human history.

The human who cultivates his or her moral capacities can be said to be a good human. But it is of course not that simple. Always lurking in the wings of our nature are what we can call moral susceptibilities, susceptibilities that emanate from the robust self-interest that we also bring to the project of being alive.

Here I will briefly consider six moral capacities that undergird our ability to flourish in human community, namely, strategic reciprocity, humaneness, fair-mindedness, courage, reverence, and mindfulness. I will argue that these have arisen during our evolutionary history and have acquired vast additional import and complexity in the context of our human mentality, a mentality that allows us to engage in symbolic language and hence to formulate abstractions. These moral capacities stand in tension with our susceptibilities to greed, hubris, self-absorption, fearfulness, xenophobia, and prejudice, behaviors that overwhelm us in the face of prolonged stress when we hunker down and engage not in community but in self-interested survival patterns, the default behavior of all creatures.

Strategic reciprocity

We can begin with the capacity for strategic reciprocity, which is a salient behavior in social primates and also, curiously, in vampire bats, but undescribed in other social animals. Strategic reciprocity, also known as reciprocal altruism, refers to behavior that we can summarize as 'I'll scratch your back if you scratch mine'. Self-interest remains paramount – my back will be scratched, my coat will be groomed, my status in the social hierarchy will be protected – and in exchange I will groom you and form an alliance to protect your social status. The cultivation of strategic reciprocity entails elaborate acts of cognition – I must remember who reciprocates and who cheats or defects, I must burnish my reputation for being a cooperator, and so on – and humans are astoundingly good at it. Our economic, political, and legal systems are heavily grounded in strategic reciprocity, and it is of vast importance in structuring communities that flourish. But in the end, strategic reciprocity is a game, a calculus, and indeed computers can be programmed to be astoundingly good at it as well. After we finish teaching our children that they should be good at strategic reciprocity if they are to flourish in community, it feels like we still have much left to say to them about morality.

The virtues

So we can next turn to four moral capacities which, when cultivated, acquire the status we often call virtues. Two of these we can designate as pro-social or valenced virtues in the sense that their cultivation assures the flourishing of community. The first is humaneness, which generates such responses as compassion, agape, benevolence, and charity, and the second is fair-mindedness, which generates such responses as justice, honesty, and trustworthiness. Primatologists have documented manifestations of these traits in nonhuman primates, who are observed to engage in consolation, in reconciliation, and in affection for one another and for one another's offspring. I also find most attractive the thesis, argued by Geoffrey Miller in his book *The Mating Mind* (2000), that just as we favor humaneness and fair-mindedness in our choice of mates, so did both capacities come to be reinforced by sexual selection during the 5 million years of hominid evolution. Importantly, our ability to form abstract concepts, which develops with maturation and education, allows us to enlarge these capacities such that we come to extend humaneness and fair-mindedness to other human groups, thereby tempering our susceptibility to xenophobia, and then as well to other species, to ecosystems, to the planet itself. We come to care about suffering and injustice in all its manifestations. There are no more promising antidotes than these for our susceptibilities to greed and hubris.

The other two cardinal virtues – courage and reverence – are more complicated. First let's consider what they are.

When we speak of courage, as opposed to reflexive acts of self-defense or defense of kin, we are speaking of the capacity to hold a large idea, a large passion, as being more important than one's own safety. So – the mountain climber is courageous because conquering the mountain trumps her fear of falling; Martin Luther was courageous because his religious conviction trumped his fear of papal authority. Courage, I believe, is essential to human creativity: the passion to break new ground, solve a problem, write a poem, is fueled by courage and defeated by fearfulness.

When we speak of reverence, which is celebrated in a book of that title by philosopher Paul Woodruff (2001), we are speaking of the capacity to carry the sense that there are entities larger than the human being, and hence larger than the self, to which one accords awe and gratitude and to which one develops obligation and commitment. Theistic persons traditionally offer reverence towards a supernatural deity or deities, whereas the nontheistic religious naturalist locates reverence in the natural world, the material world, in all its wondrous manifestations and evolutionary history. We speak of reverent family life, reverent leadership, reverent community. Reverence, in whatever context, endows us with humility and hence defeats our susceptibility to self-absorption.

The reason that courage and reverence are complicated virtues is that they are inherently neutral, inherently unvalenced. Courage can be displayed in the name of any ideal, and reverence can be held for any ideal, as we so tragically witnessed on September 11. Courage and reverence can make bounteous contributions to the flourishing of community, but they can also sabotage community and hijack the good.

This dilemma brings me to the final moral capacity on my list, the capacity for mindfulness.

Mindfulness

Mindfulness represents the human capacity to take in understandings of reality without the distortions introduced by need, bias, and prejudice. Rigidity, dogmatism, and fundamentalism are antonyms to mindfulness – mindfulness is constantly evolving, ready for surprise.

Wisdom and knowledge are entailed by mindfulness, but mindfulness demands more of us. It is knowledge or wisdom that pulls the mind-and-heart of the knower towards a connection with the way things are in all their exciting particularity. You cannot be mindful and know things in a purely academic way; as you become mindful of something, your feelings and your behavior towards it are transformed.

Mindfulness is a central concept in Buddhism, where it is lifted up both as a mental state and as a practice. The mindful person, Buddhism tells us, assumes the attitude of pure observation, freed from all false views, and apprehends a reality that is not only objective but also becomes subjective. The mindful person really, really sees.

Mindfulness is also described as a path, a work in progress, rather than an endpoint or achievement. This is because the mindful person is prepared to perceive each particular situation in its uniqueness and respond to it appropriately.

In the broadest and deepest sense, the 'naturalism' part of religious naturalism is all about mindfulness. Scientists, trained in a particular kind of 'pure observation', have provisioned us with stunning understandings of the natural world, and these understandings then provision the religious naturalist with countless substrates for mindful apprehension. So, for example, mindfulness of the body is no longer just about breathing and walking as in the original Buddhist practice; we are now able to contemplate as well the molecular and genetic underpinnings of the body and its evolution from simpler forms.

The religious naturalist is called to be mindful of the following understandings from biology:

- Mindful of our place in the scheme of things.
- Mindful that life evolved, that humans are primates.
- Mindful of the dynamics of molecular life and its emergent properties.
- Mindful of the fragility of life and its ecosystems.
- Mindful that life and the planet are wildly improbable.
- Mindful that all of life is interconnected.
- Mindful of the uniqueness of each creature.
- Mindful of future generations.

And from psychology and anthropology:

- Mindful that our thoughts and feelings are neural.
- Mindful of the evolutionary continuity between our minds and other animals' minds.
- Mindful of human diversity, including diversity of temperament.
- Mindful of human creativity and its wondrous manifestations.
- Mindful of the influence of ethnic and family roots and tribal connection.
- Mindful that children best flourish when loved and nurtured.
- Mindful of the human need for personal wholeness and social coherence.

Similar lists can be drawn from the physical sciences and the earth sciences, from cultural history and imaginative literature, and so on. All such lists are expected to be incomplete and open-ended. They are offered to remind us of what is at stake.

And now, a central claim. I would suggest that virtues, and particularly the neutral virtues, will generate flourishing communities only to the extent that they are mindful virtues. Mindfulness is a precondition for virtue and hence for morality, or, rather, the cultivation of mindfulness and the cultivation of virtue must go together as an essential collaboration if we are to attain moral maturity. The attacks of September 11 may have been executed in the name of reverence and courage, but it was neither mindful reverence nor mindful courage.

Moral susceptibilities

We can conclude by circling back to our moral susceptibilities. How do we go about stacking the decks of our psyches, and our children's psyches, so that mindfulness trumps fundamentalism, mindful courage trumps fearfulness, humaneness trumps hubris and xenophobia, fair-mindedness trumps greed, and mindful reverence trumps self-absorption?

One way to stack the deck is through mindful moral education. From my perspective, this is robustly feasible in the context of religious naturalism. Nor is the project defeated by the naturalistic fallacy: our 'Is' is that we are social animals; our 'Ought' is that we be good social animals. Importantly, religious naturalists are not constrained to describing and celebrating moral concepts in the context of evolutionary biology alone. The moral capacities and susceptibilities of which I speak are, needless to say, embedded in the stories and rituals of all the major traditions – indeed, their universality is yet another testimonial to their centrality to human nature – and there are many ways to convey the rich meanings of these traditions to ourselves and our children in naturalistic contexts.

A second way to stack the deck, obviously, is to ameliorate the conditions wherein humans are physically or emotionally impoverished, threatened, defeated, abused, humiliated, lonely, and insecure. Such conditions of prolonged stress induce us to hunker down and render us vulnerable to fundamentalisms that promise deliverance.

Hope

Hope is another one of those complicated human capacities, complicated in that it can so often be elicited by false promise. But mindful hope, if we can speak of such a thing, is perhaps what we most need in these times of ours.

Notes

1 http://religious-naturalist-association.org.

2 The wasps and ants are an informative exception. An ant colony can be analogized to a multicellular organism, such as a human, where individual worker ants are, to a first approximation, the equivalent of individual somatic cells. The ants, and the cells, are genetically identical and individually sterile; their mandate is to cooperate in ensuring the viability and reproductive success of the queen/germ line. A self-interested cell in a human, focused only on its own replication, might generate a malignancy, but not another human. A human has far more tenuous obligations to cooperate with other humans in her/his community than a cell (or ant) has to cooperate with other cells (ants): human self-interest has not been discarded in the name of sociality.

References

Arnhart, L. (1998) *Darwinian Natural Right: The Biological Ethics of Human Nature.* Albany NY: State University of New York Press.

Foot, P. (2001) *Natural Goodness.* New York: Oxford University Press.

Goodenough, U. (1998) *The Sacred Depths of Nature.* New York: Oxford University Press.

Goodenough, U. (2001) 'Causality and Subjectivity in the Religious Quest', *Zygon*, Vol. 36, pp. 725–34.

Goodenough, U. (2003) 'Religious Naturalism and Naturalizing Morality', *Zygon*, Vol. 38, pp. 101–09.

Goodenough, U. (2009) 'Ecomorality: Toward an Ethic of Sustainability', in L. Mazur (ed.), *A Pivotal Moment: Population, Justice, and the Environmental Challenge.* Washington D.C.:, Island Press, pp. 372–82.

Goodenough, U. and T. Deacon (2003) 'From Biology to Consciousness to Morality', *Zygon*, Vol. 38, pp. 801–19.

Goodenough, U. and T. Deacon (2007) 'The Sacred Emergence of Nature', in P. Clayton (ed.), *The Oxford Handbook of Religion and Science.* Oxford University Press, pp. 854–71.

Goodenough, U. and P. Woodruff (2001) 'Mindful Virtue, Mindful Reverence', *Zygon*, Vol. 36, pp. 585–95.

Haught, J. (2001) 'Religion and the Brain', *Religion and Ethics Newsweekly,* www.pbs.org/wnet/religion and ethics/week510/cover.html.

Hursthouse, R. (1999) *On Virtue Ethics.* New York: Oxford University Press.

Miller, G.F. (2000) *The Mating Mind: How Sexual Choice Shaped the Evolution of Human Nature.* New York: Doubleday.

Woodruff, P. (2001) *Reverence: Renewing a Forgotten Virtue.* New York: Oxford University Press.

25

MAKING LEGAL USE OF THE VALUATION OF NATURE

Colin T. Reid

Nature can be valued in many ways, but legal relationships reflect only some of these. Western legal systems have traditionally taken into account only some of the ways in which humanity can benefit from nature, predominantly where the benefit is based on physical possession of land or on tangible produce which has a commercial market. Other benefits have not been captured by legal rights and therefore fall wholly outwith the range of legal concern. More recently, an increasing range of values and interests have been brought within the sphere of legal consideration, notably the intellectual property matters discussed elsewhere in this book. Further possibilities, however, arise from the increasing identification of the many ecosystem services provided to us by the natural world and emerging appreciation of what they are worth (TEEB, 2010). The recognition of such services and their valuation creates the opportunity for new approaches to conserving and enhancing these services and the benefits they provide. There are considerable challenges in translating these ideas into enduring and enforceable legal mechanisms and this chapter explores how new legal rights and relationships might be used to limit and reverse the degradation of biodiversity which threatens us all. After some consideration of the role of ecosystem services, the chapter examines different mechanisms based on the "beneficiary pays" or "polluter pays" principles, before considering some common challenges to be faced and the potential for new approaches to conservation.

Ecosystem services and conservation

Humanity enjoys, and indeed depends on, many benefits provided by the ecosystems in which we live. Work on identifying and valuing these ecosystem services endeavours to ensure that these are not overlooked as we take strategic and operational decisions on how land and other resources are to be used. Studies at an international level have identified many such services, which can be divided into four categories (UNEP, 2005):

- Provisioning services in the form of products obtained from ecosystems, including food, fresh water, fuelwood, fibres, biochemicals and genetic resources.
- Regulating services obtained from the regulation of ecosystem processes, including climate regulation, disease regulation, water regulation, water purification and pollination.

- Cultural services in the form of non-material benefits obtained from ecosystems, including spiritual, religious, aesthetic and educational benefits, recreation and tourism, and cultural heritage.
- Supporting services which are necessary for the production of all other ecosystem services, including soil formation and nutrient cycling.

Identifying these services reveals that what might conventionally be regarded as unproductive land (since it is not being put to a commercial use) may actually be providing something very valuable to society, something which deserves to be recognized and indeed rewarded. Attempts at precise valuations are difficult and contested, both in terms of technical methodology and over the anthropocentric focus involved. Nevertheless, it is widely agreed that it would be ruinously expensive if these services were not available and had to be replaced; for example it has been estimated that the direct value of pollination services to UK agriculture (without including the vital contribution of pollination to supporting other ecosystem services) was £603 million in 2010, rising to £1.9 billion if the cost of replacing insect pollination with artificial means is included (EAC, 2013). Whether or not a monetary valuation is adopted, the scale of the contribution and the desirability of ensuring it is not wholly overlooked are clear. Work at the broad level is one thing, but it is a further challenge to proceed to the more precise identification, separation and valuation of ecosystem services that are pre-requisites of using such services as the basis for mechanisms for attributing costs and benefits between individual parties. This is not least because the services are often overlapping and hard to disentangle; maintaining a healthy peat bog provides benefits for both biodiversity and carbon storage, while contributing with many other elements in the wider catchment to the quality and quantity of water resources. Moreover there may be disagreement over how immediate, as opposed to indirect, the links with human benefits must be before they are taken into account. The services are also likely to be provided by areas defined by their ecosystem functions, such as stream catchments, not by discrete plots defined by the legal boundaries of ownership, which are often on a completely different scale. Moreover although legal boundaries may respect ecological features, for example where the watershed is a boundary between two estates, they are as likely to cut across them, for example by using a stream rather than a watershed as the boundary. So long as the limitations are recognized, however, the absence of precision may not be a fatal problem if the alternative is the status quo which contributes to the loss of biodiversity by completely failing to recognize and reward the provision of ecosystem services (Reid and Nsoh, 2016).

The conventional approach to conserving biodiversity is to rely on "command and control" techniques. These prohibit activities which are harmful to the particular species and habitats designated as being of particular value, restricting the landowners' rights to use their land as they wish and penalizing the killing and taking of protected plants and animals. The prohibitions often extend to include connected activities such as the possession, sale and transfer of specimens, both because of the difficulty of obtaining evidence of direct wrongdoing in often remote areas and in order to suppress the trade which may be fuelling the damaging exploitation of rare species.

Such an approach will always have a major part to play in conserving biodiversity, but the focus on specific sites and specific elements of the fauna and flora ignores the wellbeing of those that are not designated for protection, and is at odds with the wide acceptance that an ecosystem approach is needed to stop and reverse the devastating loss of biodiversity that we have seen in the past two centuries. The Convention on Biological Diversity defines an ecosystem as "a dynamic complex of plant, animal and micro-organism communities and their non-living environment interacting as a functional units" and the ecosystem approach involves "a strategy for the integrated management of land, water and living resources that promotes conservation and sustainable use in an equitable way", with a priority being "the conservation of ecosystem structure and functioning, in order to maintain ecosystem services" (CBD-COP 5). This calls for "adaptive management to deal with the complex and dynamic nature of ecosystems", requiring attention to ecosystems at all scales, "a grain of soil, a pond, a forest, a biome or the entire biosphere" (CBD-COP 5). Conservation laws must endeavour to devise and apply mechanisms which can provide this more holistic approach.

By recognizing the ways in which human activities are affecting the ecosystem and the services it provides, a basis exists for developing new means of reflecting and influencing our impact on biodiversity, predominantly applying various forms of economic instruments. The effect of our activities can be positive or negative, and both aspects can be incorporated into mechanisms, rewarding or penalizing activities as appropriate. There are thus two classes of intervention designed to influence our impacts on ecosystem services. Where there is a positive impact, the provider of this should be rewarded by the beneficiaries and this is reflected in schemes of Payment for Ecosystem Services, with the state often identified as the beneficiary making the payment on behalf of the wider community. Examples include agri-environment schemes giving financial rewards to farmers for adopting biodiversity-friendly methods. The alternative is based on the "polluter pays" principle and endeavours to make sure that those who cause losses to biodiversity are required to compensate for this in some way. Examples include biodiversity offsetting schemes where those whose activities will cause loss in one place are allowed to proceed only if they provide equivalent gains elsewhere. There are technical challenges with both approaches, as well as concerns that they inappropriately treat biodiversity as a commodity, but both may have a role as a supplement to more direct conservation measures.

Beneficiary pays – payment for ecosystem services

Although not always fully expressed as such, payment for ecosystem services schemes are far from uncommon (Reid and Nsoh, 2016, ch. 3). In Europe, the most obvious example is within the European Union's Common Agricultural Policy, where agri-environment schemes have become increasingly significant, offering financial support linked not to agricultural production but to management activities that benefit the environment (European Commission, 2005). Since the 1980s these have developed from allowing special payments to farmers adopting specific environmentally sensitive practices in designated areas to the stage where compliance with a certain level of environmental stewardship is a pre-requisite for receiving any support payments. Farmers are thus rewarded for their role in protecting and managing the rural environment, maintaining and enhancing the ecosystem services it provides. Other schemes operate at varying levels of generality. Some are narrowly targeted, such as the New York City Catskill Watershed Scheme, designed to ensure that a high quality of drinking water is available through the management of activities within the catchment, rather than the more expensive option of purifying a lower quality of water once it has been taken into the supply system (Kenny, 2006). Some are linked to other regulatory controls, such as within biodiversity conservation schemes within the UK, where although there are in the background legal restrictions on the use of land within the various categories of sites designated for conservation purposes, the preference is to achieve the conservation goals through management agreements with the occupiers of the land, supported by payments for the requisite management operations (Reid, 2009). Other schemes take a more broad-brush approach such as the original form of the payment for ecosystems services programme in Costa Rica designed to protect the forests there, which operated on a first-come-first served basis rather than being more precisely targeted (Porras *et al.*, 2013). In relation to watershed services alone, one estimate is that globally over 115 million hectares of land are managed under agreements designed for that purpose (Bennett, Carroll and Hamilton, 2013).

There are several challenges in linking those providing ecosystem services with those who are benefiting from them, calling for the identification of what exactly is being paid for, who is entitled to payment and who should be paying. One possibility, therefore is to avoid such difficulties and simply accept that the recognition of service provision should be applied at the macro level, as a justification for decisions on national taxation and expenditure rather than as a matter of individual transactions. Subsidies paid to farmers struggling to make a living through non-intensive farming on marginal lands can be justified by pointing out that this is not giving them "something for nothing" but represents recompense for the valuable services their care of the land is providing. These benefits are enjoyed by the population as a whole who can therefore be expected to contribute to them through funds raised by taxation. The same can apply to other ways in which support is given to maintain land in a state where environmental benefits are provided, even though more intensive or industrial uses would be more commercially profitable. Where more finely tuned mechanisms are attempted, endeavouring to make a direct connection between individual beneficiaries and providers, it may be accepted that a fairly rough and ready scheme is appropriate to capture the essence and key benefits of the approach, rather than being thwarted by the obstacles in the way of a perfectly refined solution.

In the first place, it must be ascertained what services are to be paid for. As noted above this is complicated by the fact that many services overlap and are difficult to disentangle, and this creates difficulty in dealing with the range of services involved. Where the owners who are maintaining an active peat bog are rewarded for its role in absorbing carbon from the atmosphere, it may seem inappropriate for them also to be rewarded for its role as a reservoir for water resources, a service which requires no additional effort on their part. Yet being in a position to sell more than one product arising from the same form of land management is not uncommon; a farmer from the single activity of growing barley is able to sell both the grain and the straw, usually to different buyers, and there is no reason why those producing a range of ecosystem services should not be rewarded for each.

It may be possible to separate each distinct service and sell them individually to different buyers (known as "layering" or "stacking"), but the difficulties of disentangling and valuing each service can be avoided in two ways. One is to accept the interconnected nature of the services being provided and offer them as a single multi-faceted "bundle", so that buyers take the complete package, attracted either by the bundle as a whole or by one particular aspect which is of importance to them (for example those in search of a carbon sink to offset their emissions might be competing with a water company keen to preserve its catchment). The other is to identify a single lead service and to recognize and sell it alone, while recognizing that in practice there will be further services which are "piggy-backing" on this. A study of payments based on water resources revealed that nearly a third included some form of stacking or bundling (Bennett, Carroll and Hamilton, 2013), while virtually every other case will in practice have produced benefits beyond those the water-resource-based ones explicitly addressed.

A further issue is that for pragmatic reasons, schemes may not attempt to quantify and pay for the actual service provided, but rather pay for the management activities (or absence of damaging interventions) which it is hoped will lead to such provision. Given the uncertainties of a dynamic environment, and risk of future policy changes, there is a risk that such payments may not in fact result in actual benefits, especially where there is a time-lag between the activity and the benefits, such as with planting woodland. Nevertheless, such "input" rather than "output" payments provide support at the time when most expense is being incurred and fit the short-term financial arrangements likely to affect both the service provider and those providing payment, such as public bodies which are reluctant or unable to commit funds to be paid decades in the future when the biodiversity benefits of actions taken today may be realized. This does, of course, raise the issue that the person being paid is not the person responsible for the service being provided at that time, but their predecessor who started the ball rolling.

This temporal issue is one aspect of the second question, of who should receive payment (Nsoh and Reid, 2013). A spatial issue is raised by the fact that services are provided by areas of land based on ecosystem functioning, not the boundaries of legal landownership, and separating the contribution of any single plot may be impossible. Even where the land providing the service can be isolated, further issues arise from the layers of interests held by the different parties with some legal rights in the land, such as between owner and tenants or holders of mineral rights, or where there are elements of shared interests, such as in common land. In some parts of the world there can be more basic problems in identifying those with legal rights to the land, or in ensuring that the legal position matches the interests of those actually occupying, managing and using the land. Ensuring that the right person gets paid may therefore not be straightforward. In policy terms there are also questions of whether payments should be targeted at areas where a comparatively healthy environment already exists, rewarding the more eco-friendly management in the past, or at those where there is most scope for improvement, even though this may mean favouring those who have previously adopted damaging practices.

The third element is identifying who is going to pay for the service. As noted above, since many of the benefits are enjoyed by the community at large, it may be appropriate for the state in some form to take responsibility. In the United Kingdom it is the statutory nature conservation agencies that make the payments under the management agreements serving to conserve protected areas. For example, Scottish Natural Heritage has the power to enter management agreements and make payments under the National Parks and Access to the Countryside Act 1949, section 16, the Countryside (Scotland) Act 1967, section 49A and the Countryside Act 1968, section 15.

Even where more specific beneficiaries are identified, such as residents in a town that will suffer less risk of flooding if upstream catchments are maintained to detain floodwater, it is still likely that payments will be directed through public sector bodies. However, in such cases if there were a collective insurance agreement, then those concerned in that might be the appropriate funder.

There are, though, cases where individual beneficiaries are involved. Those whose business depends on the continuing supply of an ecosystem service may wish to provide funding to support this, such as water companies seeking to protect their sources, as in the Catskill example mentioned above. Other demand can be created by schemes where parties are required to show that they are limiting the environmental harm they cause. There is a growing market in carbon offsets, under regulatory and voluntary schemes, which means that those responsible for emissions are interested in paying for the creation and maintenance of carbon sinks to balance their outputs, and to be able to demonstrate this to regulators, investors or customers. Similarly, biodiversity offsetting schemes as discussed below create service-purchasers seeking to support conservation projects to balance the adverse consequences of their activities elsewhere. It remains the case, though, that for most people, any involvement will be through collective schemes, supported by greater awareness of what is being done, rather than individual responsibility for acquiring specific services.

Polluter pays – liability, offsetting, impact fees, etc.

While payment for ecosystem services is based on rewarding those who undertake the activities we wish to encourage, the polluter pays principle is based on trying to ensure that those who cause harm have to bear the full cost of that. Conventionally, legal systems have recognized only harm in the form of physical injury or damage to property, so that the adverse effects on the wider community caused by diffuse pollution or losses of biodiversity are not taken into account. Moreover, even in the case of physical injury, a considerable degree of discomfort can be caused before the threshold for actionable injury is reached. By recognizing the value of ecosystem services, it is possible to take legal account of forms of harm which currently escape attention, but at the same time thought must be given to the remedies that will be available, since the standard route of financial compensation will not guarantee that the environmental harm is actually repaired.

The recognition of ecosystem services might enable the standard rules of liability to be extended, since the loss or diminution of such services can be recognized as a type of harm which can form the basis of a legal action. Thus a pollution incident which deprives neighbours of the benefit of pollinating insects would be actionable, even though no direct physical damage to their property is caused (Reid, 2014). This idea is extended further by the EU Directive on Environmental Liability (Dir. 2004/35/EC), under which biodiversity damage is one of the heads of harm which for which liability is imposed. The challenge of identifying when relevant harm has been caused is resolved by linking the threshold to the status of species and habitats designated for protection under other EU measures (with the processes leading to designation providing the baseline data against which impacts can be measured), while the obligation to avoid or remedy harm expressly puts the focus not on financial compensation but on preventing or repairing the environmental harm. In particular the remedial action required includes not only steps to restore the habitats affected but also providing "compensatory remediation" to make up for the interim loss of natural resources and services pending the recovery of the habitat affected and "complementary remediation" where restoration of the original habitat is not possible; in such circumstances the party liable must provide a similar level of resources and services at an alternative site.

This last requirement is an example of biodiversity offsetting, which can be more directly employed to ensure that those responsible for environmental harm in one place must provide compensating gains elsewhere to ensure that no net loss to biodiversity results overall (Reid and Nsoh, 2016 ch. 4). Offsetting has been used in other contexts, such as in relation to carbon emissions, as a means of allowing the negative effects of development which results in environmental harm to be coupled with positive gains which it is hoped will balance the position overall. In relation to biodiversity, though, the equation is much more challenging than for greenhouse gases. The elements of biodiversity are not fungible in the way that greenhouse gases are – a coastal sand dune is not a direct equivalent of a woodland – and even the same type of habitat in different locations will have different characteristics and values depending on local conditions and its connectivity with other habitat. Moreover there are significant temporal issues since a newly created or restored habitat will take years before it can provide the same range of ecosystem services and functions as one that has been undisturbed – centuries, in the case of woodland by the time that a varied mix of young, mature, senescent and dead trees is in place. To secure the long-term commitment that the land will continue to be managed so as to provide the biodiversity benefits, recourse can be had to enduring property law mechanisms in the form of conservation covenants or easements (Reid and Nsoh, 2016 ch. 5). In all cases the measuring of equivalences to determine whether or not there is a "net loss" is difficult, with a choice to be made between thorough but slow and complex approaches and much more simple (if not simplistic) ones.

Various offsetting schemes are in operation, notably the US Wetland Mitigation Scheme (Womble and Doyle, 2012) and schemes in Australia (Coggan *et al.*, 2013), and they again provide a means of ensuring that adverse impacts which are simply not regarded by conventional legal approaches are not overlooked. Where offsetting is mandatory, it ensures that the losses to biodiversity caused by development are recognized, both providing an incentive to minimize such loss and requiring compensatory measures for such loss as is unavoidable. The demand for such compensatory action can in turn provide an incentive for managing land so as to produce the sort of benefits which can be sold to fulfil the offset requirements; as opposed to being unrewarded, delivering such conservation benefits becomes an income-generating use of the land, since there will be a pool of purchasers willing to pay for the conservation activities which can offset the harm which they are causing. The relationships between the party required to provide an offset, the party providing it (assuming that, as is likely to be the case, the developer requiring the offset does not have the capacity or desire to do so directly) and any regulatory body requiring the offset or certifying its validity and scale, can take many forms and a complex web of arrangements can be created using contractual, property and public law devices.

A simpler approach to seeking compensation for harm cause by land management decisions is not to look for steps that directly provide compensatory environmental benefits but to require that some form of impact fee (or charge or levy or tax) is paid (Reid and Nsoh, 2016 ch. 7.4). This operates on the basis of a recognition that environmental harm is being caused which generates an obligation to make some financial contribution in return. As with beneficiary pays mechanisms, this can operate at a very general level through the national taxation system, reflecting the many ways in which our economic activities impose burdens on the natural world (both in general or more specifically as with fuel or vehicle taxes that vary according to the environmental impact caused), but more direct linkages can be created. When permission is given for certain types of development (or at the time when the development is complete, which may offer a better fit with the cash-flow of development projects) a fee can be charged which to some extent reflects the losses arising and which is then used for purposes that in some way compensate for that. The first of two key challenges is identifying the scale of any fee that is to be paid, which requires both an assessment of the impact being caused and then the adoption of a methodology by which this is converted into a financial equivalent, with options for a more or less sophisticated and complex approach to be taken. The second is determining who is to receive the fee and the range of purposes to which it can be put. Within governmental and public sector structures, the identity of the recipient may be significant, since payment to a statutory body dedicated to nature conservation will provide greater certainty as to the eventual use of the sums than if they are paid to a government department with much wider responsibilities.

In comparison with schemes where a more direct offset is required, the financial form of the compensatory measure is simpler and allows more flexibility, with potentially greater room for manoeuvre in terms of supporting projects which are ecologically important but may be less closely related to the specific impacts being felt and their locations. As usual, such flexibility has advantages and disadvantages. One underlying fear with relying on impact fees is that despite the best intentions initially, over time the financial receipts will either be submerged in the general national finances or used to replace existing expenditure rather than actually offering something extra to balance the additional losses caused by the projects in question. More fundamentally, such fees can be seen as the epitome of the idea, anathema to some, that biodiversity can be treated as a commodity to be disposed of at human hands, to be taken into account for its monetary worth alone, not any inherent value.

Common potential and challenges

This brief look at some means by which the value of ecosystem services can be utilized to support conservation shows the potential of this approach. Adverse impacts on the environment are one of the great externalities not accounted for in conventional economic and commercial systems and paying attention to ecosystem services opens the door to ways of enabling them to be taken into account. The degradation of our ecosystems has been largely the cumulative result of countless aspects of the way in which we live our lives rather than the product of major incidents which can be directly regulated. Ensuring that habitat is cared for in a way that services are maintained and enhanced involves costs, both direct costs of management and the lost opportunities of converting land to intensive agriculture or undertaking development of various sorts. Moreover, the long-term commitment and the very many variables involved mean that conserving a healthy environment is not well-suited to a legal approach based purely on prohibiting specific harmful acts. The willing co-operation of those occupying the land is necessary and this can be encouraged and confirmed through the making of payments that recognize the benefits being provided to society, encouraging a wide range of positive actions while maintaining legal prohibitions to prevent the most obviously damaging conduct.

In terms of land use, by providing not just rhetorical support but a tangible income-stream for conservation-friendly uses of land, the pressure towards environmentally destructive but commercially profitable uses is eased as these no longer offer the only ways of generating an income. Landowners with areas suitable for providing ecosystem services are provided incentives for doing just that. Farmers whose fields provide the natural storage of floodwater to the benefit of towns downstream could be paid for maintaining this potential, rather than constructing a raised river-bank to protect their crops and therefore their own income from that land, regardless of the cost to others in the form of much more expensive flood damage downstream. Those with land that provides valuable habitat, or who could increase the biodiversity value of their land, could be rewarded by such means. The schemes may also encourage wider participation in conservation programmes, bringing in money from the private sector as opposed to conservation being an enterprise funded by, and therefore largely directed by, the public sector.

There is growing experience of mechanisms which can be used to give effect to our appreciation of the ecosystem services. These may require the creation or refinement of legal rights and relationships, but the existing "tool-kit" has a lot to offer. Simply recognizing that ecosystem services are valuable enables them to become the subject of contractual and other legal arrangements. Obligations under public and private law can require parties to make payments or to manage land in certain way (or ensure that others do this on their behalf) so as to deliver specified benefits. Such obligations can be linked to existing procedures, for example as a condition of obtaining planning permission for new buildings. Property law can also be used as a vehicle for the long-term arrangements required to secure enduring benefits to make up for the loss.

There are, however, significant challenges to be faced. At the technical level the brief discussion above has hinted at the difficulties of valuing ecosystem services and determining equivalences, whether or not the further step is taken of converting values into monetary terms. With some services, such as the Catskill Scheme, some key calculations can be straightforward, setting the cost of building and running water purification plant against the cost of managing the catchment to ensure a high-quality supply such that the purification costs can be avoided (Kenny, 2006). In other cases, though, the calculations are much more difficult, especially where biodiversity is involved in view of its fundamentally non-fungible nature and the significance of locational and temporal issues (Salzman and Ruhl, 2000). The need to co-ordinate action over time and space may also have to be taken into account, with effective conservation requiring coherent long-term programmes rather than relying on the cumulative outcome of individual transactions for various disaggregated ecosystem services.

There is also great uncertainty at many levels. We often simply do not have the basic knowledge and understanding to appreciate fully the current state of the environment, the services it provides and the relationship between different elements that might enhance or diminish these. The dynamic nature of the environment, all the more so in an age of climate change, means that we cannot say with certainty whether measures to restore or recreate habitats will succeed. A further issue is ensuring that where offsetting or payment for enhanced services is involved, what is being delivered is truly additional to the status quo, providing value for the credit recognized or money spent.

In addition there is a more generic and deep ethical concern about the valuation of ecosystems and the employment of mechanisms such as those discussed above. From various perspectives this can be regarded as ethically unacceptable (Reid and Nsoh, 2016, ch. 8). Such views would reject outright the notion that the natural world should be valued simply for the services (tangible and intangible) it provides for humans and can be seen as amenable to being shaped by humankind for its own benefit. On this basis, all members of the Earth community deserve to be respected, and it is wrong to regard nature as something which we can buy and sell, to focus on ideas such as a "net loss" to biodiversity as if the elements of nature are interchangeable at the whim of humans. We progressed from accepting slavery on the basis of the economic and other benefits it produced for some humans to the position of respecting the rights of all humans; similar respect, it can be argued, should be paid to the non-humans with which we share the planet.

Conclusion

Recognizing the value of ecosystem services enables them to be brought into consideration in a way which too often does not happen today. While policy at international, regional and national levels may call for the interests of biodiversity to be included as decisions are taken, the continuing loss of biodiversity reveals that too little is being done. There are some bright spots in relation to particular protected species and habitats, but the overall decline continues. The future of humanity depends on the continued functioning of the ecosystem, yet the interests of nature are simply ignored. This is not so much a conscious choice as the result of the fact that biodiversity does not feature in the decision-making processes which determine how we use our planet. Wild animals and plants do not have legal rights which can compete against those of individuals, landowners or investors and without a recognized value biodiversity does not feature in the financial calculations which determine how resources are to be deployed. Identifying the valuable services provided by the natural world provides a means for bringing nature into our considerations.

Various mechanisms can be used to give legal shape to the relationships needed to reflect and protect the value and interests of the ecosystem. There are technical challenges to be overcome in devising workable and effective systems and ethical questions over the acceptability of such an approach. Nevertheless, the potential is there for progress to be made. Biodiversity conservation cannot be achieved without continuing reliance on command-and-control regulation, but mechanisms based on ecosystem services can provide a useful supplement, influencing behaviour across a wider front and involving a wider range of parties. The fact that a perfect system cannot be found should not prevent us from thinking about deploying the imperfect ones we have when the alternative is that nature continues to be ignored and to suffer.

References

Bennett, G., Carroll, N. and Hamilton, K. (2013) *Charting New Waters: State of Watershed Payments 2012*, Forest Trends, Washington D.C.

CBD-COP 5 Fifth Ordinary Meeting of the Conference of the Parties to the Convention on Biological Diversity, 15–26 May 2000, Nairobi, Kenya (CBD-COP 5), Decision V/6, para. 5, Principle 5.

Coggan, A., Buitelaar, E., Bennett, J. and Whitten, S.M. (2013) "Transferable Mitigation of environmental impacts of development: two cases of offsets in Australia" *Journal of Environmental Policy and Planning*, Vol. 15, p. 303.

EAC (2013) House of Commons Environmental Audit Committee, *Pollinators and Pesticides* (Seventh Report of 2012–13) 2012–13 HC 668.

European Commission (2005) Directorate General for Agriculture and Rural Development), *Agri-environment Measures Overview on General Principles, Types of Measures, and Application*, European Commission, Brussels.

Kenny, A. (2006) "Ecosystem Services in the New York City Watershed" *Ecosystem Marketplace*, www.ecosystemmarketplace.com/articles/ecosystem-services-in-the-new-york-city-watershed-1969-12-31-2/.

Nsoh, W. and Reid, C.T. (2013) "Privatisation of Biodiversity: Who can sell ecosystem services?" *Environmental Law and Management*, Vol. 25, p. 12.

Porras, I., Barton, D.N., Miranda, M. and Chacón-Cascant, A. (2013) *Learning from 20 years of Payments for Ecosystem Services in Costa Rica*, International Institute for Environment and Development (UK), London.

Reid, C.T. (2009) *Nature Conservation Law* (3rd edn), W Green, Edinburgh.

Reid, C.T. (2014) "Taking Account of Environmental Damage – a brief overview" *Environmental Law and Management*, Vol. 26, p. 164.

Reid, C.T. and Nsoh, W. (2016) *The Privatisation of Biodiversity? New Approaches to Conservation Law*, Edward Elgar, Cheltenham.

Salzman, J. and Ruhl, J.B. (2000) "Currencies and Commodification of Environmental Law" *Stanford Law Review*, Vol. 53, p. 607.

TEEB (2010) The Economics of Ecosystems and Biodiversity (TEEB), *Mainstreaming the Economics of Nature: A Synthesis of the Approach, Conclusions and Recommendations of TEEB*, http://teebweb.org.

UNEP (2005) *UNEP Millennium Ecosystem Assessment, Ecosystems and Human Well-Being: Synthesis*, Island Press, Washington D.C.

Womble, P. and Doyle, M. (2012) "The Geography of Trading Ecosystem Services: A Case Study of Wetland and Stream Compensatory Mitigation Markets" *Harvard Environmental Law Review*, Vol. 36, p. 229.

26

BOUNDED OPENNESS AS THE MODALITY FOR THE GLOBAL MULTILATERAL BENEFIT-SHARING MECHANISM OF THE NAGOYA PROTOCOL

Joseph Henry Vogel, Klaus Angerer, Manuel Ruiz Muller and Omar Oduardo-Sierra

Introduction

'Access to genetic resources' and 'the fair and equitable sharing of benefits' (ABS) has beleaguered the thirteen Conferences of the Parties (COP) to the 1992 United Nations Conference on Biological Diversity (CBD). Despite the expectation that COP10 would achieve a workable policy, the negotiations resolved none of the many contentious issues (Kamau *et al.*, 2010) and even added a new one (West, 2012). One might say that had the discipline of economics not been studiously ignored since 1992 (Oduardo-Sierra *et al.*, 2012), a 'fair and equitable' policy would have arisen as early as COP1 in 1994. As will be elaborated here, the field of economics appropriate for ABS is the economics of information, which recognizes that '[i]nformation is a fundamentally different commodity from normal goods. It is costly to produce but cheap to reproduce' (Samuelson and Nordhaus, 2010, pp. 222–23). The asymmetry means that government should either finance information or grant the 'owner special protection against the material's being copied and used by others without compensation' (Samuelson and Nordhaus, 2010, p. 223).

The failure of the COPs to achieve an economically sound policy for ABS can be traced to the definition of 'genetic resources' as 'material' in Article (2) of the CBD (CBD Secretariat, 1992). Over the years and decades, the definition has become the apotheosis of error, as accusations of 'digital biopiracy' attest (ETC Group, 2010). The correction is to re-define 'genetic resources' as 'natural information'. The term captures the essence of the object of access for research and development (R&D) while suggesting discrimination from its complement, viz., artificial information.[1] By applying the economics of information to natural information, a strong policy implication emerges for the modality of the 'Global Multilateral Benefit-Sharing Mechanism' (GMBSM), which is the title of Article 10 of the Nagoya Protocol (NP). The second section of this chapter will explain that modality as 'bounded openness'. The third section will contrast the economic implications of genetic resources as material with that of genetic resources as natural information.

Understanding how bounded openness achieves fairness and equity, requires a command of economics no more advanced than that found in an introductory course, as evidenced by the Online Discussion Groups on Article 10 of the Nagoya Protocol, held in 2013. Only one of the 143 experts was an economist. Nevertheless, experts from various disciplines participated in threads that explored the economics of the GMBSM. The fourth section of this chapter reproduces passages from those comments which, when considered together, show how easily bounded openness can be assimilated.

At the close of the 2013 Online Discussion, the Secretariat expunged the economics of 'genetic resources' as 'natural information' in its official Synthesis of the Online Discussion (CBD Secretariat, 2013a). Similar efforts were evidenced in other fora thereafter. The penultimate section reviews the measures taken against the economics-of-information solution. The bureaucratic resistance integrates with broad patterns about policymaking and invites an interdisciplinary enquiry about group behavior. One promising approach is to 'think [not just] like an economist' (Siegfried *et al.*, 1991), but also like an ethologist and contemplate the role of leadership in rational argumentation. Heeding the generic advice of Noam Chomsky, the Conclusion and Recommendations 'face honestly and realistically the question of how policy decisions are made' (Chomsky, 2016, p. 161).

Bounded openness

Core concepts from the economics of information have been applied to genetic resources since 1992 (Vogel, 1992; Swanson *et al.*, 1994; Stone, 1995; Kagedan, 1996), the same year that the draft text of the CBD was presented for signature at the Earth Summit in Rio de Janeiro. However, the best term to capture the policy implication appeared only in 2010: 'bounded openness' (May, 2010, pp. 142–46). Chris May launched the neologism with respect to artificial information and later endorsed its extension to natural information.[2] The following wording compresses the breadth and depth of May's thesis into one sentence:

> Bounded openness: Legal enclosures which default to, yet depart, from *res nullius* [property of no one] to the extent the departures enhance efficiency and equity, which must be balanced when in conflict.
>
> (Peruvian Society of Environmental Law, 2016, p. 2, fn 2)

Succinct definitions require unpacking. The seemingly unnecessary 'default to' stresses that access to information is open unless specifically addressed. The legal bounds are departures from the default position and are imposed to enhance efficiency and equity. The clause 'when in conflict' is also necessary because economists often leave the wrong impression that efficiency and equity are always in conflict (see, for example, Samuelson and Nordhaus, 2010, p. 323).

By the criteria of efficiency and equity, bounded openness emerges as the optimal modality for the 'transboundary situations' referenced in Article 10 of the NP, the GMBSM (CBD Secretariat, 2010). Such situations implicitly assume that genetic resources are information inasmuch as material cannot be in two jurisdictions at the same time, quantum physics notwithstanding. As the modality for the GMBSM, natural information would flow unencumbered for R&D, which is the 'openness' in 'bounded openness'; the 'bounded[ness]' is a significant royalty on any subsequent intellectual property to be shared among the countries of origin, proportional to habitat. For globally dispersed species, which comprise the majority of those accessed for R&D (Oldham *et al.*, 2013, p. 6), the sharing would be simple: none, as the royalties would finance the fixed costs of the infrastructure necessary for classification.[3] Compliance is also simple. In the case of patents, applications for protection would be amended to include a box to tick on whether or not natural information was utilized. Only successful commercializations of processes and products protected by intellectual property rights would trigger follow-up disclosures about the biological samples accessed, thereby facilitating determination of the geographic dispersion of that species.[4] Chemical analysis would also be required to determine the dispersion of the natural information across taxa.

By leaving genetic resources unencumbered in R&D, bounded openness would enable the 'creative destruction' (Schumpeter, 1950, p. 83) of game-changing technologies that derive from utilization. Because the potential of diverse sectors of the economy is in the balance, an insight from Nobel Memorial Laureate James M. Buchanan is most germane: 'Until and unless we begin to take the long-term perspective in our private and in our public capacities ...we are doomed to remain mired in the muck of modern politics' (Buchanan, 1987, p. 126). Buchanan encourages economists to explore 'analysis of the working properties of alternative sets of constraining rules' (Buchanan, 1986). The advice can be applied to the reception of bounded openness by the experts convened in the aforementioned 2013 online discussion. All were painfully aware that under the current 'set of constraining rules', few bilateral Material Transfer Agreements (MTAs) have ever been concluded (Carrizosa *et al.*, 2004) and of those few concluded and divulged, the royalty percentages are scandalously low (Markandya and Nunes, 2012).

The above thumbnail sketch reflects the power of economic abstraction, where the solution has been compacted into a few paragraphs of an introductory statement. Bounded openness is so commonsensical that its essence was even perceived without formal economics and before the drafting of the CBD in 1992. Antecedents lie in the notion of 'Range States' by Cyril de Klemm (1989). Nevertheless, recourse to abstract economics was required to keep the idea as simple as possible. Manuel Ruiz Muller has documented the twenty-five-year trajectory in *Genetic Resources as Natural Information: Implications for the Convention on Biological Diversity and the Nagoya Protocol* (2015), the preface of which examines the prickly question of provenance (Vogel, 2015). Although some scholars have indeed insisted that biodiversity is 'code by any other name' (Chen, 2004, p. 497) and made the 'argument for a more open system' (Safrin, 2004, p. 668), many more have just ignored the fact that genetic resources are information and the economic implications thereof.[5] To date, the latter group has held sway with the COPs and Secretariat.

In 2007 the environment ministers from the G8+5 recognized that the application of economics may be useful for the objectives of the CBD. A non-economist was commissioned to lead the drafting of a series of reports,[6] which would address, *inter alia*, ABS. The hope was that the resulting analyses would help inform decisions in the COPs. To contextualize the receptivity of the experts in the 2013 Online Discussion to bounded openness, a critique of that literature seems prudent.

From studied to selective ignorance

At COP10, much fanfare accompanied the launch of *The Economics of Ecosystems and Biodiversity* (TEEB), whose Advisory Board includes the Executive Secretary of the CBD. The webpage of TEEB opens unabashedly 'The TEEB study is underpinned by an assessment of state-of-the-art science and economics' (TEEB, 2016). Nevertheless, the application of the economics of information to ABS does not appear once in any of the TEEB publications even though state-of-the-art economics includes the economics of information, as evidenced by the Nobel Memorial Prizes awarded in 1974, 1982 and 2001 to pioneers of the field.[7] To understand how TEEB morphs 'studied ignorance' into 'selective ignorance', one must first examine its methodology.

Chapter 5 of *TEEB Ecological and Economic Foundations* is titled 'The economics of valuing ecosystem services and biodiversity' (de Groot, 2010). The overall approach has much intuitive appeal:

> The concept of total economic value (TEV) of ecosystems and biodiversity is used throughout this chapter. It is defined as the sum of the values of all service flows that natural capital generates both now and in the future – appropriately discounted.
>
> (Pascual and Muradian, 2010, p. 6)

In the taxonomy of service flows, the benefits from access are both use and non-use values. The TEEB authors set broad criteria for any market-like solution of the summed worth:

> [M]arket creation process requires three main stages: demonstration of values, appropriation of values and sharing the benefits from conservation … Demonstration refers to the identification and measurement of the flow of ecosystem services and their values.
>
> (Pascual and Muradian, 2010, p. 5)

To satisfy the criteria for access, the TEEB authors of Chapter 5 quote another TEEB report, coincidentally also a Chapter 5, just a few clicks away. In *TEEB for National and International Policy Making*, the value of genetic resources for bioprospecting is 'demonstrat[ed through] measurement', with an ostensible precision that goes down to a dime per hectare:

> To the dismay of those who believe that genetic resources are a global resource of high value, these estimates come out rather low. A key early study [(Simpson *et al.*, 1996)] calculated values of genetic resources in 1996 prices at between US$ 0.2/hectare (California), and US$20.6/hectare (Western Ecuador) and argued that these estimates could be on the high side. Other studies making the same point include Barbier and Aylward [(1996)] and Firn [(2003)].
>
> (ten Brink, 2009, p. 35)

The discerning reader will be perplexed. Table 5.2 of *TEEB for National and International Policy Making* identifies six 'Market Sectors Dependent on Genetic Resources' from the year 2006: (1) pharmaceutical, (2) biotechnology, (3) crop protection products, (4) agricultural seeds, (5) ornamental horticulture, (6) personal care, botanical, food and beverage industries (ten Brick, 2009, p. 34). Total annual sales compute to US$827 billion, of which pharmaceuticals lead at US$640 billion. In a near trillion-dollar per annum market, how can estimates for the value of genetic resources 'come out rather low'? *TEEB Ecological and Economic Foundations* does not waver:

> Reasons identified for values being so low included the **high cost of developing the final goods and bringing them to market, the long time lags involved and inefficiencies in the systems for exploiting genetic resources.**
>
> (ten Brink, 2009, p. 3; bold in original)

The explanation has a familiar ring. Jack R. Kloppenburg identified it more than twenty years earlier in the first edition of *First the Seed: The Political Economy of Biotechnology*: 'Curiously, this argument relies implicitly on a [Marxian] labor theory of value. It is asserted that only the application of scientists' labor adds value to the natural gift of germplasm' (Kloppenburg, 2004 [1988], p. 185).

The non-Marxian reason for the low market values has been explained repeatedly in the literature: competition in the provision of a good that is intangible (see, for example, Vogel, 1991). Although the TEEB authors provide evidence that they were exposed to the economic explanation (ten Brink, 2009, p. 63, fn 24), they did not identify it among the other 'reasons identified for values being so low'. The omission becomes conspicuous as they boldly broach, literally and typographically, the issue of rents:

> A key question in the ABS context is **how much of the value of final products is attributable to genetic *material* and how much to other factors of production (labour, capital, local knowledge *et al.*).** To answer this question we need to distinguish between what a producer of drugs or other products has to pay to obtain the genetic *material*; and what the *material* is worth to the producer (i.e., the maximum that a company would pay). The difference between this maximum payment and the cost of obtaining the genetic *material* is called its 'rent'.
>
> (ten Brink, 2009 pp. 34–35; italics added, bold in original)

Although the above definition of 'rent' is correct, the acceptance of its elimination through competition is as hypocritical as it is wrong. What is of interest to R&D is not the genetic material per se but the natural information contained therein. As an intangible, a justification for rent exists in the opportunity costs of habitat conservation. Without rules constraining competition, rents are eliminated as the market price falls to the marginal cost of collection, viz., what 'a producer of drugs or other products has to pay [in a competitive market] to obtain the genetic material'. The justification for a rent lies in the intangible nature of genetic resources as information, which is the same justification for rents that underpins all monopoly intellectual property rights.

'State-of-the-art' TEEB ignored not only the academic literature which justifies rents for genetic resources, but also the official enquiry in COP9 Decision IX/12, which:

> *Requests* the Executive Secretary to invite, in consultation with the Co-Chairs of the Working Group, relevant experts to address the Working Group on Access and Benefit-sharing, at the appropriate time, on the following issues: Should economic rent be charged for access to genetic resources and what is the justification for such a rent or against such a rent? What should be the basis for the valuation of such rent?
>
> (CBD Secretariat, 2008)

The Ecuadorian delegation introduced the question about rents at COP9 in the hope of eliciting an official recognition of the relevant economics. Unfortunately, the question was left unanswered en route to COP10. Fortunately, an answer can be gleaned from the TEEB reports. One can safely say that any significant royalty on an annual trillion-dollar market would generate more revenue, by orders of magnitude, than the budget needed for the incremental infrastructure to enable the 'market-like' solution for ABS.[8]

The justification for rents coheres with the provocative call for a 'biodiversity cartel',[9] which the authors of TEEB did address:

> Suggestions include forming a cartel to negotiate on behalf of all owners of such resources i.e. like the Organization of the Petroleum Exporting Countries (OPEC) on behalf of global oil producers … Like all cartels, this might succeed in obtaining a higher share of the rent from exploitation of genetic *materials* but also be unstable with strong incentives to undercut the agreed price (e.g. the price of crude oil has fluctuated since 1974, when the cartel started restricting supply).
>
> (ten Brink 2009, p. 39, italics added)

The presumed instability of the proposed biodiversity cartel hinges on the mistake of categorizing a sequence of nucleotide bases in DNA with crude oil. The correct analogy for natural information would have been the state-sanctioned monopoly protection of artificial information. With beguiling deference, all the authors of the TEEB reports enjoy a blanket disclaimer in Chapter 1: 'In the TEEB assessment, we largely follow the definitions of the United Nations 1992 Convention on Biological Diversity' (de Groot, 2010, p. 15).

Delegates in the COP who look to TEEB for any guidance on ABS should note well that in 2015, Sarah Winands-Kalkuhl and Karin Holm-Müller published the peer-refereed 'Bilateral vs. Multilateral? On the economics and politics of a global mechanism for genetic resource use'. The extensive article, some 26,000 words in total, does not cite any of the TEEB reports. Winands-Kalkuhl and Holm-Müller 'find that the economically preferable instrument of a comprehensive global mechanism is politically not feasible any time soon due to path dependencies and an arguably narrow understanding of national sovereignty' (Winands-Kalkuhl and Holm-Müller, 2015, p. 305).

Receptivity of 2013 online discussion groups to bounded openness

How did the 143 experts convened in the 2013 online discussions react to the economics of information and its implication of bounded openness? The comments are retrievable through the search key from the intralink in the ABS Clearing-House Mechanism (CBD Secretariat, 2013b). Four rounds with thematic questions ran, from 8 April to 24 May 2013. In the light of the economics of information, one can examine the strength and acceptance of its application by the progression of thought in the commentary.[10]

Ruiz Muller was one of the first to respond the thematic question of Round 1: What could be the 'transboundary situations' covered by Article 10 of the NP that are within the scope of the Protocol?

> I would argue that even though as some other colleagues have stated, there may be political and negotiation difficulties on a number of fronts, the interesting thing about this online discussion is that we can actually try and think a little outside the box and see whether there may be some alternative options… [#4773].

A multitude of opportunities arose to think outside the box. Over the four rounds, Vogel would make thirty-eight interventions, responding as early as possible in the posting of each round. In choosing what to reference, Vogel followed the recommendation of the Executive Secretary for 'Credible sources of information, preferably from peer-reviewed articles' (CBD Secretariat, 2013c). Vogel's first post occurred on the second day of the Discussion:

> A 'transboundary situation' arises when natural information (e.g., a gene, a metabolite, a design from biomimicry, and so on) is diffused over species and the species, over jurisdictions. 'Natural information' is not a 're-definition' of 'genetic resources' but a simple acknowledgement of the commonplace in molecular biology (e.g., genomics, proteomics, metabolomics, and so on).
>
> Seen thus, 'transboundary situations' raise questions that lend themselves to economic analysis. Why should the first-Party-to-contract receive all the benefits from the conservation of the natural information? Moreover, competition among providing Parties means that the price contracted will fall to the marginal cost of access, essentially nothing. The implications are an application of the 'economics of information', studiously ignored in the COP (see Oduardo-Sierra *et al.* [2012]).
>
> Economists have long argued that one cannot put a fence around (artificial) information; hence, exclusionary mechanisms must be institutional, viz., intellectual property rights. For natural information, something similar could be achieved through the CBD. Openness would be bounded by simple disclosure requirements in patent applications for species accessed in R&D. Only with commercial success of the patent would it be worthwhile to determine the countries of origin (N.B. plural). The International Barcode of Life [iBOL] could facilitate that determination. Distribution of revenues from royalties could be proportional to habitat size, which seems the fairest and most equitable arrangement [#4788].

Mr Marco D'Alessandro of the Federal Office for the Environment, Switzerland, was swift to counter:

> In my understanding, in the context of the Nagoya Protocol, we are however neither talking about species, subspecies or any other taxonomic entity, nor are we talking about information. In the context of the Nagoya Protocol we are talking about genetic resources. And a genetic resource is defined as genetic material of actual or potential value according to Article 2 of the Convention. It is hard to imagine how a genetic resource as material can occur in two or more countries at the same time. I believe that in principle it should always be possible to determine the source of a specific material, and therefore the bilateral approach is probably applicable in most cases, no matter whether material with similar properties is found in different countries [#4953].

Vogel responded to D'Alessandro's intervention and that of other experts who raised similar points:

> Thomas Nickson of Monsanto comments that 'the scope and fundamental elements of the Nagoya Protocol must be respected in any decision on the need for and modalities of a GMBSM' [#4973]. Perhaps the most 'fundamental element' of the Protocol is conveyed in the two adjectives that modify the 'Sharing of Benefits' in the full title 'The Nagoya Protocol on Access to Genetic Resources and the Fair and Equitable Sharing of Benefits Arising from their Utilization to the Convention on Biological Diversity'.
>
> The inherent unfairness and inequity in bilateral negotiations of transboundary resources are compounded by Mutual Agreed Terms (MATs) which stipulate 'confidential business information' over the royalty negotiated. The beguiling term 'Mutually Agreed Terms' appears 25 times in the Nagoya Protocol and transparency, thrice.

> Mr Nickson continues that 'much more thought needs to be given to how a GMBSM could or could not give rise to monetary and non-monetary benefits, and affect the conservation and sustainable use of biodiversity'. Much thought over two decades has been given to a multilateral benefit-sharing mechanism; the results appear in the refereed literature. A recent application of the 'economics of information' to bioprospecting has generated an unambiguous conclusion: almost all of the contentious issues well analyzed by Kamau *et al.* [(2010)] can be resolved with a system of 'bounded openness'.
>
> (Vogel *et al.*, 2011, p. 59)

The forum was off to a somewhat pugnacious start. Participants could expect that Vogel and Ruiz Muller would answer questions from each round in the light of the appropriate economics. To facilitate referencing, Vogel would quote the participants and identify the number of their comment in his response.

By the second week of the Discussion, the term 'natural information' had already taken hold. Vogel notes:

> Mr Ibañez de Novion cites 'natural information' ten times in his comment (#5000). The term even becomes the referent by its third usage: 'Since there is natural information subject to access…'. The term 'natural information' invites the application of 'the economics of information' with policy implications diametrically opposed to the 'bilateral approach'. In other words, any argument that integrates the term 'natural information' into the 'bilateral approach' is not coherent. This is the unspeakable economics of ABS.
>
> (Vogel, 2008 [#5009])

Others took issue. Vogel quotes one such comment as he opens his response:

> 'The suggestion that you can share benefits based on the percentage of a habitat or ecosystem a country has is not yet practical and has yet to be shown to be scientifically appropriate'. Contrary to Dr Smith's assertion [#4845] and Ms. Muñoz' affirmation [#4856], the approach is extremely practical and also scientifically appropriate. Through bounded openness to genetic resources, one can eliminate the transaction costs that have beleaguered ABS for twenty years. Only species from commercially successful patents would be examined. No costs would be incurred for those with dead-end R&D. N.B. Such a system will be fiercely resisted. As Nobel Memorial Laureate in Economics Joseph E. Stiglitz writes: '[f]or lawyers, transaction costs are a benefit, because they are a source of their income' [(2008, p. 1706)].
>
> For the rare commercial successes, one does not need to know the exact boundaries of the habitat to share the royalties, fairly and equitably. One simply needs to apply the best estimate of boundaries given the state of science. As iBOL advances, the estimate will become ever more precise. For the cases where the origin is ubiquitous (e.g., many microorganisms), the royalties collected should defray the fixed costs of the infrastructure which drives the system, viz., iBOL [#4869].

Perhaps because the disagreements were profound, Ruiz Muller urged open-mindedness:

> That the politics for implementing a 'new' approach may be hard – indeed. But the technical, economic and scientific evidence seems – if we are careful to read it and

> process it – overwhelmingly in favour and supportive of a new approach if these foundations are recognized and accepted. As a Turkish proverb says 'No matter how far you have gone on the wrong road, turn back'. This may be required in ABS discussions in the near future [#4872].

Turning back would not be easy and the likelihood of intransigence was not lost on Mr Joe Ripley of the US Department of State:

> The points made so far reflect serious and deep divisions surrounding the interpretation of Article 10 and the entire Protocol. Some have raised questions about the temporal and geographic scope of the Protocol, about core definitions, about whether the Protocol is access driven or use driven, and a few have challenged the sovereign basis that underpins the Convention on Biological Diversity. Even the science behind the Protocol is under debate.
>
> How Parties to the Protocol resolve these questions will have a tremendous effect on those who desire to conduct biological research of any kind. What we have seen among U.S. researchers at this time is considerable uncertainty surrounding the implementation of the Protocol and the expectation that transaction costs (including litigation costs) will only rise. This has caused many researchers to not conduct the research that they otherwise would have [#5017].

Were participants other than Vogel and Ruiz Muller perceiving bounded openness as a bridge over the 'serious and deep divisions' perceived by Ripley? The parenthetical remark by Pierre du Plessis at the end of Round 2 is hopeful:

> A GMBSM will deliver most value if it is created as soon as possible and implemented alongside the bilateral system as a complementary option. It would also be useful not to prejudge the exact modalities of the GMBSM and to remember that it could encompass sub-sections, such as ring-fenced regional, sub-regional or sectoral arrangements; it could also (especially if implemented in a phased, stepwise approach aimed at developing an SMTA [standard Material Transfer Agreement] with sectoral model clauses) accommodate different ABS models, such as 'bounded openness over natural information' (the logic of which, to my mind at least, seems fairly unassailable in cases where the subject matter is indeed strictly natural information)… [#5191].

Similar to du Plessis' parenthetical praise, Bienvenu Célestin expressed gratitude in Comment [#5138]. At the end of the rounds, Prudence Tangham Galega of Cameroon presented the summary for the African Group:

> This is a concluding submission which issues from an African Group consultation. It cuts across all key questions on the on-line discussion and is without prejudice to African positions in future negotiations on Article 10…
>
> *Response to emerging issues…*
>
> *'Bounded openness'*
>
> The 'bounded openness' idea is interesting to Africa, but it is not clear how the transition from the current bilateral ABS system can be made in practice. We are open to further discussions along these lines and including this approach in the GMBSM on a trial basis, possibly on a sub-regional level. Africa would require substantial capacity

> building and technology transfer, especially in taxonomy and compiling biodiversity inventories, to make this work [#5298, italics added].

Although the expression of interest in bounded openness by Africa is heartening, even more encouraging is a subtle veer in course among the experts who had long ignored the economics. They were now engaged in the debate, albeit begrudgingly. Ripley argued:

> The administration of a single flat royalty rate would be short sighted, as it does not take into consideration different industry cost structures or market prices. The imposition of a flat rate, or the wrong flat rate, for example, could well determine which industries undertake R&D on genetic resources and which do not [#5164].

Vogel responded:

> Flat or standard royalty rates are found across large swaths of the economy, as a quick Google search will reveal (580,000 hits at 11:00 PM GMT). The economic justification for a flat rate lies in a simple question: Are the transaction costs of negotiating royalties case by case less than the value of deals that would have only been consummated below that [flat] rate? The scarcity of MTAs negotiated case by case demonstrates that the transaction costs of the bilateral approach are so high as to be prohibitive (for empirical evidence, see Santiago Carrisoza *et al.* 'Accessing Biodiversity and Sharing Benefits: Lessons from Implementing the Convention on Biological Diversity') [(2004)]. A standard royalty under 'bounded openness over natural information' would greatly reduce these costs and thereby facilitate R&D. Contrary to Ripley's assertion, such a flat rate is far-sighted [#5166].

Could Ripley and Vogel be both right and wrong? The pharmaceutical industry has long been the most profitable beneficiary of genetic resources, especially for drugs that have no other indicated therapy.[11] By synthesizing the exchange of Ripley and Vogel, one sees how they are simultaneously right and wrong. The GMBSM could impose a flat rate, but not the same flat rate across the six aforementioned market sectors identified by TEEB.

Why discriminate royalty rates solely by sector? Many other characteristics are also relevant.[12] Simple math quantifies just how many combinations would have to be negotiated. A quick review of the website from the World Intellectual Property Organization indicates that as many as ten specific types of intellectual property might be relevant to ABS over the six sectors.[13] There is also the presence or absence of substitutability of inputs in R&D or production, and directness in derivative use or indirectness in research streams. So, at least 240 distinct combinations (6x10x2x2) exist for which the COPs must negotiate flat royalty rates. Thinking like an economist, negotiation should begin over the percentage royalties for the combinations which hold the highest mathematical expectation, i.e., the probability of the event multiplied by its value. Undoubtedly, the highest expectations will be those combinations which include the pharmaceutical sector. To prevent the summed royalties from utilization of multiple ensembles of information becoming prohibitively costly, the royalties for each ensemble would have to be weighed proportional to some cap.

An example may make the math more comprehensible. Suppose a blockbuster medicine is developed which utilizes four distinct ensembles of natural information. Suppose further that the COP decides on a royalty of 5% for the combination 'pharmaceutical/patent/non-substitutable input/directness in research stream' and 2% for 'pharmaceutical/patent/

non-substitutable input/indirectness in research stream', with a cap of 15% on the sum of all royalties paid on any patented product. Of the four distinct ensembles of natural information utilized, imagine further that ensembles one to three correspond to the first combination, and the ensemble of the fourth, to the second combination. Without the cap, the total royalty would be 17% ((3x5%) + (1x2%)) and Ripley's complaint would gain traction. To stay within a cap of 15%, the percentage long advocated by Vogel for pharmaceuticals (1994, p. 37), the weighted rate would be (3x15/17x5%) + (1x15/17x2%). The sharing among countries of origin would be proportional to the geographic range of the species which harbor the natural information accessed.

Bureaucratic resistance

Twelve weeks after the Discussion Groups closed, the Secretariat published the 'Synthesis of the Online Discussions on Article 10 of The Nagoya Protocol on Access and Benefit-sharing' (CBD Secretariat, 2013a). Despite forty-seven interventions that referred to economics and twenty-two interventions, to the economics of information, neither term was cited once. Identification of the category mistake in the definition of 'genetic resource' was also omitted despite multiple interventions which stressed the point. Although 'natural information' was cited, the concept was nonetheless misconstrued: 'A representative of an NGO expressed the view that transboundary resources are the norm if genetic resources are understood as natural information' (p. 26). The sentence is a non sequitur inasmuch as genetic resources are information for R&D whether or not they are transboundary.

Beyond the omissions in the official Synthesis lies an overarching flaw: it does not synthesize. 'Synthesis' is the 'combination of parts or elements so as to form a whole' (*Merriam-Webster*, 2016). The text is a 'classification', defined as 'an arrangement of people or things into groups based on ways that they are alike' (*Merriam-Webster*, 2016). In essence, the Secretariat classified the comments without the light of any theoretical framework, reminiscent of Theodosius Dobzhansky's famous remark about biology without evolution: 'a pile of sundry facts some of them interesting or curious but making no meaningful picture as a whole' (Dobzhansky, 1973, p. 129).

The lacunae did not escape the invitees to the 'Expert Meeting on Article 10 of the Nagoya Protocol' held at the Secretariat from 17 to 19 September 2013. In Point 15 of the Report about the Official Synthesis:

> Experts identified three points that were raised during the online discussions but that they considered had not been adequately reflected in the synthesis:
> (a) The concept of the 'economics of information';
> (b) The question of the definition of genetic resources from the Convention;
> (c) The issue of transaction costs, including transaction cost [sic] that may be appropriate.
> (CBD Secretariat, 2013d, p. 3)

In the closing point (26) 'Areas for Further Examination':

> The experts also suggested that it would be useful if Parties and others could be invited to provide possible scenarios on modalities for a GMBSM as well as information regarding the implications of these scenarios (p. 5).

Two years later, any hope engendered by Points 15 and 26 was dashed by Notification 2015–023, which requested written submissions of views on the very same themes vetted in the 2013 Online Discussion (CBD Secretariat, 2015a). Volunteer participation was lower by almost an order of magnitude.[14] Nevertheless, a twelve-member group of experts, under the acronym of ABSSG, submitted a comprehensive view. Its introduction is reminiscent of point 15(b) above, which in turn reflects some twenty-five years of economic analysis on genetic resources as information:

> A globally functional legal/commercial regime can only function on the basis of clear and unambiguous international consensus on critical components. Legal principles can only apply if the parties, negotiators, arbitrators and courts are able to clearly determine whether or not a particular resource is included within the ABS regime, and how the 'resource' in question is characterized – in this case as a chemical, a piece of encoded information, or as a hybrid embracing both aspects.
>
> (Young and Minnis, eds, 2015, p. 1)

Despite the prominence that ABSSG placed on resolving the definition of genetic resources – appearing on page one of the twenty-two-page document – the Secretariat once again expunged the issue from its subsequent official Synthesis. This time the point was not to be restored in the subsequent Expert Report (CBD Secretariat, 2016a). Curiously, unlike the experts who met in 2013 to evaluate the official Synthesis of the Online Discussions, those chosen in 2016 did not find the official Synthesis of the written submissions non-representative of what was submitted.

The category mistake in Article (2) of the CBD is no longer confined to ABS and the GMBSM. The damage reverberates in the cutting-edge and high-tech field of 'synthetic biology', which was the new and emerging issue for COP11. The 'Updated report and synthesis of views' on synthetic biology shows little agreement as to its definition (CBD Secretariat, 2015c), thereby frustrating any regulation of technologies which may be 'creatively destructive', not just economically, but also biophysically. The Secretariat is in a real pickle. Any proposed definition which deploys 'genetic resources' as 'material' will controvert the very essence of synthetic biology. Enzymes, genes, proteins and other metabolites can be disembodied into data which become the object of R&D. Databases exist which obviate the need to access the material source.

A serendipitous opportunity arose for the Peruvian Society for Environmental Law (known by its acronym in Spanish, SPDA) to suggest a definition which not only allows inclusion and exclusion criteria, but also integrates with bounded openness as the modality for the GMBSM:

> **Synthetic Biology:** the extremely intensive use of artificial information in the manipulation of natural information.
>
> (Peruvian Society for Environmental Law, 2016, p. 3)

Did the position of any of the other thirty-two submitted views about the 'Updated report and synthesis of views' on synthetic biology cohere with that of the SPDA, i.e., with bounded openness? Analyzing each, Vogel perceived common ground between the submission of the SPDA and that of New Zealand, which he communicated to the Lead Adviser of the New Zealand Ministry of Foreign Affairs and Trade (Vogel, 2016). The response was as unexpected as it was unequivocal: the Adviser asked Vogel not to quote the New Zealand submission and formally requested that the Secretariat delete it. The Secretariat obliged not only to un-publish what had been in the public record but also to upload a replacement file (New Zealand National Focal Point, CBD Secretariat, 2016b), submitted more than a month after the deadline (CBD Secretariat, 2016). In other words, the view which most cohered with bounded openness over natural information was now no longer held, nor even readily accessible.

Which of the thirty-three submitted views showed the greatest bureaucratic resistance to bounded openness? No opposition was more clearly enunciated than that of the non-party:

> The United States notes that digital genetic information is not covered under the Convention or the Nagoya Protocol and do not agree that there could be a shift in understanding of what constitutes a genetic resource. We therefore do not support calls to set up mechanisms for 'clarifying the issue of *digital genetic resource information* as it relates to access and benefit-sharing', and we stress the need to adhere to topics that are within the scope of the CBD and Nagoya Protocol. Additionally, we do not support including digital information (e.g., sequence data) under a definition of synthetic biology.
>
> (USA National Focal Point, 2016; italics added)

Deploying transitivity is fruitful in logic. Inasmuch as 'genetic resources' are defined as 'material', 'digital genetic resource information' becomes 'digital material information'. The illogic of the juxtaposed adjectives pales, however, against the false premise by which the USA 'do[es] not agree that there could be a shift in understanding of what constitutes a genetic resource'. The CBD is a framework treaty and as such, 'shift[ing]' from a mistake to a correction should not only be possible but wholeheartedly welcome.

Beyond the contradictions and the faulty arguments, lies an absurdity. The USA is defending the conceptual status quo of a convention that it has not ratified.

Conclusion and recommendations

That politics is illogical requires no comment. That policymaking is likewise, does. The explanation is as obvious as it is dispiriting: policymaking is driven as much by politics as by expertise. Any delusion to the contrary risks the 'tragedy of unpersuasive power' (Vogel, 2013), whereby experts persevere in rational argumentation while the object of discussion collapses. To avert the tragedy, one must recognize that logic and evidence will not be sufficient to change policy. Experts must address the necessity of leadership in making their case. In the controversial language of human ethology, a spokesperson must be found who exhibits 'the broadest capacity to induce a following response in the entire group' (Ardrey, 1970, p. 124).

Until such a leader appears, what can be done? Stakeholders from academia, government and NGOs are under tremendous pressure to conclude MTAs, which are intrinsically unfair and inequitable. Is there a stopgap? Participants to a symposium about ABS held in Quito, Ecuador in 2013 contemplated how model MTAs could be amended to safeguard any future GMBSM from the invocation of retroactivity for genetic resources already transacted. Participants discussed a clause to the effect that:

> Once a multilateral system is established for access and benefit-sharing that reflects the transboundary nature of genetic resources and their essence as natural information, the agreements and bilateral contracts will be subject to the principles and rules of this new international system.[15]

The wording was soon put to the test in Puerto Rico in 2014, where a Steering Committee was formed for an initiative that would become the Puerto Rico Center for Tropical Diversity and Bioprospecting (2016). Vogel, then a member of the Committee, suggested that the clause above be foundational for the whole endeavor. After elaborating the rationale over several email exchanges, he was euphemistically 'excused' from said Committee. The story is referenced in 'Preventing jurisdiction shopping for transboundary resources in a non-party: The case of Puerto Rico' (CBD Secretariat, 2015d), which was the only 'new and emerging issue' submitted for COP13. Despite being the only issue, the Executive Secretary recommended that the Subsidiary Body on Scientific, Technical and Technological Advice (SBSTTA) not adopt it. The Secretary argued that 'the crux of the issue' submitted was 'the lack of universal membership of the Convention on Biological Diversity', which lay outside the purview of SBSTTA (CBD Secretariat, 2016c, p. 2). Although the argument is a straw man, the SBSTTA accepted the recommendation without comment (CBD Secretariat, 2016d).

Parties to the CBD and NP should take stock. With the meteoric rise of synthetic biology since 2010, the entry into force of the NP in 2014, and the institutionalization of tropical transboundary bioprospecting in a megadiverse non-party in 2016, the stars are aligning for bounded openness. Any articulate advocate may more easily release the ethological 'following response' than at any other time in the history of the CBD. Potential leaders exist among the many consummate politicians who populate the COP delegations, North and South. The cynic may even suspect that successful leadership will hail from quarters once associated with MTAs. Such is the nature of politics. Experts vested in MTAs should likewise be fluid. They may take stock in a remark that the British economist Joan Robinson attributed to her mentor, John Maynard Keynes, the indisputable Darwin of economics: 'When someone persuades me that I am wrong, I change my mind. What do you do?' (Harcourt, 1986, p. 99, fn 3).

Notes

1 The adjective 'artificial' implicitly assumes the adverb 'intentionally'. For example, drug-resistant pathogens can be considered natural information subject to ABS (Vogel *et al.*, 2013).

2 On the book jacket of *Genetic Resources as Natural Information* (Ruiz Muller, 2015), May writes: 'The authors operationalize my idea of "bounded openness" for genetic resources as natural information. They show how radically different sectors of intellectual property can learn much from one another'.

3 Data mined on patents indicate that origins 'limited to one or a very small number of countries are likely, on the basis of available distribution data, to be exceptions rather than the rule' (Oldham *et al.*, 2013, p. 6).

4 In the 1980 Budapest Treaty on the International Recognition of the Deposit of Microorganisms for the Purposes of Patent Procedure, 'three quarters of applicants [sampled] do not disclose the country of origin of genetic resources that appear in their claims' (Hammond, 2014, p. 1).

5 For example, E.C. Kamau and G. Winter contrast the 'material substratum' of the genome with its '"intellectual" component' (2013, p. 108), ironically and wrongly preferring a metaphor which implies artificial information.

6 Pavan Sukhdev, the TEEB Study Leader, is a 'career banker' and also CEO of 'GIST Advisory', which 'provides sustainability consultancy services to enable governments, corporations, civil society organisations, banks and financial institutions to discover, measure, value and manage their impacts on natural, social and human capital', http://gistadvisory.com.

7 In the sequence of the years cited, the award was conferred on Friedrich August von Hayek, George J. Stigler and George A. Akerlof, A. Michael Spence, and Joseph E. Stiglitz, www.nobelprize.org/nobel_prizes/economic-sciences/laureates/index.htmls.

8 Under bounded openness, the only relevant question is: 'Does probable cause exist to justify public investment in the infrastructure needed to enable a market in genetic resources?' (Vogel, 2007, p. 49).

9 The title of the anthology *The Biodiversity Cartel* (Vogel, 2000) heeds the enduring advice of John Maynard Keynes: 'We do not distinguish, at first, between the color of the rhetoric with which we have won a people's assent and the dull substance of the truth of our message. There is nothing insincere in the transition. Words ought to be a little wild for they are the assault of thoughts on the unthinking' (Keynes, 1933, p. 761).

10 Biologists may perceive a template for the phrase in Dobzhansky's most famous essay (1973).

11 Profit maximization by a monopolist occurs at the price where marginal revenue equals marginal cost, which is located in the elastic region of the demand curve. Any pricing in the inelastic region is paradoxical but explicable. Full exercise of monopoly power risks a public backlash and price regulation. Nevertheless, restraint in pricing can break down as newcomers to the industry perceive the potential for windfall profits, as has occurred in the USA (see, for example, Cha, 2015). For ABS, the traditional inelasticity of demand in pharmaceuticals means that industry could have passed on any significant royalty to consumers. So, an unresolved paradox is industry reluctance to embrace bounded openness.

12 India has adopted differentiation in flat rates in its *Guidelines for Access and Benefit-sharing* (Mazoomdaar, 2014). Brazil also appears to be moving toward a table of rates, with some sectors charging as low as one-tenth of 1% (Paquin-Jaloux, 2015). Users of transboundary resources can be expected to source genetic material in the Provider country with the lowest flat rate.

13 The number is larger than may be expected as distinctions exist within broad categories of intellectual property (e.g., patents can be further classified as utility, design or plant patents). See www.wipo.org.

14 Sixteen submissions can be counted (CBD Secretariat, 2015b) versus the 143 participants to the 2013 Online Discussions (CBD Secretariat, 2013b).

15 Based on discussion from the Workshop Symposium 'Taller Distribución de Beneficios, Sociedad Peruana de Derecho Ambiental', held in Quito, Ecuador on 8 March 2013.

References

Ardrey, R. (1970) *The Social Contract*, Dell Publishing Co., Inc., New York.

Barbier, E.B. and Aylward, B. (1996) 'Capturing the pharmaceutical value of biodiversity in a developing country', *Environmental and Resource Economics*, Vol. 8, pp. 157–81.

Buchanan, J.M. (1986) 'James M. Buchanan Jr. – Prize Lecture: The constitution of economic policy', www.nobelprize.org/nobel_prizes/economic-sciences/laureates/1986/buchanan-lecture.html.

Buchanan, J.M. (1987) *Comment: Constitutional Strategy and the Monetary Regime*, University of Chicago Press, Chicago.

Carrizosa, S., S.B. Brush, B.D. Wright, and P. McGuire (2004) *Accessing Biodiversity and Sharing Benefits: Lessons from Implementing the Convention on Biological Diversity*, IUCN, Gland, Switzerland and Cambridge, UK, http://era-mx.org/biblio/Carrizosa_et_al_2004.pdf.

CBD Secretariat (1992) Text of The Convention, www.cbd.int/convention/text/default.shtml.

CBD Secretariat (2008) COP9 Decision IX/12: Access and benefit-sharing, www.cbd.int/decision/cop/default.shtml?id=11655.

CBD Secretariat (2010) Text of the Nagoya Protocol on Access and Benefit-Sharing, www.cbd.int/abs/.

CBD Secretariat (2013a) 'Synthesis of the online discussions on Article 10 of The Nagoya Protocol on Access and Benefit-sharing', www.cbd.int/doc/meetings/abs/absem-a10-01/official/absem-a10-01-02-en.pdf.

CBD Secretariat (2013b) ABS Clearing-House: Pilot Phase, https://bch.cbd.int/abs/art10_groups/searchforum/.

CBD Secretariat (2013c) Notification No. 2013–018, www.cbd.int/doc/notifications/2013/ntf-2013-018-emerging-issues-en.pdf.

CBD Secretariat (2013d) 'Report of the Expert Meeting on Article 10 of the Nagoya Protocol on Access and Benefit-Sharing', www.cbd.int/doc/?meeting=ABSEM-A10-01.

CBD Secretariat (2015a) 'Submission of views in preparation for the Expert Meeting on the need for and modalities of a global multilateral benefit-sharing mechanism and the first meeting of the Compliance Committee of the Nagoya Protocol', www.cbd.int/doc/notifications/2015/ntf-2015-049-abs-en.pdf.

CBD Secretariat (2015b) Submissions on Article 10 of the Nagoya Protocol pursuant to decision NP-1/10, www.cbd.int/abs/submissions.shtml.

CBD Secretariat (2015c) 'Updated report and synthesis of views in response to paragraph 7(b) of Decision XII/24 on new and emerging issues: Synthetic biology', http://bch.cbd.int/synbio/peer-review.

CBD Secretariat (2015d) New and Emerging Issues, www.cbd.int/emerging/.

CBD Secretariat (2016a) Expert group meeting on Article 10 of the Nagoya Protocol on Access and Benefit-sharing, Montreal 1–3 February, www.cbd.int/doc/?meeting=ABS-A10EM-2016-01.

CBD Secretariat (2016b) Peer-review of the outcomes of the process in response to decision XII/24 on synthetic biology, http://bch.cbd.int/synbio/peer-review.

CBD Secretariat (2016c) New and emerging issues relating to the conservation and sustainable use of biological diversity: Note by the Executive Secretary, www.cbd.int/kb/record/meetingDocument/106931.

CBD Secretariat (2016d) Meeting documents, www.cbd.int/doc/?meeting=sbstta-20.

Cha, A.E. (2015) 'Drug and biotech industry trade groups give Martin Shkreli the boot', *The Washington Post*, 24 September, www.washingtonpost.com/news/to-your-health/wp/2015/09/24/why-big-pharma-hates-martin-shkreli-too/.

Chen, J. (2004) 'Webs of life: Biodiversity conservation as a species of information policy', *Iowa Law Review*, Vol. 89, pp. 495–608.

Chomsky, N. (2016) *Who Rules the World?*, Metropolitan Books Henry Holt and Company, New York.

de Groot, R. (2010) 'Integrating the ecological and economic dimensions in biodiversity and ecosystem service valuation', *The Economics of Ecosystems and Biodiversity: The Ecological and Economic Foundations*, www.teebweb.org/2013-08-30_archive/wp-content/uploads/Study and Reports/Reports/Ecological and Economic Foundations/TEEB Ecological and Economic Foundations report/TEEB Foundations.pdf.

de Klemm, C. (1989) 'Migratory species in international law', *Natural Resources Journal*, Vol. 9, no. 4, pp. 935–78, http://lawschool.unm.edu/nrj/volumes/29/4/02_klemm_migratory.pdf.

Dobzhansky, T. (1973) 'Nothing in biology makes sense except in the light of evolution', *The American Biology Teacher*, Vol. 35, pp. 125–29.

ETC Group (2010) 'Synthetic biology: Creating life forms', www.etcgroup.org/files/publication/pdf_file/ETC_COP10SynbioBriefing081010.pdf.

Firn, R.D. (2003) 'Bioprospecting–why is it so unrewarding?', *Biodiversity and Conservation*, Vol. 12, pp. 207–16.

Hammond, E. (2014) 'Patent claims on genetic resources of secret origin', *Third World Network*, www.twn.my/title2/biotk/misc/budapest_final_21%20Feb2014.pdf.

Harcourt, G.C. (1986) 'On the influence of Piero Sraffa on the contributions of Joan Robinson to economic theory', *The Economic Journal*, Vol. 96, pp. 96–108.

Kagedan, B.L. (1996) 'The Biodiversity Convention, intellectual property rights, and the ownership of genetic resources: International developments prepared for the Intellectual Property Policy Directorate Industry Canada, http://iatp.org/files/Biodiversity_Convention_Intellectual_Property_.pdf.

Kamau, E.C., B. Fedder and G. Winter (2010) 'The Nagoya Protocol on Access to Genetic Resources and Benefit-sharing: What is new and what are the implications for provider and user Countries and the scientific community?', *Law, Environment and Development Journal*, Vol. 6, no. 3, p. 246, www.lead-journal.org/content/10246.pdf.

Kamau, E.C. and G. Winter (2013) 'An introduction to the international ABS regime and a comment on its transposition by the EU', *Law, Environment and Development Journal*, Vol. 9, no. 2, www.lead-journal.org/content/13106.pdf.

Keynes, J. M. (1933) 'National self-sufficiency', *The Yale Review*, Vol. 22, no. 4, pp. 755–69.

Kloppenburg Jr., J.R. (2004[1988]) *First the Seed: The Political Economy of Plant Biotechnology*, 2nd edn, University of Wisconsin Press, Madison, Wisconsin.

Markandya, A. and P.A.L.D. Nunes (2012) 'Is the value of bioprospecting contracts too low?', *International Journal of Ecological Economics and Statistics*, www.feem.it/userfiles/attach/20101213150154NDL2010-154.pdf.

Mazoomdaar, J. (2014) 'Centre sits on royalty slabs for bio resources, loses Rs 25,00 cr a year', *NATION*, 19 November, http://indianexpress.com/Article/india/india-others/centre-sits-on-royalty-slabs-for-bio-resources-loses-rs-25000-cr-a-year/.

May, C. (2010) *The Global Political Economy of Intellectual Property Rights*, 2nd edn, Routledge, London.

Merriam-Webster Dictionary (2016), www.merriam-webster.com/.

Oduardo-Sierra, O., J.H. Vogel and B.A. Hocking (2012) 'Monitoring and tracking the economics of information in the Convention on Biological Diversity: Studied ignorance (2002–2011)', *Journal of Politics and Law*, Vol. 5, no. 2, pp. 29–39, http://dx.doi.org/10.5539/jpl.v5n2p29.

Oldham, P., S. Hall and O. Forero (2013) 'Biological Diversity in the patent system', *PLoS ONE*, Vol. 8, no. 11, p. 6, http://dx.doi.org/10.1371/journal.pone.0078737.

Paquin-Jaloux, S. (2015) 'Brazilian revised ABS law goes back the Deputy chamber', 16 April, http://biodivsourcing.com/brazilian-revised-abs-law-goes-back-the-deputy-chamber/.
Pascual, U. and R. Muradian (2010) 'Chapter 5: The economics of valuing ecosystem services and biodiversity', www.teebweb.org/wp-content/uploads/2013/04/D0-Chapter-5-The-economics-of-valuing-ecosystem-services-and-biodiversity.pdf.
Peruvian Society for Environmental Law (SPDA) (2016) 'Submitted view for the Updated report and synthesis of views in response to paragraph 7(b) of Decision XII/24; and Report of the Meeting of the Ad Hoc Technical Expert Group on Synthetic Biology', http://bch.cbd.int/synbio/peer-review.
Puerto Rico Center for Tropical Diversity and Bioprospecting (2016) http://prsciencetrust.org/puerto-rico-center-for-tropical-biodiversity-and-bioprospecting/.
Robinson D.F. (2014) *Biodiversity, Access and Benefit-Sharing: Global Case Studies*, Routledge, Oxon.
Ruiz Muller, M. (2015) *Genetic Resources as Natural Information: Implications for the Convention on Biological Diversity and Nagoya Protocol*, Routledge, London.
Safrin, S. (2004) 'Hyperownership in a time of biotechnological promise: the international conflict to control the building blocks of life', *The American Journal of International Law*, Vol. 98, no. 4, pp. 641–85.
Samuelson, P.A. and W.D. Nordhaus (2010), *Economics*, 19th edn, McGraw-Hill Irwin, New York.
Schumpeter, J.A. (1950) *Capitalism, Socialism, and Democracy*, 3rd edn, Harper and Brothers, New York.
Siegfried, J.J., R.L. Bartlett, W.L. Hansen, A.C. Kelley, D.N. McCloskey and T.H. Tietenberg (1991) 'The status and prospects of the economics major', *The Journal of Economic Education*, Vol. 22, no. 3, pp. 97–224, www.tandfonline.com/doi/citedby/10.1080/00220485.1991.10844710?scroll=top&needAccess=true.
Simpson, R.D., R.A. Sedjo and J.W. Reid (1996) 'Valuing biodiversity for use in pharmaceutical research', *The Journal of Political Economy*, Vol. 104, no. 1, pp. 163–85.
Stiglitz, J.E. (2008) 'Economic foundations of intellectual property rights', *Duke Law Journal*, Vol. 57, pp. 1693–1724, http://scholarship.law.duke.edu/dlj/vol57/iss6/3/.
Stone, C.D. (1995) 'What to do about biodiversity, property rights, public goods and the Earth's biological riches', *Southern California Law Review*, no. 68, pp. 577–605.
Swanson, T.M., D.W. Pearce and R. Cervigni (1994) 'The appropriation of the benefits of plant genetic resources for agriculture: An economic analysis of the alternative mechanism for biodiversity conservation', Secretariat of the FAO Commission on Plant Genetic Resource, Rome.
TEEB (2016) *The Economics of Ecosystems and Biodiversity*, www.teebweb.org/our-publications/teeb-study-reports/ecological-and-economic-foundations/.
ten Brink, P. (2009) 'Chapter 5: Rewarding benefits through payments and markets', *TEEB – The Economics of Ecosystems and Biodiversity for National and International Policy Makers*, www.cbd.int/doc/case-studies/inc/cs-inc-teeb.Chapter%205-en.pdf.
USA National Focal Point (2016) 'Submitted view for the Updated report and synthesis of views in response to paragraph 7(b) of Decision XII/24; and Report of the Meeting of the Ad Hoc Technical Expert Group on Synthetic Biology', http://bch.cbd.int/synbio/peer-review.
Vogel, J.H. (1991) 'The intellectual property of natural and artificial information', *CIRCIT Newsletter*, Melbourne, Australia, June, p. 7.
Vogel, J.H. (1992) *Privatisation as a Conservation Policy*, CIRCIT, Melbourne, Australia.
Vogel, J.H. (1994) *Genes for Sale*, Oxford University Press, New York.
Vogel, J.H. (ed.) (2000) *The Biodiversity Cartel*, CARE, Quito, Ecuador.
Vogel, J.H. (2007) 'Reflecting financial and other incentives of the TMOIFGR: The biodiversity cartel', in M. Ruiz and I. Lapeña (eds) *A Moving Target: Genetic Resources and Options for Tracking and Monitoring their International Flows*, IUCN, Gland, Switzerland, pp. 47–74, http://data.iucn.org/dbtw-wpd/edocs/EPLP-067-3.pdf.
Vogel, J.H. (2008) 'The unspeakable economics of ABS', *Bridges*, Vol. 12, no. 4, International Centre for Sustainable Trade and Development, http://ictsd.net/i/news/bridges/27572/.
Vogel, J.H. (2013) 'The Tragedy of unpersuasive power: The Convention on Biological Diversity as exemplary', *International Journal of Biology*, Vol. 5, no. 4, pp. 44–54, www.ccsenet.org/journal/index.php/ijb/Article/view/30097/18019.
Vogel, J.H (2015) 'Foreword: On the Silver Jubilee of "Intellectual Property and Information Markets: Preliminaries to a New Conservation Policy"', in M. Ruiz Muller (ed.) *Genetic Resources as Natural Information: Implications for the Convention on Biological Diversity and Nagoya Protocol*, Routledge, London, pp. xii–xxv.
Vogel, J.H. (2016) 'Common ground in bounded openness: Analysis of the submissions from New Zealand and Sociedad Peruana de Derecho Ambiental (Peruvian Society of Environmental Law) regarding the

CBD peer-review of synthetic biology outcomes' (paper distributed under creative commons license and on file with author).

Vogel, J.H., N. Álvarez-Berrío, N. Quiñones-Vilche, J.L. Medina-Muñiz, D. Pérez-Montes, A.I. Arocho-Montes, N.Vale-Merniz, R. Fuentes-Ramirez, G. Marrero-Girona, E.Valcárcel Mercado and J. Santiago-Rios (2011) 'The economics of information, studiously ignored in the Nagoya Protocol on Access and Benefit-sharing', *Law, Environment and Development Journal*, Vol. 7, no. 1, pp. 51–65, www.lead-journal.org/content/11052.pdf.

Vogel, J.H., C. Fuentes-Rivera, B.A. Hocking, O. Oduardo-Sierra and A. Zubiaurre (2013) 'Human pathogens as capstone application of the economics of information to Convention on Biological Diversity', *International Journal of Biology*, Vol. 5, no. 2, pp. 121–34, www.ccsenet.org/journal/index.php/ijb/article/view/22760.

West, S. (2012) 'Institutionalised exclusion: the political economy of benefit-sharing and intellectual property', *Law, Environment and Development Journal*, Vol. 8, no. 1, p. 19, http://lead-journal.org/content/12019.pdf.

Winands-Kalkuhl, S. and K. Holm-Müller (2015) 'Bilateral vs. multilateral? On the economics and politics of a global mechanism for genetic resource use', *Journal of Natural Resources Policy Research*, Vol. 7, no. 4, pp. 305–22, http://dx.doi.org/10.1080/19390459.2015.1097022.

Young, T.R. and A. Minnis (eds), K. Angerer, L. Benjamin, E.C. Kamau, G. Dutfield, C.H.C. Lyal, E. Mawal, S. Peña Moreno, M. Ruiz Muller, T.T. Huong Trang and J.H. Vogel (2015) 'Submission of views in preparation for the Expert Meeting on the need for and modalities of a Global Multilateral Benefit-sharing Mechanism of the Nagoya Protocol'. Collective submission of the IUCN Joint SSC-WCEL Global Specialist Group on ABS, Genetic Resources and Related Issues (ABSSG), www.cbd.int/abs/submissions.shtml.

27

THE IPBES, BIODIVERSITY AND THE LAW

Design, functioning and perspectives of the Intergovernmental Science-Policy Platform on Biodiversity and Ecosystem Services

Guillaume Futhazar, Denis Pesche and Sandrine Maljean-Dubois

Even though the need for a global science-policy interface has been identified as a recurring gap in the governance of biodiversity, establishing such an institution has not been trouble free. The creation of the Intergovernmental Science-Policy Platform on Biodiversity and Ecosystem Services (IPBES) was in fact the outcome of a relatively long and laborious international negotiation process during which several interests and world views had to be carefully conciliated. The complex array of rules and procedures that governs IPBES is the result of this slow genesis.

The purpose of this chapter is to provide a general overview of this young institution in order to make sense of its mandate, structure, rules of procedure and general place in the complex institutional landscape of biodiversity governance.

Firstly, the early history of the Platform will be summarized in order to put into context its mandate and functioning. Secondly, and most importantly, its institutional characteristics will be detailed so as to decipher its day-to-day work. Finally, we will offer a broader legal perspective of this institution by briefly discussing its relation with other important actors in international environmental law.

Setting up the IPBES

A tortuous path[1]

Since its emergence in the early 1990s, the Convention on Biological Diversity (CBD) has been confronted with a lack of scientific knowledge on how to tackle the loss of biodiversity. This situation led, *inter alia*, to the establishment of the CBD's *Subsidiary Body on Scientific Technical and Technological Advice* (SBSTTA) and the elaboration of the *Global Biodiversity Assessment* (GBA) in 1995, which did not receive the expected attention from policymakers. Drawing on this initial experience, the Millennium Ecosystem Assessment (MA) was launched by UN Secretary General Kofi Annan in June 2001. The primary aim of the MA was to meet the needs for environmental assessments of the CBD and other biodiversity-related Multilateral Environmental Agreements such as the Convention to Combat Desertification (UNCCD), the Convention on Migratory Species (CMS) or the Ramsar Convention on Wetlands. The MA ran from 2001 to 2005 and involved more than 1,400 scientists. Following its publication in 2005, the 60th session of the UN General Assembly recognized the need for strengthened scientific knowledge for its activities and a follow-up process started working on how to continue the MA experience.

Parallel to this, an international conference on biodiversity was held in 2005, in Paris. This event is often referred to the other starting point of the negotiation process that led to the establishment of the Platform. Following the call by former French president Jacque Chirac to establish a mechanism similar to the Intergovernmental Panel on Climate Change (IPCC) but for biodiversity, an international consultation was launched: the International Mechanism of Scientific Expertise on Biodiversity (IMoSEB). Both processes, the IMoSEB and the MA follow-up, merged as the negotiation entered an intergovernmental phase under the supervision of the United Nations Environment Programme (UNEP) in 2008. From this point on, *ad hoc* intergovernmental and multi-stakeholder meetings were held annually over the course of three years in Putrajaya, Nairobi and Busan (Earth Negotiation Bulletin – ENB, 2008; ENB, 2009; ENB, 2010).

During this three year period, States and stakeholders had heated debates on the purpose and the structure of this possible new science-policy interface (SPI) and even on the need for establishing such an institution. Two important elements allowed the negotiations to carry on during this uncertain period. Firstly, UNEP's gap analysis on SPIs for biodiversity (UNEP/IPBES, 2009) contributed to building a consensus among participants as to what shortcomings the future institution should address. For instance, the insufficiency of knowledge about biodiversity or the ineffectiveness of communication between scientists and decision-makers. Secondly, the 2010 international year for biodiversity kept the Platform on the political agenda as biodiversity was a central topic of global governance at this time. As a result, the third *ad hoc* meeting held in Busan ended on the agreement that the IPBES "should be established" and provided indications on its exact mission and possible future structure.

The negotiations then entered a second stage where, even though the question of the functions and structure of the Platform were agreed upon, participants still had to determine how to formally establish the IPBES.

In 2011, the Busan outcome was submitted to the 65th session of the United Nations General Assembly (UNGA) which adopted a resolution calling for UNEP to organize a plenary meeting in order to operationalize the Platform as soon as possible. However, following this resolution, negotiators felt uncertain about whether or not the Platform had been formally established. During the meeting held under UNEP's auspices, in October 2011 (ENB, 2011), the UN Office of Legal Affairs shed light on this existential limbo by declaring that there had been no decision formally establishing the new institution but merely recommendations inviting States to do so (UNEP/IPBES, 2011). Moreover, the 2011 meeting, even though it was convened as a "plenary", did not have the mandate to establish the Platform, but only to determine the modalities and institutional arrangements of the Platform.

Several options were available for the formal establishment of the Platform, and States decided, a year later, on April 21, 2012, in Panama (ENB, 2012), to adopt a resolution to this end during the second "plenary" (UNEP/IPBES, 2012). Thus the IPBES was created.

The IPBES is not a new international organization in the strict legal sense. The members of the Platform were adamant in stressing that the IPBES had no international legal capacity. Moreover, they also emphasized that "any future decisions of the Platform have a legally non-binding nature". Also, even though it has strong ties with UN bodies, the IPBES was not attributed a UN status, the Platform is therefore only an "independent intergovernmental body".

An ambitious purpose

As stated in its "policies and procedures" documents, the IPBES' objective is to "strengthen the science-policy interface for biodiversity and ecosystem services for the conservation and sustainable use of biodiversity, long-term human well-being and sustainable development, with the following functions".

But what exactly is a science-policy interface? The notion has several definitions, both in official documents, such as the Gap Analysis, and academic papers (Van den Hove, 2007). Broadly speaking, SPIs encompass "structures and process" that allow exchanges between policymakers and scientists in order to improve the governance of specific subjects. SPIs do not promote a unilateral relation where scientists simply provide policymakers with information, they favour a co-productionist approach whereby both actors participate in the construction of knowledge relevant for decision-making. For instance, this logic entails that priorities are jointly identified by both scientists and policymakers. It also means that specific procedures and institutional features are set up in order to promote co-operation and coordination between the different actors involved.

In fulfilling its role as an SPI, the Platform performs regular assessments of knowledge on biodiversity and ecosystem services. It also identifies priorities in terms of key scientific information needed for policymakers and in terms of capacity-building needs in order to improve the science-policy interface. The Platform assists policy-making by identifying the appropriate tools and methodologies arising from its assessments and by making them broadly available. The Platform does not conduct any new research by fulfilling these commitments and therefore should not be considered as a "scientific research" institution. Also, all of the products of IPBES are meant to be policy relevant but not policy prescriptive.

The exact content of these activities is based on the needs of governments and priorities identified by the Plenary. While only States are able to make requests to the Platform – autonomously or through the governing bodies of other multilateral environmental agreements related to biodiversity – it also welcomes suggestions and inputs from other non-State actors, such as UN bodies, relevant stakeholders, indigenous and local communities or even the private sector.

Here, the Platform distances itself from its acknowledged and obvious model, the Intergovernmental Panel on Climate Change (IPCC). Indeed, the IPCC only focuses on the production of periodic assessments which have greatly contributed in keeping the climatic issue in the global political agenda. The identification of research priorities, though implicit in its assessments has never been formally recognized as a function of the Panel (IPCC, 2013).

A universal but state-centric composition

Having a global scope, the Platform aims at achieving a universal membership and is open to any United Nations member State which formally expresses the will to be a part of it. As of February 2017, 126 States were members of the Platform.

The first phase of the negotiation process leading to the establishment of the IPBES was generally open to all stakeholders and, at this point, the intergovernmental nature of the Platform was not self-evident. However, the Busan outcome (2010) conveyed a not unusual approach by restricting the membership to United Nations member States and regional economic integration organizations.[2] Therefore, as an intergovernmental body, the decision-making powers remain solely in the hands of States.

Though state-centric in its decision-making process, the Platform nevertheless strives to engage with the community of relevant stakeholders. It does so in two ways: through observer status and the Stakeholder Engagement Strategy (IPBES, 2015a).

Observer status has a broad scope and is open to any State not a member of the Platform and any non-State actor qualified in matters covered by the IPBES. However, the admission procedure for new observers is still subject to debate. Consensus on this issue has been postponed at each plenary of the Platform since 2013. As a consequence of this diplomatic stalemate, interim procedures are still in place. For the time being, observers can be admitted to the Plenary after review by the Bureau and only if no member expresses opposition to the admission.

Stakeholders are acknowledged as key partners of the Platform and are expected to play a major role in the diffusion of the IPBES' products. The Stakeholder Engagement Strategy offers a framework for this collaboration, but its exact implementation is still subject to discussion.

Institutional characteristics of the platform

In this section, we will present the structure and the different process and procedures of the Platform.

IPBES' structure

The Platform is composed of a Plenary, two subsidiary bodies, the Bureau and the Multidisciplinary Expert Panel (MEP), and a Secretariat. The choice of a dual structure for IPBES' subsidiary bodies was made at the Panama meeting in 2012. Indeed, some countries initially wanted a single subsidiary body – similar to what can be seen in the IPCC context – while other participants stressed the fact that a dual structure would allow for greater scientific independence by having a body devoted to administrative and political questions. The latter opinion prevailed and has shaped the current structure of the Platform.

The Plenary

The Plenary is the decision-making body of the Platform. Its rules of procedure were set at the Panama meeting (2012) and are based on the procedures of the UNEP governing council. It makes its decisions by consensus or, if not possible, by a two-thirds vote of present and voting Platform members.[3] The presence of a majority of members participating in the session is required for any decision to be made.

The Plenary meets on an annual basis for several days during which intense negotiations take place on different topics: financial and procedural arrangements, adoption of the Platform's work programme, review of the progress on its deliverables, nomination of MEP and Bureau members, and, acceptance, adoption and approval of its reports (we will come back to these three different terms in the next section of the chapter).

Before each plenary session, "stakeholders' days" are organized in order to keep the community of stakeholders up to date on the work of the Platform and provide opportunity for engagement in its work by contributing as an expert or by diffusing the needs and results of the Platform across different networks.

The Bureau

The Bureau is the subsidiary body in charge of the administrative functions of the Platform and is in charge of supervising the conduct of the Platform's activities. This general function leads it to preside over the plenary meetings, contribute to the elaboration of draft decisions concerning procedural and financial arrangements, review the requests for observer status or even identify potential strategic partnerships for the Platform.

The Bureau is composed of ten officers and includes a Chair and four Vice-Chairs. They are elected for a three years by the Plenary among a list of candidates proposed by the members of the Platform. Each UN region[4] has a list of nominees among which two officers are elected. The Chair and Vice-Chairs are selected with due consideration to scientific and technical expertise and the Chair is rotated among the five UN regions every three years.

Currently, IPBES's chair is Sir Robert Watson who succeeded to Professor Abdul Hamid Zakri at the end of the fourth plenary meeting in 2016. The fact that Sir Robert is the current Chair of IPBES, highlights the numerous connections between this organization and other science policy interfaces, such as the MA and the IPCC in which he was also involved.

The Multidisciplinary Expert Panel

The MEP is the subsidiary body of the Platform in charge of overseeing its scientific and technical elements. It is composed of twenty-five members, five per UN region, elected by the Plenary based on regional lists comprising nominations by member States. The MEP members are selected on the basis of their expertise and their ability to work in international scientific and policy processes.

The MEP plays a crucial role in the supervision of the different assessments of the Platform. As such, its exact composition has been the subject of lengthy discussions between members and stakeholders as it evidently appeared as a crucial element in ensuring the credibility, relevance and legitimacy of the Platform.

One of the first questions relating to this composition was on the geographical structure of the MEP. While it was agreed without much debate that the Bureau should have divisions reflecting the UN structure, reaching a consensus on the MEP division was less straightforward. The fact that IPBES is concerned with biodiversity and ecosystem services prompts the question of having a biogeographical division rather than a political division which does not reflect the physical realities of the globe. Even though this argument was duly taken into account by the Platform, the UN division was nevertheless kept for the MEP for practical purposes, while highlighting that assessments will take into account biogeographical divisions.

The nomination and election of MEP members is an exclusive prerogative of member States, though the possibility of letting the stakeholders play a role in the nomination of MEP members was mentioned during the second plenary of the Platform. This strict governmental overview of the process is similar to the procedure in force in the IPCC context.

While nominating members of the MEP, States are encouraged to ensure a geographical, gender and disciplinary balance. Indeed, it has been demonstrated that this balance was a key feature for environmental assessments to achieve a higher impact (Agrawala, 1998; Biermann, 2002). However, this balance has yet to be achieved. The first members of the MEP were predominantly male and from the natural sciences (Morin *et al.*, 2016), and even if its new composition is more balanced, there is still room left for improvement (Montana and Borie, 2015).

The Secretariat

During the first phases of the negotiation process, the interim Secretariat of the IPBES was located in the headquarters of UNEP in Nairobi. It was then moved to Bonn since 2013, following the Panama meeting. The Secretariat benefits from an institutional partnership between the Platform and the UNEP, the United Nations Educational, Scientific and Cultural Organization (UNESCO), the Food and Agriculture Organization (FAO) and the United Nations Development Programme (UNDP). Basically, the Secretariat is under the administrative responsibility of UNEP while the other organizations can provide staff support. However, even under the administrative responsibility of UNEP, the Secretariat remains "solely accountable to the IPBES plenary on policy and programmatic matters".

IPBES's work programme

In this section we will provide an overview of the current work programme, the existing procedures for its implementation and its financial requirements.

The first work programme: setting the stage for IPBES's first achievements

IPBES's first work programme is the result of an innovative process that is absent from the IPCC framework (Futhazar, 2016). In January 2013, in Bonn, governments were invited to formulate requests to the Platform on issues that they felt to be of significant importance for the governance of ecosystems and biodiversity. Additionally, other actors, such as UN bodies or stakeholders, could formulate suggestions and inputs. Based on the information thus received, the MEP and the Bureau established priorities based on the relevance, urgency, geographic scope and level of complexity of the different requests. On the basis of this synthesis, in which the priority was given to governments' requests, a draft of the first work programme was submitted to member States in December 2013 in Antalya. After lengthy discussions, the first work programme of the IPBES was thus adopted (IPBES, 2013a), and is composed as follows.

The first objective of the work programme aims to "**strengthen the capacity and knowledge foundations of the SPI to implement key functions of the Platform**". Four deliverables were agreed upon for this objective. The first two are related to capacity-building needs with the aim of identifying capacity-building needs clearly linked to achieving the Platform work programme (deliverable 1a.) and to integrate the capacity-building needs into activities. The third deliverable aims to produce procedures and approaches for participatory processes working with indigenous and local knowledge systems. And the fourth one aims to identify and prioritize knowledge and data needs for policy making.

The second and third objectives are related to the production of environmental assessments. The second aims to "**strengthen the SPI on biodiversity and ecosystem services at and across subregional, regional and global levels**". Three deliverables were programmed for this objective. The first one is methodological and deals with the issue of producing and integrating assessments across different scales while also taking into account different visions, approaches and knowledge on biodiversity. The second one is a set of four regional assessments based on a biogeographical division of the planet. Finally, the last one is the global assessment of biodiversity and ecosystem services that will draw on the previous assessments while also taking into account elements that were not taken into account by the regional approach, such as oceans and the Polar regions.

The third objective aims to "**strengthen the SPI on biodiversity and ecosystem services with regard to thematic and methodological issues**". Six different assessments are supposed to be produced for this objective and two of them, the assessment of pollinators, pollination and food production and the methodological assessment of scenarios and models of biodiversity and ecosystem services, were formally endorsed during the 2016 plenary in Kuala Lumpur. The four remaining thematic and methodological assessments are meant to provide comprehensive reviews on land degradation and restoration, invasive alien species, sustainable use and conservation of biodiversity, and the conceptualization of values of biodiversity.

Finally, the fourth objective is dedicated to the communication of the Platform's activities and their evaluation. Five different deliverables are dedicated to this objective: a catalogue of relevant assessments; an information and data management plan; a catalogue of policy support tools and methodologies; a set of communication, outreach and engagement strategies, products and process; and finally a review of the Platform's activities and impact so as to contribute to its future development and adjustments.

As illustrated by the broad scope of this first work programme, the Platform intends to produce far more than "simple" assessments. And to do so, it has adopted a complex array of procedures in order to ensure the credibility, legitimacy and relevance of its products.

Procedural complexity: a safeguard for credibility, legitimacy and relevance

Literature on SPIs stresses the importance of ensuring that Interfaces are perceived by all actors as credible, legitimate and relevant (Koetz, Farrel and Bridgewater, 2012). Generally speaking, credibility is the perceived validity of information, methods and procedures of an SPI; legitimacy is the perceived fairness and political balance of an SPI; and relevance reflects the extent to which the work carried out within a SPI is responsive to the conditions and needs of the policy process (Koetz, Farrel and Bridgewater, 2012).

The preparation process can be divided into four stages. Firstly, the scoping process for each assessment: what topics will be addressed by an assessment? Secondly, the expert selection process: who will be in charge of drafting the deliverables? Thirdly, the drafting and review process: how will the deliverables be elaborated and their content reviewed? Lastly, the clearance process: how will reports be adopted by the plenary? This division into four stages is similar to the process that has been used for decades within the IPCC.

Scoping

The scoping process of assessments is conducted by a small group of experts,[5] selected by the MEP on the basis of governments and stakeholders' nominations. The scoping submitted to the Plenary contains several scientific and technical elements but also procedural and administrative aspects related to the realization of the assessment. It gives the member States an overview of what can be expected from the assessment and how its realization is expected to unfold. The scoping process is therefore a safeguard for the assessments' relevance. After a review of the scoping document and its acceptance, the expert selection process for the assessment can begin.

Selecting experts

Several rules govern the selection of experts by the MEP (IPBES, 2013b). Firstly, while both States and stakeholders can nominate experts for any assessments, the MEP has to respect a clear ratio in its selection: "experts selected from those presented by relevant stakeholders should not exceed twenty per cent". This rule, also applicable to the selection of experts in charge of the scoping process, illustrates the willingness of States to preserve a high degree of control over the process (Futhazar, 2016) and was the result of a compromise during the second plenary. The 80/20 rule combined with the state-centric election of the MEP in charge of selecting experts, thus gives a relatively narrow path for stakeholder nominated experts to engage in the process. Also, when selecting the experts, the MEP has to give careful consideration to gender, geographical and disciplinary balance.

These demanding procedures make it difficult for the MEP to satisfy all criteria, and the Platform's Secretariat has pointed out that the 80/20 should be eased in order to give the MEP more flexibility in selecting experts (IPBES, 2015b). The gap in expertise nevertheless remains a recurring issue, especially concerning experts for Eastern European Countries and social scientists (IPBES, 2016; IPBES, 2017a). Still, through the use of a gap filling procedure, the Platform has continuously solicited governments and managed to reach its threshold in order to initiate the drafting process. If these difficulties persist over time, they might prove to be an issue for the credibility and legitimacy of the Platform.

Another aspect of the selection process is the management of conflict of interests. Each nominated expert, for the MEP or for any assessment, has to provide a declaration of interests which will be reviewed by a conflict of interests committee. The adoption of this procedural framework during the third plenary echoed the concerns that were being raised concerning the presence of two scientists from pesticide-producing companies in the group in charge of the assessment on pollinators, pollination and food production (Duperray, Hrabanski and Oubenal, 2016). Nevertheless, as shown by the clearance of this assessment in 2016, it seems that the Platform was able to circumvent suspicions on the credibility and legitimacy of its first deliverable.

Drafting the assessments

Once all experts have been selected and roles attributed, the drafting process begins. To support this long process, several technical support units (TSU) are established in order to provide technical and administrative assistance to the experts' team. Once a first draft of the assessment is established, its review is generally open to all experts. On the basis on the received comments, a second draft is presented as well as a first version of the summary for policy makers. These documents are open to comments from experts and governments. The final drafts of the assessment and summary for policymakers are then prepared by taking into account all the relevant comments. At the end of this process, and after validation by the MEP, they are submitted to the Plenary for a very specific clearance phase.

The assessments' final steps

Once the drafting process is over, the assessments and their summaries for policy makers are presented to the plenary. There, it can accept, adopt and approve the products it is presented with. This precise terminology is important as each word is related to a different product. On the one hand, the plenary accepts "the full scientific, technical and socioeconomic assessment reports". This acceptance means that the member States formally approve of the assessments as a whole. On the other hand, the plenary approves the summary for policy makers (SPM). As SPMs provide a brief and non-technical overview of the assessment's mains finding, they are key and strategic documents that potentially concentrate the most policy relevant points. As such, they are approved line by line during a long process where the MEP and experts who were

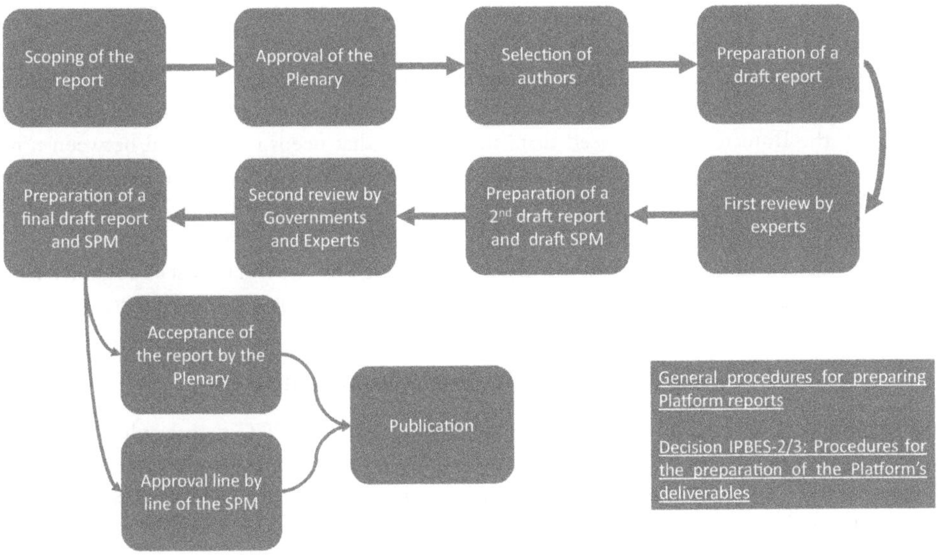

Figure 27.1 General procedures for preparing Platform reports

involved in the elaboration process mediate claims by State to change the wording of the SPM. Here, the balance is subtle as States may try to change the general meaning of an SPM while experts will ensure that its content is coherent with the finding of the assessment to which it is linked. This type of line by line negotiation is a hallmark of the IPCC where the debates on SPM could often go on for several days. Two SPM were adopted during the fourth plenary, and the pollination SPM certainly had contentious points concerning neonicotinoids and bee mortality. Nevertheless the approval process did not prove to be particularly difficult, and a general satisfaction with both assessments was expressed by delegates and stakeholders at the end of the meeting (ENB, 2016).

The adoption procedure, which has not yet been used by the Platform, concerns synthesis reports that combine materials drawing from assessment reports and are written in a non-technical style. These reports, that also have SMP, are subject to a section-by-section (and not line by line) agreement by the Plenary.

Figure 27.1 offers a schematic overview of the drafting procedures currently implemented by the Platform.

Budgetary influence and financial shortcomings

A closer look at the allocated budget for each of the objectives of the Platform is a good indicator of the priorities of the Platform's members. The estimated cost of the first work programme indicates that 44.5% of the estimated budget will be allocated to objective 2, with a predominant allocation to the global assessment. In comparison 25% will be spent on all of the methodological and thematic assessments planned in objective 3. It would seem that even though the work programme wants to provide States with more than a simple global assessment, members of the Platform appear to be more favourable to this type of product (Brooks, Lamoreux and Soberon, 2014).

However, the mandate of the IPBES has met with financial difficulties. For its activities, the IPBES has a core trust fund that can "receive voluntary contributions" (IPBES, 2013c).

Contributions can be made both in cash and in kind (such as the detachment of staff to support the work of TSUs of the Secretariat). In 2015, the Chair of IPBES highlighted that contributions were insufficient in comparison to the expected budget of the first work programme which was estimated at US$40.8 m. In 2017, the financial situation was still described as being critical by the Bureau which stressed that "the funding that needs to be raised between now and the seventh session of the Plenary … to complete the first work programme amounts to $10,755,622, corresponding to the difference between the total cost ($40,506,766) and the total resources received or pledged to date ($29,751,144)" (IPBES, 2017b).

Having described the history and the structure of the IPBES, we can now seek to understand its relations with other institutions operating for the global governance of biodiversity.

Navigating a complex institutional landscape

An important point of distinction between the IPBES and the IPCC is that, while the latter was established before the adoption of the United Nations Framework Convention on Climate Change (UNFCCC), the Platform came into being in an institutional and legal environment where conventions and organizations had been in place for several decades. This situation has raised several questions on how to manage the numerous possible overlaps between the mandate of the IPBES and other organizations.

Well-established relations with the UNEP, UNESCO, FAO and UNDP

With regards to IPBES' relations with the UNEP, UNESCO, FAO and UNDP, the situation was addressed during the first and second plenaries of the Platform which led to the adoption of decision IPBES-2/8 on Collaborative partnership arrangement to establish an institutional link between the Plenary and UNEP, UNESCO, the FAO and UNDP. In the end, the framework of co-operation between IPBES and these other institutions is fairly straightforward: while the UNEP is the only organization in charge of administering the Platform's Secretariat, all organizations assist the Platform by participating in the implementation of its work programme, sharing information, attending meetings or even providing staff.

The situation is, however, less clear with regards to the links between the Platform and the other biodiversity-related conventions.

Towards co-operation dynamic with other Multilateral Environmental Agreements (MEA)

One of the many questions regarding the IPBES during its conceptualization was what the relation between the CBD and the Platform should be. Some even raised the possibility of establishing the Platform within the framework of the CBD, an option that could not be accepted by the USA as they were not party to the CBD (Vadrot, 2014). Therefore, the Platform was established as a SPI independent of any conventional framework. Consequently, practices were established in order to allow coherent action between the new institution and the MEA bodies.

Firstly, MEA decision-making bodies are able to convey requests from their members to the Platform. For instance, during the elaboration of the first work programme, the Secretariat received requests from the CBD, the Convention on International Trade in Endangered Species of Wild Fauna and Flora (CITES), the CMS and the UNCCD (IPBES, 2013d). The current content of the work programme illustrates the relevance of IPBES activities for these conventions.

The assessments on land degradation and restoration will evidently provide useful information for the UNCCD and the CBD intends to draw on IPBES's findings for the preparation of the next Global Biodiversity Outlook.

Currently the Platform's relations with the other MEA are set within a wide range of decisions all adopted by MEAs Conferences of the Parties (COP). For instance, decisions 16.13 to 16.15 of the Convention on International Trade in Endangered Species of Wild Fauna and Flora (CITES) call for increased communication between the bodies of the CITES and the Platform. This is achieved through the participation of CITES representatives as observers during the plenaries but also MEP meetings. Moreover, reports are regularly produced in order to keep both parties and conventional bodies up to date on the relevant activities of the Platform. This approach is generally the same as those adopted in the context of the other MEAs.

However, the CBD differentiates itself from the other MEAs as it seems, at least for the moment, to have a higher degree of interaction with the Platform. For instance, though anecdotic in terms of practical effects, the CBD COP was the first one to have a memorandum of understanding with the IPBES. Moreover, during the last COP in Cancun, parties to the Convention formally endorsed the conclusions and recommendations of the assessment on pollinators, pollination and food production (CBD, 2016).

Though it might be difficult to precisely pin down the legal consequences of an endorsement by a COP (Brunée, 2002) of a non-prescriptive document, it nevertheless has a legal significance. The dual distinction between law and non-law might be less relevant than the idea of normative strength (Thibierge, 2009), and one could suggest that by being endorsed in the context of an international agreement, IPBES's assessment has gained in normative strength. Is it the path towards legal translation of IPBES production? Will we witness similar endorsement for other assessments but in different legal context? By having such strong links with other MEAs, IPBES is, once more, differentiating itself from the IPCC that is still trying to enhance its co-operation with the UNFCCC bodies (IPCC, 2015).

Conclusion

IPBES is taking its first steps and much more remains to be done for such a young institution to achieve the same degree of prestige as its older sister, the IPCC. One thing is certain however: though it might have been elaborated on the model of the IPCC, it will not follow the same path. The end of its work programme, coinciding with the end of the strategic plan for biodiversity 2010–20, will be an important landmark in any assessment of the Platform's chances of becoming a key player in the global governance of biodiversity. Its first assessments have proved to be promising, but time alone will tell whether the Platform does in fact manage to effectively "strengthen the science-policy interface for biodiversity and ecosystem services".

Notes

1 For a more detailed account on the negotiation process that led to the establishment of the Platform, see Vadrot (2014, 2016).
2 For the moment, no such organization is a member of the Platform. The European Union takes part in the work of the Platform but only has the observer status.
3 The first four plenaries never had to rely on voting procedures in order to adopt any decision.
4 The five UN regions are the African group, the Asia-Pacific group, the Eastern European group, the Latin American and Caribbean Group, the Western European and Other group.
5 Twenty experts for the global assessment scoping.

References

Agrawala S. (1998) "Structural and process history of the IPCC." *Climate Change*, Vol. 39, pp. 621–42.

Biermannn F. (2002) Institutions for scientific advice: global environmental assessments and their influence in developing countries. *Global Governance*, Vol. 8, No. 2, pp. 195–219.

Brooks T., Lamoreux J. and Soberon J. (2014) IPBES ≠ IPCC. *Trends in Ecology and Evolution*, Vol. 29, pp. 543–45.

Brunée J. (2002) COPing with consent: law-making under multilateral environmental agreements. *Leiden Journal of International Law*, Vol. 15, No. 1, pp. 1–52.

CBD. (2016) Decision XIII/15, Implications of the IPBES assessment on pollinators, pollination and food production for the work of the Convention.

Duperray F., Hrabanski M. and Oubenal M. (2016) First thematic assessment on pollination: Between the legitimization of IPBES and tensions regarding the selection of knowledge and experts. *In*: Hrabanski M. and Pesche D. (eds) *The Intergovernmental Platform on Biodiversity and Ecosystem Services (IPBES) Meeting the challenges of biodiversity conservation and governance*, London and New York: Routledge.

ENB. (2008) Report of the Ad Hoc Intergovernmental Multi-Stakeholder Meeting on an Intergovernmental Science-Policy Platform for Biodiversity and Ecosystem Services (IPBES), Vol. 158, No. 1.

ENB. (2009) A Summary Report of the Second Ad Hoc Intergovernmental and Multi-stakeholder Meeting on an Intergovernmental Science-Policy Interface on Biodiversity and Ecosystem Services, Vol. 158, No. 6.

ENB. (2010) A Summary Report of the Third Ad Hoc Intergovernmental and Multi-Stakeholder Meeting on an Intergovernmental Science-Policy Platform on Biodiversity and Ecosystem Services, Vol. 158, No. 11.

ENB. (2011) Summary of the First Session of the Plenary Meeting on the Intergovernmental Science-Policy Platform on Biodiversity and Ecosystem Services, Vol. 16, No. 94.

ENB. (2012) Summary of the Second Session of the Plenary Meeting on the Intergovernmental Science-Policy Platform on Biodiversity and Ecosystem Services, Vol. 16, No. 104.

ENB. (2016) Stakeholder Days and Fourth Session of the Plenary of the Intergovernmental Panel on Biodiversity and Ecosystem Services, Vol. 31, No. 29.

Futhazar G. (2016) "From climate to biodiversity Procedural transcriptions and innovations within IPBES in the light of IPCC practices". *In*: Hrabanski M. and Pesche D. (eds) *The Intergovernmental Platform on Biodiversity and Ecosystem Services (IPBES) Meeting the challenges of biodiversity conservation and governance*, London and New York: Routledge.

IPBES. (2013a) Decision IPBES-2/5: Work programme for the period 2014–2018.

IPBES. (2013b) Decision IPBES-2/3: Procedures for the preparation of Platform deliverables.

IPBES. (2013c) Decision IPBES-2/7: Financial and budgetary arrangements.

IPBES. (2013d) Supporting documentation on the prioritization of requests, inputs and suggestions put to the Platform, IPBES/2/INF/9, 9–14 December 2013.

IPBES. (2015a) Decision IPBES-3/4: Communication, stakeholder engagement and strategic partnership, annex II.

IPBES. (2015b) Implementation of the Work Programme for 2014–2018, IPBES/3/2, 12–17 January 2015.

IPBES. (2016) Implementation of the Work Programme for 2014–2018, IPBES/4/2, 22–28 February 2016.

IPBES. (2017a) Implementation of the Work Programme 2014–2018, IPBES/5/2, 7–10 March 2017.

IPBES. (2017b) Financial and budgetary arrangements for IPBES, IPBES/5/10, 7–10 March 2017.

IPCC (2005) Chairman's Vision Paper on the Future of the IPCC, IPCC-XLI/Doc. 5, 24–27 February 2015.

IPCC. (2013) Principles governing IPCC work.

Koetz T., Farrell K. and Bridgewater P. (2012) "Building better science-policy interfaces for international environmental governance: assessing the potential within the IPBES". *International Environmental Agreements: Politics, Law and Economics*, Vol. 12, No. 1, pp. 1–21.

Montana J. and Borie M. (2015) "IPBES and biodiversity expertise: regional, gender and disciplinary balance in the composition of the interim and 2015 Multidisciplinary Expert Panel". *Conservation Letters*, pp. 1–5.

Morin J.-F., Louafi S., Orsini A. and Oubenal M. (2016) "Boundary organizations in regime complexes. A social network Assessment of IPBES". *Journal of International Relations and Development*, pp. 1–32.

Thibierge C. (ed.) (2009) *La force normative – Naissance d'un concept*. Paris: LGDJ.

UNEP/IPBES. (2009) Gap analysis for the purpose of facilitating the discussions on how to improve and strengthen the science-policy interface on biodiversity and ecosystem services, UNEP/IPBES/2/INF/1, 5–9 October 2009.

UNEP/IPBES. (2011) Legal opinion of the secretariat concerning certain legal issues relating to the establishment and operationalization of the platform, UNEP/IPBES.MI/INF/9, 3 October 2011.
UNEP/IPBES. (2012) Annex I, Resolution, Intergovernmental Science-Policy Platform on Biodiversity and Ecosystem Services, UNEP/IPBES.MI/2/9, 16–21 April 2012.
Vadrot A. (2014) *The Politics of Knowledge and Global Biodiversity*. London and New York: Routledge.
Vadrot A. (2016) "The birth of a science-policy interface for biodiversity The history of IPBES". *In*: Hrabanski M. and Pesche D. (ed.) *The Intergovernmental Platform on Biodiversity and Ecosystem Services (IPBES) Meeting the challenges of biodiversity conservation and governance*, London and New York: Routledge.
Van den Hove S. (2007) "A rationale for science-policy interfaces". *Futures*, Vol. 39, No. 7, pp. 807–26.

INDEX

Italic page numbers indicate figures; **bold** numbers indicate tables.

For Product Safety Concerns and Information please contact our EU
representative GPSR@taylorandfrancis.com
Taylor & Francis Verlag GmbH, Kaufingerstraße 24, 80331 München, Germany

www.ingramcontent.com/pod-product-compliance
Ingram Content Group UK Ltd.
Pitfield, Milton Keynes, MK11 3LW, UK
UKHW052055230926
475702UK00031B/2811
* 9 7 8 0 3 6 7 5 0 5 2 4 0 *